Vitamins and Neurodegeneration

This book examines the neurobiological impact of vitamins in the pathophysiology and prevention of neurodegenerative diseases, offering an integrative perspective on nutritional modulation of brain health. It explores the antioxidant, anti-inflammatory, and neurotrophic mechanisms of vitamins, alongside their pharmacokinetics, bioavailability, and blood–brain barrier transport. With a focus on age-associated deficiencies, impaired absorption, and cognitive decline, it highlights mechanistic links between micronutrient status and neuronal integrity. It also investigates synergistic interactions with cofactor minerals, polyunsaturated fatty acids, and polyphenols, as well as the influence of gut microbiota on vitamin-mediated neuroprotection. Specific emphasis is placed on the therapeutic implications of vitamins in Alzheimer's disease, Parkinson's disease, amyotrophic lateral sclerosis (ALS), Huntington's disease, and multiple sclerosis (MS).

Key Features:

- Explores the neuroprotective and cognitive benefits of essential vitamins, including A, C, E, B complex, D, and K.
- Examines pharmacokinetics, bioavailability, and blood–brain barrier transport mechanisms relevant to vitamin efficacy.
- Analyzes the epidemiology of vitamin deficiency, aging-related malabsorption, and their link to cognitive impairment.
- Highlights nutrient–nutrient and vitamin–drug interactions in neurological health and therapy.
- Provides disease-specific insights into the role of vitamins in Alzheimer's disease, Parkinson's disease, ALS, Huntington's disease, and MS.

This book is intended for researchers and students of neuroscience and nutritional sciences.

Vitamins and Neurodegeneration

Mechanisms and Therapeutic Insights

Edited by
Sachchida Nand Rai

CRC Press
Taylor & Francis Group
Boca Raton London New York

CRC Press is an imprint of the
Taylor & Francis Group, an **informa** business

Designed cover image: Shutterstock

First edition published 2026
by CRC Press
2385 NW Executive Center Drive, Suite 320, Boca Raton FL 33431

and by CRC Press
4 Park Square, Milton Park, Abingdon, Oxon, OX14 4RN

CRC Press is an imprint of Taylor & Francis Group, LLC

ISBN: 9781041168508 (hbk)
ISBN: 9781041168515 (pbk)
ISBN: 9781003686675 (ebk)

DOI: 10.1201/9781003686675

Typeset in Times
by codeMantra

Dr. Rai lovingly dedicates this book to the cherished memory of his late father, Dr. Ravindra Rai, whose unwavering faith, guidance, and encouragement have been a constant source of inspiration throughout his life. His father's steadfast belief in his potential, combined with his enduring support and motivation, provided the foundation upon which this work was conceived and completed. This book stands as a tribute to his father's enduring influence, values, and the profound impact he had on Dr. Rai's personal and professional journey.

Contents

Preface

One of the most difficult and urgent problems facing modern medicine is neurodegenerative illnesses. Conditions like Alzheimer's disease, Parkinson's disease, Huntington's disease, multiple sclerosis, and amyotrophic lateral sclerosis have a significant impact on patients, families, and healthcare systems. They are characterized by progressive loss of neurons, functional decline, and impairments in cognition or movement. Their complex nature, which is fueled by oxidative stress, mitochondrial dysfunction, inflammation, and altered nutritional signaling, needs a deep comprehension of the underlying biology as well as preventive and therapeutic approaches.

The importance of nutrition and vitamins for brain health has garnered a lot of attention in recent years. Beyond their traditional roles as metabolic cofactors, vitamins are now known to enhance neurotransmitter production, adjust oxidative equilibrium, affect gene expression, and even control epigenetic processes. The need for a resource that combines molecular insights with clinical and translational viewpoints is highlighted by the mounting evidence connecting vitamin deficits to neurodegenerative diseases. It is the goal of this book, *Vitamins and Neurodegeneration: Mechanisms and Therapeutic Insights.*

This book provides a thorough and comprehensive understanding of how vitamins improve cognitive health, prevent neuronal damage, and enhance brain function. It provides researchers, clinicians, and students with a multifaceted perspective by bridging basic science, clinical research, and translational applications.

The landscape of neurodegenerative diseases is presented in Chapter 1, which also highlights the fundamental role that nutrition plays in maintaining brain health. It provides an overview of the developing discipline of nutritional neuroscience and its significance for comprehending illness processes and potential treatments.

The processes by which vitamins function in the neurological system are covered in detail in Chapter 2. It looks at every vitamin, from fat-soluble to B complex, emphasizing their neurotrophic, anti-inflammatory, and antioxidant functions and how they maintain the integrity and functionality of neurons.

The pharmacokinetics, bioavailability, and blood–brain barrier transit of vitamins are examined in Chapter 3, along with how these aspects affect their therapeutic utility and effectiveness. How vitamins enter the central nervous system and work is explained in this chapter.

The epidemiology of vitamin deficiencies and their connections to cognitive deterioration are discussed in Chapter 4. It highlights the significance of nutritional surveillance and public health measures for older persons by drawing on worldwide data, age-related absorption trends, and population-specific risk factors.

Important vitamin groups are covered in Chapters 5 and 6. While Chapter 6 offers a thorough explanation of B vitamins in methylation, mitochondrial energy production, neurotransmitter synthesis, and cognitive health, Chapter 5 looks at antioxidant vitamins A, C, and E and their combined neuroprotective effects.

In Chapter 7, vitamins D and K are highlighted, and their functions in immunological regulation, vascular defense, and the preservation of brain health during aging are examined.

The topic of nutrient interactions is expanded in Chapter 8, which looks at how vitamins interact with minerals, polyunsaturated fatty acids (PUFA), and polyphenols. The notion of nutritional synergy is presented, highlighting the ways in which mixtures of nutrients can improve neuroprotection.

The gut–brain axis is the subject of Chapter 9, which looks at how vitamins and microbes interact to affect neurophysiology. Probiotics, prebiotics, and other microbiome-targeted tactics are discussed as potential new intervention possibilities.

The interconnections between vitamins, drugs, and nutrients in neurological diseases are examined in Chapter 10, which offers helpful advice for safe and efficient therapeutic use.

The therapeutic implications of vitamins for various neurodegenerative illnesses are summarized in Chapters 11 and 12. Alzheimer's and Parkinson's illnesses are covered in Chapter 11, and ALS, Huntington's disease, and multiple sclerosis are covered in Chapter 12, with a focus on translational potential, clinical trial results, and new prospects.

When taken as a whole, this book provides a comprehensive understanding of the ways in which vitamins interact with neurological health, disease pathology, and therapeutic innovation, acting as both a practical guide and a scientific reference. It emphasizes the potential of vitamins as modulators of the course of neurodegenerative diseases by integrating knowledge from biochemistry, neuroscience, nutrition, and clinical medicine.

As editors and contributors, we believe that *Vitamins and Neurodegeneration: Mechanisms and Therapeutic Insights* will be a useful tool for students, doctors, pharmacologists, nutritionists, and neuroscientists. Our goal is to stimulate additional research, promote interdisciplinary cooperation, and ultimately aid in the creation of practical, evidence-based approaches for brain health and cognitive longevity.

Sachchida Nand Rai

Acknowledgments

First and foremost, I express my deepest gratitude to the Almighty for guiding me throughout this journey and for bestowing upon me the strength, perseverance, and wisdom to undertake this challenging endeavor.

I dedicate this work to my late father, Dr. Ravindra Rai, whose constant inspiration, guidance, and unwavering belief in me have been a cornerstone throughout my academic and professional journey. I am equally grateful to my mother, Asha Rai, for her unconditional support, encouragement, and motivation at every step of my life.

I would also like to acknowledge my beloved wife, whose presence, patience, and support have been invaluable during the writing and editing of this book. Her encouragement has been a constant source of strength. I am deeply thankful to my brother, Ashwini Rai, and sister, Reena Rai, for their moral, ethical, and thoughtful guidance, as well as their unasked yet invaluable suggestions that have enriched my work.

It gives me immense pleasure to acknowledge Bharat Ratna Mahamana Pt. Madan Mohan Malviya Ji, the visionary founder of Banaras Hindu University, Varanasi, Uttar Pradesh, India, whose legacy continues to inspire scholars worldwide.

I, Sachchida Nand Rai, sincerely acknowledge the Indian Council of Medical Research (ICMR), Department of Health Research, Ministry of Health and Family Welfare, Government of India, for their support through the Research Associate (ICMR-RA) Award (3/1/3/196/Neuro/2021-NCD-I), which has been instrumental in enabling the completion of this work.

I am also profoundly grateful to all the contributors who have shared their expertise, research, and insights for the preparation of this book. My heartfelt thanks go to the publishing team at CRC Press for their professionalism, enthusiasm, and steadfast belief in this project. Their expertise in design, production, and marketing has been essential in bringing this book to fruition.

To all the contributors from India and across the globe, your knowledge, passion, and dedication have greatly enriched the pages of *Vitamins and Neurodegeneration: Mechanisms and Therapeutic Insights*. Your diverse perspectives, groundbreaking research, and tireless efforts have been pivotal in enhancing the depth, comprehensiveness, and impact of this work.

I remain deeply grateful to everyone who has supported and contributed to the creation of this seminal volume.

Sachchida Nand Rai

Editor Biography

Sachchida Nand Rai is currently an assistant professor at the University Centre for Research & Development (UCRD), Chandigarh University, Mohali, Punjab, India. He earned his Ph.D. in Biochemistry from BHU in 2018. His research primarily focuses on the anti-Parkinsonian potential of medicinal plants such as *Mucuna pruriens*, *Tinospora cordifolia*, and *Withania somnifera*, as well as their bioactive constituents including ursolic acid, betulinic acid, and chlorogenic acid, particularly in toxin-induced Parkinsonian mouse models. Prior to his current role, Dr. Rai completed a 3-year postdoctoral fellowship at the Centre of Biotechnology, University of Allahabad, Prayagraj, from January 27, 2020, to January 26, 2023. He has made substantial contributions to neuroscience research, with numerous publications in high-impact journals. He serves as an academic editor and editorial board member for several reputed journals. Additionally, he is a member of the advisory board for *Heliyon Neuroscience*. His ongoing research investigates the neuroprotective effects of commonly used Indian medicinal plants in toxin-induced Parkinsonian models, aiming to uncover novel therapeutic strategies for neurodegenerative disorders.

1 Neurodegenerative Diseases and the Role of Nutrition

Falguni Goel[1], *Vipin Kumar Garg*[2], *Kajal Kumari*[1], *Swapnil Sharma*[1], *Payal Singh*[3], *and Sachchida Nand Rai*[3]

[1]Department of Pharmacy, Banasthali Vidyapith, Banasthali, India
[2]Department of Pharmaceutical Technology, Meerut Institute of Engineering & Technology (MIET), Meerut, India
[3]University Centre for Research & Development (UCRD), Chandigarh University, Mohali, Punjab, India

ABBREVIATIONS

AD	Alzheimer's disease
ALS	Amyotrophic lateral sclerosis
BBB	Blood–brain barrier
BDNF	Brain-derived neurotrophic factor
CNS	Central nervous system
CR	Caloric restriction
DASH	Dietary Approaches to Stop Hypertension
DHA	Docosahexaenoic acid
EPA	Eicosapentaenoic acid
GABA	Gamma-aminobutyric acid
HD	Huntington's disease
IF	Intermittent fasting
MIND	Mediterranean-DASH Intervention for Neurodegenerative Delay
PD	Parkinson's disease
ROS	Reactive oxygen species

1.1 INTRODUCTION

The structure and function of the central nervous system are impacted by neurodegenerative disorders, which comprise a broad category of progressive illnesses. These diseases present in the clinic as cognitive decline, mobility impairment, memory loss, psychiatric disturbances, and, in the end, severe disability (McWhirter, Ritchie, Stone, & Carson, 2020). They are distinguished by changes in brain architecture, synaptic disruption, and gradual neuronal death. These include Alzheimer's disease (AD), which causes progressive dementia; Parkinson's disease (PD), which causes tremor, rigidity, and bradykinesia; Huntington's disease (HD), a hereditary disease that manifests as both a movement disorder and psychiatric changes; and amyotrophic lateral sclerosis (ALS), which causes muscle weakness and paralysis while sequentially tearing down motor neurons (Tröster, Pulaski, & Woods, 2019). Even though each of these illnesses has unique pathophysiological characteristics, they all share elements such as chronic neuroinflammation, oxidative stress, mitochondrial dysfunction, misfolded protein accumulation, and disruption of neurotransmitter pathways. The aforementioned parallels point to several common molecular pathways that are crucial in the emergence of neurodegenerative illnesses. This suggests that there may be particular prophylactic measures that can be used for many diseases at the same time (Pathak et al., 2022).

Demographic shifts are largely to blame for the increase in neurodegenerative disease diagnoses globally. The World Health Organization cautions that the number of dementia patients could triple by 2050, placing an unmanageable strain on unpaid caregivers and eventually health systems (Gitlin & Hodgson, 2018). Over 55 million people worldwide currently suffer from AD alone, which has significant financial ramifications due to lost productivity, family grief, and emotional challenges in addition to the expense of treatment. Over the next 20 years, the rate of development of PD, the second most common neurodegenerative disorder in the world's populations, is predicted to double and that is just one ND (Obeso, Monje, & Matarazzo, 2022). These figures are not the only lived realities that patients and their families deal with: emotional distress, a decreased sense of independence and quality of life, etc. Neurodegenerative illnesses are therefore urgent clinical, societal, and economic issues that demand immediate attention (Costa, Adduono, & Jones-Bonofiglio, 2024).

Nutrition has become the most important modifiable factor for brain health and disease in light of this expanding catastrophe. Dietary intake is practical and actionable for illness prevention and management in a way that makes genetic predispositions, aging, and other factors that are only biologically based unmanageable (Abi-Dargham et al., 2023). Nutritional neuroscience has shown significant evidence in the last two decades that nutrients are imperative for maintaining neuronal resilience and functionality (Yousafzai, Rasheed, & Bhutta, 2013). Select vitamins can provide antioxidant defenses, aid in neurotransmitter synthesis, and regulate homocysteine metabolism. Specific polyunsaturated

DOI: 10.1201/9781003686675-1

fatty acids, particularly omega-3 fatty acids, can also support anti-inflammatory signaling pathways, synaptic function, and membrane fluidity. The polyphenols and flavonoids present in fruit, vegetables, herbs, etc., can also influence signaling pathways that enhance neurogenesis, ameliorate oxidative stress, and favor neuroprotection (Moosavi, Hosseini, Saso, & Firuzi, 2015). Diminished risk of neurodegenerative diseases, delayed cognitive decline, and reduced susceptibility to oxidative damage are often associated with adequate intakes of these nutrients (Bianchi, Herrera, & Laura, 2021).

Dietary patterns have emerged as factors affecting brain health, as well as individual nutrients. The Mediterranean diet, which emphasizes whole grains, legumes, fish, olive oil, fruit, and vegetables, has been positively associated with a reduced risk of AD and cognitive decline (Fekete et al., 2025). The Mediterranean-Dietary Approaches to Stop Hypertension (DASH) Intervention for Neurodegenerative Delay (MIND) diet, which incorporates the Mediterranean and DASH diets, consistently slows age-related cognitive decline. The ketogenic diet may benefit patients with epilepsy and neurodegeneration by providing alternative energy sources for glucose-starving neurons, according to preliminary studies. Intermittent fasting (IF) and calorie restriction also provide protective effects in animal studies through augmented autophagy, reduced oxidative stress, and improved mitochondrial function (Vergara Nieto, Diaz, Hernández, & Sagredo, 2025). Taken together, these findings suggest that nutrition interacts actively with the processes leading to neurodegeneration via metabolic, molecular, and epigenetic pathways, in addition to promoting general health (Nicolia, Lucarelli, & Fuso, 2015).

The association between nutrition and the gut–brain axis is an emerging area of interest. The human microbiome affects brain health by producing metabolites such as short-chain fatty acids, modifying immune responses, and regulating the integrity of the blood–brain barrier (BBB) (Parker, Fonseca, & Carding, 2020). Diet has a pronounced impact on the composition and function of gut microbiota, and dysbiosis has been associated with the development of both PD and AD. High-fiber and polyphenol diets promote beneficial microbes that produce anti-inflammatory metabolites (Aziz, Hussain, Hameed, & Lin, 2024). High-fat and high-sugar diets correlate with pro-inflammatory microbiota profiles. Because of this relationship, there are opportunities for new nutritional approaches that enhance cognitive resilience through restoration of microbiome status (Brewer-Smyth, 2022).

From these perspectives, nutrition can be thought of as an essential component of neurodegenerative disease management and prevention. There is accumulating evidence of the utility of a nutritional strategy together with pharmaceutical interventions, lifestyle practice exercise, and cognitive stimulation even though it cannot yet be described as a cure on its own. For individuals who are at risk of or are already undergoing neurodegeneration, the combination of personalized nutrition offers great potential for customizing dietary interventions (Milošević, Arsić, Cvetković, & Vučić, 2021).

1.2 PATHOPHYSIOLOGY OF NEURODEGENERATIVE DISEASES

Numerous interconnected factors contribute to the complex and multifaceted pathophysiology of neurodegenerative disorders, which eventually manifest as increasing neuronal malfunction and death. Human neurodegenerative illnesses share common underlying mechanisms that contribute to the neurodegenerative process, despite the fact that each ailment has unique clinical symptoms and pathological factors (Zaib et al., 2023).

One of the key factors across these diseases is oxidative stress resulting from a stage where the generation of reactive oxygen and nitrogen species exceeds the capacity of antioxidant defenses. Due to the brain's high oxygen consumption, neuronal membranes' high levels of polyunsaturated fatty acids, and its low store of antioxidants, the effect of oxidative injury can impact neuronal function (Stachowicz, 2023). Damaged lipids, proteins, and nucleic acids by excessive free radicals contribute to the cell's integrity and function. Mitochondrial dysfunction is closely related to oxidative signaling. Neurons are highly dependent on energy and rely on mitochondrial oxidative phosphorylation to survive (Trigo, Avelar, Fernandes, Sá, & da Cruz e Silva, 2022). When mitochondrial complexes of the electron transport chain are flawed, ATP output declines while reactive oxygen species (ROS) output increases, resulting in a feedback loop that increases neuronal injury. Disruption in mitochondrial function has been shown to affect initiation and progression in diseases such as PD, which have mutations in the mitochondrial quality control genes PINK1 and Parkin (Wang et al., 2021). In Alzheimer-type dementia, there is in vivo evidence that amyloid-β accumulates in mitochondria and impairs cellular metabolism while increasing free radical activity (Ekundayo et al., 2024).

Another significant aspect is neuroinflammation, which is the inability to regulate immune responses in the central nervous system (CNS). Glial cells including microglia and astrocytes act to maintain neuronal viability by eliminating debris, providing trophic factors, and maintaining homeostasis (Lopez-Ortiz & Eyo, 2024). Chronic activation of glial cells causes a shift toward pro-inflammatory phenotype that leads to the release of pro-inflammatory cytokines at sustained release. This inflammatory response conveys no protection; rather, persistently active glia cause continuous oxidative injury, synaptic dysfunction, and neuronal apoptosis (Zhu et al., 2022). In many neurodegenerative diseases, the BBB breaks down, allowing entry of blood-derived immune cells into the CNS. This further activates neuroinflammation while creating a noxious environment that may promote disease propagation (Sabahi et al., 2021).

A common factor of many neurodegenerative diseases is the imbalance between cellular proteostasis and misfolded proteins, leading to excess accumulation. The key pathological feature consists of amyloid-β plaques located in the extracellular space and hyperphosphorylated tau forming neurofibrillary tangles in neurons (Gallardo & Holtzman,

2020). These aggregates block synaptic communication, disrupt axonal transport, and cause cascades of oxidative and inflammatory injury. In PD, misfolded α-synuclein aggregates are packaged together by the cell and accumulate in Lewy bodies, blocking synaptic vesicle trafficking and mitochondrial function (Bhushan, Goel, & Singh, 2025). In HD, large polyglutamine repeats in the huntingtin protein can lead to toxic protein aggregation and neuronal dysfunction globally. The failure of the ubiquitin proteasome system and cellular autophagy–lysosomal degradation systems worsens the accumulation of toxic proteins and leads to progressive loss of neurons (Goel, Dobhal, & Singh, 2025).

Neurotransmission and synaptic plasticity are also deeply impaired during neurodegenerative diseases. The brain's ability to process, encode, and retrieve information relies on neurotransmitter signaling and synaptic plasticity (Chaudhary, 2025). Cholinergic deficits are present and are perceived to contribute to memory and cognition impairment due to the loss of basal forebrain cholinergic neurons. Excitotoxicity, which refers to excessive glutamate release, results in further losses as overstimulation of NMDA receptors results in calcium overload, leading to cellular death (Goel, Dobhal, Kumar, Rai, & Yadav, 2025). In PD, the pattern of degeneration of dopaminergic neurons in the substantia nigra disrupts basal ganglia circuits that results in motor symptoms of tremor, rigidity, and bradykinesia. In terms of upper motor neuron degeneration, in ALS, neurons with excitotoxic damage contribute to worsening paralysis. Reduced integrity of synapses and impaired long-term potentiation reduce the ability for plasticity of learning and memories that further add to outcomes of neuronal degeneration (Bin Ibrahim, Benoy, & Sajikumar, 2022).

The gut–brain axis has been implicated as an important modulator of neurodegenerative pathology. The gut microbiome interacts with the central nervous system through neural, endocrine, immune, and metabolic pathways (Loh et al., 2024). Altered composition of the microbiota dysbiosis has been found to occur in AD and PD. In PD, gastrointestinal symptoms, such as constipation, commonly precede motor symptoms by several years, indicating that gut pathology may contribute to early mechanisms of the disease (Warnecke, Schäfer, Claus, Del Tredici, & Jost, 2022). Dysbiosis may lead to systemic inflammation, increased intestinal permeability, and the translocation of metabolites and amyloids from gut microbial communities to the central nervous system. Pro-inflammatory signatures of microbiota correlate with higher levels of neuroinflammation and amyloid deposition, while diets containing higher amounts of fiber and polyphenols support communities of bacteria that produce short-chain fatty acids and may provide neuroprotective benefits (Mafe & Büsselberg, 2025).

The pathophysiology seen in neurodegenerative diseases is a constellation of overlapping processes. Oxidative stress and mitochondrial dysfunction can affect neuronal energy and redox homeostasis, neuroinflammation can cause a self-perpetuating toxic state, protein misfolding is a perturbation of intracellular architecture, poor neurotransmission may affect plasticity and communication, and dysfunction of the gut–brain connection will spread pathology outside of the central nervous system (Korczowska-Łącka et al., 2023). These processes are feedback loops that may act in synergy, creating added vulnerability in neurons and expediting the disease process. Acknowledging these related and overlapping pathophysiology processes is an essential platform to form vehicle and preventive approaches, such as nutritional approaches, that attenuate pathological cycles and support neuronal health (Mishra et al., 2025) (Figure 1.1).

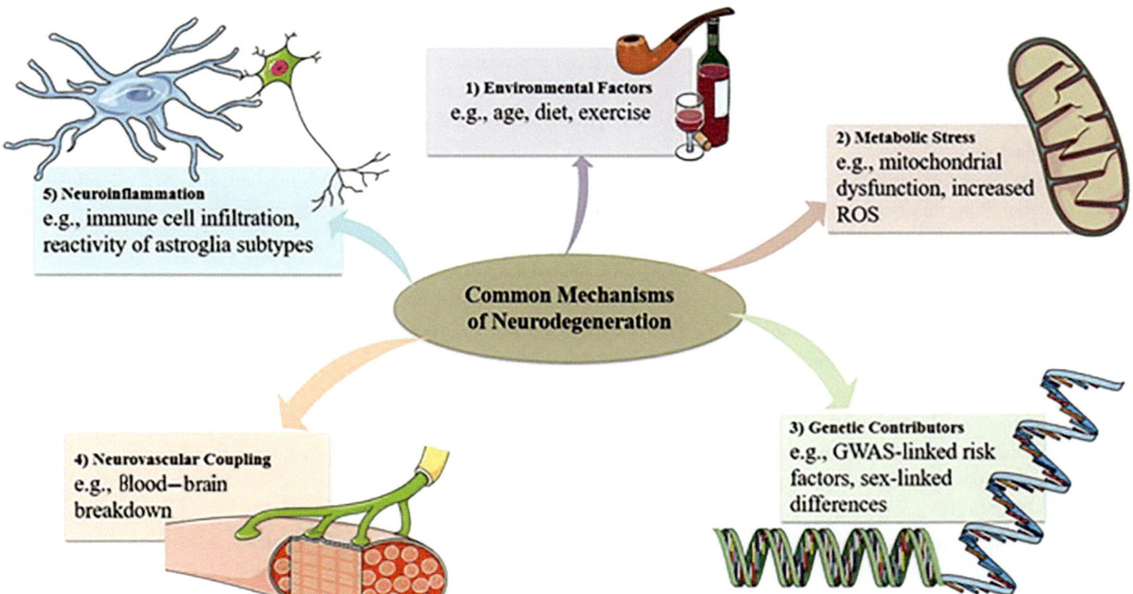

FIGURE 1.1 This figure shows that neuronal loss and cognitive decline are caused by integrated pathways of neurodegeneration that include oxidative stress, inflammation, protein misfolding, and disruption of the gut–brain axis.

1.3 NUTRITIONAL NEUROSCIENCE: MECHANISTIC INSIGHTS

Nutritional neuroscience has begun to reveal that diet is essential for brain health and resilience to neurodegenerative disease. Diet affects the nervous system not just by providing general support to the metabolic processes but also by impacting cellular pathways involved in oxidative balance, inflammation, synaptic plasticity, neurotransmitter creation, and neuronal viability. Mechanistic insight is essential for channeling dietary actions to prevent and manage neurodegenerative diseases (Nazam et al., 2021).

Because nutrients modulate processes that are central to neuronal survival and cognition, they are also inherently important to neuroprotection. Antioxidant nutrients, vitamins, fatty acids, and amino acids may serve alternately as cofactors in enzyme activity, structural components of membranes, and as signaling pathways within neural circuitry (Zhang, Liu, Gao, Sehgal, & Wu, 2022). For example, if there are adequate antioxidants, there will not be sufficient excess ROS built up to cause oxidative damage and outdate active pathways. Excess oxidative damage to neurons accelerates protein misfolding, mitochondrial dysfunction, and synaptic degeneration (Jurcau, 2021). Likewise, the nutrients involved in methylation (e.g., folate, vitamin B12) signal the epigenetic mechanisms controlling gene expression for neuronal repair and synaptic plasticity. Thus, dietary components can determine resilience in cellular pathways or initiate endpoints leading to degenerative processes (Balasubramanian et al., 2024) (Table 1.1).

Vitamins and minerals have unique and non-substitutable actions in the brain. The B vitamins, such as B6, B9 (folate), and B12, play fundamental roles in the one-carbon cycle that governs methylation reactions and homocysteine metabolism (Lyon, Strippoli, Fang, & Cimmino, 2020). By modifying metabolic pathways, achieving adequate vitamin levels with B vitamins can lower the risk of neurodegeneration. Because vitamin D regulates calcium homeostasis, immunoinflammatory signaling, and the expression of neurotrophic factors, particularly brain-derived neurotrophic factor (BDNF), which stimulates synaptic growth and memory, it also has significant neuroprotective effects (Khairy & Attia, 2021). Vitamin C mainly increases catecholamine production and antioxidant activity in aqueous conditions, while vitamin E protects neuronal membranes from lipid peroxidation as a lipid-soluble antioxidant. Characteristic toxicities or deficiencies are neuroinconsistent, and trace elements are necessary including zinc, iron, copper, and selenium (Skalny, Chernova, & Skalnaya, 2023).

The relationship between polyunsaturated fatty acids and cognitive health has long been researched, with a focus on the omega-3 family. A crucial structural lipid in neuronal membranes, docosahexaenoic acid (DHA) ensures the fluidity and integrity of synapses, which are critical for effective neurotransmission (Hashimoto, Hossain, Al Mamun, Matsuzaki, & Arai, 2017). By shifting the balance from pro-inflammatory to anti-inflammatory eicosanoids and new classes of specialized pro-resolving mediators, eicosapentaenoic acid (EPA) also promotes omega-3 activities and affects inflammatory signaling.

TABLE 1.1
Nutritional Neuroscience—Mechanistic Insights

Nutrient	Mechanism of Action	Cognitive Impact	Key Examples	References
Vitamins (B-complex, C, D, E)	Antioxidant activity, regulation of neurotransmitter synthesis, myelin maintenance, modulation of neurotrophic factors	Enhanced neuronal survival, reduced oxidative damage, improved memory and cognition	Vitamin B12, folate, vitamin D3, vitamin E	Orywal et al. (2025)
Minerals (Mg, Zn, Se, Fe, Cu)	Cofactors for enzymatic activity, synaptic signaling, antioxidant defense (selenoproteins), mitochondrial support	Regulation of synaptic plasticity, prevention of neuroinflammation, maintenance of neuronal integrity	Magnesium, zinc, selenium, iron	Solovyev (2015)
Polyunsaturated fatty acids	Membrane fluidity, modulation of neuroinflammation, promotion of neurogenesis, mitochondrial bioenergetics	Improved cognitive performance, reduced risk of Alzheimer's and Parkinson's diseases, better synaptic connectivity	Omega-3 (DHA, EPA), omega-6 fatty acids	Yan, Zhang, Liu, Li, and Zheng (2025)
Polyphenols and flavonoids	Activation of antioxidant pathways (Nrf2), inhibition of neuroinflammation, modulation of signaling cascades (MAPK, PI3K/Akt)	Protection against oxidative stress, enhanced memory, delay in cognitive decline	Resveratrol, quercetin, curcumin, EGCG	Ali et al. (2018)
Amino acids	Precursors for neurotransmitters (serotonin, dopamine, GABA, glutamate), regulation of excitatory/inhibitory balance	Improved mood, cognitive regulation, reduction in neurodegenerative progression	Tryptophan, tyrosine, glutamine	Zhang et al. (2023)
Nutraceuticals and functional foods	Multifactorial modulation of oxidative, inflammatory, and apoptotic pathways	Synergistic neuroprotection and cognitive enhancement	Ginkgo biloba, ashwagandha, ginseng, *Bacopa monnieri*	Kadiri, Tiwari, Roy, and Bhowmik (2025)

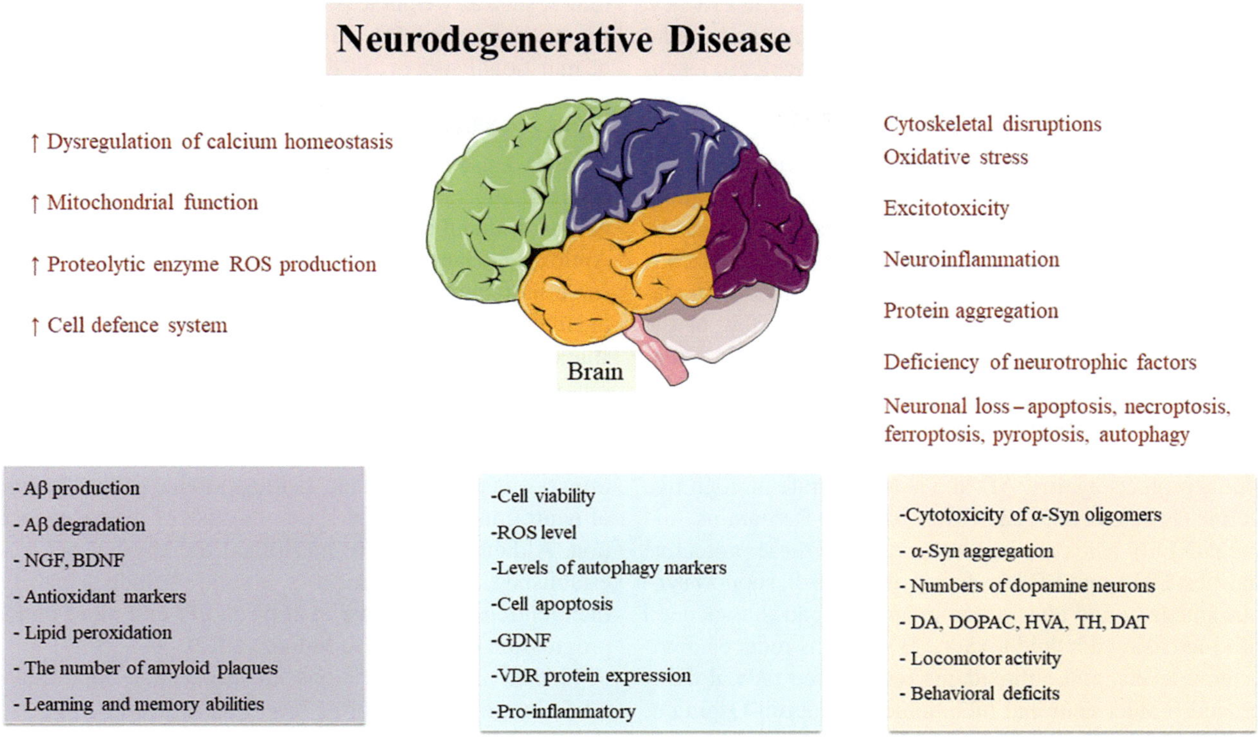

FIGURE 1.2 The main dietary components that support brain health are shown in the schematic that illustrates the molecular insights of nutritional neuroscience.

The diet that contains omega-3 fatty acids inhibits the long-term neuroinflammatory reactions that characterize most neurodegenerative illnesses (Thomas, Thomas, Radcliffe, & Itsiopoulos, 2015). Evidence from both experimental models and human clinical studies suggests that higher intakes of omega-3 fatty acids contribute to memory improvements, decreases in amyloid burden, and slower cognitive decline. Omega-6 fatty acids and trans fats promote inflammation and oxidative stress, leaving open a line of evidence regarding the effects of diet from the point of view of the quality of dietary fat and brain health (Garbicz et al., 2021).

Polyphenols and flavonoids of plant-based foods have been demonstrated as functional antioxidants and signaling molecules. Most antioxidants found in foods are not able to activate endogenous defense mechanisms (Abeyrathne, Nam, Huang, & Ahn, 2022). Many of the functional effects of these polyphenols are mediated by activating endogenous mechanisms. Resveratrol, epigallocatechin gallate, and curcumin are agents that activate the Nrf2 signaling pathway, leading to the up-regulation of antioxidant defense and detoxification phase II genes (Shahcheraghi et al., 2021). These molecules also inhibit canonical NF-κB signaling, leading to decreased expression of pro-inflammatory cytokines. Polyphenols possess numerous neuroprotective properties besides modulation of oxidative stress and inflammatory stress, including enhancement of synaptic plasticity, stimulation of neurogenesis, and mediation of protection against excitotoxicity (Shi et al., 2024).

Amino acids, which serve as the building blocks of proteins, are also precursors for important neurotransmitters. Their availability is important for neurochemical homeostasis. For example, serotonin synthesis occurs from tryptophan, which impacts mood, cognition, and circadian regulation (Ghallab & Elassal, 2024). In the instance of dopamine, norepinephrine, and epinephrine, both tyrosine and phenylalanine act as precursors, which are neurotransmitters that impact attention, motivation, and motor control. Both glutamate and gamma-aminobutyric acid (GABA), the predominant inhibitory neurotransmitter, are derived from amino acids, and damage will occur due to excitotoxicity without their careful balance (Dhurandhar, Tomar, Namdeo, & Bodakhe, 2025). Disruption of amino acid metabolism will therefore ultimately lead to neurotransmitter imbalances observed in depression, PD, and AD. Nutritional support using adequate amino acids from dietary intake will facilitate neurotransmitter synthesis and synaptic communication and provide a buffer for deficits due to neurodegeneration (Samanta, Chakraborty, & Bagchi, 2024) (Figure 1.2).

1.4 ROLE OF NUTRITION IN SPECIFIC NEURODEGENERATIVE DISEASES

Diet might affect neuronal survival, synaptic function, and disease progression, and their interest in the role of nutrition in neurodegenerative illnesses is growing. Dietary treatments are becoming powerful adjuvant strategies that can impact disease pathology, increase treatment efficacy, and improve overall quality of life, even if pharmaceutical approaches are still the mainstay of managing neurodegenerative disorders (Pandey et al.). Although each neurodegenerative disease has its own set of difficulties, there are commonalities among neurological disorders, including oxidative stress, mitochondrial dysfunction, inflammation, and protein aggregation, all of which have mechanisms through which nutrition might be beneficial (Goel, 2025).

Amyloid-β buildup, tau hyperphosphorylation, oxidative stress, and neuroinflammation are only a few of the pathogenic processes in AD that are significantly influenced by nutritional status. There is significant epidemiological evidence that the Mediterranean diet, which is rich in fruits, vegetables, nuts, seafood, olive oil, and whole grains, protects against AD and slows the rate of cognitive decline (Fekete et al., 2025). Antioxidants, B vitamins, and omega-3 fatty acids are important parts of these protective qualities. DHA and EPA, as found in fatty fish, support synaptic integrity and alter inflammatory cascades associated with increasing amyloid toxicity. B vitamins reduce homocysteine levels, and antioxidants reduce free radical damage and protect neuronal membranes (Calderón-Ospina & Nava-Mesa, 2020). Polyphenols exert their neuroprotective effects by activating Nrf2-mediated antioxidant pathways and suppressing inflammatory responses. The MIND diet has recently been evaluated as a novel protective dietary pattern context that could be beneficial to cognitive health; the MIND diet has also been associated with much lower incidence of AD, further highlighting the importance of targeted dietary patterns to reduce the current burden of AD (Kheirouri & Alizadeh, 2022).

In the incidence of PD, in which dopaminergic neurons' degeneration leads to motor symptoms and also, in most cases, non-motor symptoms, many diet interventions may have an impact on the disease course. Omega-3 fatty acids have been shown to regulate dopamine neurotransmission and lower the effects of neuroinflammation, therefore providing symptom-alleviating effects and possibly disease-modifying effects; antioxidants, including vitamin E and vitamin C, and coenzyme Q10, are sought for roles in limiting aspects of oxidative stress, which is important to note since this is a leading influencer in dopaminergic neuronal demise (Li & Song, 2022). Moreover, research into flavonoids, notably from tea, citrus fruits, and cocoa, suggests their protective effects on nigrostriatal pathways and possible effects in lowering risk and severity of PD due to having those polyphenols in the diet. The gut–brain axis leading to dysbiosis is important to note in which a shift toward an inflammatory response and α-synuclein pathology occurs. An intervention-like dietary plan full of fiber and fermented sources, which support beneficial gut microbiota, could lead to the avoidance of a dysbiotic gut and decrease inflammatory response (Dogra, Mani, & Katare, 2022).

The huntingtin gene's enlarged repeats are a hallmark of HD, an inherited condition. It causes some nutritional issues, such as weight loss, hypermetabolism, and an apparent catabolic condition (Singh & Agrawal, 2022). The goal of nutritional support in these patients is not only to slow neuronal degeneration but also to support body weight and energy balance. Increased caloric intake, often with nutrient-dense diets, is required to counteract the increase in metabolic rate. Omega-3 fatty acids may help regulate neuronal membrane function and reduce excitotoxicity. Antioxidant nutrients such as vitamins C and E provide some protection from oxidative stress, which is increased in Huntington's pathology (Samanta et al., 2024). Emerging studies have also indicated that creatine and coenzyme Q10 supplementation may provide some metabolic support, since they are involved in energy metabolism and mitochondrial stabilization. Adequate intake of vitamins and minerals is also important, as they provide proper function of neurotransmitters and lessen the risk of neurodegeneration. Although definitive nutritional therapy has yet to be established, dietary guidance should include supportive interventions to improve quality of life and slow disease progression (Bernstein & Munoz, 2012).

Nutritional considerations are especially important in ALS since individuals often face dysphagia and progressive muscle weakness that can result in malnutrition, speeding disease progression and worsening prognosis (Pal, Goel, & Garg, 2025). Appropriate caloric consumption is essential because deficits in energy can exacerbate muscle wasting and respiratory muscle weakness (Tramontano & Palange, 2023). High-fat diets, as well as ketogenic diets, have been studied as potential diets to provide energy substrates for diseased neurons under metabolic stress. There is potential benefit from foods and supplements that make up antioxidants as they are presumed to lower oxidative stress. Also, omega-3 fatty acids provide anti-inflammatory benefits (Homroy et al., 2024).

Diet affects immune cell signaling and the preservation of neuronal structural capital in multiple sclerosis (MS), a disease where autoimmune-mediated demyelination and neuroinflammation are key pathological features. Reduced inflammatory activity and improved functional results are linked to diets that contain more omega-3 fatty acids, fruits, vegetables, and whole grains and fewer saturated fats (Mariamenatu & Abdu, 2021). The Mediterranean diet enhances quality of life and lowers relapse rates in MS patients. Given that a lack of vitamin D has been closely linked to an increased risk and severity of MS, its function in regulating immune cells may be notable (Haghmorad et al., 2024). Therefore, getting enough vitamin D from food and/or supplements may have both therapeutic and preventive benefits. Polyphenols such as curcumin, resveratrol, and flavanols from green tea have been studied for their potential to modulate inflammatory signaling and promote remyelination. Probiotics and prebiotics may also be beneficial to the balance of gut microbiota that can influence immune modulation in MS (Yousefi et al., 2023).

The role of nutrition in neurodegenerative diseases is more than just nourishing our bodies, and it can also actively

TABLE 1.2
Role of Nutrition in Specific Neurodegenerative Diseases

Neurodegenerative Disease	Dietary Components	Mechanistic Role in Disease Modulation	Functional Outcomes	References
Alzheimer's disease	Omega-3 fatty acids, B vitamins (B6, B12, folate), vitamin D, vitamin E, polyphenols (curcumin, resveratrol), Mediterranean diet	Reduction in amyloid-β aggregation, attenuation of tau pathology, decreased oxidative stress and neuroinflammation, improved cholinergic transmission	Improved memory, reduced cognitive decline, slower disease progression	Sultana, Hia, and Reddy (2024)
Parkinson's disease	Polyunsaturated fatty acids (DHA, EPA), vitamin D, coenzyme Q10, polyphenols (green tea catechins), antioxidants, ketogenic diet	Protection against dopaminergic neuronal loss, reduction in oxidative stress, mitochondrial support, modulation of α-synuclein aggregation	Improved motor function, reduced tremors, better mitochondrial energy metabolism	Rahimmi, Tozandehjani, Daraei, and Khademerfan (2022)
Huntington's disease	High-calorie diets, creatine supplementation, polyphenols, omega-3 fatty acids, amino acids (tryptophan, glutamine)	Support of neuronal energy metabolism, attenuation of excitotoxicity, reduction in neuroinflammation, improvement in neurotransmitter balance	Stabilization of weight loss, improved energy status, modest delay in motor and cognitive decline	Shirodkar and Babre (2025)
Amyotrophic lateral sclerosis	Antioxidants (vitamin E, C, selenium), polyunsaturated fatty acids, ketogenic diet, functional foods (CoQ10, curcumin)	Reduction in oxidative damage, modulation of glutamate excitotoxicity, improved mitochondrial bioenergetics	Slower progression of muscle weakness, preservation of motor function, improved survival in some cases	Newton (2020)
Multiple sclerosis	Vitamin D, omega-3 fatty acids, polyphenols (resveratrol, curcumin), Mediterranean and anti-inflammatory diets, probiotics	Regulation of immune response, reduction in demyelination, gut microbiota modulation, decreased oxidative and inflammatory damage	Reduced relapse rate, improved mobility and fatigue management, enhanced quality of life	Mandato et al (2023)

alter pathological processes. Modification of dietary intake will alter amyloid and tau pathology related to AD; it will support dopaminergic neurons and gut health in PD; it will address the problems associated with hypermetabolism in HD; it can provide metabolic resilience in ALS; and it can regulate inflammatory responses in MS (Thapa et al., 2025). Although nutritional approaches cannot yet be used as a curative modality, there is increasing evidence for their use as a complementary treatment approach. Therefore, a usable form of nutritional therapy could overlap with pharmacological and lifestyle interventions and may ultimately help achieve a more global way to either slow down progression of disease or improve management of symptoms, thereby improving quality of life, for people suffering from these debilitating progressive conditions (Goel, Kumar, & Sharma, 2025) (Table 1.2).

1.5 NUTRITIONAL INTERVENTIONS AND THERAPEUTIC APPROACHES

The nutritional approaches have transitioned from a broad population health promotion approach to deliberate exploration as therapeutic options for neurodegenerative diseases. The focus has shifted beyond just preventing nutrient deficiencies to attempting to create nutritional implementations and dietary patterns that influence cellular metabolism, reduce neuroinflammation, and modulate neurodegenerative pathways (Shirodkar & Babre, 2025). Insights from preclinical and clinical research support the potential of specific diets and nutraceutical supplements to prevent cognitive decline, help alleviate motor symptoms, and enhance the effectiveness of conventional therapeutic interventions (Pal, Goel, Sharma, & Garg, 2025).

Caloric restriction (CR) and IF are two of the most researched interventions for lifespan and brain health. Both CR and IF increase lifespan by stimulating autophagy, mitochondrial biogenesis, and antioxidant activities to decrease the damaging effects of oxidative stress. CR slows the progression of age-related changes by minimizing the accumulation of damaged proteins and dysfunctional organelles, which are key processes associated with diseases such as AD and PD (Frankowska, Lisowska, & Witkowski, 2022). These processes may be minimized with IF by cycling between feeding and fasting periods, thereby permitting metabolic switching from glucose to ketone metabolism to build energy efficiency and fortify resiliency in response to energy stress, enhancing brain health. Evidence of IF enhancing synaptic plasticity and memory has occurred in animal models of AD and HD (Kumar, Goel, & Rai, 2025).

The ketogenic diet originally emerging as a treatment for epilepsy has also been studied for the neuroprotective effects it seems to have on the brain. The diet seems to utilize ketone bodies as an alternative source of energy, thus bypassing abnormal glucose metabolism often found in people with AD and PD (Jensen, Wodschow, Nilsson, & Rungby, 2020). Ketones appear to augment mitochondrial functioning while decreasing ROS and potentially stabilizing neurocytoarchitecture with increased GABAergic neurotransmission. Ketogenic diets appear to reduce motor deficit and improve motor performance in animal models of PD, while improving memory and reducing the amyloid-β burden in animal models of AD (Tao, Leng, & Zhang, 2022). Though clinical studies remain limited, some reports indicate that a ketogenic or modified Atkins diet may improve motor function and, in some patients, cognitive function in patients suffering from neurodegenerative disorders (Bohnen, Albin, & Bohnen, 2023).

The Mediterranean diet, which is high in fruits, vegetables, whole grains, olive oil, legumes, fish, and moderate wine consumption, has been extensively researched as a dietary pattern associated with less cognitive decline. It is possible that the high levels of polyphenols, omega-3 fatty acids, and monounsaturated fats confer anti-inflammatory and antioxidant effects that combat or protect against synaptic injury as well as vascular impairment (Oppedisano et al., 2020). Observational studies have consistently demonstrated a decreased risk of AD and PD disease in individuals who follow a Mediterranean diet pattern (Maraki et al., 2023). Although the Mediterranean diet pattern is high in an extremely diverse array of nutrients and the interactions of these nutrients together synergistically support neuronal survivability and vascular health, the diet also addresses two of the major risk factors for neurodegeneration: chronic inflammation and impaired cerebral blood flow (Franco, Interdonato, Cordaro, Cuzzocrea, & Di Paola, 2023).

The DASH and MIND diets are similar to the Mediterranean diet and promote generally plant-based foods, healthy fats, limited amounts of red meat, and refined sugars. The DASH diet was created to decrease hypertension but also involves optimizing vascular health and decreasing systemic inflammation, which indirectly benefits brain health (Soltani, Chitsazi, & Salehi-Abargouei, 2018). The MIND diet, derived from both the Mediterranean and DASH diets and made specifically for neuroprotection, reduced the incidence of AD and delayed cognitive decline. The MIND diet does not directly limit CR, unlike other diets, but encourages specific brain-healthy foods, like leafy greens, berries, nuts, and whole grains, and limits consumption of butter, cheese, and sweets (Kheirouri & Alizadeh, 2022). Longitudinal studies suggested that adherence to and moderation of the MIND diet had important reductions in dementia risk, which adds a vital aspect in considering its therapeutic relevance (de Crom, Mooldijk, Ikram, Ikram, & Voortman, 2022).

Nutraceuticals, functional foods, and herbal therapies increasingly will be seen as adjunctive approaches to influence neurodegenerative processes. Nutraceuticals, such as curcumin, resveratrol, and omega-3 fatty acid supplements, have strong antioxidant and anti-inflammatory properties and target some of the pathways that contribute to neurodegeneration (Chiu, Venkatakrishnan, & Wang, 2020). For example, curcumin can inhibit the aggregation of amyloid-β and can modulate inflammatory pathways. Resveratrol has the potential to activate the sirtuin pathways that increase

TABLE 1.3
Nutritional Interventions and Therapeutic Approaches in Neurodegenerative Diseases

Dietary Approach	Mechanistic Basis	Therapeutic Outcomes	Key Evidence	References
Caloric restriction and intermittent fasting	Enhance autophagy, improve mitochondrial function, reduce oxidative stress and inflammation, upregulate neurotrophic factors	Delayed neurodegeneration, improved cognitive performance, increased neuronal resilience	Preclinical and human studies show benefits in AD and PD models	Vergara Nieto et al. (2025)
Ketogenic diet	Promote ketone body utilization, enhance mitochondrial bioenergetics, reduce glutamate excitotoxicity, modulate oxidative stress	Improved motor function, reduced seizures, better energy metabolism in neurons	Effective in epilepsy, studied in AD, PD, and ALS	Korczowska-Łącka et al. (2023)
Mediterranean diet	Rich in polyphenols, PUFAs, vitamins, and minerals; reduce oxidative stress and systemic inflammation	Lower risk of cognitive decline and Alzheimer's disease, better memory and executive function	Observational studies and clinical trials show protective effects	Fekete et al. (2025)
DASH and MIND diets	Focus on fruits, vegetables, whole grains, healthy fats; reduce hypertension and vascular risk factors	Improved vascular health, reduced dementia risk, enhanced cognitive outcomes	MIND diet shown to slow cognitive aging by ~7.5 years in large cohorts	Kheirouri and Alizadeh (2022)
Nutraceuticals and functional foods	Provide concentrated bioactive compounds that target oxidative stress, inflammation, protein aggregation, and mitochondrial dysfunction	Neuroprotection, enhanced cognition, slower progression of neurodegenerative diseases	Curcumin, resveratrol, ginkgo biloba, omega-3 supplements, coenzyme Q10	Makkar et al. (2020)
Herbal interventions	Modulation of neurotransmission, antioxidant and anti-inflammatory pathways, neurogenesis stimulation	Cognitive enhancement, neuroprotection, reduced progression of pathology	Ashwagandha, *Bacopa monnieri*, *Panax ginseng*, *Rhodiola rosea*	Kadiri et al. (2025)

mitochondrial functioning and therefore longevity. Omega-3 supplementation has beneficial effects on synaptic functioning and also reduces inflammatory signaling (Mora, Arola, Caimari, Escoté, & Puiggròs, 2022). Functional foods that are enriched with prebiotics and probiotics also display potential in modulating the gut–brain axis, affecting immune modulation and neurochemical balance similar to that of curcumin, resveratrol, and omega-3 supplements, and may be particularly beneficial for PD and MS. Numerous herbal therapies have been studied for their cognitive-enhancing and antioxidant effects, including *Ginkgo biloba* and *Bacopa monnieri*; studies on these have produced mixed results, although with some promising outcomes based on clinical studies (Lorca et al., 2023) (Table 1.3).

1.6 EMERGING TRENDS AND TRANSLATIONAL PERSPECTIVES

Rapid change is occurring in the field of nutritional neuroscience, as new trends shift away from broad dietary recommendations and toward more individualized, mechanistic, focused therapies. We are making more progress in understanding the intricate interactions between diet, genetics, epigenetics, and other environmental factors that affect brain health and disruptive neurodegenerative processes thanks to systems biology, omics-based technologies, and computational models (Haripriyaa & Suthindhiran, 2024). The need to create translational ways to transition from laboratory to clinical settings is growing due to the increasing frequency of neurodegenerative disorders worldwide (Memudu, Oviosun, Nwanama, & Enyioma-Alozie, 2025).

As part of precision medicine, tailored nutrition is one of the most crucial avenues. While standard dietary guidelines may employ a "one-size-fits-all" approach, personalized nutrition takes into consideration intra-individual heterogeneity in metabolic responses, genetics, and lifestyle factors (Pointner & Haslberger, 2022). Variations in the genes that code for oxidative stress management, vitamin transport, and lipid metabolism impact the likelihood of neurodegeneration following food consumption. Precision nutrition may refer to the use of genetic testing, metabolomics, and microbiome testing; however, in the future, it will include many individualized food and drink options to account for our individual vulnerabilities to improve treatment outcomes (Lagoumintzis & Patrinos, 2023). For example, in AD, individuals carrying the APOE ε4 allele have a different response rate to dietary fat intake than non-carriers, reinforcing the need for recommendations that are based on an individual's genotype. A second example is recommending diets that contain high levels of polyunsaturated fatty acids or nutraceuticals that target the mitochondria for patients diagnosed with PD and carry certain mitochondrial haplotypes (Tragni et al., 2022).

The emerging scientific fields of nutrigenomics and epigenetic regulation of neurodegeneration are closely associated with individualized nutrition. Simply defined, nutrigenomics studies how foods affect gene expression, while epigenetics studies how heritable changes in gene activity take place without modifications to the DNA sequence (Nasir et al., 2022). The epigenetic markers of deoxyribonucleic acid (DNA) methylation, histone changes, and non-coding ribonucleic acid (RNA) can be altered by nutrients such as folate, vitamin B12, polyphenols, and omega-3 fatty acids. This may boost pathways associated with neuronal survival, synaptic plasticity, and inflammation (Golia et al., 2019). Polyphenols such as resveratrol and curcumin activate the Nrf2 and sirtuin pathways, which regulate genes related to neuroprotection and oxidative stress. The proper DNA methylation patterns, which balance the aberrant expression of genes linked to neuron degeneration illnesses including HD and AD, depend on methyl donors like choline and folate (Mohd Murshid, Aminullah Lubis, & Makpol, 2022). Even if mechanistic insights are still in their infancy, there is promise for developing diets or nutraceutical preparations that enable the restoration of improved epigenetic states to reverse the load of the neurodegenerative condition (Martínez-Iglesias, Naidoo, Carrera, Corzo, & Cacabelos, 2023).

Another arena of research has shifted to the gut–brain axis and the contribution of microbiota playing a role in the nutrition–brain connection. The two-way communication between the gut microbiome and the CNS is being more widely recognized as important for brain health (Yehualashet, Melaku, Gebreyes, Wassie, & Sahilu, 2025). Dysbiosis, due to a dysregulated microbial state as a result of increased intestinal permeability, systemic inflammation, and pathologic microbial metabolites including short-chain fatty acids, designates the microbiome to be an agent of neurodegeneration. Nutritional interventions that can rehabilitate the microbial composition, such as prebiotics, probiotics, synbiotics, and high-fiber diets, have been shown to be promising in preclinical and in our preliminary clinical studies (Yang, Sun, Cai, Chen, & Gu, 2020). The probiotics' strains have also been shown to be positively correlated with memory performance and reduce neuroinflammation and effects on neurotransmitter biosynthesis. Polyphenols, and omega-3 fatty acids, also changed components of microbiota, resulting in greater bioavailability of neuroprotective metabolites. Future-gazing suggests that these nutrition and microbiology interventions on the gut–brain axis can also lead to therapeutic approaches for nutritional interventions in disorders such as AD, PD, and MS (Abubakar et al., 2025).

Nutritional therapy is a difficult process to promote from the results of experimental studies into patient care. A significant barrier that is ubiquitous across the therapeutic care for chronic diseases is the heterogeneity in clinical trials with respect to study design trials with different populations, trials using different biomarkers and outcomes, and trials with additional environmental factors, including active comparator group interventions (Fountzilas, Tsimberidou, Vo, & Kurzrock, 2022). The bioavailability and pharmacokinetics of several nutraceuticals have been poorly characterized and sometimes do not even make a therapeutic impact. Standardization or pharmacological advice does not occur when brand-name nutraceuticals are provided alongside nutraceutical medication therapy, according to translational and clinical perspectives (Shah, Banerjee, Singh, Kulhari, & Anand Saharan, 2025). Another obstacle to successfully implementing nutritional strategies in chronic diseases is the logistical difficulty of recruiting large numbers of patients and sites for trials to assess the uptake of nutritional strategies into mainstream practice. Funding-related regulatory barriers also limit the size of consortium trials (Asgary, Garland, Ro, Stribling, & Waldman, 2022).

1.7 FUTURE DIRECTIONS

Nutrition and neurodegenerative illness continue to expand, and future research avenues will focus on integrating dietary-related strategies into complete models of neurodegenerative disease prevention and therapy. Even if pharmacological therapies remain the cornerstone of treatment, there is a need to reconsider healthcare systems in light of the growing body of research demonstrating the neuroprotective effect of food intake and nutrients (Ghallab & Elassal, 2024). These days, nutrition is seen to have a more iterative purpose, including illness-modifying processes that may delay onset, slow progression, and enhance quality of life in addition to supportive mechanisms for disease therapy (Pope, 2021).

Integrating nutrition into daily life will be a major area of attention for future research. Even if the therapies are evidence-based, safe, practical, and affordable, clinical practice guidelines and recommendations typically disregard food in favor of medication, rehabilitation, and psychosocial support based on a biological model (Marx et al., 2023). It may be possible to identify dietary risk factors and deficiencies even earlier in the course of the disease by methodically including nutritional screening into what are currently standard neurological clinical evaluations (Boy et al., 2023). For example, screening could evaluate the patients' antioxidant, omega-3, vitamin D, and vitamin B12 status. These lifestyle factors may offer supplementary therapies. Neurologists, registered dietitians, and clinical nutritionists can collaborate to create customized food programs that support prescription pharmaceutical treatments as part of multidisciplinary care teams (Short et al., 2025). The promise of incorporating personalized dietary recommendations based on each person's unique profile will not only have a therapeutic impact on a patient's medication but also a dietary management strategy once precision medicine reaches the proper level of maturity (Franz, Boucher, & Evert, 2014).

Another option is to use nutritional intervention as a preventive measure early in life. It is believed that many neurodegenerative illnesses, including PD and AD, develop slowly over a lengthy period of time with few or no clinical symptoms until manifesting (Pathak et al., 2022). Therefore, throughout the disease's preclinical window, it is crucial to be prepared to take preventive measures to address modifiable risk factors. By preserving metabolic homeostasis, reducing oxidative stress, and preserving synaptic integrity, nutritional treatments throughout a person's preclinical years can have a significant long-term risk impact (Shohag et al., 2022). In large cohort studies, dietary patterns like the Mediterranean diet or the MIND diet are consistently linked to a decreased incidence of dementia and cognitive decline, according to a wealth of research with solid evidence. Addressing dietary conditions from childhood to adulthood has the potential to be a public health strategy that acknowledges and responds to the substantial preclinical neurodegenerative burden that is common in the community (Gillette-Guyonnet, Secher, & Vellas, 2013). To help people better understand the connection between a person's daily behavior, including their dietary habits, and the possible long-term threat to their brain health, public health initiatives that lessen the population's overall neurodegenerative burden should be combined with educational and awareness campaigns. This could include leveraging community programs, workplace wellness programs, or country health promotion programs to widen the reach (Arena et al., 2013).

Future neurodegenerative care should also be multimodal, including medication, lifestyle, and diet. Research indicates that although food is helpful in and of itself, it is unlikely to be sufficient to halt the progression of disease (Arena et al., 2013). Nutrition will complement other components of a care continuum, such as tailored medication therapies, physical activity, and cognitive stimulation. For example, regular mindfulness and stress reduction can further strengthen the anti-inflammatory effects of an antioxidant-rich diet, while exercise promotes neuroplasticity and will complement a person's neuroprotective intake of daily dosages of omega-3 fatty acids (Clemente-Suárez, Martín-Rodríguez, Curiel-Regueros, Rubio-Zarapuz, & Tornero-Aguilera, 2025). Nutrition will allow for embedded opportunities to improve tolerability, which opens up options for combining treatment programs, or to increase the effectiveness of specific pharmaceuticals. Technology such as new digital health applications, increasingly complex wearable medical devices, and artificial intelligence (AI)–based dietary monitoring platforms will enable a future in which real-time, individualized multimodal approaches to neurodegeneration treatment are possible (Mathur, Bhattacharjee, Sehgal, & Shekhar, 2024).

1.8 CONCLUSION

As the globe ages and there are few effective treatments for neurodegenerative disorders, these conditions rank among the most important health issues facing contemporary society. There was sufficient evidence in the previous discussion to support the importance of diet in the pathogenesis, development, and treatment of neurodegenerative disorders. The process of nutrition in neurodegeneration is intricate and varied. Through a complex network that includes oxidative stress, mitochondrial dysfunction, neuroinflammation, protein misfolding, neurotransmission, synaptic plasticity, and gut–brain interactions, nutrition is not merely a confounding factor to take into account; it is the primary determinant of neuronal health. The foundation for neurodegenerative disease risk reduction is nutrition, which combines specific nutrients with comprehensive dietary patterns to create risk reduction contexts and slow the progression of neurodegenerative illnesses.

The potential benefits of diet in the treatment of neurodegenerative illnesses could be both immediate and long-term, according to clinical research. Individually, well-designed nutritional therapies may result in improved quality of life, reduced symptom burden, and preservation of cognitive and motor abilities. Early adoption of brain-healthy diets may reduce the prevalence and postpone the onset of neurodegenerative diseases like PD and AD in the general population, potentially alleviating the financial and social strain on our healthcare systems and caregivers. Research-based knowledge transfer to clinical practice will necessitate a major public health initiative that supports food access, nutrition education, and nutrition literacy.

The goal of future research is to establish standard dietary practices by connecting molecular understanding to clinical relevance through large, longitudinal, well-controlled trials. Advances in nutrigenomics, metabolomics, microbiome research, and other fields that combine an individual's unique genetic factors and metabolic response will likely lead to a greater usage of precision nutrition. In order to tailor the trip to each patient, we will also observe the integration of digital health tools, medications, and lifestyle modifications with nutrition as components of comprehensive multimodal solutions. Making evidence-based nutrition interventions flexible and sustainable for patients, care partners, and communities around the world will be a scientific and logistical challenge.

Neurodegenerative disease prevention and treatment will surely involve nutrition. Overall, there are enough data to suggest that dietary interventions will contribute to future chances for brain health, even though there will still be obstacles to clinical implementation and evaluation. This is an exciting time because it offers the chance to combine evidence from personalized medicine, neuroscience, and nutrition with public health efforts to envision a future in which neurodegenerative diseases are described not only as well-managed but also as prevented through early, accessible, and sustainable nutrition for health.

ACKNOWLEDGMENT

None.

CONFLICT OF INTEREST

None.

REFERENCES

Abeyrathne, E. D. N. S., Nam, K., Huang, X., & Ahn, D. U. (2022). Plant-and animal-based antioxidants' structure, efficacy, mechanisms, and applications: a review. *Antioxidants, 11*(5), 1025.

Abi-Dargham, A., Moeller, S. J., Ali, F., DeLorenzo, C., Domschke, K., Horga, G., … Rubio, J. M. (2023). Candidate biomarkers in psychiatric disorders: state of the field. *World Psychiatry, 22*(2), 236–262.

Abubakar, M., Giri, A., Goel, F., Khan, M., Gupta, J., Kumar, D., … Kumar, N. (2025). Diabetes, Alzheimer's disease risk factors, and the cafeteria diet: a comprehensive review. *Current Neuropharmacology*. doi:10.2174/011570159X384737250626094315.

Ali, T., Kim, T., Rehman, S. U., Khan, M. S., Amin, F. U., Khan, M., … Kim, M. O. (2018). Natural dietary supplementation of anthocyanins via PI3K/Akt/Nrf2/HO-1 pathways mitigate oxidative stress, neurodegeneration, and memory impairment in a mouse model of Alzheimer's disease. *Molecular Neurobiology, 55*(7), 6076–6093.

Arena, R., Guazzi, M., Briggs, P. D., Cahalin, L. P., Myers, J., Kaminsky, L. A., … Babu, A. S. (2013). Promoting health and wellness in the workplace: a unique opportunity to establish primary and extended secondary cardiovascular risk reduction programs. *Mayo Clinic Proceedings, 88*(6), 605–617.

Asgary, R., Garland, V., Ro, V., Stribling, J. C., & Waldman, R. (2022). A systematic review of effective strategies for chronic disease management in humanitarian settings; opportunities and challenges. *Preventive Medicine, 161*, 107154.

Aziz, T., Hussain, N., Hameed, Z., & Lin, L. (2024). Elucidating the role of diet in maintaining gut health to reduce the risk of obesity, cardiovascular and other age-related inflammatory diseases: recent challenges and future recommendations. *Gut Microbes, 16*(1), 2297864.

Balasubramanian, P., Kiss, T., Gulej, R., Nyul Toth, A., Tarantini, S., Yabluchanskiy, A., … Csiszar, A. (2024). Accelerated aging induced by an unhealthy high-fat diet: initial evidence for the role of Nrf2 deficiency and impaired stress resilience in cellular senescence. *Nutrients, 16*(7), 952.

Bernstein, M., & Munoz, N. (2012). Position of the academy of nutrition and dietetics: food and nutrition for older adults: promoting health and wellness. *Journal of the Academy of Nutrition and Dietetics, 112*(8), 1255–1277.

Bhushan, R., Goel, F., & Singh, S. (2025). Beyond dopamine: novel therapeutic pathways for Parkinson's disease through receptor signaling. *CNS & Neurological Disorders-Drug Targets*, *24*(6), 434–451.

Bianchi, V. E., Herrera, P. F., & Laura, R. (2021). Effect of nutrition on neurodegenerative diseases: a systematic review. *Nutritional Neuroscience, 24*(10), 810–834.

Bin Ibrahim, M. Z., Benoy, A., & Sajikumar, S. (2022). Long-term plasticity in the hippocampus: maintaining within and 'tagging' between synapses. *The FEBS Journal, 289*(8), 2176–2201.

Bohnen, J. L., Albin, R. L., & Bohnen, N. I. (2023). Ketogenic interventions in mild cognitive impairment, Alzheimer's disease, and Parkinson's disease: a systematic review and critical appraisal. *Frontiers in Neurology, 14*, 1123290.

Boy, N., Mühlhausen, C., Maier, E. M., Ballhausen, D., Baumgartner, M. R., Beblo, S., … Heringer-Seifert, J. (2023). Recommendations for diagnosing and managing individuals with glutaric aciduria type 1: third revision. *Journal of Inherited Metabolic Disease, 46*(3), 482–519.

Brewer-Smyth, K. (2022). *Adverse Childhood Experiences: The Neuroscience of Trauma, Resilience and Healing throughout the Life Course* (pp. 303–357). Springer. doi:10.1007/978-3-031-08801-8_12.

Calderón-Ospina, C. A., & Nava-Mesa, M. O. (2020). B Vitamins in the nervous system: current knowledge of the biochemical modes of action and synergies of thiamine, pyridoxine, and cobalamin. *CNS Neuroscience & Therapeutics, 26*(1), 5–13.

Chaudhary, U. (2025). *Expanding Senses Using Neurotechnology: Foundation of Brain-Computer Interface Technology* (Vol. 1, pp. 225–245). Springer.

Chiu, H.-F., Venkatakrishnan, K., & Wang, C.-K. (2020). The role of nutraceuticals as a complementary therapy against various neurodegenerative diseases: a mini-review. *Journal of Traditional and Complementary Medicine, 10*(5), 434–439.

Clemente-Suárez, V. J., Martín-Rodríguez, A., Curiel-Regueros, A., Rubio-Zarapuz, A., & Tornero-Aguilera, J. F. (2025). Neuro-nutrition and exercise synergy: exploring the bioengineering of cognitive enhancement and mental health optimization. *Bioengineering, 12*(2), 208.

Costa, I. G., Adduono, K., & Jones-Bonofiglio, K. (2024). Biopsychosocial needs and the intersection of mental health and aging. In: I. G. Costa & K. Jones-Bonofiglio (Eds.), *Supporting Older Persons on Their Aging Journey: An Emancipatory Approach to Nursing Care* (p. 163). Canadian Scholars.

de Crom, T. O., Mooldijk, S. S., Ikram, M. K., Ikram, M. A., & Voortman, T. (2022). MIND diet and the risk of dementia: a population-based study. *Alzheimer's Research & Therapy, 14*(1), 8.

Dhurandhar, Y., Tomar, S., Namdeo, K. P., & Bodakhe, S. H. (2025). Excitatory amino acids as therapeutic agents: reversing neurodegenerative trajectory by tackling excitotoxicity. *Neurological Sciences, 46*(2), 549–560.

Dogra, N., Mani, R. J., & Katare, D. P. (2022). The gut-brain axis: two ways signaling in Parkinson's disease. *Cellular and Molecular Neurobiology, 42*(2), 315–332.

Ekundayo, B. E., Obafemi, T. O., Adewale, O. B., Obafemi, B. A., Oyinloye, B. E., & Ekundayo, S. K. (2024). Oxidative stress, endoplasmic reticulum stress and apoptosis in the pathology of Alzheimer's disease. *Cell Biochemistry and Biophysics, 82*(2), 457–477.

Fekete, M., Varga, P., Ungvari, Z., Fekete, J. T., Buda, A., Szappanos, Á., ... Godos, J. (2025). The role of the Mediterranean diet in reducing the risk of cognitive impairment, dementia, and Alzheimer's disease: a meta-analysis. *Geroscience, 47*, 3111–3130.

Fountzilas, E., Tsimberidou, A. M., Vo, H. H., & Kurzrock, R. (2022). Clinical trial design in the era of precision medicine. *Genome Medicine, 14*(1), 101.

Franco, G. A., Interdonato, L., Cordaro, M., Cuzzocrea, S., & Di Paola, R. (2023). Bioactive compounds of the Mediterranean diet as nutritional support to fight neurodegenerative disease. *International Journal of Molecular Sciences, 24*(8), 7318.

Frankowska, N., Lisowska, K., & Witkowski, J. M. (2022). Proteolysis dysfunction in the process of aging and age-related diseases. *Frontiers in Aging, 3*, 927630.

Franz, M. J., Boucher, J. L., & Evert, A. B. (2014). Evidence-based diabetes nutrition therapy recommendations are effective: the key is individualization. *Diabetes, Metabolic Syndrome and Obesity: Targets and Therapy, 7*, 65–72.

Gallardo, G., & Holtzman, D. M. (2020). Amyloid-β and Tau at the crossroads of Alzheimer's disease. *Tau Biology*, 187–203. doi:10.1007/978-3-030-28288-3_12.

Garbicz, J., Całyniuk, B., Górski, M., Buczkowska, M., Piecuch, M., Kulik, A., & Rozentryt, P. (2021). Nutritional therapy in persons suffering from psoriasis. *Nutrients, 14*(1), 119.

Ghallab, Y. K., & Elassal, O. S. (2024). Biochemical and neuropharmacology of psychiatric disorders. In: W. Mohamed & F. Kobeissy (Eds.), *Nutrition and Psychiatric Disorders: An Evidence-Based Approach to Understanding the Diet-Brain Connection* (pp. 25–47). Springer.

Gillette-Guyonnet, S., Secher, M., & Vellas, B. (2013). Nutrition and neurodegeneration: epidemiological evidence and challenges for future research. *British Journal of Clinical Pharmacology, 75*(3), 738–755.

Gitlin, L. N., & Hodgson, N. A. (2018). *Better Living with Dementia: Implications for Individuals, Families, Communities, and Societies.* Academic Press.

Goel, F. (2025). Exploring the therapeutic role of Moringa oleifera in neurodegeneration: antioxidant, anti-inflammatory, and neuroprotective mechanisms. *Inflammopharmacology*, 1–17. doi:10.1007/s10787-025-01794-y.

Goel, F., Dobhal, V., & Singh, S. (2025). Protective effects of diacerein against quinolinic acid–induced Huntington's disease–like symptoms in adult zebrafish by targeting GSK-3β signalling. *Naunyn-Schmiedeberg's Archives of Pharmacology*, 1–23. doi:10.1007/s00210-025-02701-2.

Goel, F., Dobhal, V., Kumar, D., Rai, S. N., & Yadav, D. K. (2025). Evidence based molecular pathways, available drug targets, pre-clinical animal models and future disease modifying treatments of Huntington's disease. *Molecular Biology Reports, 52*(1), 1–17.

Goel, F., Kumar, D., & Sharma, A. (2025). Impact of corticoid receptors on Alzheimer's disease: a neuroendocrine perspective. *Inflammopharmacology*, 1–16. doi:10.1007/s10787-025-01795-x.

Golia, M. T., Poggini, S., Alboni, S., Garofalo, S., Albanese, N. C., Viglione, A., ... Limatola, C. (2019). Interplay between inflammation and neural plasticity: both immune activation and suppression impair LTP and BDNF expression. *Brain, Behavior, and Immunity, 81*, 484–494.

Haghmorad, D., Soltanmohammadi, A., Jadid Tavaf, M., Zargarani, S., Yazdanpanah, E., Shadab, A., & Yousefi, B. (2024). The protective role of interaction between vitamin D, sex hormones and calcium in multiple sclerosis. *International Journal of Neuroscience, 134*(7), 735–753.

Haripriyaa, M., & Suthindhiran, K. (2024). The important role of systems biology in neuroscience and neurology and its associated diseases. In S. Joshi, R. R. Ray, M. Nag & D. Lahiri (Eds.), *Systems Biology Approaches: Prevention, Diagnosis, and Understanding Mechanisms of Complex Diseases* (pp. 343–392). Springer.

Hashimoto, M., Hossain, S., Al Mamun, A., Matsuzaki, K., & Arai, H. (2017). Docosahexaenoic acid: one molecule diverse functions. *Critical Reviews in Biotechnology, 37*(5), 579–597.

Homroy, S., Chopra, R., Singh, P. K., Dhiman, A., Chand, M., & Talwar, B. (2024). Role of encapsulation on the bioavailability of omega-3 fatty acids. *Comprehensive Reviews in Food Science and Food Safety, 23*(1), e13272.

Jensen, N. J., Wodschow, H. Z., Nilsson, M., & Rungby, J. (2020). Effects of ketone bodies on brain metabolism and function in neurodegenerative diseases. *International Journal of Molecular Sciences, 21*(22), 8767.

Jurcau, A. (2021). Insights into the pathogenesis of neurodegenerative diseases: focus on mitochondrial dysfunction and oxidative stress. *International Journal of Molecular Sciences, 22*(21), 11847.

Kadiri, S. K., Tiwari, P., Roy, S. P., & Bhowmik, S. (2025). *Herbal Medicine and Neurological Diseases Herbal Pharmacopeia* (pp. 287–304). CRC Press.

Khairy, E. Y., & Attia, M. M. (2021). Protective effects of vitamin D on neurophysiologic alterations in brain aging: role of brain-derived neurotrophic factor (BDNF). *Nutritional Neuroscience, 24*(8), 650–659.

Kheirouri, S., & Alizadeh, M. (2022). MIND diet and cognitive performance in older adults: a systematic review. *Critical Reviews in Food Science and Nutrition, 62*(29), 8059–8077.

Korczowska-Łącka, I., Hurła, M., Banaszek, N., Kobylarek, D., Szymanowicz, O., Kozubski, W., & Dorszewska, J. (2023). Selected biomarkers of oxidative stress and energy metabolism disorders in neurological diseases. *Molecular Neurobiology, 60*(7), 4132–4149.

Kumar, D., Goel, F., & Rai, S. N. (2025). The role of ABCA12 in neurodegenerative diseases: a review of molecular mechanisms and potential therapeutic implications. *European Journal of Clinical and Experimental Medicine, 23*(2), 468–481.

Lagoumintzis, G., & Patrinos, G. P. (2023). Triangulating nutrigenomics, metabolomics and microbiomics toward personalized nutrition and healthy living. *Human Genomics, 17*(1), 109.

Li, P., & Song, C. (2022). Potential treatment of Parkinson's disease with omega-3 polyunsaturated fatty acids. *Nutritional Neuroscience, 25*(1), 180–191.

Loh, J. S., Mak, W. Q., Tan, L. K. S., Ng, C. X., Chan, H. H., Yeow, S. H., … Khaw, K. Y. (2024). Microbiota–gut–brain axis and its therapeutic applications in neurodegenerative diseases. *Signal Transduction and Targeted Therapy, 9*(1), 37.

Lopez-Ortiz, A. O., & Eyo, U. B. (2024). Astrocytes and microglia in the coordination of CNS development and homeostasis. *Journal of Neurochemistry, 168*(10), 3599–3614.

Lorca, C., Mulet, M., Arévalo-Caro, C., Sanchez, M. Á., Perez, A., Perrino, M., … Gallart-Palau, X. (2023). Plant-derived nootropics and human cognition: a systematic review. *Critical Reviews in Food Science and Nutrition, 63*(22), 5521–5545.

Lyon, P., Strippoli, V., Fang, B., & Cimmino, L. (2020). B vitamins and one-carbon metabolism: implications in human health and disease. *Nutrients, 12*(9), 2867.

Mafe, A. N., & Büsselberg, D. (2025). Could a Mediterranean diet modulate Alzheimer's disease progression? The role of gut microbiota and metabolite signatures in neurodegeneration. *Foods, 14*(9), 1559.

Makkar, R., Behl, T., Bungau, S., Zengin, G., Mehta, V., Kumar, A., … Arora, S. (2020). Nutraceuticals in neurological disorders. *International Journal of Molecular Sciences, 21*(12), 4424.

Mandato, C., Colucci, A., Lanzillo, R., Staiano, A., Scarpato, E., Schiavo, L., … Napoli, J. S. (2023). Multiple sclerosis—related dietary and nutritional issues: an updated scoping review with a focus on pediatrics. *Children, 10*(6), 1022.

Maraki, M. I., Yannakoulia, M., Xiromerisiou, G., Stefanis, L., Charisis, S., Giagkou, N., … Sakka, P. (2023). Mediterranean diet is associated with a lower probability of prodromal Parkinson's disease and risk for Parkinson's disease/dementia with Lewy bodies: a longitudinal study. *European Journal of Neurology, 30*(4), 934–942.

Mariamenatu, A. H., & Abdu, E. M. (2021). Overconsumption of Omega-6 polyunsaturated fatty acids (PUFAs) versus deficiency of Omega-3 PUFAs in modern-day diets: the disturbing factor for their "balanced antagonistic metabolic functions" in the human body. *Journal of Lipids, 2021*(1), 8848161.

Martínez-Iglesias, O., Naidoo, V., Carrera, I., Corzo, L., & Cacabelos, R. (2023). Natural bioactive products as epigenetic modulators for treating neurodegenerative disorders. *Pharmaceuticals, 16*(2), 216.

Marx, W., Manger, S. H., Blencowe, M., Murray, G., Ho, F. Y.-Y., Lawn, S., … Ruusunen, A. (2023). Clinical guidelines for the use of lifestyle-based mental health care in major depressive disorder: World Federation of Societies for Biological Psychiatry (WFSBP) and Australasian Society of Lifestyle Medicine (ASLM) taskforce. *The World Journal of Biological Psychiatry, 24*(5), 333–386.

Mathur, S., Bhattacharjee, S., Sehgal, S., & Shekhar, R. (2024). Role of artificial intelligence and internet of things in neurodegenerative diseases. In L. Gaur, A. Abraham & R. Ajith (Eds.), *AI and Neuro-Degenerative Diseases: Insights and Solutions* (pp. 35–62). Springer.

McWhirter, L., Ritchie, C., Stone, J., & Carson, A. (2020). Functional cognitive disorders: a systematic review. *The Lancet Psychiatry, 7*(2), 191–207.

Memudu, A. E., Oviosun, A., Nwanama, E. K., & Enyioma-Alozie, S. (2025). Introduction to translational neuroscience of neurodegenerative disorders. In: I. Chatterjee & N. Moradikor (Eds.), *Integrating Neuroimaging, Computational Neuroscience, and Artificial Intelligence* (pp. 1–16). CRC Press.

Milošević, M., Arsić, A., Cvetković, Z., & Vučić, V. (2021). Memorable food: fighting age-related neurodegeneration by precision nutrition. *Frontiers in Nutrition, 8*, 688086.

Mishra, S., Singh, A., Naqvi, W., Garg, P., Srivastava, N., & Srivastava, P. (2025). Implications of systems biology in understanding the pathophysiology of neurological diseases. In: B. Sokouti (Ed.), *Systems Biology and In-Depth Applications for Unlocking Diseases* (pp. 245–257). Elsevier.

Mohd Murshid, N., Aminullah Lubis, F., & Makpol, S. (2022). Epigenetic changes and its intervention in age-related neurodegenerative diseases. *Cellular and Molecular Neurobiology, 42*(3), 577–595.

Moosavi, F., Hosseini, R., Saso, L., & Firuzi, O. (2015). Modulation of neurotrophic signaling pathways by polyphenols. *Drug Design, Development and Therapy, 10*, 23–42.

Mora, I., Arola, L., Caimari, A., Escoté, X., & Puiggròs, F. (2022). Structured long-chain omega-3 fatty acids for improvement of cognitive function during aging. *International Journal of Molecular Sciences, 23*(7), 3472.

Nasir, A., Bullo, M. M. H., Ahmed, Z., Imtiaz, A., Yaqoob, E., Safdar, M., … Yaqoob, S. (2022). Nutrigenomics: epigenetics and cancer prevention: a comprehensive review. *Critical Reviews in Food Science and Nutrition, 60*(8), 1375–1387.

Nazam, F., Shaikh, S., Nazam, N., Alshahrani, A. S., Hasan, G. M., & Hassan, M. I. (2021). Mechanistic insights into the pathogenesis of neurodegenerative diseases: towards the development of effective therapy. *Molecular and Cellular Biochemistry, 476*(7), 2739–2752.

Newton, C. (2020). Amyotrophic lateral sclerosis (ALS): the application of integrative and functional medical nutrition therapy (IFMNT). In: D. Noland, J. A. Drisko & L. Wagner (Eds.), *Integrative and Functional Medical Nutrition Therapy: Principles and Practices* (pp. 863–912). Springer.

Nicolia, V., Lucarelli, M., & Fuso, A. (2015). Environment, epigenetics and neurodegeneration: focus on nutrition in Alzheimer's disease. *Experimental Gerontology, 68*, 8–12.

Obeso, J. A., Monje, M. H., & Matarazzo, M. (2022). Major advances in Parkinson's disease over the past two decades and future research directions. *The Lancet Neurology, 21*(12), 1076–1079.

Oppedisano, F., Macrì, R., Gliozzi, M., Musolino, V., Carresi, C., Maiuolo, J., … Guarnieri, L. (2020). The anti-inflammatory and antioxidant properties of n-3 PUFAs: their role in cardiovascular protection. *Biomedicines, 8*(9), 306.

Orywal, K., Socha, K., Iwaniuk, P., Kaczyński, P., Farhan, J. A., Zoń, W., … Mroczko, B. (2025). Vitamins in the prevention and support therapy of neurodegenerative diseases. *International Journal of Molecular Sciences, 26*(3), 1333.

Pal, A., Goel, F., & Garg, V. K. (2025). From genetics to function: the role of ABCA12 in autism neurobiology. *Journal of Molecular Neuroscience, 75*(2), 67.

Pal, A., Goel, F., Sharma, A., & Garg, V. K. (2025). Oxidative stress and antioxidant therapeutics in autism spectrum disorder: a biochemical and structure–activity relationship perspective. *Molecular Diversity*, 1–19. doi:10.1007/s11030-025-11274-8.

Pandey, P. K., Shrivastava, A., Srivastava, A., Shukla, S., Jain, A., Venkatesh, V., … Jamal, F. (2025). Holistic management of diverse neurodegenerative diseases in different translational models. In: S. N. Rai, S. Singh & S. K. Singh (Eds.), *Neurodegenerative Diseases: Translational Models, Mechanisms, and Therapeutics* (pp. 177–205). CRC Press.

Parker, A., Fonseca, S., & Carding, S. R. (2020). Gut microbes and metabolites as modulators of blood-brain barrier integrity and brain health. *Gut Microbes, 11*(2), 135–157.

Pathak, N., Vimal, S. K., Tandon, I., Agrawal, L., Hongyi, C., & Bhattacharyya, S. (2022). Neurodegenerative disorders of Alzheimer, parkinsonism, amyotrophic lateral sclerosis and multiple sclerosis: an early diagnostic approach for precision treatment. *Metabolic Brain Disease, 37*(1), 67–104.

Pointner, A., & Haslberger, A. G. (2022). Personalized nutrition for healthy aging, a review. In A. G. Haslberger (Ed.), *Advances in Precision Nutrition, Personalization and Healthy Aging* (pp. 97–143). Springer.

Pope, D. B. (2021). *Health Related Quality of Life (HRQoL) in Adults Diagnosed with Multiple Sclerosis.* Capella University.

Rahimmi, A., Tozandehjani, S., Daraei, M., & Khademerfan, M. (2022). The neuroprotective roles of dietary micronutrients on Parkinson's disease: a review. *Molecular Biology Reports, 49*(8), 8051–8060.

Sabahi, M., Joshaghanian, A., Dolatshahi, M., Jabbari, P., Rahmani, F., & Rezaei, N. (2021). Modification of glial cell activation through dendritic cell vaccination: promises for treatment of neurodegenerative diseases. *Journal of Molecular Neuroscience, 71*(7), 1410–1424.

Samanta, S., Chakraborty, S., & Bagchi, D. (2024). Pathogenesis of neurodegenerative diseases and the protective role of natural bioactive components. *Journal of the American Nutrition Association, 43*(1), 20–32.

Shah, P., Banerjee, S., Singh, A., Kulhari, H., & Anand Saharan, V. (2025). From traditional to modern medicine: the role of herbs and phytoconstituents in pharmaceuticals, nutraceuticals, and cosmetics. In: A. Singh, H. Kulhari, & V. A. Saharan (Eds.), *Formulating Pharma-, Nutra-, and Cosmeceutical Products from Herbal Substances: Dosage Forms and Delivery Systems* (pp. 3–73). Wiley.

Shahcheraghi, S. H., Salemi, F., Peirovi, N., Ayatollahi, J., Alam, W., Khan, H., & Saso, L. (2021). Nrf2 regulation by curcumin: molecular aspects for therapeutic prospects. *Molecules, 27*(1), 167.

Shi, R., Gao, D., Stoika, R., Liu, K., Sik, A., & Jin, M. (2024). Potential implications of polyphenolic compounds in neurodegenerative diseases. *Critical Reviews in Food Science and Nutrition, 64*(16), 5491–5514.

Shirodkar, S. S., & Babre, N. (2025). The role of nutrition in neurodegeneration. In T. S. Koduru, R. A. M. Osmani, E. Singh & S. Dutta (Eds.), *The Neurodegeneration Revolution* (pp. 167–202). Elsevier.

Shohag, S., Akhter, S., Islam, S., Sarker, T., Sifat, M. K., Rahman, M. M., … Sharma, R. (2022). Perspectives on the molecular mediators of oxidative stress and antioxidant strategies in the context of neuroprotection and neurolongevity: an extensive review. *Oxidative Medicine and Cellular Longevity, 2022*(1), 7743705.

Short, E., Akers, L., Callahan, E. A., Allen, C. C., Crespo-Bellido, M., Deuman, K., … Steeves, E. A. (2025). The role of registered dietitian nutritionists within food is medicine: current and future opportunities. *Journal of the Academy of Nutrition and Dietetics, 125*(8), 1075–1084.

Singh, A., & Agrawal, N. (2022). Metabolism in Huntington's disease: a major contributor to pathology. *Metabolic Brain Disease, 37*(6), 1757–1771.

Skalny, A., Chernova, L., & Skalnaya, A. (2023). Essential trace elements in neurodevelopment: an updated narrative. In: C. R. Martin, V. Patel & V. R. Preedy (Eds.), *Vitamins and Minerals in Neurological Disorders* (pp. 673–690). Elsevier.

Solovyev, N. D. (2015). Importance of selenium and selenoprotein for brain function: from antioxidant protection to neuronal signalling. *Journal of Inorganic Biochemistry, 153*, 1–12.

Soltani, S., Chitsazi, M. J., & Salehi-Abargouei, A. (2018). The effect of dietary approaches to stop hypertension (DASH) on serum inflammatory markers: a systematic review and meta-analysis of randomized trials. *Clinical Nutrition, 37*(2), 542–550.

Stachowicz, K. (2023). The role of polyunsaturated fatty acids in neuronal signaling in depression and cognitive processes. *Archives of Biochemistry and Biophysics, 737*, 109555.

Sultana, O. F., Hia, R. A., & Reddy, P. H. (2024). A combinational therapy for preventing and delaying the onset of Alzheimer's disease: a focus on probiotic and vitamin co-supplementation. *Antioxidants, 13*(2), 202.

Tao, Y., Leng, S. X., & Zhang, H. (2022). Ketogenic diet: an effective treatment approach for neurodegenerative diseases. *Current Neuropharmacology, 20*(12), 2303–2319.

Thapa, K., Khan, H., Chahuan, S., Dhankhar, S., Kaur, A., Garg, N., … Singh, T. G. (2025). Insights into therapeutic approaches for the treatment of neurodegenerative diseases targeting metabolic syndrome. *Molecular Biology Reports, 52*(1), 260.

Thomas, J., Thomas, C., Radcliffe, J., & Itsiopoulos, C. (2015). Omega-3 fatty acids in early prevention of inflammatory neurodegenerative disease: a focus on Alzheimer's disease. *BioMed Research International, 2015*(1), 172801.

Tragni, V., Primiano, G., Tummolo, A., Cafferati Beltrame, L., La Piana, G., Sgobba, M. N., … Volpicella, M. (2022). Personalized medicine in mitochondrial health and disease: molecular basis of therapeutic approaches based on nutritional supplements and their analogs. *Molecules, 27*(11), 3494.

Tramontano, A., & Palange, P. (2023). Nutritional state and COPD: effects on dyspnoea and exercise tolerance. *Nutrients, 15*(7), 1786.

Trigo, D., Avelar, C., Fernandes, M., Sá, J., & da Cruz e Silva, O. (2022). Mitochondria, energy, and metabolism in neuronal health and disease. *FEBS Letters, 596*(9), 1095–1110.

Tröster, A. I., Pulaski, S. J., & Woods, S. P. (2019). Neuropsychology of movement disorders and motor neuron disease: Parkinson's disease, progressive supranuclear palsy, essential tremor, Huntington's disease, and amyotrophic lateral sclerosis. In: C. L. Armstrong & L. Morrow (Eds.), *Handbook of Medical Neuropsychology: Applications of Cognitive Neuroscience* (pp. 415–440). Springer.

Vergara Nieto, Á. A., Diaz, A. H., Hernández, M., & Sagredo, D. (2025). A narrative review about metabolic pathways, molecular mechanisms and clinical implications of intermittent fasting as autophagy promotor. *Current Nutrition Reports, 14*(1), 78.

Wang, X.-L., Feng, S.-T., Wang, Z.-Z., Yuan, Y.-H., Chen, N.-H., & Zhang, Y. (2021). Parkin, an E3 ubiquitin ligase, plays an essential role in mitochondrial quality control in Parkinson's disease. *Cellular and Molecular Neurobiology, 41*(7), 1395–1411.

Warnecke, T., Schäfer, K. H., Claus, I., Del Tredici, K., & Jost, W. H. (2022). Gastrointestinal involvement in Parkinson's disease: pathophysiology, diagnosis, and management. *NPJ Parkinson's Disease, 8*(1), 31.

Yan, Y., Zhang, Y., Liu, M., Li, L., & Zheng, Y. (2025). Neuroprotection vs. Neurotoxicity: the dual impact of brain lipids in depression. *International Journal of Molecular Sciences, 26*(6), 2722.

Yang, H., Sun, Y., Cai, R., Chen, Y., & Gu, B. (2020). The impact of dietary fiber and probiotics in infectious diseases. *Microbial Pathogenesis, 140*, 103931.

Yehualashet, A. S., Melaku, E. E., Gebreyes, D. S., Wassie, T. A., & Sahilu, B. Y. (2025). The bilateral cross communication in microbiota-gut-brain axis as a promising therapeutic target for Alzheimer's disease: a focus on neuroinflammation. *Discover Medicine, 2*(1), 33.

Yousafzai, A. K., Rasheed, M. A., & Bhutta, Z. A. (2013). Annual research review: improved nutrition–a pathway to resilience. *Journal of Child Psychology and Psychiatry, 54*(4), 367–377.

Yousefi, B., Babaeizad, A., Banihashemian, S. Z., Feyzabadi, Z. K., Dadashpour, M., Pahlevan, D., ... Eslami, M. (2023). Gastrointestinal tract, microbiota and multiple sclerosis (MS) and the link between gut microbiota and CNS. *Current Microbiology, 80*(1), 38.

Zaib, S., Javed, H., Khan, I., Jaber, F., Sohail, A., Zaib, Z., ... Ogaly, H. A. (2023). Neurodegenerative diseases: their onset, epidemiology, causes and treatment. *ChemistrySelect, 8*(20), e202300225.

Zhang, M., Liu, L.-Y., Xu, Y., Wang, W.-Z., Qiu, N.-Z., Zhang, F.-F., ... Xu, X.-Y. (2023). Imbalance of multiple neurotransmitter pathways leading to depression-like behavior and cognitive dysfunction in the triple transgenic mouse model of Alzheimer disease. *Metabolic Brain Disease, 38*(7), 2465–2476.

Zhang, T., Liu, Q., Gao, W., Sehgal, S. A., & Wu, H. (2022). The multifaceted regulation of mitophagy by endogenous metabolites. *Autophagy, 18*(6), 1216–1239.

Zhu, G., Wang, X., Chen, L., Lenahan, C., Fu, Z., Fang, Y., & Yu, W. (2022). Crosstalk between the oxidative stress and glia cells after stroke: from mechanism to therapies. *Frontiers in Immunology, 13*, 852416.

2 Mechanisms of Vitamin Action in the Nervous System

Antioxidant, Anti-inflammatory, and Neurotrophic Functions

Prabhakar Singh[1], *Saqib Hassan*[1], *Mohammad Murtaza Mehdi*[2], *and V. Ramesh Kumar*[1]

[1]Department of Biotechnology, Sathyabama Institute of Science and Technology, Chennai, India
[2]Department of Biochemistry, School of Bioengineering and Biosciences, Lovely Professional University, Phagwara, India

2.1 INTRODUCTION

Vitamins are essential micronutrients that are vital for preserving the nervous system's structural and functional integrity. Recent research has demonstrated their diverse activities as antioxidants, inflammatory modulators, and enhancers of neurotrophic signaling, in addition to their traditional metabolic roles. Because of its high metabolic activity and high lipid content, the nervous system is susceptible to oxidative stress (OS), inflammation, and decreased neurotrophic support, which, in turn, lead to neurodegenerative diseases and cognitive loss. OS, which is caused by the uncontrolled generation of reactive oxygen species (ROS), has been detected to be a characteristic shared by neurodegenerative disorders (Barnham et al., 2004; Onose et al., 2022). Numerous significant biological processes are known to be connected to ROS. Overproduction of ROS can cause oxidative degradation of molecules implicated in aging progression and a number of ailments, including cancer, cardiovascular disease, and neurological diseases. Naturally, the body contains a number of systems, including antioxidants that may balance the effects of OS. These antioxidants, which can be generated internally or obtained externally, can identify ROS and lessen the oxidation of the cellular components that are impacted. Due to high oxygen consumption by the brain, the cells in the brain are vulnerable to ROS and the oxidative damage that follows to proteins, lipids, and nucleic acids (Teleanu et al., 2022).

Natural antioxidants include vitamins like A, C, and E and their derivatives and several B complex vitamins. Vitamin C scavenges free radicals and regenerates oxidized vitamin E, preserving the redox equilibrium, while vitamin E, which is lipid-soluble, localizes inside neuronal membranes to prevent lipid peroxidation. B vitamins like riboflavin and niacin work as cofactors in redox processes necessary for mitochondrial activity. Vitamin A and carotenoids quench singlet oxygen and lower oxidative load. These vitamins operate in concert to create a synergistic antioxidant network that protects neurons and delays the onset of aging. The central nervous system's (CNS) inflammatory reaction to events that disrupt homeostasis is known as neuroinflammation. This reaction involves a variety of CNS cell types, including microglia and astrocytes. This reaction can be seen in ischemia, viral, and traumatic neurological events, in addition to neurodegenerative diseases. Understanding the interactions between the immune system and the CNS is essential because it plays a significant role in the onset and development of various disorders (Solleiro-Villavicencio & Rivas-Arancibia, 2018).

It is important to note that neuroinflammation and OS are two different disease processes. However, when the disease worsens, these two occurrences influence and/or contribute to one another. Consequently, one of them may be suppressed if the other is inhibited (Gao et al., 2014). The activation of microglia is the primary cause of neuroinflammation. When disease, trauma, stress, or pathology induces abnormalities in brain homeostasis, microglia react quickly. Several inflammatory and cytotoxic substances that cause neuroinflammation and neurodegeneration are released when microglia are activated. However, microglia also play advantageous functions in reorganizing neural networks and preserving CNS homeostasis. For example, microglia provide some neuroprotection in Alzheimer's disease (AD) by phagocytically clearing Aβ. Neuronal injury, synapse loss, and inflammation result when microglia cease to perform their advantageous roles. Furthermore, it is shown

DOI: 10.1201/9781003686675-2

that the activation of proinflammatory signaling pathways in AD increases inflammation, OS, and neurodegeneration in the absence of the clearance of accumulated Aβ (Bisht et al., 2018; Gao et al., 2014).

Vitamins have important impacts on neuroinflammatory pathways in addition to their antioxidant properties. The pathophysiology of diseases, including multiple sclerosis (MS), Parkinson's disease (PD), and AD, is significantly influenced by neuroinflammation, which is defined by microglial activation and the production of proinflammatory cytokines. Vitamins have significant immunomodulatory effects by blocking the NF-κB pathway to restrict inflammatory cascades, downregulating inflammatory cytokines, and increasing anti-inflammatory responses. Vitamins maintain synaptic integrity and shield neurons from long-term harm by reducing neuroinflammation. Vitamins' neurotrophic properties, which support neuronal survival, development, and repair, are equally significant. Vitamins increase synaptic plasticity and neural resilience by increasing the expression of neurotrophic factors such as brain-derived neurotrophic factor (BDNF) and nerve growth factor (NGF). Additionally, they control the expression of genes essential for synaptic remodeling, axonal guidance, and neuronal development. These neurotrophic effects demonstrate how vitamins can support both protection and healing in damaged brain systems (Teleanu et al., 2022).

The comprehensive character of vitamin activity in the neurological system is reflected in the interaction of neurotrophic, anti-inflammatory, and antioxidant pathways. While anti-inflammatory effects provide an environment that is conducive to neurotrophic signaling, antioxidant defenses reduce OS and, consequently, inflammatory triggers. When combined, these mechanisms improve cognitive function, preserve neuronal health, and postpone the development of neurodegenerative diseases. Dementia, depression, MS, peripheral neuropathy, and developmental defects have all been closely linked to vitamin deficiencies. Vitamins integrate growth factor pathways, immunological modulation, and redox balance to function as multitargeted modulators of brain health. They are positioned as potential agents for therapeutic interventions in neurological diseases as well as preventative nutrition due to their varied modes of action (Teleanu et al., 2022).

2.2 VITAMINS AND NERVOUS SYSTEM

Vitamins are a class of organic nutrients that must be obtained from food since the body is unable to synthesize them in large enough quantities for a range of biochemical processes. Water-soluble and lipid-soluble vitamins are the two categories into which vitamins fall. Hydrophobic substances, such as lipid-soluble vitamins, can be effectively absorbed only in the event of typical fat absorption. They are linked to certain binding proteins or carried in the circulation as lipoproteins. They should thus be provided every day. They primarily serve as cofactors for enzymes involved in the metabolism of proteins, lipids, and carbohydrates (Murray et al., 2009). Deficiency in vitamins is a common public health issue and arises from a complicated interaction of variables involving insufficient intake, poor absorption, elevated demand, and/or elevated excretion. Chronic deficiency in vitamins induces several pathological conditions, and due to their involvement in myelin production, the majority of neurological disorders linked to vitamin deficits result in peripheral neuropathy.

2.2.1 Vitamin B Complex

2.2.1.1 Thiamine (Vitamin B1)

Thiamine, commonly known as vitamin B1, is an essential nutrient crucial for energy metabolism and the proper functioning of the nervous system. Being a water-soluble vitamin, it is naturally present in foods such as nuts, beans, beef, and pork. Its biologically active form, thiamine pyrophosphate, serves as a coenzyme that plays a key role in several metabolic processes within the brain. The prevalence of vitamin B1 deficiency varies depending on the population and underlying risk factors. In developed countries, it is relatively uncommon and is most often associated with chronic illnesses, malabsorption disorders, or excessive alcohol consumption. In contrast, thiamine deficiency is more widespread in low-income regions, especially among populations whose diets rely heavily on polished rice or other processed grains that have lost their natural thiamine content (Bailey et al., 2015; Nutrients et al., 2000; Wiley & Gupta, 2019).

Neurological manifestations of thiamine deficiency include peripheral numbness, tingling, paresthesias, muscle weakness, loss of reflexes, disorientation, memory impairment, delirium, seizures, behavioral changes (such as depression, irritability, and apathy), and visual disturbances like ophthalmoplegia or nystagmus. When the deficiency primarily causes symmetrical peripheral neuropathy, it is referred to as 'dry' beriberi, which is more commonly seen in children. In adults, thiamine deficiency may present as Wernicke encephalopathy (WE), characterized by dementia-like symptoms. WE is classically defined by the clinical triad of confusion, ataxia, and nystagmus (Smith et al., 2021). Wernicke–Korsakoff syndrome (WKS) or Korsakoff syndrome might develop from WE if treatment is not received. Confabulation, altered mental state, and profound memory loss (especially anterograde memory) are the clinical triads that define the latter (Smith et al., 2021). Although early neurologic symptoms can frequently be promptly resolved with therapy, severe instances or postponed treatment result in irreversible consequences (i.e., neurological symptoms are permanent when Korsakoff syndrome is present) (Nieto-Salazar et al., 2023).

2.2.1.2 Riboflavin (Vitamin B2)

Riboflavin, also called vitamin B2, is a water-soluble vitamin in the B complex. When consumed in high quantities, it is distinguished by the distinctive brilliant yellow coloring of the urine. As a cofactor for many flavoprotein enzyme activities, riboflavin is involved in a broad range of metabolic routes and functions. Flavin mononucleotide (FMN) and flavin adenine dinucleotide (FAD) are the active forms of riboflavin. Riboflavin has been reported as a neuroprotective agent. Indeed, riboflavin has shown promise in addressing important pathogenesis-related pathways in neurological diseases, such as those linked to the pathophysiology of migraine and PD. In fact, riboflavin reduces neuroinflammation, glutamate excitotoxicity, mitochondrial dysfunction, and OS, which are linked to the etiology of a variety of neurological conditions. Furthermore, pyridoxine activation requires riboflavin. Any buildup of kynurenines or homocysteine brought on by a vitamin deficiency can have serious neurological effects. Given that 10%–15% of the global population exhibits limited riboflavin absorption and utilization, prolonged riboflavin insufficiency may contribute to the onset of various neurological disorders. This highlights the importance of maintaining adequate long-term riboflavin intake, particularly in vulnerable groups. To further establish its clinical role, large-scale, long-duration randomized double-blind placebo-controlled trials are warranted in patients with PD and migraine (Marashly & Bohlega, 2017).

2.2.1.3 Niacin (Vitamin B3)

A member of the vitamin B complex, niacin (vitamin B3) participates in a variety of metabolic processes as a coenzyme (Oliveira et al., 2019). Vitamin B3 deficiency, known as pellagra, remains most prevalent in low-income regions, particularly in parts of Africa, where the disease continues to occur more frequently than in other areas of the world (Li et al., 2016). In contrast, pellagra is relatively uncommon in Latin America, China, and India, despite maize-based diets, due to traditional food practices that improve niacin availability. In the United States, pellagra is rare owing to the fortification of foods such as wheat, and when it does occur, it is typically associated with chronic alcoholism or malabsorption syndromes rather than dietary deficiency (Berdanier, 2019). Lack of intake from dietary sources such as meats, fish, bread, cereal, and legumes is the primary cause of primary niacin insufficiency (Fukuwatari & Shibata, 2013). The most frequent causes of secondary pellagra include drug addiction, gastrointestinal disorders, and alcoholism (Badawy, 2014). Insomnia, disorientation, anxiety, dementia, delusions, encephalopathy, and muscular weakness are among its neurological symptoms and indicators, which are characterized by gastrointestinal and dermatological abnormalities. Laboratory measurement of niacin, tryptophan, nicotinamide adenine dinucleotide (NAD), and NAD phosphate (NADP) levels can confirm a suspected niacin deficit (Gasperi et al., 2019). The goal of treating niacin deficiency is to replenish the vitamin by taking oral nicotinamide supplements, which should include 250–500 mg per day. To prevent the condition from developing in the future, it is also critical to determine the origin of this deficit, treat it appropriately, and take preventative steps (Hegyi et al., 2004).

2.2.1.4 Pantothenic acid (Vitamin B5)

Vitamin B5, also known as pantothenic acid (PA), is an essential nutrient that occurs naturally in certain foods, is incorporated into fortified products, and is also available as a dietary supplement. This water-soluble B vitamin plays a central role in the biosynthesis of acyl carrier protein and coenzyme A (CoA) (Miller & Rucker, 2012). CoA is crucial for fatty acid synthesis and degradation, the transfer of acetyl and acyl groups, and multiple anabolic and catabolic processes (Faran, 2020; Nutrients et al., 2000). The primary function of the acyl carrier protein, on the other hand, is to facilitate fatty acid synthesis.

An enzyme called PA kinase is necessary for the synthesis of CoA and phosphopantetheine. CoA production is carried out by this key enzyme, which is connected to the metabolic pathway. An uncommon hereditary condition known as pantothenate kinase–associated neurodegeneration (PKAN) is brought on by mutations in the pantothenate kinase 2 (PANK2) gene. Iron buildup in the brain is linked to PKAN, a kind of neurodegeneration (Faran, 2020). Numerous mutations in PANK2 decrease the activity of pantothenate kinase 2, which may lower the amount of PA converted to CoA and, consequently, lower levels of CoA. Dystonia (contractions of opposing muscle groups), pigmentary retinopathies, and plasticity are some of the symptoms of PKAN (Faran, 2020; Hayflick, 2014). It progresses quickly, resulting in substantial impairment and functional loss (Hayflick, 2014). The main goal of treatment is to lessen symptoms (Gregory & Hayflick, 2017). Although the benefits of pantothenate supplementation in PKAN are unknown, anecdotal findings suggest that some individuals with atypical PKAN may experience a reduction in symptoms (Kurian & Hayflick, 2013).

2.2.1.5 Pyridoxine (Vitamin B6)

Vitamin B6, often known as pyridoxine, is a necessary component for several metabolic functions. For instance, it is essential for the synthesis of neurotransmitters, amino acid metabolism, and the production of red blood cells. Foods that naturally contain pyridoxine include fish, chicken, beans, almonds, and bananas. It may also be purchased as a supplement and is frequently used to treat several diseases, including premenstrual syndrome, carpal tunnel syndrome, and morning sickness (Baviera-Muñoz et al., 2022; Nieto-Salazar et al., 2023).

Pyridoxine aids in the maintenance of healthy eyes, skin, and liver and is vital for the nervous system to operate properly. It also has a part in controlling hormone levels and the immune system. Many symptoms, such as anemia, seizures, dermatitis, depression, and disorientation, can be brought on by a pyridoxine deficiency. Additionally, it may raise the risk of heart disease and several cancers (Nieto-Salazar et al., 2023).

Severe starvation, drunkenness, liver disease, pregnancy, and isoniazid therapy can all result in pyridoxine deficiency (Baviera-Muñoz et al., 2022). At the neurological level, mild deficits show themselves as sadness, irritability, disorientation, and even peripheral neuropathy. Seizures are linked to severe deficiencies. Furthermore, pyridoxine-dependent epilepsy, which manifests as refractory newborn seizures, is caused by an inborn defect in pyridoxine metabolism (Nieto-Salazar et al., 2023).

2.2.1.6 Folate (Vitamin B9)

Folate is a water-soluble vitamin found naturally in foods such as leafy green vegetables, liver, and fruits. Its synthetic counterpart, folic acid, is used in dietary supplements and fortified foods and is absorbed more efficiently by the body. Folate is crucial for DNA and RNA synthesis, red blood cell formation, and the prevention of congenital anomalies like neural tube defects. Studies from the Centers for Disease Control and Prevention (CDC) indicate that folate deficiency occurs more frequently among individuals from low-income backgrounds and elderly people in institutional care, largely due to poor nutrition and cognitive decline. Women of reproductive age, particularly non-Hispanic women, are also considered a high-risk group because of dietary insufficiency. Other contributing factors to folate deficiency in adults include celiac disease, gastric bypass surgery, malabsorption disorders, amyloidosis, inherited enzyme defects affecting folate metabolism, vitamin B12 deficiency, and the use of drugs such as methotrexate, phenytoin, sulfasalazine, and trimethoprim. Chronic alcoholism, pregnancy, hemolytic anemia, and long-term hemodialysis further increase the likelihood of deficiency.

Nowadays, folic acid flour fortification is required in the majority of industrialized nations, and folic acid supplementation is advised both before conception and throughout the first trimester of pregnancy (Khan & Jialal, 2023). The neurologic symptoms of vitamin B-12 insufficiency and folate inadequacy are comparable. These include dementia, depression, AD (accompanied by elevated homocysteine levels), cognitive impairment, coupled spinal cord degeneration, and, in certain situations, peripheral neuropathy. In certain cases, folate deficiency may worsen the underlying condition as a result of poor nutrition, apathy, or mental disorders (Reynolds, 2014). Recent research has begun to explore the potential influence of folic acid on behavioral outcomes in children, particularly in relation to autism spectrum disorder (ASD). Some studies have suggested a possible association between ASD risk and either serum autoantibodies directed against folate receptor alpha or genetic mutations in the methylenetetrahydrofolate reductase (MTHFR) enzyme. However, the evidence remains inconsistent (Reynolds, 2014; Shulpekova et al., 2021). Given the observational data supporting folate's protective effect against neural tube defects, larger clinical trials with extended follow-up periods are needed to clarify its role in neurodevelopmental outcomes.

2.2.1.7 Cobalamin (Vitamin B12)

Vitamin B12, or cobalamin, is a vital nutrient required for key metabolic processes such as DNA synthesis and myelin formation. It is naturally present in animal-derived foods, including red meat, dairy products, and eggs (Briani et al., 2013). Deficiency can occur in both men and women, with higher prevalence in individuals over the age of 60. One of the major clinical consequences of cobalamin deficiency is megaloblastic anemia, which may present with symptoms such as weakness, fatigue, and shortness of breath. Additional manifestations include skin pigmentation changes, nausea, glossitis, vomiting, and diarrhea. Since vitamin B12 is essential for nervous system maintenance, its deficiency can also give rise to a range of neurological disturbances.

Nerve injury and dysfunction may result from demyelination of the nerves. Peripheral neuropathy, which causes weakness, tingling, and numbness in the hands and feet, may arise from this. Weakness, spasticity, poor fine motor control, and trouble balancing and coordinating are some other neurologic symptoms. Severe cognitive impairment, dementia, and even mental diseases (such as depression, mania, and psychosis) may result from the deficiency (Senol et al., 2008). If cobalamin insufficiency is not addressed for an extended length of time, the neurologic effects are usually irreversible. Therefore, avoiding the emergence of severe neurologic symptoms and irreversible consequences requires early detection and treatment. Since water-soluble and fat-soluble vitamins can both produce comparable neurological symptoms, the patient's medical history, physical examination, and test results should always be included in the diagnosis and treatment process. Finding the main source of the shortfall and making sure that intake is restored are the two main steps in treating this nutritional deficit. Numerous patient characteristics influence the length of therapy and the dosage of supplements (Nieto-Salazar et al., 2023).

2.3 ASCORBIC ACID (VITAMIN C)

Ascorbic acid, referred to as ascorbate, is a vital antioxidant that is in the bloodstream and has anti-inflammatory and immune-boosting properties. The enzyme L-gulono-1,4-lactone oxidase, which is required for the synthesis of vitamin C, is absent in humans. Therefore, the body's ascorbate content is solely determined by the quantity ingested through nutrition. Under these severe conditions, vitamin C deficiency develops in critically sick patients (Doseděl et al., 2021; Spoelstra-de Man et al., 2018). According to a national study conducted in the United States in 2003–2004, 7% of people, especially smokers (who are more likely to have this deficiency), had vitamin C levels below the recommended limits. The plasma levels of vitamin C decrease in a number of diseases and other circumstances, including burns, trauma, and sepsis, usually due to the increased demand for vitamin C in these conditions (Doseděl et al., 2021). In scurvy, the majority of symptoms often appear when plasma levels fall below 0.15 mg/dL. The clinical manifestation of vitamin C insufficiency is diverse. Lethargy, irritability, and disorientation are some of the condition's early symptoms. Additional symptoms include bleeding, muscular weakness and petechial hemorrhage, tooth loss, swollen gums, spontaneous ecchymoses, poor wound healing, anemia, weakness, arthralgia, myalgia, and weight loss as the deficiency worsens. Since vitamin C is required for Schwann cells to make myelin, its epigenetic impact may be linked to peripheral neuropathy that causes calves to go numb and experience discomfort. Peripheral neuropathy is therefore believed to be caused by demethylating effects, which can influence Schwann cell differentiation, proliferation, and epigenetic activation (Valdés, 2006; Young et al., 2015). Vitamin C provides therapeutic benefits in neurodegenerative diseases and is necessary for the growth and physiological processes of neural tissue (Granger & Eck, 2018; Kocot et al., 2017; Moretti et al., 2017). Vitamin C treatment has been shown to enhance functional recovery after spinal cord damage in both clinical and animal investigations (Li et al., 2019; Murray et al., 2009). In a model with mild nerve damage, vitamin C can effectively stimulate nerve regeneration. This effect is linked to vitamin C's impact on neurons, SCs, and macrophages, all of which are important cells involved in nerve damage and repair.

2.4 RETINOL (VITAMIN A)

Retinol, retinal, and retinoic acid are the three active forms of vitamin A. They all make up a vital nutrient that is required in trace levels for the preservation of cell function for development and the proper operation of the visual system. However, the human body is unable to synthesize this essential ingredient; thus, it must be obtained from the diet. Chronically reduced food intake will result in inadequate vitamin A to support cellular functions and eyesight, which will decrease function (Günes-Bayir, 2025). The primary underlying cause of vitamin A insufficiency is inadequate vitamin intake from food, which makes it difficult to satisfy physiological demands. One of the most frequent forms of avoidable juvenile blindness, xerophthalmia, can arise from a chronic or severe deficiency (Günes-Bayir, 2025; Nieto-Salazar et al., 2023). Furthermore, decreased serum retinol levels, increased urine excretion of vitamin A, and the depletion of hepatic vitamin A supply have all been linked to viral diseases. For instance, children with low vitamin A levels are known to develop blindness as a result of conjunctival and corneal damage caused by measles virus infection.

Vitamin A insufficiency is linked to a number of clinical symptoms. First, ophthalmological symptoms, such as enlarged eyes, enlargement of the cortical disk, and vision abnormalities, particularly at night (night blindness), are among the most prevalent symptoms. Lastly, severe headaches, sleepiness, psychosis, coma, elevated intracranial pressure (pseudotumor cerebri), bulging fontanelles (in babies), mental abnormalities, cerebral edema, and irritability are examples of neurological symptoms (Vitamin, 2022). Especially in experimental models of diabetic neuropathy and peripheral nerve injury, vitamin A can boost NGF and enhance functional recovery in situations of nerve damage. It functions as a precursor to retinoic acid and affects cellular functions such as survival, proliferation, and differentiation, all of which are essential for the nervous system's growth and regeneration. Although further study is required, vitamin A is being investigated as a possible supplementary treatment for nerve injury, providing a more affordable and easily available alternative to other retinoids.

2.5 VITAMIN D (CALCIFEROL)

Vitamin D is a source of steroid hormones and a member of the fat-soluble vitamin family. It comes in two main forms: ergocalciferol (vitamin D2) and cholecalciferol (vitamin D3), both of which need to be hydroxylated in the kidney and liver in order for tissues to be biologically activated (Nikolac Gabaj et al., 2020; Ramasamy, 2020). The human body expresses the vitamin D receptor broadly (Lopez et al., 2021). Numerous biological functions are linked to vitamin D, and a lack of vitamin D is linked to several clinical disorders. Vitamin D's dietary intake effect has been found to have a positive impact on peripheral nerve regeneration. A rat model involving a 10 mm peroneal nerve transection repaired with an autograft was employed to investigate the therapeutic potential of vitamin D2 (Chabas et al., 2008). According to the study, ergocalciferol enhances axonal diameter and encourages axonal regeneration. The regulating action of vitamin D on many neurotrophic factors involved in nerve regeneration might be responsible for this improvement, with NGF showing an enhanced production (El Soury et al., 2021). Furthermore, a decrease in axon degeneration near the injury site was noted in the vitamin D–treated group, indicating a neuroprotective function for vitamin D2. Nevertheless, no discernible differences in functional recovery between the vitamin D–treated and untreated animals were found using the peroneal functional index (Chabas et al., 2008).

Vitamin D3 was shown to be more efficient than vitamin D2 when its effects on nerve regeneration were compared (Chabas et al., 2013). Cholecalciferol did, in fact, improve functional recovery more than ergocalciferol. There was a dose-dependent impact. Furthermore, compared to ergocalciferol, cholecalciferol therapy enhanced the myelination of axons and increased their width and number. Lastly, calcitriol, a physiologically active form of vitamin D3, was shown to upregulate several myelin-associated genes *in vitro* by microarray research on DRG and SCs activated with it for 24 hours (Chabas et al., 2013; Montava et al., 2015). In rabbits following facial nerve transection and autograft repair, the beneficial impact of vitamin D3 on functional recovery was also verified. When compared to vehicle therapy, weekly vitamin D3 administration enhanced myelination and decreased facial paralysis (Montava et al., 2015). In comparison with animals that were not treated, a recent study showed that either cyanocobalamin (B12) or cholecalciferol (given orally) improved the sciatic nerve functional index values and decreased the number of degenerating axons following sciatic nerve crush. Remarkably, a synergistic impact of enhanced nerve regeneration was even achieved with combination treatment (Albay et al., 2020).

Epidemiological research indicates that vitamin D insufficiency is associated with a variety of neuropsychiatric diseases and neurodegenerative diseases (Wassif et al., 2023). While vitamin D deficiency early in childhood has been associated with neuropsychiatric diseases like schizophrenia, vitamin D deficiency in adulthood has been connected to a number of detrimental brain outcomes, including PD, AD, depression, and cognitive decline. Numerous clinical and animal investigations suggest that vitamin D supplementation may have some "neuroprotective" benefits for PD, epilepsy, MS, and chronic stress (Berridge, 2017; Fenercioglu, 2024; Wassif et al., 2023).

A shift in brain activity, including the suppression of "gamma-aminobutyric acid (GABA)," an "inhibitory neurotransmitter," and the rise of "glutamate," an "excitatory neurotransmitter," leads to depression. GABA provides a relaxing and soothing impact when it attaches to GABA receptors (Berridge, 2017; Krystal et al., 2002). However, its suppression might result in despair, worry, and stress. This results from elevated calcium levels in inhibitory neurons, which trigger necrosis and neuronal death. Calcium homeostasis in neurons is regulated by vitamin D. Vitamin D controls serotonin levels, which are important in the regulation of mood and the prevention of depression by preserving calcium homeostasis and promoting tryptophan (Wassif et al., 2023). Because vitamin D affects early neurodevelopment, the immune system, and gene control mechanisms, maternal/neonatal vitamin D insufficiency has been suggested as a potential environmental risk factor for ASDs. Similar to individuals impacted by vitamin D deficiency, people with autism show immune function abnormalities, including increased levels of inflammatory cytokines and persistent inflammation in their brains (Hadi et al., 2021). Vitamin D suppresses the tryptophan hydroxylase 1 gene transcription in tissues outside the blood–brain barrier (BBB) and promotes the transcription of the serotonin-synthesizing gene tryptophan hydroxylase 2 in the brain. Serotonin's low levels in the brain and high levels in tissues outside the BBB in autism are explained by this mechanism (Patrick & Ames, 2014).

2.6 TOCOPHEROL (VITAMIN E)

Vitamin E is one of the most important lipid-soluble antioxidant nutrients. It occurs naturally in foods such as nuts, seeds, and vegetable oils and is also available as a supplement or as an additive in fortified products (Colombo, 2010). Since humans cannot synthesize vitamin E, it must be obtained through dietary sources. The vitamin E family consists of eight isomers: four tocopherols (α, β, γ, and δ) and four tocotrienols (α, β, γ, and δ) (Lešić et al., 2024; Miyazawa et al., 2019).

Recent studies suggest that vitamin E exerts multiple beneficial health effects, including anti-inflammatory, neuroprotective, anti-cancer, cardioprotective, anti-diabetic, anti-lipidemic, anti-hypertensive, and anti-atherogenic properties, as well as a potential role in modulating telomerase activity (Nieto-Salazar et al., 2023; Umer et al., 2025). Antioxidants prevent the chain reaction of lipid peroxidation by lowering the assault of free radicals. They also preserve cell membranes by replacing and repairing lipids. They are thought to have an antioxidant effect because they scavenge ROS and prevent membrane lipid peroxidation. They might also avert heart disease, cancer, and neurological conditions. Apart from its antioxidant properties, vitamin E also plays a role in cell communication, immunological function, gene expression, and other metabolic processes, as demonstrated mainly by *in vitro* cell research (Dror & Allen, 2011; Umer et al., 2025). Low circulation levels of other antioxidants, such as vitamin C and β-carotene, are frequently observed in conjunction with vitamin E deficiency, which lends credence to the idea that deficits are linked to inadequate consumption and more severe OS. The CNS, which is susceptible to the harmful effects of free radicals, can be severely impacted by a severe vitamin E deficit. Different levels of neurologic impairments may result from associated vitamin shortages in conditions such as abetalipoproteinemia, short bowel syndrome, isolated vitamin E deficiency syndrome caused by genetic anomalies in α-TTP, cystic fibrosis, chronic cholestatic liver disease, and other malabsorption syndromes (Tanyel & Mancano, 1997).

Hyporeflexia, ataxia, difficulties looking upward, and strabismus are the initial symptoms of vitamin E insufficiency. These are followed by long-tract abnormalities, severe muscular weakness, and restriction of the visual field. Infertility, dementia, heart arrhythmias, and total blindness (retinitis pigmentosa) may all occur in patients with a severe, protracted deficit (Wiley & Gupta, 2019). Individuals with autosomal recessive mutations in the α-TTP gene develop ataxia with vitamin E deficiency (AVED), which presents with neurological features such as cerebellar ataxia, loss of deep tendon reflexes, impaired vibratory sensation, dysarthria, muscle weakness, head tremors, and dystonia. The peripheral neuropathy and anemia associated with vitamin E deficiency are thought to result from excessive free radical–induced damage to erythrocyte membranes and large-caliber sensory axons. Human neurologic disorders are linked to plasma α-tocopherol concentrations less than 8 μmol/L, but *in vitro*, increased red blood cell fragility is linked to amounts less than 12 μmol/L (Dror & Allen, 2011).

The discovery of a higher number of neurons in rat DRGs under vitamin E–deficit conditions—an increase typically seen in older animals only—further supported the link between vitamin E deficiency and premature aging (Cecchini et al., 1993). The condition of vitamin E insufficiency again led to an increase in sciatic sensory neurons in the follow-up investigation. However, rats on a vitamin E–deficient diet did not differ from rats fed normally in terms of neuron loss following sciatic nerve compression damage (El Soury et al., 2021). On the other hand, it showed that rats following sciatic nerve crush damage and fed a diet lacking in vitamin E had poor nerve regeneration, as seen by fewer myelinated axons with irregular and thinner myelin compared to rats that were properly fed. OS following nerve injury can lead to myelin sheath oxidation, potentially impairing the myelination process. In recent studies, the role of vitamin E in peripheral nerve regeneration has been evaluated using a rat model with a 10 mm sciatic nerve gap repaired by a chitosan conduit (El Soury et al., 2021). The conduit was supplemented with vitamin E, pyrroloquinoline quinone (PQQ)—a redox-active antioxidant cofactor—or a combination of both compounds. Peripheral nerve regeneration was significantly improved and accelerated, as shown by functional, morphological, and electrophysiological research. This improvement was much more noticeable in mice that received topical treatment with a combination of vitamin E and PQQ. These results demonstrated that local vitamin E treatment can decrease OS, which has an impact on neuron regeneration (El Soury et al., 2021).

2.7 VITAMIN K

Vitamin K refers to a group of fat-soluble compounds characterized by a common 2-methyl-1,4-naphthoquinone ring structure. The main forms include vitamin K1 (phylloquinone [PK]), vitamin K2 (menaquinones [MKn]), and vitamin K3 (menadione). These variants differ in their biological roles, target activities, food sources, and in the nature of the side chains attached at the 3-position of the 2-methyl-1,4-naphthoquinone ring (Fusaro et al., 2017; Simes et al., 2019, 2020; Willems et al., 2014). Due to the tissues' limited capacity to retain them, vitamin K should be regularly obtained from food (Fusaro et al., 2017). Numerous physiological circumstances and biological processes are impacted by vitamin K, and a lack of it is connected to a number of pathological disorders, including pathological mineralization and inflammation (Fusaro et al., 2020). Vitamin K functions as both an antioxidant agent and an anti-inflammatory agent, providing protective effects on the CNS. While its primary clinical application is in blood coagulation, emerging evidence suggests that vitamin K may help slow the progression of MS (Lasemi et al., 2018). This is consistent with the understanding that OS contributes to neurodegenerative diseases, as vitamin K appears to protect neurons and oligodendrocytes from oxidative damage (Sakaue et al., 2011). Additionally, it may play a role in promoting remyelination following acute demyelination (Popescu et al., 2018).

Vitamin K's function in several biological processes is widely recognized, but its effects on the peripheral nervous system have not been thoroughly studied. In fact, rats that had their sciatic nerve surgically removed (neurectomy) were frequently employed as models for osteoporosis (Iwamoto et al., 2010) and to study the effects of vitamin K on bone structures, but this did not account for its role in peripheral nerve regeneration. Tsang demonstrated that vitamins K1 and K2 enhance NGF-mediated neurite outgrowth in PC12D cells *in vitro* through activation of PKA and MAPK-dependent signaling pathways (Tsang & Kamei, 2002). When combined with NGF, both K vitamins significantly increased the proportion of cells exhibiting neurite outgrowth compared to NGF treatment alone. However, vitamins K1 and K2 alone did not induce neurite outgrowth in PC12D cells. To date, there appear to be no additional studies investigating the effects of vitamin K on the peripheral nervous system (Table 2.1).

TABLE 2.1
Neuroprotective Effect of Vitamins

Vitamin	Neuroprotective Effect	Action Mechanism	References
B_1	Maintains neuronal function and prevents cognitive deterioration	Neuronal function and energy metabolism	Calderón-Ospina and Nava-Mesa (2020), Hrubša et al. (2022), and Jia et al. (2024)
B_2	Supports visual perception and cognitive processes while shielding neurons from oxidative damage	Myelin health, neurotransmitter control, antioxidant defense, and energy metabolism	Benedik (2021), Plantone et al. (2021), and Suwannasom et al. (2020)
B_3	Enhances cognitive function, lowers inflammation, shields neurons from oxidative damage, and supports mitochondrial function	Antioxidant defense, energy metabolism, neurotransmitter control, and anti-inflammatory	Benedik (2021), Liu et al. (2020), and Wuerch et al. (2023)
B_6	Reduces homocysteine, improves mood and motor coordination, and slows down cognitive decline	Synthesis of neurotransmitters (dopamine, GABA, and serotonin), metabolism of homocysteine	Calderón-Ospina & Nava-Mesa (2020) and Modica et al. (2020)
B_7	Enhances energy metabolism, neuron survival, remyelination, and inflammation reduction	Gene expression, mitochondrial energy metabolism, and myelin synthesis	Leon-Del-Rio (2019) and McCarty and DiNicolantonio (2017)
B_9	Lowers homocysteine, promotes neurodevelopment, and stops cognitive decline	Neurotransmitter production, homocysteine control, DNA synthesis, and repair	Bennett et al. (2022), Lee et al. (2024), and Shemirani et al. (2023)

(Continued)

TABLE 2.1 (*Continued*)
Neuroprotective Effect of Vitamins

Vitamin	Neuroprotective Effect	Action Mechanism	References
B_{12}	Protects myelin, lowers oxidative stress, and enhances cognitive and motor abilities	Regulation of homocysteine, myelin formation, DNA synthesis, and neurotransmitters	El-Mezayen et al. (2022), Lee et al. (2024), and Liu et al. (2023)
C	Improves cognitive function and shields neurons from oxidative damage	Neuroprotection, neurotransmitter production, and antioxidant	Chang et al. (2020), Hamid et al. (2022), and Hansen et al. (2014)
A	Lowers neuroinflammation, promotes dopamine synthesis, and lessens beta-amyloid plaques	Immune modulation, gene regulation, and synaptic plasticity	Joshi et al. (2023), Marie et al. (2021), and Nolan et al. (2022)
K	Lowers inflammation and enhances mental performance	Metabolism of sphingolipids and anti-inflammatory	Booth et al. (2022), Diachenko et al. (2024), and Regulska-Ilow et al. (2022)
D	Enhances neuronal survival, lowers neuroinflammation, supports cognitive and motor function	Neurotransmitter modulation, neuroinflammation, and calcium homeostasis	Benedik (2021), Pignolo et al. (2022), and Sangha et al. (2023)
E	Lowers inflammation, enhances synaptic function, and shields neurons from oxidative damage	Support for neurotrophic factors and antioxidants	Hao et al. (2024), Joshi et al. (2023), Nolan et al. (2022), Zaaboul and Liu (2022)

2.8 OXIDATIVE STRESS AND INFLAMMATION IN THE PATHOGENESIS OF NEUROLOGICAL DISORDERS

Neuroprotection refers to strategies aimed at preventing or minimizing damage to nerve cells and other components of the nervous system, including the brain, spinal cord, and peripheral nerves. Its primary goal is to safeguard the CNS against acute and progressive neurodegenerative disorders. OS—an imbalance between the body's antioxidant defenses and the production of ROS—plays a key role in the development of neurological diseases. The brain is particularly vulnerable to OS due to its high metabolic activity and oxygen consumption. Excessive ROS can damage essential macromolecules, leading to neurodegeneration and cellular dysfunction. OS has been implicated in a variety of neurological conditions, including ischemic stroke, MS, AD, PD, and amyotrophic lateral sclerosis (ALS) (Dash et al., 2025).

The quality of life and life expectancy of older adults with neurodegenerative diseases such as PD, AD, ALS, Huntington's disease (HD), and vascular dementia are greatly impacted by cognitive impairment and dementia. These conditions are characterized by memory loss, motor impairments, and decreased mobility. Studies have shown links between OS indicators such as glutathione peroxidase (GSH-Px), malondialdehyde (MDA), and protein carbonyls (PCs) with cognitive impairment, especially in older adults. OS is a key component in the pathogenesis of dementia. Increased GSH-Px activity has been linked to a slower rate of cognitive decline. However, ironically, higher GSH levels might slow down the decline, perhaps as a result of increased OS and decreased GSH-Px activity. Additionally, by sequestering vital proteins and RNAs, OS aids in the formation of stress granules (SGs), which impair neuronal function. OS is intricately connected to DNA damage, abnormal protein folding, and aggregation in NDDs, which exacerbates tissue damage and advances the disease. The brain's high metabolic activity and vulnerability to neuroinflammatory reactions, which are triggered by activated microglia in response to oxidative damage, make it more susceptible to oxidative damage. OS is a key component of NDDs such as AD, PD, HD, and ALS diseases, and neuroinflammation exacerbates tissue damage and advances the disease. An intriguing topic of debate that may be investigated on several levels is behavioral and cognitive impairment brought on by OS through a sequence of chain reactions in the CNS. In essence, different types of neuronal cells are vulnerable to varying degrees of OS. The amygdala, hippocampus, and cerebellar granule cells are the neuronal cells most susceptible to OS; as a result, they initially experience functional impairment (Wang & Michaelis, 2010). According to many preclinical research studies, behavioral problems and cognitive impairments are caused by three different parts of the brain: the amygdala, hippocampus, and prefrontal cortex (PFC) (Masood et al., 2008; Solanki et al., 2015).

Significant biochemical alterations in the hippocampal region provide insight into the relationships and activities of neuronal cells. Furthermore, there are structural flexibility and remodeling/regenerative capacity in the hippocampus's dentate gyrus (DG)–cornu ammonis (CA)3 system (McEwen, 2008; Sousa et al., 2000). Subsequent studies showed that pyramidal cells of CA1 have a higher level of oxidative damage than pyramidal cells of the DG, yet both are susceptible to OS (Chang et al., 2012; Huang et al., 2012; Huang et al., 2013). Increased OS in the CA regions (DG, CA1, and CA3) is critical and has significant functional implications; for example, DG is linked to memory and learning, and the ventral hippocampus is specific for anxiety and depression. The amygdala and PFC also exhibit changes in the dendritic system as a result of long-term OS. Amygdalar neurons are in charge of dendritic growth brought on by stress (Leuner & Shors, 2013; Ng et al., 2018), and the medial PFC is a key location for dendritic shrinkage. These factors can alter the prefrontal dendritic structural design and connection in PFC neuronal networks (Liston et al., 2009). Remarkably, OS in the brain includes biochemical, molecular, and structural integrity in the hippocampus and amygdala, as evidenced by a greater degree of oxidative damage and the breakdown of the antioxidant defense mechanism in these regions. Thus, OS in DG–CA function can harm neurogenesis, cell proliferation, remodeling ability, and structural plasticity, all of which can disrupt normal synaptic neurotransmission. Additionally, glutamatergic N-methyl-D-aspartate receptors are damaged by free radicals, which lower LTP and synaptic transmission (Haxaire et al., 2012; Rai et al., 2013). These incidents subsequently brought attention to the part OS plays in cognitive and behavioral deterioration.

The formation of protein aggregates, intracellular tau (τ), extracellular amyloid plaques ($A\beta$), or neurofibrillary tangles is a characteristic of AD. Higher concentrations of byproducts of oxidized lipid, protein, and DNA/RNA have been linked to Aβ-induced oxidative imbalance, according to a number of studies (Hassan et al., 2022). Additionally, the scientists noted that AD patients had lower levels of several antioxidants, including vitamin C, uric acid, and vitamin E, and antioxidant enzymes like catalase (CAT) and superoxide dismutase. Additionally, there are reports of an aberrant distribution of transition metals in AD, which might catalyze the OS. AD has also been linked to elevated RNS levels in neurons and astrocytes. It has been shown that astrocytes gradually modify the expression of cytokine-inducible (iNOS or NOS-2), neuronal (nNOS or NOS-1), and endothelial (eNOS or NOS-3) isozymes. Additionally, iNOS and eNOS are directly associated with Aβ aggregates, which further stimulates the production of nitric oxide (NO) via NO synthases (NOS). One of the main factors contributing to the production of $ONOO^-$, which damages proteins, lipids, and DNA/RNA, is elevated NO levels (Hassan et al., 2022).

Accumulating evidence indicates that the substantia nigra of PD patients contains reduced levels of glutathione (GSH) alongside increased amounts of oxidized proteins, lipids, and DNA (Hassan et al., 2022). Specifically, elevated levels of 4-hydroxynonenal (4-HNE), PCs, 8-hydroxy-deoxyguanosine, and 8-hydroxy-guanosine have been reported. Experimental studies using various OS-inducing toxins—such as 1-methyl-4-phenyl-1,2,3,6-tetrahydropyridine (MPTP), rotenone, paraquat (1,1′-dimethyl-4,4′-bipyridinium dichloride), and 6-hydroxydopamine (6-OHDA)—further support the link between OS and dopaminergic neuronal degeneration (Hassan et al., 2022). In PD patients, the ratio of reduced glutathione to its oxidized form (GSH:GSSG) is significantly lower, indicating impaired antioxidant capacity. Dopaminergic neurons may subsequently die as a result of ROS generation (Brown, 2011; Rivera-Mancía et al., 2010). Moreover, the SNpc's dopamine (DA) neurons' deterioration raises the possibility that OS originates from DA itself. In a similar vein, it is indisputable that changes in the mitochondrial electron transporter chain (METC) and catecholamine metabolism contribute to PD. There have also been reports of elevated levels of 4-hydroxy-2,3-nonenal, thiobarbituric acid (TBA) reactive compound, and MDA in PD patients.

According to reports, one of the main factors influencing the course of HD is OS. Oxidative changes to proteins, lipids, and DNA are a few examples of frequent indicators (Hands et al., 2010; Squadrone et al., 2020). In actuality, a highly sensitive biomarker of HD is 8-hydroxydeoxyguanosine (8-OHdg), an oxidation product of DNA. HD is also associated with an increased byproduct of plasma lipid peroxidation. It is impossible to overlook mitochondrial malfunction and impairment in the electron transport chain as the primary mechanisms of OS-mediated HD pathogenesis. The literature also reports elevated OS, aberrant tryptophan metabolism, and decreased GSH levels (Duran et al., 2010). Significant data suggest that OS has a role in ALS patients. Additional markers of oxidative damage in sporadic ALS (sALS) include elevated levels of PCs (Cunha-Oliveira et al., 2020; Niebrój-Dobosz et al., 2004), 8-hydroxy-2-deoxyguanosine (8-OHdG), MDA-modified proteins, 4-HNE protein adducts, and nitrotyrosine in spinal cord tissue. Furthermore, erythrocytes from sALS patients show GSH levels along with decreased activities of CAT, glutathione reductase (GR), and glucose-6-phosphate dehydrogenase, reflecting compromised antioxidant defenses (Mitsumoto et al., 2020). Mechanistically, 20% of instances of ALS have been linked to a mutation in SOD1. SOD1's primary job is to scavenge excess superoxide radicals (O^{2-}). In addition to lowering glutamate transporter (GLT) 1 expression, the mutant SOD1 may eventually cause glutamate toxicity in motor neurons (Cunha-Oliveira et al., 2020; Mitsumoto et al., 2020). Figure 2.1 shows the role of vitamins and their functions in different neurological conditions.

Another characteristic of Friedreich ataxia is the absence of the iron transporter protein frataxin. This might lead to a rise in iron levels, which encourages the Fenton reaction to convert H_2O_2 to $^{\cdot}OH$. To support this, it has been demonstrated that certain individuals with Friedreich ataxia have better energy production when they get antioxidant therapy with vitamin E and coenzyme Q10 supplements (Kim et al., 2015). Patients have been shown to have elevated amounts of serum MDA and urine 8-hydroxy-2'-deoxyguanosine, which are markers of lipid peroxidation and DNA damage, respectively (Emond et al., 2000; Schulz et al., 2000). OS is one of the main factors linked to neurodegenerative disorders and neuronal dysfunction, according to a wealth of literature. Due to its high oxygen consumption, lipid-rich organization, limited antioxidant defense mechanisms, and presence of multiple redox-active metals, OS can readily damage brain regions, resulting in neuropsychiatric and neurodegenerative disorders (Halliwell, 2006; Lin & Beal, 2006).

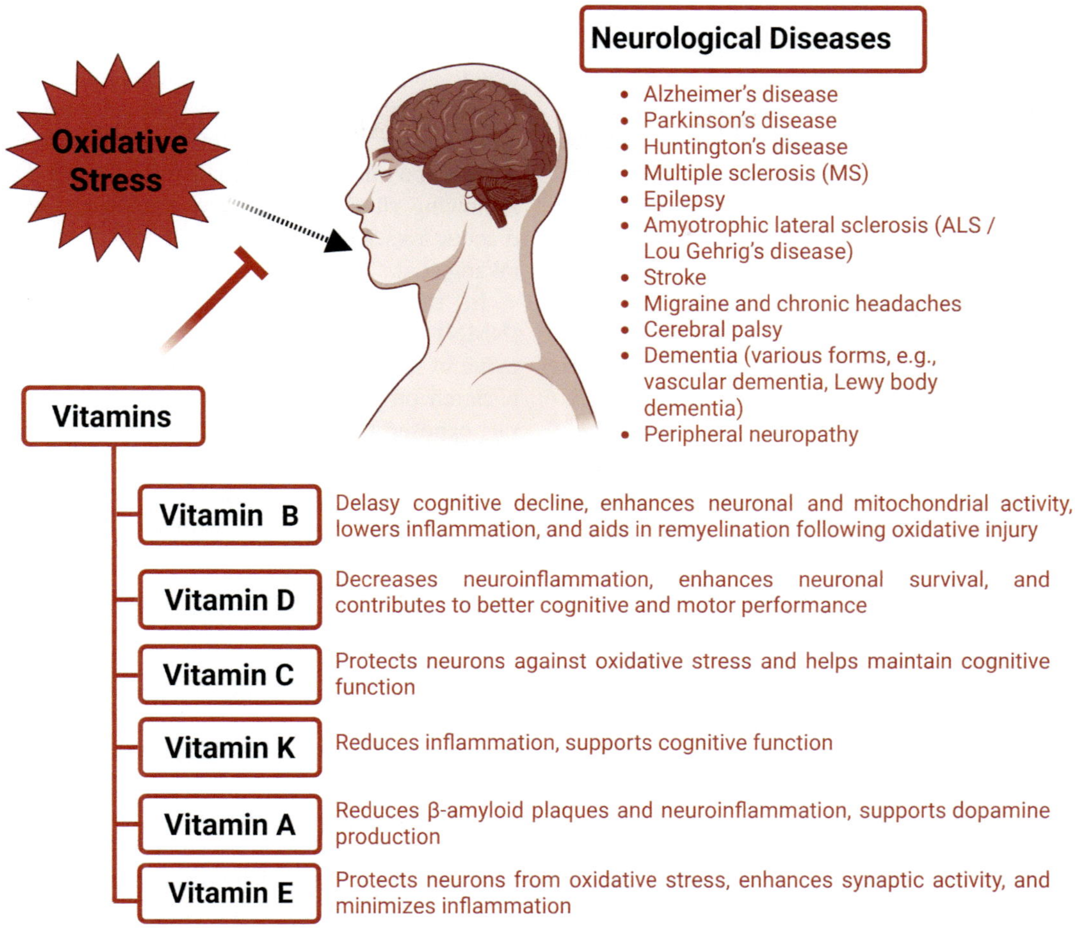

FIGURE 2.1 Vitamins and neurological functions.

2.9 ANTIOXIDANT AND ANTI-INFLAMMATORY ROLES OF VITAMINS

Many studies have explored the antioxidant properties of well-known vitamins such as vitamin E, C, and carotenoids, highlighting their importance for human health. While their antioxidant capacity is well established, the roles of several other vitamins in combating OS and the consequences of their deficiencies remain underexamined. Vitamins such as K, D, niacin, pyridoxine, and riboflavin are often overlooked but act as coenzymes in neutralizing free radicals, and their deficiencies can contribute to OS. Research has demonstrated specific effects, including vitamin B12 on retinal ganglion cells, riboflavin on lipid peroxidation and PC formation in rats (Wang et al., 2011), and vitamin K in neuroprotection (Li et al., 2003). Deficiencies in these vitamins are also associated with elevated homocysteine levels, which increase cardiovascular risk (Green et al., 2017), as well as OS-related conditions such as atherogenesis (Ohta & Foote, 2002).

2.10 VITAMIN B COMPLEX

Neuroinflammatory processes are important in the development of CNS disease, according to a growing body of evidence. A modified and/or insufficient immune response, mediated by various peripheral immune cells and CNS-resident cells, is the primary cause of these processes. One of the main immune effector cells specialized in controlling CNS homeostasis and the inflammatory response is thought to be the tissue-resident macrophage population of the CNS or microglia. In addition to being the CNS's first line of defense and helping to eradicate pathogens, microglia also help to remove dead cells and other cellular waste. In order to preserve CNS development and homeostasis, microglia collaborate closely with other glial cells, such as oligodendrocytes, astrocytes, and neuronal cells (Rakić et al., 2023). Microglia cells generate proinflammatory molecules like ROS and NO, which are produced by nitric inducible synthase (iNOS), as well as proinflammatory cytokines like tumor necrosis factor α (TNF-α), interferon (IFN)-γ, interleukin (IL)-1β, and IL-6 in response to various stimuli. Other peripheral immune cells, such as monocytes, Th1 and Th17 lymphocytes, B lymphocytes, and dendritic cells (DCs), can more easily infiltrate the site of inflammation due to the oxidative damage and disruption of the BBB caused by the unchecked microglial activation and inflammatory response in the CNS (Rakić et al., 2023).

Many neurological disorders are characterized by persistent inflammation in the CNS, which can be brought on by the dysregulation of these processes. Demyelination, increasing axonal destruction, and ultimately neurodegeneration are the hallmarks of MS, one of the most prevalent autoimmune CNS diseases. It is a complicated process that includes the production of proinflammatory mediators and interactions between several immune cells, which is followed by a loss of neuronal function and the ensuing neurodegeneration. Several studies have reported that microglial activation (antigen presentation and generation of proinflammatory molecules) exacerbated encephalomyelitis (EAE) symptoms using the animal model of MS, experimental autoimmune EAE (Rakić et al., 2023). However, it has also been suggested that microglia contribute to remyelination, T regulatory cell proliferation, and neurogenesis, which attenuates the development of EAE disease, because of their dual function in the CNS and their neuroprotective properties. According to these results, modifying microglial activation may change how the disease progresses and hasten neuroregeneration and recovery. Therefore, more research into the intricate mechanisms behind neuroinflammation processes might help develop new treatments that may slow the advancement of neurodegenerative disorders like MS and improve our knowledge of the roles played by microglia in these conditions.

According to research, treating EAE *in vivo* with the vitamin B complex (B1, B2, B3, B5, B6, and B12) improved the cellular alterations in the peripheral nerve that was impacted, the muscle that this nerve innervated, and the dysbiosis of the gut microbiota that developed as the disease progressed (Rakić et al., 2023). Consequently, in neurodegenerative diseases, VBC therapy may be a viable therapeutic approach for controlling and lowering the inflammatory response. B vitamins are essential for the metabolism of macronutrients including proteins, carbs, and fats and function as coenzymes in a variety of metabolic processes. They are recognized to be vital micronutrients for healthy brain function and are water-soluble vitamins. Neuronal malfunction and the development of neurological disorders, such as MS, have been associated with decreased levels of B vitamins (Rakić et al., 2023).

Riboflavin (vitamin B2) is often overlooked as an antioxidant nutrient, since antioxidants are usually associated with vitamins C and E and carotenoids. The glutathione redox cycle converts reduced riboflavin to its oxidized state, which confers some antioxidant properties on riboflavin (Ashoori & Saedisomeolia, 2014). In the FMN and FAD forms, riboflavin functions as a coenzyme for redox enzymes. Because the FAD coenzyme is necessary for GRx activity to convert GSSH to GSH, they function as antioxidants (Liu et al., 1993). In the intracellular environment, reduced glutathione functions as an antioxidant by neutralizing ROS while undergoing conversion to its oxidized form (Dringen et al., 2000). To restore the oxidized glutathione's antioxidant capacity, FAD was required to convert it back to its reduced state. Its capacity to deactivate peroxides, particularly hydroperoxides, depends on glutathione. Thus, it is anticipated that a lack of riboflavin may lead to an increase in lipid peroxidation. Both the intended impact of riboflavin treatment and the detrimental effects of riboflavin shortage on lipid metabolism were demonstrated in a number of animal studies (Sinbad et al., 2019). Following riboflavin delivery, a related study found that diabetic rats had lower levels of PCs and lipid peroxides (Wang et al., 2011).

In the formation of nicotinamide mononucleotide (NMN) and NAD, niacin (vitamin B3) functions as a coenzyme of redox enzymes (Long et al., 2015). According to research on the potential advantages of giving niacin to rats exposed to methyl mercury, niacin reduced the negative effects of the metal. The mechanism underlying this positive impact was proposed to be related to niacin's natural antioxidant capacity (Silva de Paula et al., 2016). In rats with alloxan-induced diabetes, niacin was also shown to ameliorate protein carbonylation, lipid peroxidation, and DNA damage brought on by ROS (Abdullah et al., 2018). In *in vitro* experiments on glycated human serum albumin, niacin also lowered DNA damage and glycation (Abdullah et al., 2017).

PA, also referred to as vitamin B5, is a component of coenzyme A (CoA), which is important for the metabolism of glucose, fatty acids, and amino acids. Coenzyme A is also involved in the tricarboxylic acid cycle, the acetylation of choline to form acetylcholine, and fatty acid biosynthesis (Kim et al., 2015). In gamma-irradiated rats, PA administration has been shown to dramatically lower OS and alleviate brain damage (Sm et al., 2018).

Although vitamin B6 (pyridoxine) is not a traditional antioxidant component, it does have some antioxidant properties. According to the paper, pyridoxine may absorb up to eight hydroxyl radicals (-OH) and functions as a powerful scavenger of -OH (Higashi-Okai et al., 2006). Vitamin B6–deficient rats developed peroxidative stress due to increased TBA in the rat heart and liver, indicating a crucial role for pyridoxine in inhibiting oxidation. An in vitro study found that vitamin B6 stopped the generation of ROS and lipid peroxidation in U937 monocytes (Kannan & Jain, 2004). Vitamin B6 deficiency affects antioxidant defense, increases lipid peroxidation, and influences the production of long-chain polyunsaturated fatty acids, all of which are linked to atherogenesis (Ohta & Foote, 2002). Furthermore, vitamin B6's antioxidant properties were demonstrated by the fact that pretreating rats with it improved antioxidant enzymes such as superoxide dismutase, CAT, GR, glutathione peroxidase, and glutathione S-transferase (GST), which helped treat chromium-induced OS (Anand, 2005). According to a theoretical study on its scavenging potential against different ROS such as hydroxyl, superoxide, and oxygen radicals, pyridoxine shows stronger reactivity with hydroxyl radicals, either through interactions with its aromatic ring or by hydrogen abstraction from substituent groups attached to the ring (Matxain et al., 2006).

Folates, commonly known as vitamin B9, are important vitamins with medicinal significance because they can lower plasma homocysteine levels, which may help prevent cardiovascular disease. In contrast, folates may have a direct antioxidant impact *in vivo* that is unrelated to their ability to reduce homocysteine. Folic acid's reduced forms were shown to possess antioxidant properties similar to those of vitamin C and vitamin E (Gliszczyńska-Świgło, 2007). While DHT is highly efficient against $ABTS^{+\cdot}$ cations, THF has the most protective effect against lipid peroxidation. Numerous studies have also shown that folate helps with neurological problems, cancer, hematological disease, and neural tube anomalies. It has been proposed that folates' antioxidant properties account for their preventive function in various diseases (Sinbad et al., 2019).

In vitro research has shown that vitamin B12 (cobalamin) may have some antioxidant properties. According to certain scientists, vitamin B12 has a direct mechanism for scavenging superoxide (Chan et al., 2018; Moreira et al., 2011). Superoxide levels in the intracellular fluid and mitochondria were reduced when cyanocobalamin was administered to human aorta cells (Moreira et al., 2011). Neuronal cells and cell-free systems showed comparable outcomes. Additionally, following vitamin B12 delivery, certain *in vivo* investigations have also documented findings with a decreased superoxide burst in the retina's ganglion cells (Chan et al., 2018). Additionally, vitamin B12's capacity to maintain glutathione is linked to its antioxidant properties. Nonetheless, research is being done on the reactions that are involved in this (Karamshetty et al., 2016; Manzanares & Hardy, 2010). Studies using rats lacking vitamin B12 showed lower levels of epidermal growth factors than controls, suggesting that vitamin B12 may prevent OS by changing the activity of nuclear factor-KB, which may modify the expression of cytokines and growth factors and prevent OS brought on by inflammation (Birch et al., 2009). A vitamin B12 shortage will result in elevated homocysteine levels because vitamin B12 promotes the conversion of homocysteine to methionine (Green et al., 2017). The body produces more ROS when homocysteine is easily converted to hydrogen peroxide (Tyagi et al., 2005). Because of its function in homocysteine metabolism, it thereby provides indirect protection against OS. The OS cycle may be significantly influenced by vitamin B12. ROS are produced when there is a subclinical vitamin B12 shortage, which hinders vitamin B12 absorption. Advanced glycated end products can be formed as a result of OS, which may decrease the absorption of vitamin B12 (Nowotny et al., 2015; Sinbad et al., 2019).

2.11 VITAMIN C

Ascorbic acid, often known as vitamin C, is a vital antioxidant that is important for brain function. Its connection to neurodegenerative diseases has drawn a lot of interest lately. This water-soluble vitamin has been shown to have a role in neuroprotection, neurotransmitter production, and OS management. In order to prevent and cure neurodegenerative diseases, the aforementioned elements are essential (Eldridge et al., 1987; Hansen et al., 2014). One of the brain's strongest antioxidants is vitamin C. It helps protect neurons from oxidative injury by neutralizing ROS and free radicals. Diseases like AD and PD may advance more slowly as a result of this preventive impact. Neurotransmitters including DA, serotonin, and noradrenaline, which are critical for cognitive processes, mood regulation, and motor control, are also produced in part by vitamin C (Harrison & May, 2009). Vitamin C also contributes to the creation of collagen, which is necessary to maintain the structural integrity of the brain's connective tissue and blood vessels. Furthermore, it maintains the BBB's integrity, which shields the brain from inflammation and toxins. Many neurodegenerative diseases have the characteristic of breaching this barrier (Chang et al., 2020). Furthermore, by aiding in neuronal repair and encouraging the production of extracellular matrix, vitamin C preserves the health of neurons. Additionally, it aids in controlling the levels of metal ions, particularly iron, in the brain. Protein aggregation and OS, which arise during neurodegenerative disorders, are linked to iron dysregulation (Choudhry et al., 2012). Vitamin C lowers oxidative damage and offers neuroprotection by reducing excessive iron buildup (Nualart et al., 2014). Additionally, it aids in combating nervous system inflammation, which is linked to the onset of AD and MS. Higher vitamin C levels may help AD patients' memory and cognitive performance by lowering OS and neuroinflammation, according to some research (Feng & Wang, 2012). Furthermore, vitamin C helps to eliminate beta-amyloid plaques, which lessens the harmful effects they have on neurons (Kook et al., 2014). According to research, consuming enough vitamin C through food or supplements can improve brain function in general and lower the risk of neurodegenerative diseases (Hamid et al., 2022).

2.12 VITAMIN A

Vitamin A, a fat-soluble vitamin that is mostly known for its effects on the immune system and eyesight, is also vital for brain health and has drawn interest for its potential to treat neurodegenerative diseases. An active form of vitamin A is essential for gene regulation, neurotransmission, synaptic plasticity, and neurodevelopment in the brain. Maintaining synaptic plasticity is essential for learning, memory, and cognitive processes because it allows synapses to change structurally and functionally in response to novel experiences. Its anti-inflammatory and antioxidant qualities further support its neuroprotective benefits, which might aid in halting neurodegenerative processes. The development of the CNS depends on retinoic acid, which also ensures appropriate brain formation by controlling the proliferation and differentiation of neurons and glial cells throughout embryonic development (Clark et al., 2020).

As an antioxidant, vitamin A aids in the neutralization of free radicals, shielding neurons from OS, a primary cause of neuronal damage and diseases such as PD and AD. Additionally, by lowering inflammation in the brain and modifying immunological responses, it may decrease the course of AD. Increased beta-amyloid plaque buildup in AD is linked to impaired retinoic acid signaling. According to some research, vitamin A supplements may help AD patients' cognitive function by lowering beta-amyloid buildup (Joshi et al., 2023). Retinoids may prevent dopaminergic neurons from degenerating in PD by controlling genes related to DA synthesis and cell survival. Vitamin A has neuroprotective qualities in PD because it can lower OS and boost DA synthesis (Marie et al., 2021). It has been demonstrated that retinoic acid lowers the activity of inflammatory T cells in MS, which leads to myelin destruction. In MS patients, this activity relieves symptoms by promoting remyelination and reducing neuroinflammation. According to Fragoso et al., vitamin A may aid in immune system regulation, encourage myelin regeneration, and lower the frequency of flare-ups. As a result, individuals with MS can benefit from improved treatment outcomes (Fragoso et al., 2014).

Although vitamin A is not a well-known antioxidant, a small number of research studies have suggested that it may have an indirect antioxidant function. According to reports, all-trans retinoic acid suppresses thioredoxin-interacting protein and lowers OS levels, which, in turn, inhibits the activation of hepatic stellate cells, an effector of hepatocellular cancer [57]. Retinoic acid, a byproduct of vitamin A, has also been found to increase the expression of genes linked to antioxidants in mature bovine oocytes cultured *in vitro* [58]. Additionally, in both healthy and varicocele sperm, all-trans retinoic acid reduced ROS and MDA while increasing glutathione transferase and superoxide dismutase activities, suggesting that retinol increases antioxidant enzyme activity [59]. Consequently, there is mounting evidence that vitamin A may help shield the body from the harm caused by OS.

2.13 VITAMIN D

In addition to being essential for bone health and calcium management, vitamin D has drawn interest for its possible neuroprotective benefits in neurodegenerative disorders. It impacts OS, immunological regulation, neuronal health, and neuroinflammation—all of which are crucial in the progression of neurodegenerative diseases (Koduah et al., 2017). Vitamin D has receptors on both neurons and glial cells in important brain regions, where it controls the synthesis of NGF and glial cell line–derived neurotrophic factor (GDNF) and affects the differentiation and maturation of neurons (Habib et al., 2020). These growth factors support the development, survival, and repair of neurons. Furthermore, vitamin D facilitates the production of critical neurotransmitters that are necessary for cognitive and motor processes, such as GABA, DA, and acetylcholine (Moretti et al., 2018). Since dysregulated calcium can result in neuronal malfunction and death, vitamin D aids in the regulation of brain calcium levels, which are essential for appropriate neuronal signaling.

Since inflammation plays a major role in the course of MS and AD, its anti-inflammatory qualities and capacity to regulate immune responses make it protective in both conditions (Caccamo et al., 2018). As an antioxidant, vitamin D also aids in the fight against OS, which is a factor in the degeneration and destruction of neurons. Cognitive decline and an elevated risk of AD have been associated with low vitamin D levels. A vitamin deficiency may exacerbate inflammation and lead to the formation of beta-amyloid plaques (Grimm et al., 2017). Additionally, vitamin D guards against neuroinflammation and promotes synaptic plasticity, which may assist AD patients in postponing cognitive decline. Vitamin D helps preserve dopaminergic neurons, which gradually disappear in PD. Higher vitamin D levels are linked to a decreased incidence of PD, according to Pignolo et al. Supplementing with this vitamin may enhance motor function and guard against dopaminergic degradation (Pignolo et al., 2022). Lower recurrence rates and a decreased risk of MS have also been associated with vitamin D's capacity to control immune responses and lessen neuroinflammation. Further evidence of vitamin D's neuroprotective properties suggests that adequate levels may decrease the course of MS by promoting remyelination and lowering inflammation (Sangha et al., 2023).

All three forms of vitamin D have been shown to reduce iron-dependent lipid peroxidation, and their antioxidant properties have been compared to those of the anticancer medication [60]. The antioxidant action of vitamin D3 (cholecalciferol) is supported by several data. According to some experimental research, giving diabetic mice vitamin D3 reduces the production of ROS by inhibiting the expression of the NADPH oxidase gene (Kono et al., 2013; Labudzynskyi et al., 2015). Additionally, it has been proposed that vitamin D administration offers substantial protection against vascular problems in diabetes that are mediated by OS (Salum et al., 2013). Furthermore, research suggested that vitamin D functions as an antioxidant due to a rise in hepatic GSH levels in rats given cholecalciferol. Pregnant women taking vitamin D supplements for nine weeks have improvements in their antioxidant indicators. Among other metabolic enzymes, it resulted in a notable rise in glutathione activity and overall antioxidant capacity (Asemi et al., 2013). In mature erythrocytes without a nucleus, vitamin D3 is shown to play a significant antioxidant role. This finding not only confirms that cholecalciferol has an antioxidant effect (Wolden-Kirk et al., 2012), but it also suggests that 1,25-dihydroxycholecalciferol may act as a direct antioxidant of the membrane by stabilizing and shielding it from lipid peroxidation through its interactions with its hydrophobic components.

2.14 VITAMIN E

As a fat-soluble antioxidant, vitamin E is essential for shielding the brain from oxidative damage. The potential advantages of this vitamin in the management of a number of neurodegenerative diseases have been investigated. Because free radicals are the primary cause of neurodegeneration, alpha-tocopherol, its most active form, is crucial for preserving brain health (Rodríguez-Lara et al., 2016). Overproduction of free radicals damages brain cells' proteins, lipids, and DNA, impairing cognitive and motor abilities. By promoting neurotrophic factors, vitamin E helps neurons survive, develop, and repair, hence reducing neurodegenerative processes.

The symptoms of neurodegenerative diseases are linked to vitamin E deficiency because it raises OS and speeds up the processes that lead to neuron death. Particularly in older adults, low vitamin E levels are linked to an increased risk of cognitive impairment, dementia, and AD (Ulatowski & Manor, 2015). The development of beta-amyloid plaques and neurofibrillary tangles in AD is accelerated by oxidative damage brought on by free radicals (Cassidy et al., 2020). OS has a major role in the destruction of dopaminergic neurons in the substantia nigra, which controls movement, in PD (Hao et al., 2024). By shielding the lipid bilayers of neuronal membranes from oxidation, vitamin E keeps polyunsaturated fatty acids from oxidizing. The integrity and functionality of nerve cells depend on this function (Rodríguez-Lara et al., 2016).

Vitamin E possesses anti-inflammatory qualities in addition to its antioxidant function, which is crucial in neuroinflammatory diseases including MS and AD (Guan et al., 2018). Vitamin E may help prevent protein aggregation (e.g., beta-amyloid and alpha-synuclein) in conditions like PD and AD, protecting neurons against toxin buildup. Higher vitamin E levels are linked to a slower rate of decline in cognitive function in AD patients (Nolan et al., 2022). According to certain clinical research, vitamin E supplements may decrease the development of AD by lowering inflammation, oxidative damage, and beta-amyloid plaque accumulation. This enhances neuroplasticity and synaptic functioning, which are essential for preserving memory and cognitive abilities (Mangialasche et al., 2010; Nolan et al., 2022). By lowering OS in dopaminergic neurons, its supplementation may therefore help to reduce motor symptoms (Hao et al., 2024; Yang et al., 2017). Although the findings of studies on vitamin E's potential to halt the course of ALS are conflicting, they point to a moderate neuroprotective impact associated with its antioxidant qualities (Xia et al., 2022). Likewise, OS has a role in demyelination in MS. Vitamin E's qualities improve myelin function and lessen oxidative damage, which helps people with MS (Miller et al., 2014). It has been discovered that genetic differences affect how well vitamin E is absorbed and metabolized, which may change its overall effects. According to recent studies, vitamin E supplements may worsen AD by encouraging the production of more amyloid β. These varied results highlight the necessity for cautious usage and perhaps account for the contradictory results regarding its supplementation (Diachenko et al., 2024; Regner-Nelke et al., 2021).

2.15 VITAMIN K

Although vitamin K is well known for its function in blood coagulation, it also has important effects on brain function and has attracted attention due to possible neuroprotective benefits. According to Diachenko et al., vitamin K may have an impact on the onset and course of a number of neurodegenerative diseases by regulating inflammation, sphingolipid metabolism, and antioxidant protection (Diachenko et al., 2024). Through its function in electron transport, vitamin K2 has been demonstrated in recent research to have potential benefits in reducing OS, microglial activation, and apoptosis in neuronal cells, especially in connection to AD (Popescu & German, 2021). Sphingolipid metabolism depends on vitamin K, which is also vital for cell communication and neuronal survival. Cognitive decline and neurodegenerative diseases, such as AD, have been linked to disturbances in sphingolipid metabolism (Denisova & Booth, 2005). Furthermore, vitamin K has antioxidant qualities that protect neurons from OS, which is a major contributor to neuronal deterioration. Its anti-inflammatory properties could also aid in lowering neuroinflammation, which might protect against the advancement of MS and AD (Lin et al., 2021).

Higher vitamin K levels may be associated with improved memory and cognitive function, according to some research. Increased vitamin K2 levels were linked to a decreased incidence of dementia, better cognitive function, less severe AD, and less phosphorylated Tau protein aggregation, according to postmortem examinations of brain samples (Booth et al., 2022). According to research, vitamin K may help nerve cells survive and flourish. Additionally, some studies have suggested that vitamin K may help prevent amyloid-beta deposition, which might prevent AD from progressing (Alam et al., 2016). Additionally, by lowering inflammation and oxidative damage in the brain, vitamin K may help shield dopaminergic neurons, just like it does in AD. Additionally, by lowering inflammation and oxidative damage in the brain, vitamin K protects dopaminergic neurons, much like its beneficial effects in AD (Booth et al., 2022). Recent research has shown that vitamin K2 has significant inhibitory effects on inflammation and the fibrillization of α-synuclein in PD, in addition to its protective effect on cognitive functioning (Yu et al., 2020). Furthermore, vitamin K has been proposed as a possible treatment option for MS patients in an effort to lessen the intensity of their symptoms or delay the course of the disease. Vitamin K may prevent nerve cell degeneration in MS by promoting sphingolipid metabolism and offering antioxidant defense. This might help with myelin repair and reduce neuroinflammation (de Oliveira et al., 2021).

2.16 CONCLUSION

In conclusion, vitamins have a critical role in both the prevention and treatment of neurodegenerative diseases. Their diverse actions, which include anti-inflammatory, antioxidant, and neurotransmitter production control, highlight their potential as supplemental treatments for neurological disorders. Key vitamins that support brain health by reducing OS, preserving cognitive function, and promoting overall neural well-being include the B complex group, along with vitamins C, D, E, A, and K. The need to consume enough vitamins to reduce the risk of neurodegenerative diseases and support individuals who have already been affected is becoming clearer as continuing research clarifies how these vitamins offer neuroprotection. Studies show that different people have different metabolisms for several vitamins; thus, more research is required. Optimizing treatment results requires a greater knowledge of the mechanisms influencing their metabolism. Although customizing vitamin supplements can be expensive, it is an interesting field for future research because of the potential advantages it may have for patients with neurodegenerative disorders and healthcare systems. The currently small number of approved health claims under Regulation 1924/2006 that relate vitamins to the prevention and treatment of neurodegenerative diseases may increase in the future as science advances, according to this mounting body of data. For those who are at risk of developing neurodegenerative disorders as well as those who already have them, focusing on a healthy diet full of vital vitamins and making other lifestyle adjustments can help preserve brain function and enhance quality of life.

ACKNOWLEDGMENT

None.

CONFLICT OF INTEREST

None.

REFERENCES

Abdullah, K., Alam, M. M., Iqbal, Z., & Naseem, I. (2018). Therapeutic effect of vitamin B3 on hyperglycemia, oxidative stress and DNA damage in alloxan induced diabetic rat model. *Biomedicine & Pharmacotherapy*, *105*, 1223–1231.

Abdullah, K., Qais, F. A., Ahmad, I., & Naseem, I. (2017). Inhibitory effect of vitamin B3 against glycation and reactive oxygen species production in HSA: An in vitro approach. *Archives of Biochemistry and Biophysics*, *627*, 21–29.

Alam, P., Chaturvedi, S. K., Siddiqi, M. K., Rajpoot, R. K., Ajmal, M. R., Zaman, M., & Khan, R. H. (2016). Vitamin k3 inhibits protein aggregation: Implication in the treatment of amyloid diseases. *Scientific Reports*, *6*(1), 26759.

Albay, C., Adanir, O., Akkalp, A. K., Dogan, V. B., Gulec, M. A., & Beytemur, O. (2020). Cyanocobalamin and cholecalciferol synergistically improve functional and histopathological nerve healing in experimental rat model. *Ideggyogyaszati Szemle*, *73*(3–4), 113–120.

Anand, S. S. (2005). Protective effect of vitamin B6 in chromium-induced oxidative stress in liver. *Journal of Applied Toxicology: An International Journal*, *25*(5), 440–443.

Asemi, Z., Samimi, M., Tabassi, Z., Shakeri, H., & Esmaillzadeh, A. (2013). Retracted: Vitamin D supplementation affects serum high-sensitivity C-reactive protein, insulin resistance, and biomarkers of oxidative stress in pregnant women. *The Journal of Nutrition*, *143*(9), 1432–1438.

Ashoori, M., & Saedisomeolia, A. (2014). Riboflavin (vitamin B2) and oxidative stress: A review. *British Journal of Nutrition*, *111*(11), 1985–1991.

Badawy, A. A.-B. (2014). Pellagra and alcoholism: A biochemical perspective. *Alcohol and Alcoholism*, *49*(3), 238–250.

Bailey, R. L., West Jr, K. P., & Black, R. E. (2015). The epidemiology of global micronutrient deficiencies. *Annals of Nutrition and Metabolism*, *66*(Suppl. 2), 22–33.

Barnham, K. J., Masters, C. L., & Bush, A. I. (2004). Neurodegenerative diseases and oxidative stress. *Nature Reviews Drug Discovery*, *3*(3), 205–214.

Baviera-Muñoz, R., Buigues-Lafuente, A., Campins-Romeu, M., Garcés-Sánchez, M., & Martinez-Torres, I. (2022). Refractory status epilepticus due to vitamin B6 deficit in a Parkinson's disease patient in treatment with levodopa/carbidopa intestinal gel. *Neurologia*, *37*(7), 608–609.

Benedik, E. (2021). Sources of vitamin D for humans. *International Journal for Vitamin and Nutrition Research*, *92*(2), 118–125.

Bennett, C., Green, J., Ciancio, M., Goral, J., Pitstick, L., Pytynia, M., Meyer, A., Kwatra, N., & Jadavji, N. M. (2022). Dietary folic acid deficiency impacts hippocampal morphology and cortical acetylcholine metabolism in adult male and female mice. *Nutritional Neuroscience*, *25*(10), 2057–2065.

Berdanier, C. D. (2019). Corn, niacin, and the history of pellagra. *Nutrition Today*, *54*(6), 283–288.

Berridge, M. J. (2017). Vitamin D and depression: Cellular and regulatory mechanisms. *Pharmacological Reviews*, *69*(2), 80–92.

Birch, C. S., Brasch, N. E., McCaddon, A., & Williams, J. H. (2009). A novel role for vitamin B12: Cobalamins are intracellular antioxidants in vitro. *Free Radical Biology and Medicine*, *47*(2), 184–188.

Bisht, K., Sharma, K., & Tremblay, M.-È. (2018). Chronic stress as a risk factor for Alzheimer's disease: Roles of microglia-mediated synaptic remodeling, inflammation, and oxidative stress. *Neurobiology of Stress*, *9*, 9–21.

Booth, S. L., Shea, M. K., Barger, K., Leurgans, S. E., James, B. D., Holland, T. M., Agarwal, P., Fu, X., Wang, J., & Matuszek, G. (2022). Association of vitamin K with cognitive decline and neuropathology in community-dwelling older persons. *Alzheimer's & Dementia: Translational Research & Clinical Interventions*, *8*(1), e12255.

Briani, C., Dalla Torre, C., Citton, V., Manara, R., Pompanin, S., Binotto, G., & Adami, F. (2013). Cobalamin deficiency: Clinical picture and radiological findings. *Nutrients*, *5*(11), 4521–4539.

Brown, D. R. (2011). Metals in neurodegenerative disease. *Metallomics*, *3*(3), 226–228.

Caccamo, D., Ricca, S., Currò, M., & Ientile, R. (2018). Health risks of hypovitaminosis D: A review of new molecular insights. *International Journal of Molecular Sciences*, *19*(3), 892.

Calderón-Ospina, C. A., & Nava-Mesa, M. O. (2020). B Vitamins in the nervous system: Current knowledge of the biochemical modes of action and synergies of thiamine, pyridoxine, and cobalamin. *CNS Neuroscience & Therapeutics*, *26*(1), 5–13.

Cassidy, L., Fernandez, F., Johnson, J. B., Naiker, M., Owoola, A. G., & Broszczak, D. A. (2020). Oxidative stress in Alzheimer's disease: A review on emergent natural polyphenolic therapeutics. *Complementary Therapies in Medicine*, *49*, 102294.

Cecchini, T., Cuppini, R., Ciaroni, S., & Grande, P. D. (1993). Increased number of dorsal root ganglion neurons in vitamin-E-deficient rats. *Somatosensory & Motor Research*, *10*(4), 433–443.

Chabas, J.-F., Alluin, O., Rao, G., Garcia, S., Lavaut, M.-N., Risso, J. J., Legre, R., Magalon, G., Khrestchatisky, M., & Marqueste, T. (2008). Vitamin D2 potentiates axon regeneration. *Journal of Neurotrauma*, *25*(10), 1247–1256.

Chabas, J.-F., Stephan, D., Marqueste, T., Garcia, S., Lavaut, M.-N., Nguyen, C., Legre, R., Khrestchatisky, M., Decherchi, P., & Feron, F. (2013). Cholecalciferol (vitamin D3) improves myelination and recovery after nerve injury. *PLoS ONE*, *8*(5), e65034.

Chan, W., Almasieh, M., Catrinescu, M.-M., & Levin, L. A. (2018). Cobalamin-associated superoxide scavenging in neuronal cells is a potential mechanism for vitamin B12–deprivation optic neuropathy. *The American Journal of Pathology*, *188*(1), 160–172.

Chang, B.-J., Jang, B.-J., Son, T. G., Cho, I.-H., Quan, F.-S., Choe, N.-H., Nahm, S.-S., & Lee, J.-H. (2012). Ascorbic acid ameliorates oxidative damage induced by maternal low-level lead exposure in the hippocampus of rat pups during gestation and lactation. *Food and Chemical Toxicology*, *50*(2), 104–108.

Chang, C.-Y., Chen, J.-Y., Wu, M.-H., & Hu, M.-L. (2020). Therapeutic treatment with vitamin C reduces focal cerebral ischemia-induced brain infarction in rats by attenuating disruptions of blood brain barrier and cerebral neuronal apoptosis. *Free Radical Biology and Medicine*, *155*, 29–36.

Choudhry, F., Howlett, D. R., Richardson, J. C., Francis, P. T., & Williams, R. J. (2012). Pro-oxidant diet enhances β/γ secretase-mediated APP processing in APP/PS1 transgenic mice. *Neurobiology of Aging*, *33*(5), 960–968.

Clark, J. N., Whiting, A., & McCaffery, P. (2020). Retinoic acid receptor-targeted drugs in neurodegenerative disease. *Expert Opinion on Drug Metabolism & Toxicology*, *16*(11), 1097–1108.

Colombo, M. L. (2010). An update on vitamin E, tocopherol and tocotrienol—perspectives. *Molecules*, *15*(4), 2103–2113.

Cunha-Oliveira, T., Montezinho, L., Mendes, C., Firuzi, O., Saso, L., Oliveira, P. J., & Silva, F. S. (2020). Oxidative stress in amyotrophic lateral sclerosis: Pathophysiology and opportunities for pharmacological intervention. *Oxidative Medicine and Cellular Longevity*, *2020*(1), 5021694.

Dash, U. C., Bhol, N. K., Swain, S. K., Samal, R. R., Nayak, P. K., Raina, V., Panda, S. K., Kerry, R. G., Duttaroy, A. K., & Jena, A. B. (2025). Oxidative stress and inflammation in the pathogenesis of neurological disorders: Mechanisms and implications. *Acta Pharmaceutica Sinica B*, *15*(1), 15–34.

de Oliveira, L. G., Angelo, Y. d. S., Iglesias, A. H., & Peron, J. P. S. (2021). Unraveling the link between mitochondrial dynamics and neuroinflammation. *Frontiers in Immunology*, *12*, 624919.

Denisova, N. A., & Booth, S. L. (2005). Vitamin K and sphingolipid metabolism: Evidence to date. *Nutrition Reviews*, *63*(4), 111–121.

Diachenko, A. I., Rodin, I. A., Krasnova, T. N., Klychnikov, O. I., & Nefedova, L. N. (2024). The role of vitamin K in the development of neurodegenerative diseases. *Biochemistry*, *89*(Suppl 1), S57–S70.

Doseděl, M., Jirkovský, E., Macáková, K., Krčmová, L. K., Javorská, L., Pourová, J., Mercolini, L., Remião, F., Nováková, L., & Mladěnka, P. (2021). Vitamin C—sources, physiological role, kinetics, deficiency, use, toxicity, and determination. *Nutrients*, *13*(2), 615.

Dringen, R., Gutterer, J. M., & Hirrlinger, J. (2000). Glutathione metabolism in brain: Metabolic interaction between astrocytes and neurons in the defense against reactive oxygen species. *European Journal of Biochemistry*, *267*(16), 4912–4916.

Dror, D. K., & Allen, L. H. (2011). Vitamin E deficiency in developing countries. *Food and Nutrition Bulletin*, *32*(2), 124–143.

Duran, R., Barrero, F. J., Morales, B., Luna, J. D., Ramirez, M., & Vives, F. (2010). Oxidative stress and plasma aminopeptidase activity in Huntington's disease. *Journal of Neural Transmission*, *117*(3), 325–332.

El Soury, M., Fornasari, B. E., Carta, G., Zen, F., Haastert-Talini, K., & Ronchi, G. (2021). The role of dietary nutrients in peripheral nerve regeneration. *International Journal of Molecular Sciences*, *22*(14), 7417.

Eldridge, C. F., Bunge, M. B., Bunge, R. P., & Wood, P. M. (1987). Differentiation of axon-related Schwann cells in vitro. I. Ascorbic acid regulates basal lamina assembly and myelin formation. *Journal of Cell Biology*, *105*(2), 1023–1034.

El-Mezayen, N. S., Abd el Moneim, R. A., & El-Rewini, S. H. (2022). Vitamin B12 as a cholinergic system modulator and blood brain barrier integrity restorer in Alzheimer's disease. *European Journal of Pharmaceutical Sciences*, *174*, 106201.

Emond, M., Lepage, G., Vanasse, M., & Pandolfo, M. (2000). Increased levels of plasma malondialdehyde in Friedreich ataxia. *Neurology-Minneapolis*, *55*(11), 1752–1752.

Faran, M. (2020). *Potential of Dairy Digestate as a Biofertilizer: Effects on Growth, Yield and Phytochemicals of Lettuce in Hydroponics*. Master's Thesis, Memorial University of Newfoundland, Memorial University Research Repository.

Fenercioglu, A. K. (2024). The anti-inflammatory roles of vitamin D for improving human health. *Current Issues in Molecular Biology*, *46*(12), 13514–13525.

Feng, Y., & Wang, X. (2012). Antioxidant therapies for Alzheimer's disease. *Oxidative Medicine and Cellular Longevity*, *2012*(1), 472932.

Fragoso, Y. D., Stoney, P. N., & McCaffery, P. J. (2014). The evidence for a beneficial role of vitamin A in multiple sclerosis. *CNS Drugs*, *28*(4), 291–299.

Fukuwatari, T., & Shibata, K. (2013). Nutritional aspect of tryptophan metabolism. *International Journal of Tryptophan Research*, *6*, IJTR. S11588.

Fusaro, M., Cianciolo, G., Brandi, M. L., Ferrari, S., Nickolas, T. L., Tripepi, G., Plebani, M., Zaninotto, M., Iervasi, G., & La Manna, G. (2020). Vitamin K and osteoporosis. *Nutrients*, *12*(12), 3625.

Fusaro, M., Mereu, M. C., Aghi, A., Iervasi, G., & Gallieni, M. (2017). Vitamin K and bone. *Clinical Cases in Mineral and Bone Metabolism*, *14*(2), 200.

Gao, H.-M., Zhou, H., & Hong, J.-S. (2014). Oxidative stress, neuroinflammation, and neurodegeneration. In: P. K. Peterson & M. Toborek (Eds.), *Neuroinflammation and Neurodegeneration* (pp. 81–104). Springer.

Gasperi, V., Sibilano, M., Savini, I., & Catani, M. V. (2019). Niacin in the central nervous system: An update of biological aspects and clinical applications. *International Journal of Molecular Sciences*, *20*(4), 974.

Gliszczyńska-Świgło, A. (2007). Folates as antioxidants. *Food Chemistry*, *101*(4), 1480–1483.

Granger, M., & Eck, P. (2018). Dietary vitamin C in human health. *Advances in Food and Nutrition Research*, *83*, 281–310.

Green, R., Allen, L. H., Bjørke-Monsen, A.-L., Brito, A., Guéant, J.-L., Miller, J. W., Molloy, A. M., Nexo, E., Stabler, S., &

Toh, B.-H. (2017). Vitamin B12 deficiency. *Nature Reviews Disease Primers*, *3*(1), 1–20.

Gregory, A., & Hayflick, S. J. (2017). Pantothenate kinase-associated neurodegeneration. In: A. M. Adam, M. P. Mirzaa, G. M. Ardinger et al. (Eds.), *GeneReviews*. University of Washington.

Grimm, M. O., Thiel, A., Lauer, A. A., Winkler, J., Lehmann, J., Regner, L., Nelke, C., Janitschke, D., Benoist, C., & Streidenberger, O. (2017). Vitamin D and its analogues decrease amyloid-β (Aβ) formation and increase Aβ-degradation. *International Journal of Molecular Sciences*, *18*(12), 2764.

Guan, J.-Z., Guan, W.-P., & Maeda, T. (2018). Vitamin E administration erases an enhanced oxidation in multiple sclerosis. *Canadian Journal of Physiology and Pharmacology*, *96*(11), 1181–1183.

Güneş-Bayir, A. (2025). Vitamin A deficiencies and impact on health. In: V. R. Preedy & V. B. Patel (Eds.), *Handbook of Public Health Nutrition: International, National, and Regional Perspectives* (pp. 1–12). Springer.

Habib, A. M., Nagi, K., Thillaiappan, N. B., Sukumaran, V., & Akhtar, S. (2020). Vitamin D and its potential interplay with pain signaling pathways. *Frontiers in Immunology*, *11*, 820.

Hadi, S., Alhadithi, H., Alkobaese, S. K., Akram, M., & Sareen, A. (2021). The effect of vitamin D3 on some metabolic processes in the human body. *IOP Conference Series: Earth and Environmental Science*, *923*(1), 012067.

Halliwell, B. (2006). Oxidative stress and neurodegeneration: Where are we now? *Journal of Neurochemistry*, *97*(6), 1634–1658.

Hamid, M., Mansoor, S., Amber, S., & Zahid, S. (2022). A quantitative meta-analysis of vitamin C in the pathophysiology of Alzheimer's disease. *Frontiers in Aging Neuroscience*, *14*, 970263.

Hands, S. L., Mason, R., Sajjad, M. U., Giorgini, F., & Wyttenbach, A. (2010). Metallothioneins and copper metabolism are candidate therapeutic targets in Huntington's disease. *Biochemical Society Transactions*, *38*(2), 552–558.

Hansen, S. N., Tveden-Nyborg, P., & Lykkesfeldt, J. (2014). Does vitamin C deficiency affect cognitive development and function? *Nutrients*, *6*(9), 3818–3846.

Hao, X., Li, H., Li, Q., Gao, D., Wang, X., Wu, C., Wang, Q., & Zhu, M. (2024). Dietary vitamin E intake and risk of Parkinson's disease: A cross-sectional study. *Frontiers in Nutrition*, *10*, 1289238.

Harrison, F. E., & May, J. M. (2009). Vitamin C function in the brain: Vital role of the ascorbate transporter SVCT2. *Free Radical Biology and Medicine*, *46*(6), 719–730.

Hassan, W., Noreen, H., Rehman, S., Kamal, M. A., & da Rocha, J. B. (2022). Association of oxidative stress with neurological disorders. *Current Neuropharmacology*, *20*(6), 1046–1072.

Haxaire, C., Turpin, F. R., Potier, B., Kervern, M., Sinet, P. M., Barbanel, G., Mothet, J. P., Dutar, P., & Billard, J. M. (2012). Reversal of age-related oxidative stress prevents hippocampal synaptic plasticity deficits by protecting d-serine-dependent NMDA receptor activation. *Aging Cell*, *11*(2), 336–344.

Hayflick, S. J. (2014). Defective pantothenate metabolism and neurodegeneration. *Biochemical Society Transactions*, *42*(4), 1063–1068.

Hegyi, J., Schwartz, R. A., & Hegyi, V. (2004). Pellagra: Dermatitis, dementia, and diarrhea. *International Journal of Dermatology*, *43*(1), 1–5.

Higashi-Okai, K., Nagino, H., Yamada, K., & Okai, Y. (2006). Antioxidant and prooxidant activities of B group vitamins in lipid peroxidation. *Journal of UOEH*, *28*(4), 359–368.

Hrubša, M., Siatka, T., Nejmanová, I., Vopršalová, M., Kujovská Krčmová, L., Matoušová, K., Javorská, L., Macáková, K., Mercolini, L., & Remião, F. (2022). Biological properties of vitamins of the B-complex, part 1: Vitamins B1, B2, B3, and B5. *Nutrients*, *14*(3), 484.

Huang, T.-T., Zou, Y., & Corniola, R. (2012). Oxidative stress and adult neurogenesis—effects of radiation and superoxide dismutase deficiency. *Seminars in Cell & Developmental Biology*, *23*(7), 738–744.

Huang, Y., Coupland, N. J., Lebel, R. M., Carter, R., Seres, P., Wilman, A. H., & Malykhin, N. V. (2013). Structural changes in hippocampal subfields in major depressive disorder: A high-field magnetic resonance imaging study. *Biological Psychiatry*, *74*(1), 62–68.

Iwamoto, J., Matsumoto, H., Takeda, T., Sato, Y., & Yeh, J. K. (2010). Effects of vitamin K2 on cortical and cancellous bone mass, cortical osteocyte and lacunar system, and porosity in sciatic neurectomized rats. *Calcified Tissue International*, *87*(3), 254–262.

Jia, W., Wang, H., Li, C., Shi, J., Yong, F., & Jia, H. (2024). Association between dietary vitamin B1 intake and cognitive function among older adults: A cross-sectional study. *Journal of Translational Medicine*, *22*(1), 165.

Joshi, P., Chia, S., Yang, X., Perni, M., Gabriel, J. M., Gilmer, M., Limbocker, R., Habchi, J., & Vendruscolo, M. (2023). Combinations of vitamin A and vitamin E metabolites confer resilience against amyloid-β aggregation. *ACS Chemical Neuroscience*, *14*(4), 657–666.

Kannan, K., & Jain, S. K. (2004). Effect of vitamin B6 on oxygen radicals, mitochondrial membrane potential, and lipid peroxidation in H_2O_2-treated U937 monocytes. *Free Radical Biology and Medicine*, *36*(4), 423–428.

Karamshetty, V., Acharya, J. D., Ghaskadbi, S., & Goel, P. (2016). Mathematical modeling of glutathione status in type 2 diabetics with vitamin B12 deficiency. *Frontiers in Cell and Developmental Biology*, *4*, 16.

Khan, K. M., & Jialal, I. (2023). Folic acid deficiency. In: *StatPearls [internet]*. StatPearls Publishing.

Kim, G. H., Kim, J. E., Rhie, S. J., & Yoon, S. (2015). The role of oxidative stress in neurodegenerative diseases. *Experimental Neurobiology*, *24*(4), 325.

Kocot, J., Luchowska-Kocot, D., Kiełczykowska, M., Musik, I., & Kurzepa, J. (2017). Does vitamin C influence neurodegenerative diseases and psychiatric disorders? *Nutrients*, *9*(7), 659.

Koduah, P., Paul, F., & Dörr, J.-M. (2017). Vitamin D in the prevention, prediction and treatment of neurodegenerative and neuroinflammatory diseases. *Epma Journal*, *8*(4), 313–325.

Kono, K., Fujii, H., Nakai, K., Goto, S., Kitazawa, R., Kitazawa, S., Shinohara, M., Hirata, M., Fukagawa, M., & Nishi, S. (2013). Anti-oxidative effect of vitamin D analog on incipient vascular lesion in non-obese type 2 diabetic rats. *American Journal of Nephrology*, *37*(2), 167–174.

Kook, S., Lee, K., Kim, Y., Cha, M., Kang, S., Baik, S., Lee, H., Park, R., & Mook-Jung, I. (2014). High-dose of vitamin C supplementation reduces amyloid plaque burden and ameliorates pathological changes in the brain of 5XFAD mice. *Cell Death & Disease*, *5*(2), e1083–e1083.

Krystal, J., Sanacora, G., Blumberg, H., Anand, A., Charney, D., Marek, G., Epperson, C., Goddard, A., & Mason, G. (2002). Glutamate and GABA systems as targets for novel antidepressant and mood-stabilizing treatments. *Molecular Psychiatry*, *7*(1), S71–S80.

Kurian, M. A., & Hayflick, S. J. (2013). Pantothenate kinase-associated neurodegeneration (PKAN) and PLA2G6-associated neurodegeneration (PLAN): Review of two major neurodegeneration with brain iron accumulation (NBIA) phenotypes. *International Review of Neurobiology*, *110*, 49–71.

Labudzynskyi, D., Zaitseva, O., Latyshko, N., Gudkova, O., & Veliky, M. (2015). Vitamin D3 contribution to the regulation of oxidative metabolism in the liver of diabetic mice. *The Ukrainian Biochemical Journal, 87*(3), 75–90.

Lasemi, R., Kundi, M., Moghadam, N. B., Moshammer, H., & Hainfellner, J. A. (2018). Vitamin K2 in multiple sclerosis patients. *Wiener Klinische Wochenschrift*, *130*(9), 307–313.

Lee, C.-Y., Chan, L., Hu, C.-J., Hong, C.-T., & Chen, J.-H. (2024). Role of vitamin B12 and folic acid in treatment of Alzheimer's disease: A meta-analysis of randomized control trials. *Aging (Albany NY)*, *16*(9), 7856.

Leon-Del-Rio, A. (2019). Biotin in metabolism, gene expression, and human disease. *Journal of Inherited Metabolic Disease*, *42*(4), 647–654.

Lešić, S., Ivanišević, Z., Špiljak, B., Tomas, M., Šoštarić, M., & Včev, A. (2024). The impact of vitamin deficiencies on oral manifestations in children. *Dentistry Journal*, *12*(4), 109.

Leuner, B., & Shors, T. (2013). Stress, anxiety, and dendritic spines: What are the connections? *Neuroscience*, *251*, 108–119.

Li, J., Lin, J. C., Wang, H., Peterson, J. W., Furie, B. C., Furie, B., Booth, S. L., Volpe, J. J., & Rosenberg, P. A. (2003). Novel role of vitamin k in preventing oxidative injury to developing oligodendrocytes and neurons. *Journal of Neuroscience*, *23*(13), 5816–5826.

Li, L., Li, Y., Fan, Z., Wang, X., Li, Z., Wen, J., Deng, J., Tan, D., Pan, M., & Hu, X. (2019). Ascorbic acid facilitates neural regeneration after sciatic nerve crush injury. *Frontiers in Cellular Neuroscience*, *13*, 108.

Li, R., Yu, K., Wang, Q., Wang, L., Mao, J., & Qian, J. (2016). Pellagra secondary to medication and alcoholism: A case report and review of the literature. *Nutrition in Clinical Practice*, *31*(6), 785–789.

Lin, M. T., & Beal, M. F. (2006). Mitochondrial dysfunction and oxidative stress in neurodegenerative diseases. *Nature*, *443*(7113), 787–795.

Lin, X., Wen, X., Wei, Z., Guo, K., Shi, F., Huang, T., Wang, W., & Zheng, J. (2021). Vitamin K2 protects against Aβ42-induced neurotoxicity by activating autophagy and improving mitochondrial function in *Drosophila*. *Neuroreport*, *32*(6), 431–437.

Liston, C., McEwen, B. S., & Casey, B. (2009). Psychosocial stress reversibly disrupts prefrontal processing and attentional control. *Proceedings of the National Academy of Sciences*, *106*(3), 912–917.

Liu, T., Soong, S., Wilson, N. P., Craig, C. B., Cole, P., Macaluso, M., & Butterworth Jr, C. E. (1993). A case control study of nutritional factors and cervical dysplasia. *Cancer Epidemiology, Biomarkers & Prevention: A publication of the American Association for Cancer Research, Cosponsored by the American Society of Preventive Oncology*, *2*(6), 525–530.

Liu, Y., Gou, M., & Guo, X. (2023). Features of plasma homocysteine, vitamin B12, and folate in Parkinson's disease: An updated meta-analysis. *Journal of Integrative Neuroscience*, *22*(5), 115.

Liu, Z., Li, C., Fan, X., Kuang, Y., Zhang, X., Chen, L., Song, J., Zhou, Y., Takahashi, E., & He, G. (2020). Nicotinamide, a vitamin B3 ameliorates depressive behaviors independent of SIRT1 activity in mice. *Molecular Brain*, *13*(1), 162.

Long, A. N., Owens, K., Schlappal, A. E., Kristian, T., Fishman, P. S., & Schuh, R. A. (2015). Effect of nicotinamide mononucleotide on brain mitochondrial respiratory deficits in an Alzheimer's disease-relevant murine model. *BMC Neurology*, *15*(1), 19.

Lopez, A.-G., Kerlan, V., & Desailloud, R. (2021). Non-classical effects of vitamin D: Non-bone effects of vitamin D. *Annales d'Endocrinologie*, *82*(2), 106–114.

Mangialasche, F., Kivipelto, M., Mecocci, P., Rizzuto, D., Palmer, K., Winblad, B., & Fratiglioni, L. (2010). High plasma levels of vitamin E forms and reduced Alzheimer's disease risk in advanced age. *Journal of Alzheimer's Disease*, *20*(4), 1029–1037.

Manzanares, W., & Hardy, G. (2010). Vitamin B12: The forgotten micronutrient for critical care. *Current Opinion in Clinical Nutrition & Metabolic Care*, *13*(6), 662–668.

Marashly, E. T., & Bohlega, S. A. (2017). Riboflavin has neuroprotective potential: Focus on Parkinson's disease and migraine. *Frontiers in Neurology*, *8*, 333.

Marie, A., Darricau, M., Touyarot, K., Parr-Brownlie, L. C., & Bosch-Bouju, C. (2021). Role and mechanism of vitamin A metabolism in the pathophysiology of Parkinson's disease. *Journal of Parkinson's Disease*, *11*(3), 949–970.

Masood, A., Nadeem, A., Mustafa, S. J., & O'Donnell, J. M. (2008). Reversal of oxidative stress-induced anxiety by inhibition of phosphodiesterase-2 in mice. *Journal of Pharmacology and Experimental Therapeutics*, *326*(2), 369–379.

Matxain, J. M., Ristilä, M., Strid, Å., & Eriksson, L. A. (2006). Theoretical study of the antioxidant properties of pyridoxine. *Journal of Physical Chemistry A*, *110*(48), 13068–13072.

McCarty, M. F., & DiNicolantonio, J. J. (2017). Neuroprotective potential of high-dose biotin. *Medical Hypotheses*, *109*, 145–149.

McEwen, B. S. (2008). Understanding the potency of stressful early life experiences on brain and body function. *Metabolism*, *57*, S11–S15.

Miller, E., Morel, A., Saso, L., & Saluk, J. (2014). Isoprostanes and neuroprostanes as biomarkers of oxidative stress in neurodegenerative diseases. *Oxidative Medicine and Cellular Longevity*, *2014*(1), 572491.

Miller, J. W., & Rucker, R. B. (2012). Pantothenic acid. In: J. W. Erdman Jr., I. A. Macdonald, & S. H. Zeisel (Eds.), *Present Knowledge in Nutrition* (pp. 375–390). Wiley.

Mitsumoto, H., Garofalo, D. C., Santella, R. M., Sorenson, E. J., Oskarsson, B., Fernandes Jr, J. A. M., Andrews, H., Hupf, J., Gilmore, M., & Heitzman, D. (2020). Plasma creatinine and oxidative stress biomarkers in amyotrophic lateral sclerosis. *Amyotrophic Lateral Sclerosis and Frontotemporal Degeneration*, *21*(3–4), 263–272.

Miyazawa, T., Burdeos, G. C., Itaya, M., Nakagawa, K., & Miyazawa, T. (2019). Vitamin E: Regulatory redox interactions. *IUBMB Life*, *71*(4), 430–441.

Modica, J. S., Bonno, D., & Lizarraga, K. J. (2020). Pearls and Oysters: Vitamin B6 deficiency presenting with new-onset epilepsy and status epilepticus in a patient with Parkinson disease. *Neurology*, *94*(24), e2605–e2607.

Montava, M., Garcia, S., Mancini, J., Jammes, Y., Courageot, J., Lavieille, J.-P., & Feron, F. (2015). Vitamin D3 potentiates myelination and recovery after facial nerve injury. *European Archives of Oto-Rhino-Laryngology*, *272*(10), 2815–2823.

Moreira, E. S., Brasch, N. E., & Yun, J. (2011). Vitamin B12 protects against superoxide-induced cell injury in human aortic endothelial cells. *Free Radical Biology and Medicine*, *51*(4), 876–883.

Moretti, M., Fraga, D. B., & Rodrigues, A. L. S. (2017). Preventive and therapeutic potential of ascorbic acid in neurodegenerative diseases. *CNS Neuroscience & Therapeutics*, *23*(12), 921–929.

Moretti, R., Morelli, M. E., & Caruso, P. (2018). Vitamin D in neurological diseases: A rationale for a pathogenic impact. *International Journal of Molecular Sciences*, *19*(8), 2245.

Murray, R. K., Kennelly, P. J., Bender, D. A., Rodwell, V. W., Botham, K. M., & Weil, P. A. (2009). *Harper's illustrated biochemistry*. McGraw-Hill Companies, Inc.

National Institutes of Health. (2022). *Carotenoids*. NIH Fact Sheet for Health. https://ods.od.nih.gov/factsheets/Carotenoids-HealthProfessional/.

Ng, L. H. L., Huang, Y., Han, L., Chang, R. C.-C., Chan, Y. S., & Lai, C. S. W. (2018). Ketamine and selective activation of parvalbumin interneurons inhibit stress-induced dendritic spine elimination. *Translational Psychiatry*, *8*(1), 272.

Niebrój-Dobosz, I., Dziewulska, D., & Kwieciński, H. (2004). Oxidative damage to proteins in the spinal cord in amyotrophic lateral sclerosis (ALS). *Folia Neuropathologica*, *42*(3), 151–156.

Nieto-Salazar, M. A., Ordóñez, K. N. A., Carcamo, Z. D. S., Cristina, A., Ordóñez, A., & Saldana, E. A. (2023). Neurological dysfunction associated with vitamin deficiencies: A narrative review. *Open Access Journal of Neurology and Neurosurgery*, *18*(1), 555979.

Nikolac Gabaj, N., Unic, A., Miler, M., Pavicic, T., Culej, J., Bolanca, I., Herman Mahecic, D., Milevoj Kopcinovic, L., & Vrtaric, A. (2020). In sickness and in health: Pivotal role of vitamin D. *Biochemia Medica*, *30*(2), 202–214.

Nolan, J. M., Power, R., Howard, A. N., Bergin, P., Roche, W., Prado-Cabrero, A., Pope, G., Cooke, J., Power, T., & Mulcahy, R. (2022). Supplementation with carotenoids, omega-3 fatty acids, and vitamin e has a positive effect on the symptoms and progression of Alzheimer's disease. *Journal of Alzheimer's Disease*, *90*(1), 233–249.

Nowotny, K., Jung, T., Höhn, A., Weber, D., & Grune, T. (2015). Advanced glycation end products and oxidative stress in type 2 diabetes mellitus. *Biomolecules*, *5*(1), 194–222.

Nualart, F., Mack, L., García, A., Cisternas, P., Bongarzone, E. R., Heitzer, M., Jara, N., Martínez, F., Ferrada, L., & Espinoza, F. (2014). Vitamin C transporters, recycling and the bystander effect in the nervous system: SVCT2 versus gluts. *Journal of Stem Cell Research & Therapy*, *4*(5), 209.

Ohta, B. K., & Foote, C. S. (2002). Characterization of endoperoxide and hydroperoxide intermediates in the reaction of pyridoxine with singlet oxygen. *Journal of the American Chemical Society*, *124*(41), 12064–12065.

Oliveira, N., Magalhaes, L., Matos, M., Aragao, G., & Bachur, T. (2019). The deficiency of thiamine and niacin as a risk factor for neurological disease. *Infarma—Ciências Farmacêuticas, 31*(2), 80–85

Onose, G., Anghelescu, A., Blendea, D., Ciobanu, V., Daia, C., Firan, F. C., Oprea, M., Spinu, A., Popescu, C., & Ionescu, A. (2022). Cellular and molecular targets for non-invasive, non-pharmacological therapeutic/rehabilitative interventions in acute ischemic stroke. *International Journal of Molecular Sciences*, *23*(2), 907.

Patrick, R. P., & Ames, B. N. (2014). Vitamin D hormone regulates serotonin synthesis. Part 1: Relevance for autism. *The FASEB Journal*, *28*(6), 2398–2413.

Pignolo, A., Mastrilli, S., Davì, C., Arnao, V., Aridon, P., dos Santos Mendes, F. A., Gagliardo, C., & D'Amelio, M. (2022). Vitamin D and Parkinson's disease. *Nutrients*, *14*(6), 1220.

Plantone, D., Pardini, M., & Rinaldi, G. (2021). Riboflavin in neurological diseases: A narrative review. *Clinical Drug Investigation*, *41*(6), 513–527.

Popescu, A., & German, M. (2021). Vitamin K2 holds promise for Alzheimer's prevention and treatment. *Nutrients*, *13*(7), 2206.

Popescu, D. C., Huang, H., Singhal, N. K., Shriver, L., McDonough, J., Clements, R. J., & Freeman, E. J. (2018). Vitamin K enhances the production of brain sulfatides during remyelination. *PLoS ONE*, *13*(8), e0203057.

Rai, S., Kamat, P. K., Nath, C., & Shukla, R. (2013). A study on neuroinflammation and NMDA receptor function in STZ (ICV) induced memory impaired rats. *Journal of Neuroimmunology*, *254*(1–2), 1–9.

Rakić, M., Lunić, T., Bekić, M., Tomić, S., Mitić, K., Graovac, S., Božić, B., & Nedeljković, B. B. (2023). Vitamin B complex suppresses neuroinflammation in activated microglia: In vitro and in silico approach combined with dynamical modeling. *International Immunopharmacology*, *121*, 110525.

Ramasamy, I. (2020). Vitamin D metabolism and guidelines for vitamin D supplementation. *Clinical Biochemist Reviews*, *41*(3), 103.

Regner-Nelke, L., Nelke, C., Schroeter, C. B., Dziewas, R., Warnecke, T., Ruck, T., & Meuth, S. G. (2021). Enjoy carefully: The multifaceted role of vitamin E in neuro-nutrition. *International Journal of Molecular Sciences*, *22*(18), 10087.

Regulska-Ilow, B., Różańska, D., Zatońska, K., & Szuba, A. (2022). Estimation of vitamin K content and its sources in the diet of the polish participants of the PURE study. *Nutrients*, *14*(9), 1917.

Reynolds, E. (2014). The neurology of folic acid deficiency. *Handbook of Clinical Neurology*, *120*, 927–943.

Rivera-Mancía, S., Pérez-Neri, I., Ríos, C., Tristán-López, L., Rivera-Espinosa, L., & Montes, S. (2010). The transition metals copper and iron in neurodegenerative diseases. *Chemico-Biological Interactions*, *186*(2), 184–199.

Rodríguez-Lara, S. Q., Cardona-Muñoz, E. G., Ramírez-Lizardo, E. J., Totsuka-Sutto, S. E., Castillo-Romero, A., García-Cobián, T. A., & García-Benavides, L. (2016). Alternative interventions to prevent oxidative damage following ischemia/reperfusion. *Oxidative Medicine and Cellular Longevity*, *2016*(1), 7190943.

Sakaue, M., Mori, N., Okazaki, M., Kadowaki, E., Kaneko, T., Hemmi, N., Sekiguchi, H., Maki, T., Ozawa, A., & Hara, S. (2011). Vitamin K has the potential to protect neurons from methylmercury-induced cell death in vitro. *Journal of Neuroscience Research*, *89*(7), 1052–1058.

Salum, E., Kals, J., Kampus, P., Salum, T., Zilmer, K., Aunapuu, M., Arend, A., Eha, J., & Zilmer, M. (2013). Vitamin D reduces deposition of advanced glycation end-products in the aortic wall and systemic oxidative stress in diabetic rats. *Diabetes Research and Clinical Practice*, *100*(2), 243–249.

Sangha, A., Quon, M., Pfeffer, G., & Orton, S.-M. (2023). The role of vitamin D in neuroprotection in multiple sclerosis: An update. *Nutrients*, *15*(13), 2978.

Schulz, J., Dehmer, T., Schols, L., Mende, H., Hardt, C., Vorgerd, M., Burk, K., Matson, W., Dichgans, J., & Beal, M. (2000). Oxidative stress in patients with Friedreich ataxia. *Neurology*, *55*(11), 1719–1721.

Senol, M., Sonmez, G., Ozdag, F., & Saracoglu, M. (2008). Reversible myelopathy with vitamin B12 deficiency. *Singapore Medical Journal*, *49*(11), e330–332.

Shemirani, F., Titcomb, T. J., Saxby, S. M., Ten Eyck, P., Rubenstein, L. M., Hoth, K. F., Snetselaar, L. G., & Wahls, T. L. (2023). Association of serum homocysteine, folate, and vitamin B12 and mood following the Swank and Wahls elimination dietary interventions in relapsing-remitting multiple sclerosis: Secondary analysis of the WAVES trial. *Multiple Sclerosis and Related Disorders*, *75*, 104743.

Shulpekova, Y., Nechaev, V., Kardasheva, S., Sedova, A., Kurbatova, A., Bueverova, E., Kopylov, A., Malsagova, K., Dlamini, J. C., & Ivashkin, V. (2021). The concept of folic acid in health and disease. *Molecules*, *26*(12), 3731.

Silva de Paula, E., Carneiro, M. F. H., Grotto, D., Hernandes, L. C., Antunes, L. M. G., & Barbosa Jr, F. (2016). Protective effects of niacin against methylmercury-induced genotoxicity and alterations in antioxidant status in rats. *Journal of Toxicology and Environmental Health, Part A*, 79(4), 174–183.

Simes, D. C., Viegas, C. S., Araújo, N., & Marreiros, C. (2019). Vitamin K as a powerful micronutrient in aging and age-related diseases: Pros and cons from clinical studies. *International Journal of Molecular Sciences*, *20*(17), 4150.

Simes, D. C., Viegas, C. S., Araújo, N., & Marreiros, C. (2020). Vitamin K as a diet supplement with impact in human health: Current evidence in age-related diseases. *Nutrients*, *12*(1), 138.

Sinbad, O. O., Folorunsho, A. A., Olabisi, O. L., Ayoola, O. A., & Temitope, E. J. (2019). Vitamins as antioxidants. *Journal of Food Science and Nutrition Research*, *2*(3), 214–235.

Sm, S., Hn, S., Na, E., & As, H. (2018). Curative role of pantothenic acid in brain damage of gamma irradiated rats. *Indian Journal of Clinical Biochemistry*, *33*(3), 314–321.

Smith, T. J., Johnson, C. R., Koshy, R., Hess, S. Y., Qureshi, U. A., Mynak, M. L., & Fischer, P. R. (2021). Thiamine deficiency disorders: A clinical perspective. *Annals of the New York Academy of Sciences*, *1498*(1), 9–28.

Solanki, N., Alkadhi, I., Atrooz, F., Patki, G., & Salim, S. (2015). Grape powder prevents cognitive, behavioral, and biochemical impairments in a rat model of posttraumatic stress disorder. *Nutrition Research*, *35*(1), 65–75.

Solleiro-Villavicencio, H., & Rivas-Arancibia, S. (2018). Effect of chronic oxidative stress on neuroinflammatory response mediated by CD4+ T cells in neurodegenerative diseases. *Frontiers in Cellular Neuroscience*, *12*, 114.

Sousa, N., Lukoyanov, N., Madeira, M., Almeida, O., & Paula-Barbosa, M. (2000). Reorganization of the morphology of hippocampal neurites and synapses after stress-induced damage correlates with behavioral improvement. *Neuroscience*, *97*(2), 253–266.

Spoelstra-de Man, A. M., Elbers, P. W., & Oudemans-Van Straaten, H. M. (2018). Vitamin C: Should we supplement? *Current Opinion in Critical Care*, *24*(4), 248–255.

Squadrone, S., Brizio, P., Abete, M. C., & Brusco, A. (2020). Trace elements profile in the blood of Huntington' disease patients. *Journal of Trace Elements in Medicine and Biology*, *57*, 18–20.

Standing Committee on the Scientific Evaluation of Dietary Reference Intakes, Institute of Medicine. (2000). Dietary reference intakes for thiamin, riboflavin, niacin, vitamin B6, folate, vitamin B12, pantothenic acid, biotin, and choline. National Academies Press.

Suwannasom, N., Kao, I., Pruß, A., Georgieva, R., & Bäumler, H. (2020). Riboflavin: The health benefits of a forgotten natural vitamin. *International Journal of Molecular Sciences*, *21*(3), 950.

Tanyel, M., & Mancano, L. (1997). Neurologic findings in vitamin E deficiency. *American Family Physician*, *55*(1), 197–201.

Teleanu, D. M., Niculescu, A.-G., Lungu, I. I., Radu, C. I., Vladâcenco, O., Roza, E., Costăchescu, B., Grumezescu, A. M., & Teleanu, R. I. (2022). An overview of oxidative stress, neuroinflammation, and neurodegenerative diseases. *International Journal of Molecular Sciences*, *23*(11), 5938.

Tsang, C. K., & Kamei, Y. (2002). Novel effect of vitamin K1 (phylloquinone) and vitamin K2 (menaquinone) on promoting nerve growth factor-mediated neurite outgrowth from PC12D cells. *Neuroscience Letters*, *323*(1), 9–12.

Tyagi, N., Sedoris, K. C., Steed, M., Ovechkin, A. V., Moshal, K. S., & Tyagi, S. C. (2005). Mechanisms of homocysteine-induced oxidative stress. *American Journal of Physiology-Heart and Circulatory Physiology*, *289*(6), H2649–H2656.

Ulatowski, L. M., & Manor, D. (2015). Vitamin E and neurodegeneration. *Neurobiology of Disease*, *84*, 78–83.

Umer, M., Shah, S. M. A., Moeckel, C., Azhar, E., Aziz, F., & Waheed, A. (2025). Global trends in vitamin D research from 1985 to 2024. *PRiMER: Peer-Reviewed Reports in Medical Education Research*, *9*, 33.

Valdés, F. (2006). Vitamina C. *Actas Dermo-Sifiliográficas*, *97*(9), 557–568.

Wang, G., Li, W., Lu, X., & Zhao, X. (2011). Riboflavin alleviates cardiac failure in Type I diabetic cardiomyopathy. *Heart International*, *6*(2), hi. 2011. e2021.

Wang, X., & Michaelis, E. K. (2010). Selective neuronal vulnerability to oxidative stress in the brain. *Frontiers in Aging Neuroscience*, *2*, 1224.

Wassif, G. A., Alrehely, M. S., Alharbi, D. M., Aljohani, A. A., & Wassif, G. (2023). The impact of vitamin D on neuropsychiatric disorders. *Cureus*, *15*(10).

Wiley, K. D., & Gupta, M. (2019). *Vitamin B1 (Thiamine) Deficiency*. StatPearls Publishing.

Willems, B. A., Vermeer, C., Reutelingsperger, C. P., & Schurgers, L. J. (2014). The realm of vitamin K dependent proteins: Shifting from coagulation toward calcification. *Molecular Nutrition & Food Research*, *58*(8), 1620–1635.

Wolden-Kirk, H., Gysemans, C., Verstuyf, A., & Mathieu, C. (2012). Extraskeletal effects of vitamin D. *Endocrinology and Metabolism Clinics of North America*, *41*(3), 571–594.

Wuerch, E., Urgoiti, G. R., & Yong, V. W. (2023). The promise of niacin in neurology. *Neurotherapeutics*, *20*(4), 1037–1054.

Xia, K., Wang, Y., Zhang, L., Tang, L., Zhang, G., Huang, T., Huang, N., & Fan, D. (2022). Dietary-derived essential nutrients and amyotrophic lateral sclerosis: A two-sample Mendelian randomization study. *Nutrients*, *14*(5), 920.

Yang, F., Wolk, A., Håkansson, N., Pedersen, N. L., & Wirdefeldt, K. (2017). Dietary antioxidants and risk of Parkinson's disease in two population-based cohorts. *Movement Disorders*, *32*(11), 1631–1636.

Young, J. I., Züchner, S., & Wang, G. (2015). Regulation of the epigenome by vitamin C. *Annual Review of Nutrition*, *35*(1), 545–564.

Yu, Y.-X., Yu, X.-D., Cheng, Q.-Z., Tang, L., & Shen, M.-Q. (2020). The association of serum vitamin K2 levels with Parkinson's disease: From basic case-control study to big data mining analysis. *Aging*, *12*(16), 16410.

Zaaboul, F., & Liu, Y. (2022). Vitamin E in foodstuff: Nutritional, analytical, and food technology aspects. *Comprehensive Reviews in Food Science and Food Safety*, *21*(2), 964–998.

3 Pharmacokinetics, Bioavailability, and Blood–Brain Barrier Transport of Vitamins

Ashish Kumar Parashar[1], Ruchika Rajani[2], Smruti Navthare[2], Chandani Prasad[3], Afrin Khan[4], and Sahil Thakre[2]

[1]Department of Pharmaceutics, Lloyd Institute of Management and Technology, Greater Noida, India
[2]Department of Pharmacy, SPTM, SVKM's NMIMS, Indore, India
[3]Department of Pharmacy, Manwatkar College of Pharmacy, Chandrapur, India
[4]Department of Pharmacy, NMT Gujarati College of Pharmacy, Indore, India

3.1 INTRODUCTION TO VITAMIN PHARMACOKINETICS

3.1.1 Overview of Vitamins

Vitamins are a group of chemical compounds that are crucial for physiological processes in the body. They are organic compounds necessary for the maintenance of regular bodily processes; however, the body cannot usually produce them. Under typical conditions, a balanced diet and proper nutrition provide the necessary vitamins; however, supplements may be required when dietary intake fails to meet the minimal nutritional needs. Vitamins play a crucial role in growth, function, and metabolism. All vitamins, except vitamin D, which is produced by the body, need to be taken from the diet. Deficiency of these vitamins is related to different diseases such as beriberi (vitamin B1 deficiency) and scurvy (vitamin C deficiency) [1]. Vitamin-related disorders can be categorized as follows.

- Avitaminosis, a complete deficiency of one or more vitamins
- Hypovitaminosis, a partial deficiency, and
- Hypervitaminosis is an excess of one or more vitamins, particularly those that are not easily excreted in the urine due to poor water solubility.

According to their solubility, vitamins can be divided into two categories:

- Water-soluble vitamins (vitamins B and C) and
- Fat-soluble vitamins (vitamins A, D, E, and K) [2].

3.2 HISTORICAL PERSPECTIVE AND VITAMIN DISCOVERY

Originally known as "vitamine" before being changed to "vitamin," these vital nutrients were found in the late 1800s and were well known in the 1920s because of Casimir Funk's research. The primary period of discovery began in the early 19th century and ended in the mid-20th century. Through the combined efforts of chemists, doctors, physiologists, and epidemiologists, the riddles surrounding each vitamin were resolved. The germ theory of disease, which holds that only four nutritional components—proteins, carbohydrates, fats, and minerals—are necessary, sparked research on vitamins associated with major ailments. Since it helps lower the time frame that people suffer from vitamin deficiencies, experimental physiology and the utilization of animal models are vital in nutritional studies [3]. Ultimately, chemists isolated various vitamins, determined their chemical structures, and developed methods for their synthesis. The discovery of vitamins has led to the introduction of numerous products into the market. Vitamin B derived from yeast is commonly available in stores under brand names, such as vitamins, double-strength yeast, and super vitamins. In 1916, Mastin's Yeast Vitamon Tablets, containing iron, calcium, and vitamins A, B, and C, as well as Nux Vomica (a homeopathic medicine for the treatment of heartburn), became popular drugs that resemble modern multivitamins. The label on the product mentioned, "This preparation contains vitamins with other ingredients which shall help in stimulating the appetite, assist digestion, treat constipation, clear the complexion, provide energy, and as a tonic, to help increase weight in weakened individuals, and treat malnutrition" [4]. Sales of vitamins A and C surged in the 1920s as the public began to believe that supplements containing vitamins also included other beneficial substances. By 1922, multivitamin-type products were extensively marketed as remedies for various ailments, including boils, blackheads, pimples, constipation, malnourishment, neurological deficiencies, physical fatigue, mental fog, weakness, and general ill-health [5].

DOI: 10.1201/9781003686675-3

3.3 CLASSIFICATION OF VITAMINS

Vitamins are organic compounds required by the body in trace amounts for growth, metabolism, protection, and health maintenance, which the body cannot synthesize. Vitamins are classified into two categories: fat-soluble and water-soluble. Humans possess 13 vitamins: nine water-soluble vitamins (eight B vitamins and vitamin C) and four fat-soluble vitamins (A, D, E, and K). Urinary output serves as a reliable indicator of vitamin intake because water-soluble vitamins dissolve readily in water and are generally excreted rapidly from the body. Consequently, consistent consumption is essential owing to its limited storage capacity. Lipids (fats) facilitate the absorption of fat-soluble vitamins through the digestive system. Accumulation of vitamins A and D in the body can result in hazardous hypervitaminosis. In cystic fibrosis, fat-soluble vitamin deficiency due to malabsorption is a significant clinical concern [6,7].

Classification of different enzymes is shown in Figure 3.1.

3.4 NOMENCLATURE OF VITAMINS

Vitamins are a collection of molecules with many common but not exclusive features: trace amounts of organic substances in natural foods, required for normal metabolism, and the absence of which causes a disease deficiency. However, this term can be applied to a wide range of substances made by the secondary metabolism of fatty acids, simple sugars, and amino acids. Vitamins can also be categorized based on their physical and chemical characteristics, such as their solubility in lipids or aqueous solvents [8]. However, the construction of a straightforward naming system is complicated by the diverse roles of the vitamins. The molecular structures of many vitamins are not well understood; therefore, a system of letter designations was developed as a simple categorization method based on physiological effects. Initially ascribed to vitamins, this method lost part of its usefulness when it was discovered that other substances, such as essential amino acids, were responsible for some of the tasks. The absence of chemical composition data also led to a complex method of defining doses as arbitrary units, with each unit being stated in conjunction with a biological phenomenon or response in a specific animal model [9] (Table 3.1).

3.5 OVERALL ASPECTS OF VITAMIN METABOLISM AND USE

Lipid-soluble vitamins are absorbed and transported by conventional lipid transport mechanisms. The relative absorption of water-soluble vitamins was partly determined by their respective aqueous solubility coefficients. The passage of water-soluble vitamins is mediated by receptor-mediated and active transport systems within the normal range of consumption. Pericellular and passive activities may be influenced by increased concentrations (5–10 times the physiological need) [10]. Absorption of most vitamins occurs in the upper gut. The main exceptions are the absorption of folic acid and vitamin B12 (also called folate or folacin), which are part of a family of folic acid derivatives, by the mid-intestine and the absorption of vitamin B12 primarily in the ileum. Therefore, a broad understanding of the variety and complexity of vitamin metabolism and processing is critical [11]. Vitamins that are soluble in lipid solvents are absorbed and transported via conventional lipid transport mechanisms. The aqueous solubility coefficients of water-soluble vitamins determine their relative absorptions. Within normal physiological limits of intake, water-soluble vitamin absorption occurs via receptor-mediated and active transport mechanisms. Pericellular and experimental physiology may also be affected by higher doses (5–10 times the physiological demand). The upper gastrointestinal system is where the majority of the vitamins are absorbed. Although vitamin B12 is primarily absorbed in the ileum, there are several noteworthy exceptions, such as the mid-intestinal absorption of vitamin B12 (also known as folate or folacin) and folic acid, which are members of the folic acid-derivative family. A comprehensive understanding of the variety and complexity of vitamin metabolism and processing is crucial [12].

3.6 PHARMACOKINETICS' SIGNIFICANCE IN VITAMIN RESEARCH

Pharmacokinetics is a discipline of pharmacology that studies the kinetics of drug absorption, distribution,

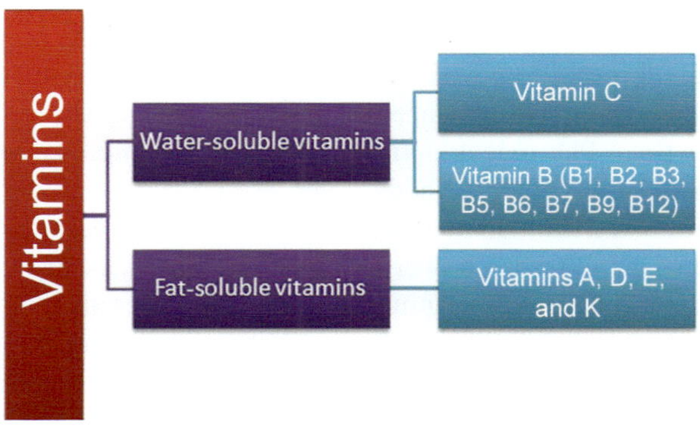

FIGURE 3.1 Classification of vitamins.

TABLE 3.1
Nomenclature, Adult Daily Needs, and Major Dietary Sources of Vitamins

Vitamins	Trivial and IUPAC Designations	Adult Daily Need (Per 1,000 Kcal or 4,180 KJ) or RDA	Major Dietary Sources
Vitamin A	Retinol IUPAC: (2E,4E,6E.8E)-3,7-dimethyl-9-(2,6,6-trimethylcyclohex-1-en-1-yl)nona-2,4,6,8-tetraen-1-ol)	600–900 μg retinol equivalents	Liver, fish liver oils, dairy products, eggs, dark green and yellow vegetables, carrots
Vitamin D	Calciferol, IUPAC: (3S,5Z,7E)-9,10-	10–15 μg (400–600 IU)	Sunlight (UV rays), fortified milk, egg yolk, fatty fish (salmon, sardines), liver
Vitamin E	α-Tocopherol (2R)-2,5,7,8-tetramethyl-2-[(4R,8R)-4,8,12-trimethyltritridecyl]-6-chromanol	8–10 mg	Vegetable oils, nuts, seeds, whole grains
Vitamin K	Phylloquinone (K1), menaquinone (K2) IUPAC: 2-methyl-3-phytylnapthalene-1,4-dione (K1)	60–120 μg	Green leafy vegetables (spinach), cabbage, broccoli, liver, gut bacteria synthesis
Vitamin B1	Thiamine IUPAC: 3[(4-amino-2-methylpyridin-5-yl]-5-(2-hydroxyethyl)-4-methylthiazol-3-ium chloride	1.1–1.2 mg	Whole grains, legumes, nuts, pork, liver
Vitamin B2	Riboflavin IUPAC: 7,8-dimethyl-10-(D-ribo-2,3,4,5-tetrahydroxypentyl) benzo[g]pteridine-2,4-dione	1.1–1.3 mg	Milk, eggs, green leafy vegetables, meat
Vitamin B3	Niacin IUPAC: Pyridine-3-carboxylic acid	14–16 mg	Meat, poultry, fish, whole grains, peanuts
Vitamin B5	Pantothenic acid IUPAC: ®-2,4-dihydroxy-N-(3-oxo-3-[-(2R)-2,4,4-trimethyl-3-oxopentanoyl] propyl)-3,3-dimethylbutamide	5 mg	Meat, whole grains, legumes, eggs
Vitamin B6	Pyridoxine, pyridoxal, pyridoxamine IUPAC: 4,5-bis (hydroxymethyl)-2-methylpyridin-3-ol (pyridoxine)	1.3–1.7 mg	Poultry, fish, fortified cereals, bananas, potatoes
Vitamin B7	Biotin IUPAC: (3as, 4S, 6aR)-2-oxohexahydrothienol[3,4-d] imidazole-4-valeric acid	30 μg	Egg yolk, nuts, legumes, liver, synthesized by intestinal bacteria
Vitamin B9	Folic acid (folate) IUPAC: (2S)-2-[4-{[2-amino-4-oxo-3,4-dihydropteridin-6-yl)methyl]amino}benzoyl)amino] pentanedioic acid	400 μg	Green leafy vegetables, legumes, citrus fruits, fortified cereals
Vitamin B12	Cobalamin IUPAC: Cyanocobalamin-C	2.4 μg	Meat, fish, eggs, milk, fortified foods
Vitamin C	Ascorbic acid IUPAC:(5R)-[(1S)-1,2-dihydroxyethyl]-3,4-dihydroxyfuran-2(5h)-one	65–90 mg	Citrus fruits, tomatoes, strawberries, broccoli, peppers

metabolism, and excretion (ADME), including the size and speed of each stage. The term "pharmacokinetics," meaning movement of a drug, is derived from the Greek words "pharmakon" (drug) and "kinesis" (movement). This refers to the mathematical study of how drug concentrations change in the body over time. Pharmacokinetics is essentially the study of how medications enter the body, travel through it, and are then eliminated through metabolism and excretion [13]. Pharmacokinetic research is crucial for pharmaceutical businesses because it determines the dosage, mode of administration, duration of action, time of peak effect, and frequency of drug administration. This critical pharmacokinetic stage influences the organs involved in metabolism and excretion as well as how drugs reach their target sites [14].

Metabolism is the process by which the chemical form of a substance is altered to make it more water-soluble for easier excretion. The process of changing a substance's chemical makeup to make it more water-soluble for easier excretion is called metabolism. In particular, "biotransformation" describes the chemical alterations that foreign substances undergo inside the body. There were two main stages of biotransformation. Phase I: Through oxidation, reduction, and hydrolysis processes, tiny polar functional groups (such as –OH and $-NH_2$) are introduced and unmasked, which change lipid-soluble drugs into more polar forms so that they can be prepared for Phase II reactions or direct excretion. Phase II involves conjugation, in which polar endogenous substances (such as glucuronic acid or sulfate) are coupled to drugs or their Phase I metabolites in Phase II. This process produces highly water-soluble conjugates that are easily eliminated from the body [15]. Excretion is a process by which drugs and their metabolites are permanently removed from the body. Except for the lungs, excretory organs are more likely to eliminate water-soluble and ionized molecules than lipid-soluble ones, which must be transformed into water-soluble forms before expulsion. The concentration of a drug in the body influences pharmacokinetic processes (ADME), and the process order explains how it affects the process rate. Important pharmacokinetic parameters can be calculated, and the temporal course of medications in the body can be quantitatively expressed using pharmacokinetic models. These three elements determine the pharmacokinetics: absorption, distribution, and elimination. Vitamins enter the bloodstream, distribute to peripheral areas, and equilibrate with different organs (such as the liver and kidneys), tissues (such as muscle and adipose tissue), and cells (such as red blood cells), whether administered orally or parenterally. The degree of affinity of a vitamin for specific internal compartments affects its redistribution rate and, consequently, the length of its terminal half-life. Certain vitamins can be explained by a two-compartment model, but the majority of vitamins, including A and C, have properties that make multi-compartment distribution more appropriate. As these vitamins pass through metabolically active organs such as the liver and kidneys, their concentration in blood plasma determines how much of them is eliminated from the circulation. The concentration of these vitamins in the blood plasma determines how much of them is eliminated from the circulation when they pass through metabolically active organs, such as the liver and kidneys. After enterohepatic reabsorption, metabolites produced in the liver travel through the biliary route to produce active vitamins including N-5, methyl tetrahydrofolate, retinoic acid, cobalamin, and 25-hydroxyvitamin D [25(OH)D] [16,17]. Certain vitamins are kept in an unmetabolized condition, such as retinol palmitate in the liver, white blood cells (such as vitamin C and folate), and adipose tissues (such as vitamins A, D, and E). Hepatic circulation transforms other vitamins, including tetrahydrofolate and folate, into active forms. Depending on their amount in free tissue and their non-protein-bound, non-ionized form, vitamins migrate between the plasma and tissue cells. The affinity of the molecule for plasma and cell surface proteins, cellular permeability parameters, and the efficiency of creating storage, such as active protein-binding sites, all affect the concentration inside the cell. Because of the intricate processes of ADME, as well as their dependence, plasma concentration alone cannot reliably predict cellular efficacy in attaining a concentration of vitamins or their active metabolites [17,18].

3.7 PHARMACOKINETIC PRINCIPLES FOR VITAMINS

3.7.1 Absorption

Intestines passively absorb various types of vitamins via diffusion, facilitated diffusion, and active transport. Fat-soluble vitamins A, D, E, and K are transported by dietary fat. Water-soluble vitamins B and C are absorbed relatively less conveniently. Determinants of absorption include intestinal pH, protein and other nutrient transport, and gastrointestinal disease. The following parameters determine vitamin absorption.

a. **Medication**: Certain medications, such as antibiotics and antacids, can affect vitamin absorption.
b. **Illness**: Conditions in the intestine, such as celiac disease or inflammatory bowel disease, can compromise the lining of the intestine and, therefore, affect vitamin absorption.
c. **Importance of Absorption**: Absorption is the key to nutrition. Inadequate vitamin intake can lead to various diseases and even perceptible signs [19].

3.8 DISTRIBUTION

Once absorbed, vitamins are distributed and transported to the blood for circulation in the body. Fat-soluble vitamins are bound to lipoproteins, whereas water-soluble vitamins circulate freely in the plasma or bind to specific carrier proteins. The distribution patterns of the vitamins are described below:

Vitamins have specialized patterns of distribution in the body that may range from a single location to many. Some organs serve as major storage sites. For example, the liver stores vitamin A predominantly and can release its fractions to the body over extended periods of time, and vitamin D is deposited in the muscle and fat tissues, contributing to the body's reserves at the time of limited sun exposure [20]. The distribution of vitamins was controlled as follows:

a. **Tissue Specificity**: Specific vitamins tend to be collected in some tissues based on their specific receptors or ability to access transport mechanisms.
b. **Carrier Proteins**: Most vitamins circulate in the blood in the form of specific binding proteins, known as carrier proteins.
c. **Metabolic Conversion**: Some vitamins undergo conversion or transport through required metabolic processes to critical destinations for significant biochemical functions.
d. The importance of delivery is as follows:
    - Maintenance of strategic health factors that allow optimal functioning.
    - To avoid vitamin deficiency.
    - To develop targeted nutritional interventions.
    - Quantitative measurement of vitamins in the body or system [21].

3.9 METABOLISM

Vitamin metabolism refers to a sequence of processes that change the structure or function of chemical substances. These are:

- **Activation**: Some vitamins must be converted into active forms, such as the conversion of vitamin D to calcitriol.
- **Conjugation**: This is the form of alteration that makes lipids more soluble or easier to eliminate.
- **Recycling**: Vitamin E illustrates the recycling of vitamin metabolites.
- **Degradation**: Destruction of surplus vitamins. This accounts for three significant facts regarding the vitamin metabolism in the human body.
  a. **Recycling**: Recycling of vitamin metabolites, such as vitamins E and B12, is an indication of the effectiveness of these nutrients in the body. Recycling of vitamin B12, for instance, reduces dietary requirements by facilitating the reabsorption of bile in the intestine.
  b. **Storage**: The human body has sophisticated systems for storing vitamins, particularly fat-soluble vitamins. Vitamins A, D, E, and K are stored in the adipose tissue and liver for several months, which provides a constant supply when intake is low. For example, vitamin D is stored in fat cells for several months, and there is a constant source of vitamin D even in the absence of sunlight.
  c. **Interactions**: Vitamins are not isolated; they interact with other nutrients that play significant roles in their absorption, metabolism, and action. For instance, vitamin C enhances the absorption of non-heme iron and calcium inhibits the absorption of iron. Some vitamins also interact synergistically; for example, vitamin D facilitates the absorption of calcium for bone health. Some minerals also influence the bioavailability and activity of the vitamins.

Recycling, storage, and interactions enable the body to use and manage vitamins efficiently. This mechanism enables access to essential vitamins, even during reduced intake, highlighting the importance of a good diet due to the complicated interrelationship among nutrients, which tends to have a strong bearing on health. An understanding of these mechanisms can be a factor for more effective strategies, as well as ensuring appropriate vitamin levels within the body [22].

3.10 EXCRETION

Vitamin excretion follows separate mechanisms for fat- and water-soluble vitamins. Fat-soluble vitamins are mainly excreted through feces and bile, whereas water-soluble vitamins are mainly excreted through the urine. Certain vitamins, such as vitamin B12, have special excretion pathways, such as enterohepatic circulation, which delay some of their excretion [23].

3.11 FACTORS AFFECTING VITAMIN PHARMACOKINETICS

The pharmacokinetics of vitamins in the elderly are influenced by a number of interdependent factors [24,25].

a. **Age**: Changes in age modify the mechanisms of absorption, metabolism, and excretion of vitamins. These changes may be the result of modifications in the digestion efficiency, metabolic processes, and organ function. For instance, the elderly can get atrophic gastritis, which diminishes gastric acid secretion and thus compromises the absorption of certain vitamins, such as vitamin B12.
b. **Sex**: Males and females differ in hormones, which affect the utilization of vitamins. For example, because of differences in biology, women require more iron during pregnancy or menstruation, whereas men require less iron. Estrogen further influences calcium absorption and utilization in women.
c. **Genetics**: Genes regulate the transport and metabolism of some vitamins. Certain genetic mutations, such as the MTHFR mutation, may render a person more susceptible to not being able to metabolize some vitamins to their active forms, similar to folate metabolism.
d. **Health Conditions**: Some illnesses may affect the metabolism of vitamins by affecting their absorption, distribution, and elimination. For example, celiac disease may lead to malabsorption of nutrients in the small intestine (SI), whereas liver or kidney disease may alter vitamin metabolism or excretion.
e. **Interactions with Drugs**: Some drugs can interfere with the metabolism or absorption of vitamins. For instance, proton-pump inhibitors (PPIs) used to treat acid reflux can interfere with the absorption of vitamin B12, and some antibiotics can affect vitamin K.
f. **Diet**: Food intake significantly contributes to the amount and metabolism of vitamins. The presence of other nutrients can promote or retard vitamin uptake. For example, the uptake of iron is improved by vitamin C, whereas the enhanced intake of fiber can inhibit the absorption of some vitamins.
g. **Lifestyle Factors**: Smoking, alcohol consumption, and exercise are some of the lifestyle factors that may affect vitamin metabolism. For example, because oxidative stress is higher among smokers, they may require additional vitamin C.

h. **Gut Microbiome**: It is an intricate bacterial population of the gut that significantly influences vitamin metabolism. Certain gut bacteria have the ability to synthesize vitamins such as K and B12, whereas others can affect nutrient absorption to some extent.
i. **Environmental Factors**: Exposure to pollutants, toxins, and ultraviolet radiation also affects vitamin metabolism. For example, excessive sun exposure enhances the production of vitamin D in the skin.
j. **Nutritional Status**: This includes the entire spectrum of nutrient consumption, and the relationship between vitamins and minerals plays a basic role in the metabolism of some vitamins. Nutrient deficiency can affect the absorption and metabolism of other nutrients.

3.12 DEFINITION AND SIGNIFICANCE OF BIOAVAILABILITY

Bioavailability is a term that is frequently used to describe drug absorption. It is defined as the relative quantity of a drug administered via a pharmaceutical product that enters the systemic circulation in its unchanged form, as well as the rate at which this process occurs. This definition inherently involves a comparative analysis. When comparing an oral formulation with an intravenous formulation of a drug, which inherently possesses 100% bioavailability, the absolute bioavailability of the drug is assessed. Conversely, when comparing two distinct oral formulations, the relative bioavailability of these formulations is evaluated. The three primary indices of drug bioavailability are typically estimated: the maximum drug concentration in plasma (C_{max}), the time required to reach this maximum (t_{max}), and the area under the plasma or serum concentration versus time curve (AUC). Generally, there is also an initial lag period (t_{lag}) before drug concentrations become measurable in plasma. There are three primary reasons why it is essential to understand its bioavailability [26].

a. **Dosage Determination**: When deciding on the right dosage of vitamins and supplements, nutritionists and medical professionals rely on bioavailability. The characteristics of the dosage form, which are partially influenced by its design and manufacturing, greatly influence a drug's bioavailability. More accurate and efficient dosage recommendations may be made by knowing how the body really absorbs and uses nutrients. This precision ensures that individuals receive the proper quantity of nutrients without having to worry about either an excess or a shortage.
b. **Nutritional Assessment**: Bioavailability can be defined as the proportion of a nutrient in a food, diet, or dietary supplement that is absorbed and used for normal body functions. This involves the absorption and transport of nutrients to the relevant body tissues and their conversion to the physiologically active compound, in such a way that the nutrient can be used to maintain normal metabolic functions. In the case of minerals, the amount available for absorption (bioaccessible) depends on the composition and physical characteristics of the diet, the content and the chemical species of the mineral, the presence of ligand promoters or inhibitors of the absorption, luminal gastrointestinal secretions, and interactions that occur as a result of the interplay of these factors.
c. **Development of Supplements**: To establish the right dosage of vitamins and supplements, doctors and nutritionists employ bioavailability. The characteristics of the dosage form, which are determined by its design and production, have a major impact on a drug's bioavailability. Making more accurate and useful dosage recommendations may be made easier with a better understanding of how the body really absorbs and uses nutrients. This precision ensures that individuals receive the proper quantity of nutrients without having to worry about either an excess or a shortage [27].

3.13 METHODS FOR ASSESSING VITAMIN BIOAVAILABILITY

Vitamin bioavailability must be evaluated because of the impact of nutrients on health, as well as the body's processes for absorbing and using them. Intricate pharmacokinetic investigations, as well as in vitro and in vivo tests, may be part of this evaluation.

a. **Pharmacokinetic Research**: Important details regarding the ADME of vitamins are provided by pharmacokinetics. During drug-kinetic studies, various modes of administration of vitamins (oral or intravenous) are preceded by systematic biological and blood sampling at predefined time intervals. Vitamin concentrations were measured using methods such as mass spectroscopy and high-performance liquid chromatography (HPLC). Data analysis encompasses the building of concentration–time curves and assessment of important parameters, including bioavailability (the proportion of dose administered that enters systemic circulation), clearance (Cl) or the rate of elimination, half-life ($t½$), or the duration taken to reduce the plasma concentration by half, and volume of distribution (Vd) or the partitioning between plasma and tissue in the body. Such studies are of great value in calculating patient-specific doses, examining nutrient–drug interactions, determining population variability, and informing formulation design [28].
b. **Depletion–Repletion Studies**: Research on Depletion and Repletion. To assess functional indicators and establish requirements, depletion-repletion studies were conducted to control

nutrient intake precisely. Traditionally, the test begins with dietary restriction that depletes a nutritional marker, followed by monitoring central recovery during the repletion stage when nutrients are added. Nutritional benefits may be measured by monitoring metabolic and physiological changes using biomarkers such as laboratory tests, including blood tests, as well as cognitive and physical assessments. This method assists in determining the functional roles of micronutrients, establishing early signs of deficiency, and facilitating the development of dietary advice and intervention methods [29].

c. **In Vitro Digestion Models**: These mimic peristaltic movement, pH and temperature fluctuations in the stomach and intestines, and chemical digestion. These are utilized to study the release and stability of nutrients, bioavailability of bioactive compounds, and impact on nutrient availability as a result of the food structure. Greater control over experimental parameters and lower cost than in vivo experiments are some benefits of using in vitro systems. Because of these advantages, these models are often used to generate the first estimates of nutrient bioavailability in drug development, as well as in food science [30].

d. **Isotope Labeling**: Stable or radioisotopes of vitamins are employed intensively to trace the metabolic pathways precisely and provide genuine information on vitamin function. Using this method, researchers can measure the efficacy of absorption, trace organ distribution, trace biochemical conversions, and kidney metabolism. Its significance is deep, as it enables higher bioavailability supplements, more personalized dietary recommendations, detection of disease- and deficiency-associated changes at an earlier time point, and better screening of nutrient–drug interactions. Isotope labeling plays a pivotal role in enabling precision nutrition by providing real-time knowledge of vitamin function and metabolism [31].

3.14 FACTORS INFLUENCING VITAMIN BIOAVAILABILITY

Vitamins are affected by a variety of physiological, nutritional, and even pharmaceutical factors that determine the bioavailability of nutrients.

a. **Physiological Factors**: Absorption of nutrients is a physiological factor. The range of gastrointestinal acidity and movement affects vitamin digestion and transport, and the overall health of the intestinal lining guarantees selective absorption and defense against pathogens. Certain vitamins are absorbed by a specific transport system, which, together with the intestinal wall, influences their absorption. The degree of function of the liver and kidneys is equally important; the two organs influence metabolism, detoxification, and elimination of vitamins. Changes that occur with age, such as reduced secretion of gastric acid, reduced blood flow in the intestines, and changes in hormone or enzyme levels, impact absorption and metabolism, which is why the elderly are more vulnerable to changes in bioavailability [32].

b. **Dietary Factors**: Food intake patterns have a significant impact on the body's use of vitamins. Both the structure of food and its fiber content can either hinder the release of certain nutrients and minerals or increase absorption. Certain nutrients also had enhanced effects on other nutrients. Vitamin C and non-heme iron absorption is one of the many examples that reinforce the significance of dietary composition. Additionally, cooking and processing methods can either improve or impair the stability of certain nutrients, particularly through heat or industrial refining, which are processes that tend to lead to nutrient loss. Anti-nutritional factors such as phytates hamper the bioavailability of certain minerals. Fat content also positively correlates with the absorption of fat-soluble vitamins, highlighting the importance of maintaining a balanced dietary fat intake [33].

c. **Pharmaceutical Formulation**: The effectiveness of vitamin administration depends on pharmaceutical formulation factors. Compliance and absorption slim down to the preferred selection of dosage forms that could be tablets, capsules, or liquids, with the latter preferred among children and the elderly. Smaller particles enhance bioavailability, and along with the grade of particles, fatty acids and surfactants also enhance absorption and permeability across intestinal membranes. Vitamins can be protected from degradation through advanced formulations such as enteric coatings and sustained-release systems. Furthermore, the formulation safeguards potency and safety in bioactive and inactive components through stability and shelf life [34].

Bioavailability encompasses vitamin absorption and reflects the host's physiology, dietary habits, and the design of the dietary formulation, which underlines the complexity and importance of the interplay between host factors and dietary formulation in developing effective supplementation strategies and optimizing therapeutic outcomes and needs across populations.

3.15 BLOOD–BRAIN BARRIER

The blood–brain barrier (BBB) is a selective, semipermeable membrane that protects the blood from the interstitial fluid of the brain, allowing the blood vessels in the brain to control the passage of molecules and ions between the blood and the brain. It is formed from endothelial cells (ECs), pericytes (PCs), the basement membrane of the capillary, and the "end feet" of astrocytes, which defend the head from poisonous materials, drain damaging elements from the brain to the blood, and nourish the brain [35]. These molecules of the BBB are tied by physical (tight junctions) and metabolic (enzymes) barriers. The structures of the central nervous system (CNS) differ in both structure and function from those of other body parts. Therefore, it is imperative that the BBB is present to maintain the structures in the CNS in a homeostatic state so that they can function without interference from other systems in the body [36].

3.16 STRUCTURE AND FUNCTION OF BLOOD–BRAIN BARRIER

Previously discussed is the structure of the BBB, which comprises a capillary basement membrane and three different types of cells: ECs, PCs, and astrocyte end feet. The BBB is pivotal for the formation and stabilization of homeostasis for neuronal functioning, defending the system from toxic damage, controlling the border between the periphery and the CNS, and providing the brain with vital elements. These functions are carried out through four main approaches: the failure of paracellular diffusion of hydrophilic compounds, active transport of nutrients toward the brain, activation of the efflux transport of hydrophobic molecules and drugs from the brain to the blood, and control of the transendothelial migration of blood cells and pathogens [37–39].

A. **Tight Junctions**: Tight junctions form highly selective diffusion barriers and can be interposed between cerebral ECs. These diffusion barriers inhibit the entry of most blood-borne substances into the brain parenchyma. TJs block the paracellular aqueous diffusion pathways between adjacent ECs. TJs are known for their adhesive properties and ability to block microvessels, preventing the passive diffusion of proteins and polar molecules into and out of the system. Role of Astrocytic End-Feet Processes: Astrocytic processes wrap around and interdigitate with the vessel wall and participate in the elaboration and preservation of the TJ barrier. However, in the case of end-feet processes, their functionality is believed not to include a barrier component in the mammalian brain.

B. **Endothelial Cells (ECs)**: Blood ECs, especially those forming the BBB, have atypical features compared to those in the rest of the body. BBB ECs, unlike other ECs, do not have fenestrae, have more extensive and coherent tight junctions, and have limited pinocytic vesicular transport. The tight junctions, the associated cross-linked ECs of the BBB, anchor the membranes to curtail paracellular transport of penetrating solutes, preventing them from passing the paracellular transport pathways. PCs, which lie within the capillary basement membrane, are crucial for angiogenesis and microvessel extension. They contain structural scaffolding and directional coordination of microvessels and form junctions with ECs.

C. **Circumventricular Organs (CVOs)**: With the exception of CVOs, which are located around the third and fourth ventricles, the BBB is present throughout the brain. Fenestrations in the blood vessels surrounding CVOs enable blood-borne molecules to diffuse across vessel walls. These exposed brain regions control the endocrine glands and the autonomic nervous system. Among the CVOs are:
   - The pineal gland
   - The median eminence
   - Organ subcommissural
   - The neurohypophysis
   - Subfornical organ
   - Postrema area
   - The lamina terminalis's organum vasculosum

3.17 MECHANISMS OF VITAMIN TRANSPORT ACROSS THE BLOOD–BRAIN BARRIER

3.17.1 Passive Diffusion

Diffusion is the transport of molecules that does not require energy. It is characterized by the movement of tiny fat-soluble particles that cross the membranes of ECs in the direction of lower concentrations. At the BBB, this route allows the passage of mostly tiny non-polar molecules (generally about 400 Da in size) and excludes nearly all water-soluble vitamins due to size and polar limitations. The tight junctions between ECs make paracellular diffusion of hydrophilic substances virtually non-existent [40].

3.18 CARRIER-MEDIATED TRANSPORT

In order to facilitate the transport of critical polar nutrients, the BBB incorporates various types of transporter proteins.

Carrier-mediated transcytosis is characterized by the presence of transporters that capture specific substrates and change their shape to transport them across the endothelium. Transporters for glucose (GLUT1) and some amino acids (via LAT1) are examples of these transporters [41].

As secondary active and facilitated transport, part of the broader solute carrier (SLC) family, certain transporters move vitamins and nucleosides tip using pre-existing ionic gradients without the direct expenditure of ATP.

These carrier systems are essential for the controlled and saturable transfer of water-soluble vitamins across the BBB [42].

TABLE 3.2
BBB Transport Mechanisms [48,49]

Mechanism	Key Features	3. Relevance to Vitamins
Passive diffusion	Transcellular movement of small lipophilic molecules (<400 Da) across membranes	Limited to small, non-polar vitamins (e.g., A, D)
Carrier-mediated transport	Saturable, facilitated, or energy-dependent SLC transporter systems	Primary route for water-soluble vitamins
Receptor-mediated transcytosis	Vesicle-mediated uptake via specific receptors (e.g., TfR, STRA6)	Critical for retinol (vitamin A); potentially others
Efflux transporters	ATP-driven extrusion of substances out of endothelial cells	Modulate CNS vitamin availability

3.19 RECEPTOR-MEDIATED TRANSCYTOSIS RMT

The binding of larger molecules or complexes to specific receptors on ECs allows RMT to mediate their transport across the BBB. During receptor–ligand engagement, vesicular endocytosis occurs, and the cargo in the vesicle is moved to the opposite end of the cell, where it is exocytosed [43].

Common transport targets include:

- Transferrin receptor (TfR)—often exploited in therapeutic delivery strategies.
- Insulin and leptin receptors, as well as other hormone and nutrient receptors.

Adsorptive-mediated transcytosis is less likely to cross the BBB owing to the lack of RMT specificity. This transcytosis is driven by positively charged cargo and electrostatic interactions with the negatively charged endothelial surface [44].

3.20 SPECIFIC BBB TRANSPORT SYSTEMS FOR VITAMINS

Particularly for the BBB, specialized routes for certain vitamins have been described in recent studies; however, general transport mechanisms are still unknown.

3.20.1 Vitamin A (Retinol)

Retinol uptake in an advanced model using human iPSC-derived brain microvascular ECs (BMECs) occurs via both lipophilic diffusion and receptor-mediated processes. Retinol trafficking is mediated by the STRA6 receptor complex with the intracellular partners LRAT and CRBP1. Notably, retinol-binding protein (RBP) and transthyretin (TTR), the principal serum transporters of retinol, do not cross the BBB; retinol must dissociate from these proteins to be internalized via STRA6, which illustrates receptor-dependent delivery [45].

- **Vitamin Transport via Carriers**
  Specific BBB transporters for other vitamins, such as B complex and C, do exist (e.g., SMVT for biotin); however, the primary function of SLC transport systems in the uptake of water-soluble vitamins has been widely documented and emphasized in the literature. For instance, vitamins and other nutrients are concentrated in the brain via carrier-mediated transport, whereas oversized molecules are included through receptors [46].
- **Efflux Mechanisms and Barrier Integrity**
  The BBB contains efflux transporters of the ABC family, such as P-glycoprotein and BCRP, which provide protective functions to the CNS by extruding various endogenous and exogenous compounds. This also affects vitamin availability in the brain [47] (Table 3.2).

3.21 PHARMACOKINETICS OF WATER-SOLUBLE VITAMINS

Water-soluble vitamins are important for various metabolic processes, acting as catalysts in cellular reactions and protective mechanisms. In contrast to fat-soluble vitamins, these vitamins are not stored in considerable amounts or have sustained retention in the body, and a continuous supply in the diet is most often required. The pharmacokinetics of these vitamins are characterized by rapid absorption in the gastrointestinal tract, considerable retention in various tissues, active transport, and elimination by the kidneys. Deficiencies are relatively easy to obtain with insufficient intake; however, in general, they are not very common and are often less severe. However, megadoses of certain vitamins can have adverse effects [50].

3.22 VITAMIN B COMPLEX

The B complex group comprises eight distinct vitamins. Each of these vitamins mainly functions as a coenzyme or precursor in metabolic processes. Despite their structural differences, most vitamins share many similarities. They are water-soluble, rely on transport systems, and are rapidly expelled through the urine [51].

3.23 THIAMINE (B1)

Thiamine absorption occurs in the SI and results from passive diffusion (at high concentrations) and active carrier-mediated (at physiological concentrations) processes. Absorbed thiamine is phosphorylated in tissues to thiamine diphosphate (a coenzyme in carbohydrate metabolism) and is the active form [52].

The vitamin is distributed widely in the body; however, thiamine body reserves, which are primarily stored in the liver, heart, and brain, are limited. Its half-life is relatively short (~10–20 days), which is why vitamin intake is continuous, as its surplus is excreted through urine [53].

3.24 RIBOFLAVIN (B2)

Riboflavin primarily enters the body through the proximal SI, which contains a saturable transport system. Riboflavin is active in its coenzyme forms, FMN and FAD, which are produced after riboflavin conversion into enterocytes.

A broad distribution is observed in the body, with high concentrations in the liver, heart, and kidneys. Riboflavin also binds tightly to proteins, prolonging its tissue retention. Excreted riboflavin, along with unabsorbed and excess riboflavin, often produces a characteristic yellow urine color [54].

3.25 NIACIN (B3)

This vitamin is synthesized from nicotinic acid and N-methyl nicotinamide. Absorption occurs through facilitated diffusion at the nutritional level and passive diffusion at pharmacological dosages. After absorption, niacin is quickly transformed into nicotinamide adenine dinucleotide (NAD^+) and NAD phosphate ($NADP^+$), which are both used in hundreds of redox reactions.

Niacin is characterized by rapid distribution and storage in tissues and is primarily disposed of through methylation in the liver and excretion in the urine. Pharmacological high dosages, commonly used in the management of dyslipidemia, significantly change the pharmacokinetics of the drug and can cause flushing syndrome, liver damage, and glucose intolerance [55].

3.26 PANTOTHENIC ACID (B5)

Through small intestinal absorption, the vitamin enters at concentrations by a sodium-linked active system and at higher intake levels through passive diffusion. At this stage, pantothenic acid is readily transformed into both coenzyme A (CoA) and acyl carrier protein, which are important for fatty acid metabolism.

However, there is a stable circulation between its widespread distribution and the absence of relative storage capacity. Moreover, the unchanged excretion of vitamins in urine supports the rapid turnover process [56].

3.27 PYRIDOXINE (B6)

Vitamin B6, or pyridoxine, can be converted to several forms, such as pyridoxal and pyridoxamine. Absorption occurs via passive diffusion in the jejunum. These forms of B6 are converted to pyridoxal 5′-phosphate (PLP). This is the active coenzyme responsible for amino acid metabolism, neurotransmitter synthesis, and hemoglobin activity.

Plasma proteins, particularly albumin, bind to pyridoxal phosphate, prolonging its half-life. Most metabolism occurs in the liver. Renal elimination primarily occurs as 4-pyridoxic acid. Very large supplemental doses can saturate the binding systems and lead to peripheral neuropathy, a rare but recognized form of toxicity [57].

3.28 BIOTIN (B7)

Biotin is absorbed in the SI. This is the only form of absorption or at least the primary form. Biotin is also synthesized by colonic bacteria; however, its quantitative significance has not yet been established.

Biotin deficiency is rare, but the most common cause is prolonged consumption of raw egg whites due to avidin binding and certain genetic defects in SMVT. Deficiency becomes common when biotin is used for metabolism by carboxylases involved in the synthesis of fatty acids, amino acid catabolism, and gluconeogenesis [58].

3.29 FOLATE (B9)

Most dietary folates are absorbed in the jejunum after undergoing deconjugation of polyglutamate forms into monoglutamates. The proton-coupled folate transporter (PCFT) is the primary pathway for folate uptake in the intestine. Folic acid, present in dietary supplements and fortified foods, is absorbed more efficiently than natural folates.

In tissues, folate is metabolized to tetrahydrofolate derivatives, which act as coenzymes in vital one-carbon transfer reactions required for DNA synthesis and certain metabolic conversions of amino acids. Folate is stored in the liver; however, the reserves are exhausted within a few months. Elimination occurs via urine and bile, and enterohepatic recycling is the major route of drug conservation [59].

3.30 COBALAMIN (B12)

B12 is the only water-soluble vitamin with unique pharmacokinetic properties. Absorption occurs only after the vitamin is bound to the intrinsic factor, a glycoprotein released by gastric parietal cells, and endocytosis of the receptor-clustered complex in the ileum. Passive diffusion contributed minimally to the absorption.

After absorption, cobalamin is widely distributed in the body and bound to transcobalamin, with transcobalamin II being the delivery vehicle for tissues. The liver, which holds immense amounts of vitamins, is the primary storage site for vitamins, explaining the delayed onset of deficiency after several years of malabsorption. The inactive forms are converted to the active coenzymes methylcobalamin and adenosylcobalamin, respectively. Of the total amount produced, only a small portion is excreted, and the primary loss occurs through bile, most of which is reabsorbed via the enterohepatic circulation [60].

3.31 VITAMIN C (ASCORBIC ACID)

Ascorbic acid is absorbed in the SI via sodium-dependent vitamin C transporters (SVCT1 and SVCT2). Regardless of the absorption method, the effectiveness of absorption decreases with an increase in intake, declining from 80% to 90% absorption at modest intakes to less than 50% absorption at high doses.

A large amount of vitamin C is stored in the adrenal glands, eyes, pituitary gland, leukocytes, and other tissues. It also acts as a potent antioxidant and cofactor in biosynthetic pathways involving oxygen and other substances [61].

Dehydroascorbic acid can be reduced to ascorbic acid or broken down to oxalate and other by-products, along with renal clearance, which is the primary route of elimination. Ascorbic acid saturation in plasma causes a significant increase in urinary acid levels. Body stores of ascorbic acid are small (1–2 g) and can be depleted quickly, causing a deficiency that eventually leads to scurvy [62].

3.32 PHARMACOKINETICS OF FAT-SOLUBLE VITAMINS

Fat-soluble vitamins have distributive metabolism, characterized by oxygen and covalent linkages to other molecules. Conjugation increases the retention time in the body, which causes an increase in imbalance and toxicity of fat-soluble vitamins [63].

3.33 VITAMIN A (RETINOL)

Vitamin A exists in several forms, including retinol, retinal, retinoic acid, and retinyl esters. It is predominantly absorbed in the upper SI, where dietary retinyl esters are soaked in pancreatic enzymes and brush border esterases. Enzymatically liberated retinol is incorporated in the presence of bile salts and lipid mixed micelles, which enter the enterocytes [64].

In intestinal cells, retinol is re-esterified and packed into chylomicrons, which are then dispersed in the lymphatic system. These lipoproteins transport retinyl esters to the liver, which is the primary storage site for vitamin A. Stellate cells in the liver serve as long-term storage cells, releasing retinol bound to RBP to the peripheral tissues as needed [65].

Metabolically, retinol and retinal undergo interchangeable processes of oxidation, whereas retinoic acid is a derivative of incessant oxidation and an active compound that binds to nuclear receptors. It is primarily lost through the biliary excretion of polar metabolites, with renal clearance being a minor pathway. Hepatic stores, which change at a relatively low rate, account for the long intervals before deficiency symptoms appear, as well as the high-dose issues, hepatotoxicity, and teratogenic side effects [66].

3.34 VITAMIN D (CALCIFEROL)

Skin can synthesize vitamin D with the help of the sun or can be consumed through the diet. Vitamin absorption occurs in the jejunum and ileum, alongside dietary fat absorption. After incorporation into vitamin D micelles, vitamin D is captured in the lymphatic vessels during the intestinal absorption of fats [67].

Vitamin D is transported in the bloodstream in the form of chylomicrons until it crosses the DBP boundary. Once the DBP boundary is crossed, the drug reaches the liver, where the first stage of metabolism occurs. 25-hydroxyvitamin D, a biomarker for vitamin D sufficiency, is synthesized in the major circulation of the body. Vitamin D is active in its form 1,25-dihydroxyvitamin D and maintains the calcium–phosphate equilibrium along with the vitamin D receptor.

The major excretory routes for vitamin D metabolites are bile and feces. Chronic toxicity can occur due to the unrestricted diffusion of 25(OH)D, which has a half-life of 2–3 weeks [68].

3.35 VITAMIN E (TOCOPHEROLS AND TOCOTRIENOLS)

Vitamin E is composed of four tocopherols and four tocotrienols. α-Tocopherol is the most active form of tocopherol in humans. Like most fat-soluble vitamins, absorption occurs during micelle formation in the SI [69].

Vitamin E becomes a part of chylomicrons and is secreted into the lymphatic system after absorption. The liver controls the distribution of vitamin E, preferentially incorporating only α-tocopherol into very low-density lipoproteins (VLDLs) via the α-tocopherol transfer protein (α-TTP). Other forms are metabolized and excreted more rapidly than the sulfate [70].

Vitamin E is found in metabolized lipid membranes and lipoproteins and works mainly as a chain-breaking antioxidant. It undergoes metabolism, which creates water-soluble carboxyethyl-hydroxychromans that are excreted in the urine, and fecal excretion also facilitates elimination [71].

α-Tocopherol has a long half-life in plasma because of selective retention, whereas the other isomers have shorter persistence. Owing to the regulatory mechanism of the body, vitamin E deficiency is unheard of, except in cases of fat malabsorption or genetically inherited disorders in α-TTP. However, toxicity from supplements is rare, especially when compared with vitamins A and D [72].

3.36 VITAMIN K

Phylloquinone (vitamin K1) and menaquinones (vitamin K2–7) are two forms of vitamin K. Phylloquinones are obtained from green leafy vegetables, whereas menaquinones are synthesized by intestinal bacteria and obtained from certain fermented foods [73].

The proximal intestine is the site of nutrient absorption. As with any other fat-soluble vitamin, absorption requires bile salts and pancreatic enzymes. Vitamin K is incorporated in chylomicrons, which are carried by the lymphatic system to the liver. Vitamin K participates in the γ-carboxylation of the glutamate residues on the liver-made and liver-stored clotting factors II, VII, IX, and X, and proteins C and S, and separate proteins [74].

Vitamin K is the exception to the rule, which states that other fat-soluble vitamins can be stored in the body. This is not the case for vitamin K. Vitamin K is swiftly transformed into various polar metabolites, which are excreted via bile and urine. The rapid transformation of vitamin K accounts for the development of vitamin K deficiency faster than other fat-soluble vitamins, especially in those with impaired fat absorption or those who have had long courses of antibiotics, which alter the gut microbiome.

Hemolytic anemia and liver toxicity are rare but are associated with use, which stresses the overuse of over-the-counter vitamins. Menadione is a K2–7 vitamin and the most commonly used synthetic vitamin K; therefore, over-the-counter vitamins are most associated with this condition [75].

3.37 ANALYTICAL METHODS FOR VITAMIN PHARMACOKINETICS

Estimating pharmacokinetics requires sensitive and reproducible methods, particularly because vitamins are present in trace amounts in biotic systems and are subject to dietary, metabolic, and genetic variations. Unlike synthetic drugs, vitamins are endogenous substances, which complicate their measurement because of high potential background levels and interferences. Analytical strategies are known for their reliability and precision in both clinical and research environments. Primary methods include extensive sample preparation, proper chromatographic separation, mass spectrometry, and immunoassay-based techniques [76].

3.38 SAMPLE PREPARATION TECHNIQUES

Analyzing biological samples for vitamins requires a complex set of techniques because the integrity of these compounds is often subject to degradation by various factors. The primary goal of sample preparation is to extract vitamins from complex biological matrices, such as plasma, serum, urine, and homogenized tissues, with minimal residual matrix effects [77].

- **Protein Precipitation (PPT)**: This mostly occurs with water-soluble vitamins (e.g., vitamins B and C). It involves the use of organic solvents, such as methanol and acetonitrile, which precipitate proteins and allow the supernatant to remain enriched with the target analytes. This is one of the fastest and easiest methods, although it is less selective.
- **Liquid–Liquid Extraction (LLE)**: Used to obtain vitamins A, D, E, and K. The solvent extraction systems of hexane or chloroform isolate and capture the analytes from the aqueous phase. These solvent systems are efficient, although emulsion formation and other factors can significantly diminish their effectiveness.
- **Solid-Phase Extraction (SPE)**: More refined than LLE, this method uses selective capture of the analytes by a C18 or polymeric resin, or ion exchange, similar to the other sorbent materials. SPE is combined with pharmacokinetic analysis to increase reproducibility and remove interfering factors, which can be achieved particularly in automated systems.
- **Derivatization**: Most vitamin D metabolites undergo chemical alterations to improve their visibility in chromatographic or spectrometric methods of analysis.
- During the various spectrophotometric analyses, BHT or ascorbic acid and other antioxidants are taken together to prevent degradation, and the process is performed in a dark environment to reduce light exposure.

3.39 CHROMATOGRAPHIC METHODS

Chromatography is central to the quantification of vitamins, as it can successfully isolate closely related compounds from one another [78].

- **High-Performance Liquid Chromatography (HPLC)**: The primary technique for vitamins soluble in fats and water. The technique used is usually reverse-phase HPLC with a column containing a C18 stationary phase. Some vitamins can be detected using UV light, but most complex mixtures/matrices require more sophisticated techniques such as fluorescence and electrochemical detectors.
- **Ultra-Performance Liquid Chromatography (UPLC)**: HPLC technique with a column containing smaller particles. It is more efficient because it results in enhanced resolution, faster separation, and reduced use of solvents, which is beneficial in larger pharmacokinetic studies.
- **Gas Chromatography (GC)**: Suitable for use with vitamins or their derivatives that can be converted into a gas at a certain temperature. In most instances, vitamins such as tocopherols and retinoids are modified (silylation) to increase their volatility and thermal stability, making them suitable for GC.
- **Capillary Electrophoresis (CE)**: Not as popular as the other techniques, but applicable to small, positively charged, water-soluble vitamins. It uses small sample volumes; however, the analysis is slower than that of more commonly used techniques. HPLC is considered superior because of its consistent analyses.

Most textbooks teach the basic principles of coupling certain chromatographic methods with sensitive detectors or even tandem mass spectrometry to augment analysis.

3.40 MASS SPECTROMETRY

Mass spectrometry, especially when combined with liquid chromatography (LC–MS/MS), is the most commonly used tool for vitamin pharmacokinetics [79].

- **Sensitivity and Selectivity**: MS can detect certain measurements at concentrations of picomolar to femtomolar, which is critical for the detection of certain vitamins existing in trace concentrations, such as 25-hydroxyvitamin D.
- **Quantitative Performance**: Tandem MS utilizing multiple reaction monitoring (MRM) can provide highly precise quantification of certain pre-designated ions by identifying specific precursor and product ions.
- **Metabolite Identification**: Within the field of vitamin studies, high-range MS (HRMS) is particularly useful for defining the metabolites of certain vitamins and helps identify some of their isomers.
- **Stable Isotope Dilution**: The current best practice involves the use of external standards containing isotope-labeled internal standards to minimize the suppressive effect of ions and attenuate the variation in reproducibility.

MS provides unpaired accuracy and information but is limited by expensive instruments, specialized training, and insufficient sample preparation, which may lead to matrix effects.

3.41 IMMUNOASSAYS

Immunoassays offer a unique set of functionalities, especially in clinical settings and large epidemiological studies where rapid testing is required [80].

- **Enzyme-Linked Immunosorbent Assay (ELISA)**: Assays for vitamins B12 and D, which are simple and inexpensive.
- **Radioimmunoassay (RIA)**: Vitamin D metabolite analysis, which is less popular because of the indiscriminate handling of radiation.
- **Chemiluminescent Immunoassays (CLIAs)**: Increasingly preferred in clinical settings because of their sensitivity, level of automation, and faster turnaround times.

The primary deficiency of immunoassays is the diminished level of cross-reactivity, which is lower than that of the more commonly used mass spectrometry methods. Closely related isoforms may not be distinguished, which is critical for pharmacokinetic analyses.

3.42 CLINICAL IMPLICATIONS OF VITAMIN PHARMACOKINETICS

Vitamin pharmacokinetics, which is the study of how vitamins are absorbed, distributed, metabolized, and excreted, is very important for getting the best therapeutic results. Knowing how these processes work can help you give the right dose, deal with drug interactions, and avoid both deficiencies and toxicities [81].

3.43 DOSING STRATEGIES

When deciding how much of a vitamin to take, you need to think about whether it is fat-soluble (A, D, E, K) or water-soluble (B complex, C). You don't need to take fat-soluble vitamins as often because they build up in the body. Water-soluble vitamins, on the other hand, need to be taken regularly because they are quickly excreted. The dose must also be adjusted based on things like the patient's age, how well their kidneys or liver are working, their nutritional status, and their disease conditions. In some cases, like malabsorption or a serious illness, you may need higher or parenteral doses. Vitamin solubility is one of the most important factors in determining the right dose because it affects absorption, storage, excretion, and the risk of toxicity [82]. Vitamin dosage consideration has been given in Figure 3.2.

3.44 SOLUBILITY AND ITS IMPACT ON DOSING

Vitamins A, D, E, and K are fat-soluble vitamins that are absorbed along with dietary fats and stored in the body's liver and adipose tissue. These vitamins can be stored in the body, so you don't have to take them every day [83]. Depending on the clinical situation, you can take them less often, like once a week or once a month. This same property, though, makes it more likely to be toxic if you take too much of it, especially if you take it for a long time or in high doses [4].

Water-soluble vitamins, such as vitamin C and the B complex group (B1, B2, B6, B12, folate), on the other hand, are not stored in appreciable quantities. They must be consistently consumed, frequently daily, to maintain appropriate plasma levels because they are quickly eliminated in the urine. Although some B vitamins (like B6 in high doses) can still be toxic, the risk of toxicity is typically lower for water-soluble vitamins.

3.45 PATIENT-SPECIFIC FACTORS

Dosing strategies must be **individualized** to the patient [84]. Key considerations include:

- **Age**: The dietary needs and metabolic rates of infants, kids, adults, and senior citizens vary. For instance, because their skin synthesis is diminished, older adults may require more vitamin D.

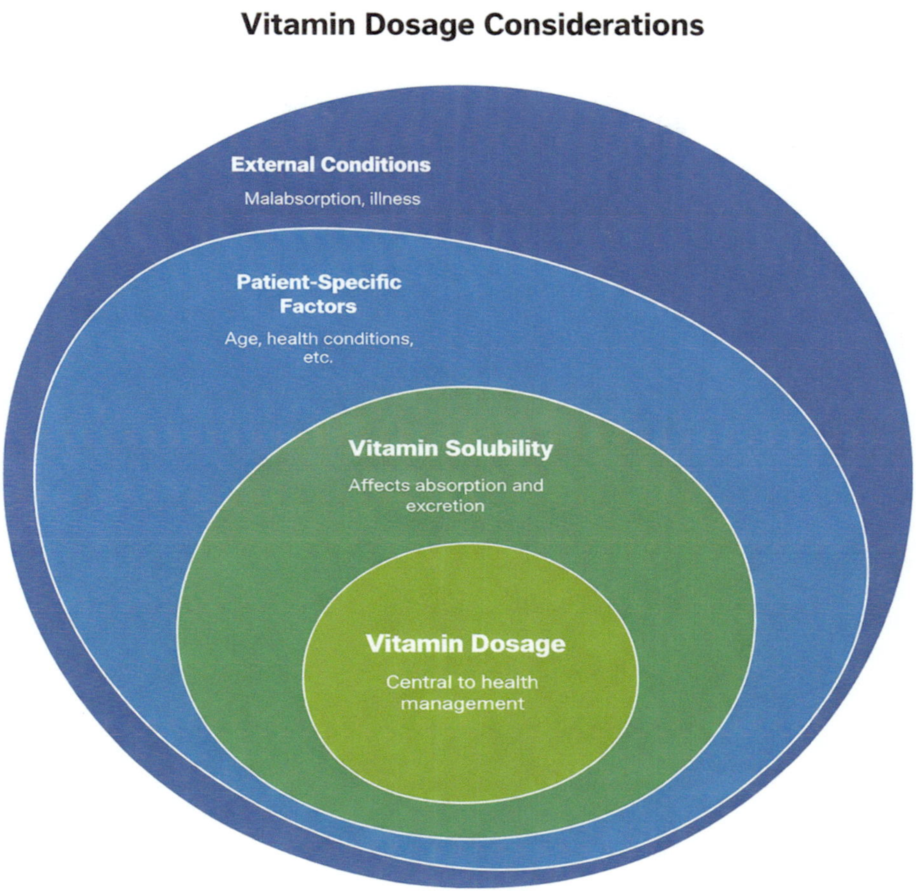

FIGURE 3.2 Vitamin dosage consideration.

- **Renal and Liver Function**: The metabolism and excretion of vitamins depend heavily on these organs. The pharmacokinetics of both water- and fat-soluble vitamins can be changed by impaired function. Patients with renal dysfunction may require adjusted dosages of vitamin C or folate, while those with liver disease may accumulate toxic levels of vitamins A and D.
- **Nutritional Status**: Those who have eating disorders, restrictive diets, or inadequate food intake are more likely to experience deficiencies and may need to take supplements more frequently or in greater amounts.
- **Medical Conditions**: The absorption of both fat- and water-soluble vitamins can be hampered by certain illnesses, such as malabsorption syndromes (e.g., celiac disease, Crohn's disease, bariatric surgery). Increased metabolic demands and altered absorption are common in critically ill patients, particularly those in intensive care units, which calls for higher dosages or parenteral (intravenous or intramuscular) supplementation.
- **Drug Interactions**: Vitamin metabolism or absorption may be hampered by certain drugs. PPIs, for example, can decrease the absorption of vitamin B12 over time, and anticonvulsants like phenytoin can alter the metabolism of vitamin D. Vitamin dosage adjustments based on patient-specific health factors are given in Figure 3.3.

3.46 ROUTE OF ADMINISTRATION

- **Oral Route**: Due to their affordability and ease of use, oral supplements are usually recommended for the majority of patients [84].
- **Parenteral Route**: Patients with severe deficiencies, malabsorption, or situations requiring quick replenishment are the only ones eligible for enteral routes (intramuscular or intravenous). For instance, high-dose IV thiamine is necessary in Wernicke's encephalopathy, and injectable B12 is used in pernicious anemia.

3.47 MONITORING AND SAFETY

Particularly with fat-soluble vitamins, continuous monitoring is necessary to evaluate effectiveness and prevent toxicity. This could entail clinical assessment of symptoms or blood tests (e.g., serum 25-hydroxyvitamin D, B12 levels). To make necessary dosage adjustments, adherence, food consumption, and co-existing medical conditions should be evaluated on a regular basis [85].

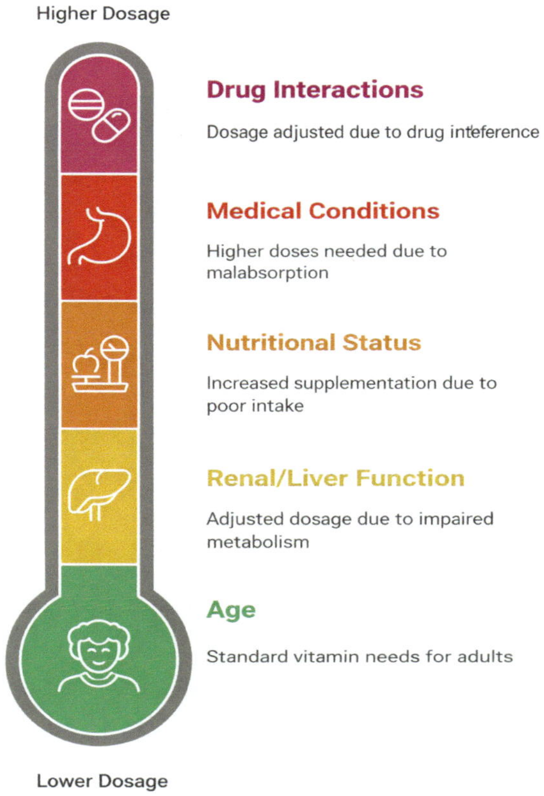

FIGURE 3.3 Vitamin dosage adjustments based on patient-specific health factors.

3.48 DRUG–VITAMIN INTERACTIONS

Vitamins and drugs can interact when vitamins change the pharmacological effects of drugs or when drugs affect how vitamins are absorbed, metabolized, distributed, or excreted. Vitamin excess or deficiency, as well as changes in the toxicity or efficacy of medications, can result from these interactions. Optimizing patient outcomes requires an understanding of these relationships, particularly in populations with polypharmacy, chronic illnesses, or higher nutritional needs [86].

3.49 TYPES OF DRUG–VITAMIN INTERACTIONS

3.49.1 Drug-Induced Vitamin Deficiency

Many drugs interfere with vitamin status by reducing intestinal absorption, increasing metabolism, or enhancing excretion [87]. Common examples include:

- **Vitamin B12 and Proton-Pump Inhibitors (PPIs)**: The prolonged administration of PPIs (e.g., omeprazole) or H2 receptor antagonists (e.g., ranitidine) results in reduced gastric acid secretion, thereby impairing the release of vitamin B12 from dietary proteins. This condition may precipitate vitamin B12 deficiency, particularly among the elderly population.
- **Vitamin D and Anticonvulsants**: Medications such as phenytoin, carbamazepine, and phenobarbital induce hepatic enzyme activity, thereby accelerating the catabolism of vitamin D. If unaddressed, this can lead to osteomalacia or secondary hyperparathyroidism.
- **Folate and Methotrexate or Trimethoprim**: Methotrexate and trimethoprim act as inhibitors of dihydrofolate reductase, thereby disrupting folate metabolism and potentially resulting in megaloblastic anemia. Co-administration of folinic acid (leucovorin) may be employed to mitigate toxicity.
- **Vitamin B6 and Isoniazid or Hydralazine**: These pharmacological agents form complexes with vitamin B6 (pyridoxine), leading to deficiency and peripheral neuropathy. Supplementation is frequently recommended during treatment.
- **Vitamin K and Broad-Spectrum Antibiotics**: Extended use of antibiotics, particularly cephalosporins such as ceftriaxone, can suppress intestinal microbiota responsible for synthesizing vitamin K, thereby causing hypoprothrombinemia and an elevated risk of bleeding.

3.50 VITAMIN EFFECTS ON DRUG EFFICACY OR TOXICITY

Conversely, some vitamins can **modify the action of drugs**, either enhancing or diminishing their effects [88]:

- **Vitamin K and Warfarin**: Vitamin K counteracts the anticoagulant effect of warfarin by facilitating the synthesis of clotting factors. Abrupt increases in dietary or supplemental vitamin K can diminish the efficacy of warfarin, potentially resulting in therapeutic failure.
- **Vitamin C and Iron Absorption**: Vitamin C enhances the absorption of non-heme iron, which may improve outcomes in patients with iron-deficiency anemia. However, in conditions such as hemochromatosis, excessive vitamin C intake may exacerbate iron overload.
- **Vitamin E and Anticoagulants/Antiplatelets**: High doses of vitamin E can exert antiplatelet effects, potentially increasing the risk of bleeding when used in conjunction with aspirin, clopidogrel, or warfarin.
- **Vitamin B6 and Levodopa**: Pyridoxine enhances the peripheral metabolism of levodopa, thereby reducing its availability in the brain. This interaction is particularly significant in patients taking levodopa without a dopa decarboxylase inhibitor, such as carbidopa.

3.51 RISK FACTORS FOR DRUG–VITAMIN INTERACTIONS

- **Older Individuals**: More prone to using several medications and experiencing changes in nutrient absorption [89].
- **Chronic Illnesses**: Disorders such as those affecting the liver, kidneys, or gastrointestinal system heighten susceptibility.
- **Poor Nutrition**: Existing nutrient shortages worsen the impact of drug-related depletions.
- **Taking Multiple Drugs**: The use of various medications raises the chances of interactions.
- **Extended Treatment Periods**: The longer a drug is administered, the greater the potential effect on vitamin levels.

3.52 CLINICAL IMPLICATIONS AND MANAGEMENT

- **Monitoring**: Clinicians are advised to conduct regular reviews of patient medications and evaluate for indications of vitamin deficiencies, particularly in cases of prolonged therapy. Laboratory evaluations, such as serum B12, folate, vitamin D, and INR, may be necessary for patients identified as high risk [90].
- **Supplementation and Prevention**: Prophylactic vitamin supplementation may be appropriate for patients undergoing specific medication regimens.
  For example:
  (1) Folic acid with methotrexate
  (2) Vitamin B6 with isoniazid
  (3) Calcium and vitamin D with corticosteroids or anticonvulsants
- **Patient Education**: Patients should be educated regarding potential interactions between vitamins and pharmaceuticals. Individuals on anticoagulants, such as warfarin, should maintain a consistent dietary intake of vitamin K and report any alterations in supplements or diet.

3.53 VITAMIN DEFICIENCIES AND TOXICITIES

Vitamins are indispensable micronutrients that fulfill vital functions in metabolism, immune response, cellular operations, and tissue preservation. While sufficient intake promotes health and development, both deficiencies and excesses of vitamins can result in significant clinical outcomes. A comprehensive understanding of the etiology, manifestations, and ramifications of vitamin imbalances is essential for accurate diagnosis, effective treatment, and prevention [90].

3.53.1 Vitamin Deficiencies

Vitamin deficiencies arise when the intake, absorption, metabolism, or storage of vitamins is inadequate to fulfill physiological requirements. These deficiencies can be classified as primary, resulting from insufficient dietary intake, or secondary, stemming from conditions that hinder absorption or utilization.

3.53.1.1 Common Causes of Vitamin Deficiencies

- Inadequate dietary intake, which may result from factors such as poverty, restrictive diets, or alcoholism, can lead to nutritional deficiencies [91].
- Malabsorption syndromes, including conditions like celiac disease, Crohn's disease, or the aftermath of bariatric surgery, also contribute to nutrient deficiencies.
- Additionally, certain physiological states, such as pregnancy, infection, or chronic illness, can increase nutritional requirements.
- Drug interactions, such as isoniazid depleting vitamin B6 or phenytoin affecting folate and vitamin D levels, further complicate nutrient absorption.
- Aging is another factor, as reduced gastric acid production in the elderly can impair vitamin B12 absorption.

3.53.1.2 Clinical Manifestations by Vitamin

- Vitamin A deficiency can lead to night blindness, xerophthalmia (dry eyes), compromised immune function, and dermatological changes [92].
- A deficiency in vitamin B1 (thiamine) is associated with beriberi, which presents with neurological and cardiovascular symptoms, and Wernicke–Korsakoff syndrome, particularly in individuals with alcohol use disorder.

- Insufficient vitamin B2 (riboflavin) may result in angular stomatitis, glossitis, and seborrheic dermatitis.
- Vitamin B3 (niacin) deficiency manifests as pellagra, characterized by dermatitis, diarrhea, and dementia. A lack of vitamin B6 (pyridoxine) can cause peripheral neuropathy, irritability, and anemia.
- Vitamin B9 (folate) deficiency is linked to megaloblastic anemia and neural tube defects during pregnancy.
- Vitamin B12 (cobalamin) deficiency can lead to pernicious anemia, neurological impairments, and glossitis.
- Vitamin C deficiency results in scurvy, which includes symptoms such as bleeding gums, petechiae, fatigue, and impaired wound healing.
- Vitamin D deficiency is associated with rickets in children, osteomalacia in adults, muscle weakness, and bone pain.
- A deficiency in vitamin E can cause neuropathy, ataxia, and hemolytic anemia, particularly in premature infants.
- Vitamin K deficiency may result in prolonged bleeding, easy bruising, elevated INR, and hemorrhagic disease in newborns.

3.54 VITAMIN TOXICITIES

Hypervitaminosis, or vitamin toxicity, arises when the consumption or buildup of vitamins surpasses the body's capacity to process or eliminate them, particularly in the case of fat-soluble vitamins. This condition is typically caused by taking too many supplements rather than from food sources [93].

3.54.1 Causes of Toxicity

- Excessive consumption of multivitamins or single-vitamin supplements [94]
- Administration of high-dose parenteral vitamins
- Utilization of megadoses in alternative or unregulated therapies
- Accidental ingestion by children

3.54.2 Clinical Manifestations by Vitamin

- **Vitamin A (Hypervitaminosis A)**: Symptoms include nausea, liver damage, bone pain, blurred vision, and headaches; during pregnancy, it can have teratogenic effects [95].
- **Vitamin D**: It can lead to hypercalcemia, nausea, kidney stones, and vascular calcification.
- **Vitamin E**: It may increase bleeding risk due to its antiplatelet effects and cause gastrointestinal upset.
- **Vitamin K**: Although rare, synthetic forms in high doses might cause hemolytic anemia or jaundice in infants.

Water-soluble vitamin toxicities are less common but can occur:

- **Vitamin B6**: Chronic intake of high doses exceeding 100 mg per day may result in neuropathy.
- **Niacin (B3)**: At pharmacological doses, it can cause flushing, hepatotoxicity, and hyperglycemia.
- **Vitamin C**: It may lead to gastrointestinal distress and, in predisposed individuals, the formation of kidney stones.

3.55 RISK POPULATIONS

Certain groups are more susceptible to vitamin imbalances [96]:

A. **Children and infants** (e.g., risk of vitamin D inadequacy from fortified diets or lack of sunshine)
B. **Women who are pregnant** (e.g., higher demands for vitamin D, iron, and folate)
C. **Elderly** (e.g., polypharmacy, low absorption, decreased appetite)
D. **Patients with malabsorptive diseases** or chronic illnesses (such as inflammatory bowel disease, chronic renal disease)
E. **Alcoholics** (such as those with deficiencies in thiamine, folate, vitamin B6, or vitamin C)
F. **Vegetarians and vegans** (risk of B12 deficiency, for example).

Vitamin imbalance risk spectrum across different population groups is shown in Figure 3.4.

3.55.1 Diagnosis and Monitoring

Diagnosis involves a combination of [97]:

A. Medical history, including food habits and symptoms
B. Examine physical symptoms such as skin changes, neurological indications, and mucosal findings.

Laboratory tests:

A. Serum levels of vitamins (e.g., 25(OH)D for vitamin D, B12, folate)
B. Functional indicators (e.g., methylmalonic acid for B12 status, prothrombin time for vitamin K)

3.55.2 Prevention and Treatment

A. Promoting a healthy diet that includes fruits, vegetables, whole grains, lean meats, and fortified foods [98].
B. Supplementation: Based on professional standards and individual risk factors, such as prenatal vitamins, vitamin D for osteoporosis, and B12 for vegans.

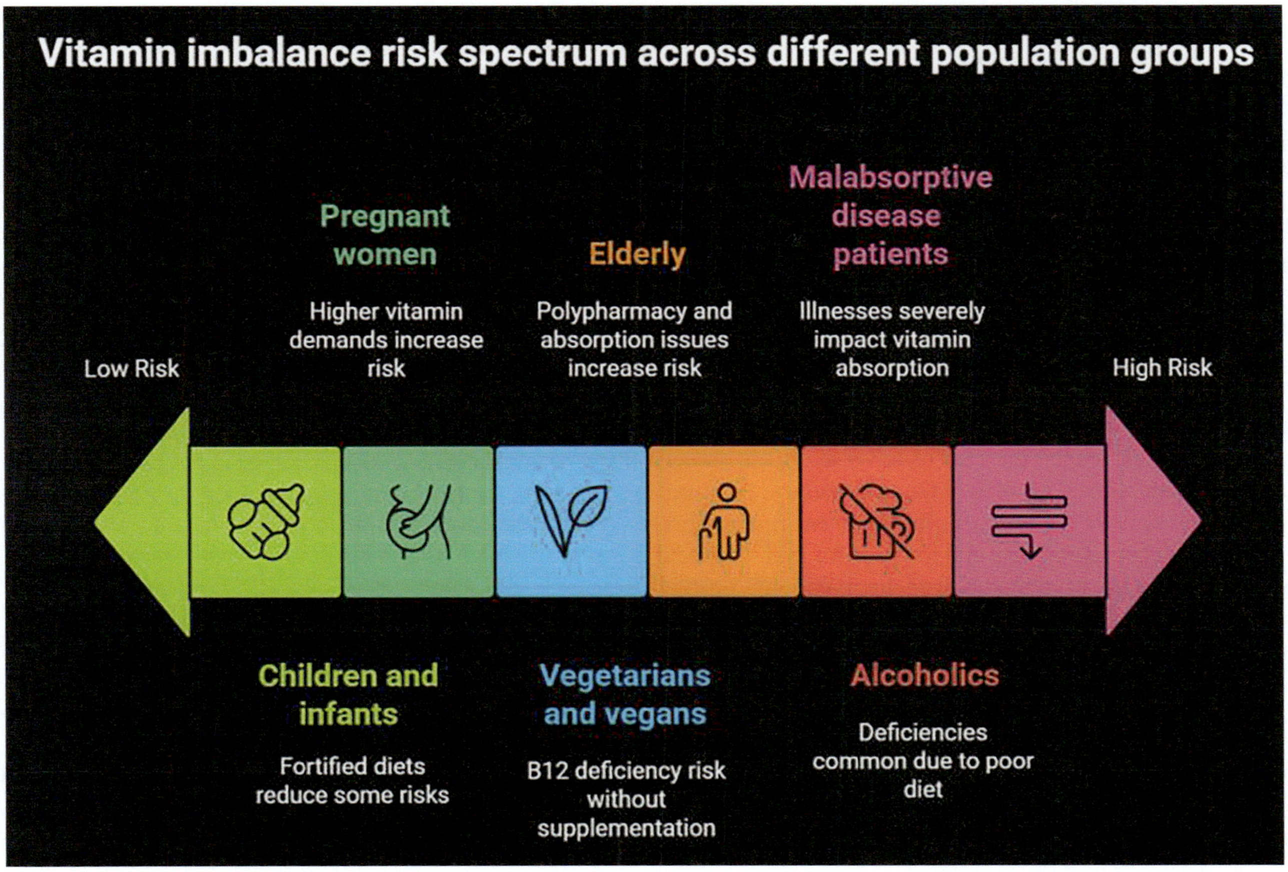

FIGURE 3.4 Vitamin imbalance risk spectrum across different population groups.

C. Parenteral treatment is used for severe shortages or malabsorption, such as IM B12 in pernicious anemia and IV thiamine in Wernicke's encephalopathy.
D. Supplements should not be taken over the suggested amounts unless medically necessary and under supervision.

3.56 CHALLENGES AND FUTURE DIRECTIONS

3.56.1 Current Limitations in Vitamin Pharmacokinetic Research

In the domain of vitamin pharmacokinetics, which examines the ADME of vitamins within the human body, several significant challenges persist. A primary challenge is the inter-individual variability, which complicates standardization efforts and is influenced by factors such as genetic makeup, age, sex, microbiome composition, and overall health status. Furthermore, much of the existing research relies on in vitro systems or animal models, which may not accurately reflect human physiological processes. The lack of standardized assays and biomarkers for evaluating vitamin status further impedes data comparability across studies. Additionally, the dose–response relationships for many vitamins remain inadequately understood, particularly for fat-soluble vitamins, where excessive intake may lead to toxicity. Moreover, the interactions between nutrients and drugs, as well as between different nutrients, which can significantly alter vitamin kinetics, are not yet well comprehended [99].

3.57 EMERGING TECHNOLOGIES AND APPROACHES

New developments provide encouraging ways to get around these restrictions. The intricate biological networks that affect vitamin metabolism are being clarified with the use of omics technologies, such as microbiomics, proteomics, metabolomics, and genomics. Real-time monitoring of vitamin distribution and function in the body is becoming better thanks to non-invasive biomarkers and advanced imaging tools [100].

Simultaneously, complex information can now be integrated to anticipate individual reactions to vitamin consumption thanks to machine learning and AI-driven modeling. Moreover, delivery methods based on nanotechnology are being investigated to enhance vitamin bioavailability and tailored administration, especially for unstable or poorly absorbed vitamins. By decreasing the need for animal testing and increasing translational relevance, the creation of organ-on-a-chip and three-dimensional tissue models also offers preclinical vitamin research more physiologically appropriate platforms.

3.38 PERSONALIZED VITAMIN SUPPLEMENTATION

Vitamin supplementation is increasingly moving toward personalized nutrition, wherein recommendations are tailored to an individual's health status, dietary preferences, lifestyle, and genetic makeup. This shift is driven by the recognition that a universal approach is inadequate and may lead to either excessive or insufficient supplementation. Technologies such as nutrigenomic testing, microbiome analysis, and digital health monitoring (e.g., wearables and applications) are facilitating the development of individualized nutritional profiles. Personalized supplementation has the potential to optimize health outcomes, enhance disease prevention, and reduce the risk of adverse effects associated with generalized supplementation strategies. However, challenges remain regarding cost-effectiveness, data privacy, ethical considerations, and the need for robust clinical evidence to support individualized interventions. Further research and regulation are necessary to ensure safe, effective, and equitable implementation [101].

FUNDING

None.

CONFLICT OF INTEREST

None.

REFERENCES

1. Combs Jr GF, McClung JP. *The vitamins: fundamental aspects in nutrition and health*. Academic press; 2016 Dec 15.
2. Moini J, Ferdowsi K. *Handbook of Nutritional Disorders*. CRC Press; 2024 Dec 10.
3. Semba RD. The discovery of the vitamins. *International Journal for Vitamin and Nutrition Research*, 2012; 82(5): 310–315.
4. Schalch W, Dostálová L. *Vitamins* (pp. 657–690). Springer, 1983.
5. Mousa Kareem M, S Jabir M. *Vitamins*. Intechopen, 2020.
6. Ming NC, Khalil MI, Gan SH. *Vitamins* (pp. 66–79). Wiley, 2023.
7. Borowitz D, Baker RD, Stallings V, Consensus report on nutrition for pediatric patients with cystic fibrosis. *Journal of Pediatric Gastroenterology and Nutrition*, 2002; 35(3): 246–259. doi:10.1097/00005176-200209000-00004.
8. Analyst. Nomenclature of vitamins. *Analyst*, 1953; 78(922): 72.
9. Cheldelin VH. Nomenclature of the vitamins. *Nutrition Reviews*, 2009; 9(10): 289–292.
10. Kallner A. *Quantitative Aspects of Vitamin Metabolism* (pp. 53–72). Springer, 1980.
11. Johnsen S. *Absorption*. Princeton University, 2011.
12. Watanabe F, Nakano Y. Vitamin B12. *Nihon Rinsho Japanese Journal of Clinical Medicine*, 1999; 57(10): 2205–2210.
13. Bonate PL. Pharmacokinetics. *WIREs Computational Stats*, 2011; 3(4): 332–342.
14. Boarder M, Dixon J, Zetterström T, Navti P, Newby D. *Pharmacokinetics*. Oxford University, 2016.
15. Bertol E, Favretto D. *Pharmacokinetics* (pp. 133–178). Crc, 2022.
16. Godfrey KR. Pharmacokinetics. *Transactions of the Institute of Measurement and Control*. 1982; 4(4): 213–223.
17. Johnson TB, Endicott MM. Vitamin research. *Science*, 1939; 89(2309): 297–298.
18. Brown J, Crison J, Timmins P. Predicting feasibility and characterizing performance of extended-release formulations using physiologically based pharmacokinetic modeling. *Therapeutic Delivery,* 2012 Sep 1; 3(9): 1047–1059.
19. Said HM, Kumar C. Intestinal absorption of vitamins. *Current Opinion in Gastroenterology*. 1999; 15(2): 172.
20. Nordlund M. *Distribution*. Edinburgh University, 2017.
21. Li X, Lv H, Luo W, Yang W, Kong L, Zhu Q, Zeng L. Recent advances in detection techniques for vitamin analysis: A comprehensive review. *Food Chem X,* 2025 Jan 31; 26: 102226. doi:10.1016/j.fochx.2025.102226. PMID: 39995404; PMCID: PMC11848456.
22. Audisio M, Fidanza A. Vitamins and lipid metabolism. *Acta vitaminologica et enzymologica*. 1982; 4(1–2): 105–114.
23. Wellhöner HH. *Vitamin B12* (pp. 175–177). Springer, 1975.
24. Sugandhi VV, Pangeni R, Nangare S, Patravale V, Qin C, Jadhav K et al. Pharmacokinetics of vitamin dosage forms: A complete overview. *Food Science & Nutrition*, 2023; 12(1): 48–83.
25. Iznaga-Escobar NN, Perez-Rodriguez RR, Mishra AK. Factors affecting pharmacokinetics of monoclonal antibodies: A review article. *Methods and Findings in Experimental and Clinical Pharmacology*, 2004; 26(2): 123.
26. Ball GFM. *Bioavailability and Analysis of Vitamins in Foods*. Springer, 1998.
27. Amore BM. Bioavailability and bioequivalence. In: AV Lyubimov (Ed.), *Encyclopedia of Drug Metabolism and Interactions* (pp. 1–39). Wiley, 2012.
28. Deguchi Y. The blood-brain barrier as a focus of drug metabolism and pharmacokinetics research. *Drug Metabolism and Pharmacokinetics*, 2011; 26(3): 219.
29. Terada R. Repletion. *Qui Parle*, 2016; 24(2): 15–43.
30. Kaushik D, Verma R. *In Vitro Digestion Models and Pharmacokinetic Aspects of SMEDDS* (pp. 169–186). Bentham Science, 2021.
31. Bluck LJ. Recent progress in stable isotope methods for assessing vitamin metabolism. *Current Opinion in Clinical Nutrition and Metabolic Care*. 2009; 12(5): 495–500.
32. Nature. Vitamins and anti-vitamins. *Nature*, 1948; 162(4130): 985–986.
33. Mills MGL, Mills MEJ. *Diet*. Oxford University, 2017.
34. Usman S, Fawzy KA, Usman A, Beevi R. *Basis of Pharmaceutical Formulation* (pp. 281–319). CRC Press, 2019.
35. Wevers NR, De Vries HE. Morphogens and blood-brain barrier function in health and disease. *Tissue Barriers*, 2015; 4(1): e1090524.
36. Kazemi-Lomedasht F, Moosavi SA, Behmardi A, Hemati E, Pooresmaeil F, Dabiri M et al. Nanoparticles as powerful tools for crossing the blood-brain barrier. *CNS & Neurological Disorders – Drug Targets*, 2023; 22(1): 18–26.
37. Schlosshauer B. The blood-brain barrier: Morphology, molecules, and neurothelin. *BioEssays*, 1993; 15(5): 341–346.

38. Abbott NJ, Friedman A. Overview and introduction: The blood–brain barrier in health and disease. *Epilepsia*, 2012; 53(s6): 1–6.
39. Stamatovic SM, Johnson AM, Keep RF, Andjelkovic AV. Junctional proteins of the blood-brain barrier: New insights into function and dysfunction. *Tissue Barriers*, 2016; 4(1): e1154641.
40. Ganong WF. Circumventricular organs: Definition and role in the regulation of endocrine and autonomic function. *Clinical and Experimental Pharmacology and Physiology*, 2000; 27(5–6): 422–427.
41. Nałęcz KA. Solute carriers in the blood-brain barrier: Safety in abundance. *Neurochemical Research*, 2016; 42(3): 795–809.
42. Yazdani S, Jaldin-Fincati JR, Pereira RVS, Klip A. Endothelial cell barriers: Transport of molecules between blood and tissues. *Traffic*, 2019; 20(6): 390–403.
43. Ruan S, Xiao W, Sun X, Yu W, Qin L, Hu C et al. Acid-responsive transferrin dissociation and GLUT mediated exocytosis for increased blood–brain barrier transcytosis and programmed glioma targeting delivery. *Advanced Functional Materials,* 2018; 28(30): 1802227.
44. Lu W. Adsorptive-mediated brain delivery systems. *Current Pharmaceutical Biotechnology*, 2012; 13(12): 2340–2348.
45. Berry DC, Vreeland AC, O'Byrne SM, Blaner WS, Noy N. Cross talk between signaling and vitamin A transport by the retinol-binding protein receptor STRA6. *Molecular and Cellular Biology*, 2012; 32(15): 3164–3175.
46. Geier EG, Papp AC, Sadee W, Yee SW, Webb A, Giacomini KM et al. Profiling solute carrier transporters in the human blood–brain barrier. *Clinical Pharmacology & Therapeutics,* 2013; 94(6): 636–639.
47. On N, Miller D. Transporter-based delivery of anticancer drugs to the brain: Improving brain penetration by minimizing drug efflux at the blood-brain barrier. *Current Pharmaceutical Design,* 2014; 20(10): 1499–1509.
48. Malcor JD, Khrestchatisky M, David M, Martinez J, Payrot N, Jacquot G et al. Chemical optimization of new ligands of the low-density lipoprotein receptor as potential vectors for central nervous system targeting. *Journal of Medicinal Chemistry,* 2012; 55(5): 2227–2241.
49. Said HM. Recent advances in carrier-mediated intestinal absorption of water-soluble vitamins. *Annual Review of Physiology,* 2004; 66(1): 419–446.
50. Waagbø R. Water-soluble vitamins in fish ontogeny. *Aquaculture Research*, 2009; 41(5): 733–744.
51. Ali MA, Hafez HA, Kamel MA, Ghamry HI, Shukry M, Farag MA. Dietary vitamin B complex: Orchestration in human nutrition throughout life with sex differences. *Nutrients*, 2022; 14(19): 3940.
52. Sambon M, Wins P, Bettendorff L. Neuroprotective effects of thiamine and precursors with higher bioavailability: Focus on benfotiamine and dibenzoylthiamine. *International Journal of Molecular Sciences,* 2021; 22(11): 5418.
53. Elmadfa I, Genser D, Rust P, Majchrzak D. The thiamine status of adult humans depends on carbohydrate intake. *International Journal for Vitamin and Nutrition Research*, 2001; 71(4): 217–221.
54. Balasubramaniam S, Yaplito-Lee J. Riboflavin metabolism: Role in mitochondrial function. *Journal of Translational Genetics and Genomics,* 2020; 4(4). doi:10.20517/jtgg.2020.40.
55. Figge HL, Souney PF, Mutnick AH, Sacks F, Figge J. Nicotinic acid: A review of its clinical use in the treatment of lipid disorders. *Pharmacotherapy*, 1988; 8(5): 287–294.
56. Jansen PAM, Van Vlijmen-Willems IMJJ, Rutjes FPJT, Zeeuwen PLJM, Hermkens PHH, Blaauw RH et al. Discovery of small molecule vanin inhibitors: New tools to study metabolism and disease. *ACS Chemical Biology*, 2013; 8(3): 530–534.
57. Vrolijk MF, Hageman GJ, Van De Koppel S, Van Hunsel F, Bast A. Inter-individual differences in pharmacokinetics of vitamin B6: A possible explanation of different sensitivity to its neuropathic effects. *PharmaNutrition*, 2020; 12: 100188.
58. Agrawal S, Agrawal A, Said HM. Biotin deficiency enhances the inflammatory response of human dendritic cells. *American Journal of Physiology-Cell Physiology*, 2016; 311(3): C386–C391.
59. Wright AJA, Dainty JR, Finglas PM. Folic acid metabolism in human subjects revisited: Potential implications for proposed mandatory folic acid fortification in the UK. *British Journal of Nutrition*, 2007; 98(4): 667–675.
60. Burger R, Mehlman C, Allen R. Human plasma R-type vitamin B12-binding proteins. I. Isolation and characterization of transcobalamin I. TRANSCOBALAMIN III. and the normal granulocyte vitamin B12-binding protein. *Journal of Biological Chemistry*, 1975; 250(19): 7700–7706.
61. Gerster G. High-dose vitamin C: A risk for persons with high iron stores? *International Journal for Vitamin and Nutrition Research*, 1999; 69(2): 67–82.
62. Nualart FJ, Rivas CI, Montecinos VP, Godoy AS, Guaiquil VH, Golde DW et al. Recycling of vitamin C by a bystander effect. *Journal of Biological Chemistry*, 2003; 278(12): 10128–10133.
63. Maqbool A, Stallings VA. Update on fat-soluble vitamins in cystic fibrosis. *Current Opinion in Pulmonary Medicine*, 2008; 14(6): 574–581.
64. Goodman D. Overview of current knowledge of metabolism of vitamin A and carotenoids. *Journal of the National Cancer Institute*, 1984; 73(6): 1375–1379.
65. Blomhoff R, Wake K. Perisinusoidal stellate cells of the liver: Important roles in retinol metabolism and fibrosis. *FASEB Journal*, 1991; 5(3): 271–277.
66. Seitz HK. Alcohol and retinoid metabolism. *Gut,* 2000; 47(6): 748–750.
67. Mansbach CM, Gorelick F. Development and physiological regulation of intestinal lipid absorption. II. Dietary lipid absorption, complex lipid synthesis, and the intracellular packaging and secretion of chylomicrons. *American Journal of Physiology-Gastrointestinal and Liver Physiology*, 2007; 293(4): G645–G650.
68. Garbossa SG, Folli F. Vitamin D, sub-inflammation and insulin resistance: A window on a potential role for the interaction between bone and glucose metabolism. *Reviews in Endocrine and Metabolic Disorders,* 2017; 18(2): 243–258.
69. Schmölz L. Complexity of vitamin E metabolism. *World Journal of Biological Chemistry,* 2016; 7(1): 14.
70. Kono N, Satow Y, Ohto U, Uchida Y, Hiramatsu T, Arai H et al. Impaired α-TTP-PIPs interaction underlies familial vitamin E deficiency. *Science*, 2013; 340(6136): 1106–1110.
71. Gee PT. Unleashing the untold and misunderstood observations on vitamin E. *Genes & Nutrition,* 2010; 6(1):5–16.

72. Kaempf-Rotzoll DE, Traber MG, Arai H. Vitamin E and transfer proteins. *Current Opinion in Lipidology*, 2003; 14(3):249–254.
73. Bus K, Szterk A. Relationship between structure and biological activity of various vitamin K forms. *Foods*, 2021; 10(12): 3136.
74. Zhou S, Mehta BM, Feeney EL. A narrative review of vitamin K forms in cheese and their potential role in cardiovascular disease. *International Journal of Dairy Technology*, 2022; 75(4): 726–737.
75. Shearer MJ, Newman P. Recent trends in the metabolism and cell biology of vitamin K with special reference to vitamin K cycling and MK-4 biosynthesis. *Journal of Lipid Research*, 2014; 55(3): 345–362.
76. Zelzer S, Goessler W, Herrmann M. Measurement of vitamin D metabolites by mass spectrometry, an analytical challenge. *Journal of Laboratory and Precision Medicine*, 2018; 3: 99.
77. Jemal M. High-throughput quantitative bioanalysis by LC/MS/MS. *Biomedical Chromatography*, 2000; 14(6): 422–429.
78. Cherdak MA. Brain aging. *Problems of Geroscience*, 2023; 2: 71–79.
79. Becher F, Ezan E, Clement G, Pruvost A, Tabet JC. Quantification of small therapeutic proteins in plasma by liquid chromatography–tandem mass spectrometry: Application to an elastase inhibitor EPI-hNE4. *Analytical Chemistry*, 2006; 78(7): 2306–2313.
80. Sakamoto S, Morimoto S, Tanaka H, Putalun W, Vimolmangkang S, Shoyama Y et al. Enzyme-linked immunosorbent assay for the quantitative/qualitative analysis of plant secondary metabolites. *Journal of Natural Medicines*, 2017; 72(1): 32–42.
81. Cloyd JC, Remmel RP. Antiepileptic drug pharmacokinetics and interactions: Impact on treatment of epilepsy. *Pharmacotherapy*, 2000; 20(8 Pt 2): 139S–151S.
82. Paparella R, Micangeli G, Panvino F, Rasio D, Niceta M, Tarani L et al. Water-soluble vitamins: Hypo- and hypervitaminosis in pediatric population. *Pharmaceutics*, 2025; 17(1): 118.
83. Fletcher RH, Fairfield KM. Vitamins for chronic disease prevention in adults: Clinical applications. *JAMA*, 2002; 287(23):3127.
84. Lešić S, Ivanišević Z, Špiljak B, Tomas M, Šoštarić M, Včev A. The impact of vitamin deficiencies on oral manifestations in children. *Dentistry Journal*, 2024; 12(4): 109.
85. Heinz J, Dierkes J, Domröse U, Westphal S, Neumann KH, Luley C. Washout of water-soluble vitamins and of homocysteine during haemodialysis: Effect of high-flux and low-flux dialyser membranes. *Nephrology*, 2008; 13(5): 384–389.
86. Mccabe BJ. Prevention of food-drug interactions with special emphasis on older adults. *Current Opinion in Clinical Nutrition and Metabolic Care*, 2004; 7(1): 21–26.
87. Dibble J, Sheridan P, Losowsky M. A survey of vitamin d deficiency in gastrointestinal and liver disorders. *QJM: An International Journal of Medicine*, 1984; 53(209): 33–42.
88. Giner-Soriano M, Molina A, Vilaplana-Carnerero C, Vallano A, Ouchi D, Morros R et al. Effectiveness and safety of oral anticoagulants for non-valvular atrial fibrillation: A population-based cohort study in primary healthcare in Catalonia. *Frontiers in Pharmacology*, 2023; 14: 1198765.
89. Roberts JW, Mouradian MM, Cora-Locatelli G, Amantea MA, Bravi D, Chase TN. Catechol-O-methyltransferase inhibitor tolcapone prolongs levodopa/carbidopa action in parkinsonian patients. *Neurology*, 1993; 43(12): 2685.
90. Sander J, Torensma B, Siepe J, Schorp T, Schulte T, Schmeer C et al. Assessment of preoperative multivitamin use on the impact on micronutrient deficiencies in patients with obesity prior to metabolic bariatric surgery. *Obesity Surgery*, 2025; 35(5): 1818–1826.
91. Palermo NE, Holick MF. Vitamin D, bone health, and other health benefits in pediatric patients. *Journal of Pediatric Rehabilitation Medicine*, 2014; 7(2): 179–192.
92. Darnton-Hill I. Public health aspects in the prevention and control of vitamin deficiencies. *Current Developments in Nutrition*, 2019; 3(9): nzz075.
93. Farris WA. Protracted hypervitaminosis a following long-term, low-level intake. *JAMA*, 1982; 247(9): 1317.
94. Nekoukar Z, Zakariaei Z, Manouchehri A. Accidental vitamin D3 overdose in a young man. *International Journal for Vitamin and Nutrition Research*, 2023; 94(2): 82–85.
95. Aberger S, Schreiber N, Pilz S, Eller K, Rosenkranz AR, Kirsch AH. Targeting calcitriol metabolism in acute vitamin D toxicity: A comprehensive review and clinical insight. *Nutrients*, 2024; 25(18): 10003.
96. Pawlak R, Lucus D, Cullum-Dugan D, Parrott SJ, Raj S. How prevalent is vitamin B12 deficiency among vegetarians? *Nutrition Reviews*, 2013; 71(2): 110–117.
97. Woo KS, Kwok TCY, Celermajer DS. Vegan diet, subnormal vitamin B12 status and cardiovascular health. *Nutrients*, 2014; 6(8): 3259–3273.
98. Thurman JE, Mooradian AD. Vitamin supplementation therapy in the elderly. *Drugs & Aging*, 1997; 11(6): 433–449.
99. Ruskovska T, Milenkovic D, Maksimova V. Polyphenols in human nutrition: From the in vitro antioxidant capacity to the beneficial effects on cardiometabolic health and related inter-individual variability: An overview and perspective. *British Journal of Nutrition*, 2019; 123(3): 241–254.
100. Gharibzahedi SMT, Moghadam M, Amft J, Tolun A, Hasabnis G, Altintas Z. Recent advances in dietary sources, health benefits, emerging encapsulation methods, food fortification, and new sensor-based monitoring of vitamin B12: A critical review. *Molecules*, 2023; 28(22): 7469.
101. Sugandh F, Memon UA, Chandio M, Karishma F, Varrassi G, Khuwaja S et al. Advances in the management of diabetes mellitus: A focus on personalized medicine. *Cureus*, 2023; 15(8): e43697.

4 Epidemiology of Vitamin Deficiency, Age-Related Absorption, and Cognitive Decline

Riya Sarkar[1], *Rajat Kanti Ghoshal*[2], *Supriya Kumar Moulick*[3], *Md. Abubakar*[4], *Abhishek Ambawatiya*[4], *Nitesh Kumar*[4], *and Krishnendu Adhikary*[5]

[1]Department of Medical Laboratory Technology, Dr. B. C. Roy Academy of Professional Courses, Fuljhor, Durgapur, India
[2]Department of Applied Microbiology, Vellore Institute of Technology, Tiruvalam Rd, Katpadi, Vellore, India
[3]Department of Tea Science, University of North Bengal, Darjeeling, India
[4]Department of Pharmacology and Toxicology, National Institute of Pharmaceutical Education and Research, Hajipur, India
[5]Department of Medical Laboratory Technology, Paramedical College, Durgapur, India

4.1 INTRODUCTION

Vitamin deficiencies significantly contribute to the global disease burden, particularly in low- and middle-income countries. Vitamin deficiencies have a profound impact on healthy aging in older adults. Malnutrition and micronutrient insufficiencies are increasingly recognized as modifiable risk factors for neurodegenerative conditions, highlighting their relevance in healthy aging (Dror et al., 2014). A study of generally healthy older adults found that 45% had inadequate nutrient and micronutrient deficiencies (MNDs), with every 1% drop in total calorie intake linked to about 100 g of weight loss. Undernutrition is also worryingly common, affecting roughly 5%–10% of older adults living in the community, 50% of those in rehabilitation facilities, 20% in residential care, and 40% of hospitalized older adults (Corish & Bardon, 2019). Numerous age-related disorders can directly or indirectly compromise nutrition by reducing nutrient absorption or limiting access to food, leading to deficiencies that may initiate or aggravate disabilities. Frailty, characterized by a decline in functional capacity, is strongly associated with malnutrition in its onset and progression. Moreover, aging leads to alterations in body composition that contribute to the development of sarcopenia (Norman et al., 2021). Together, frailty, aging, and sarcopenia significantly increase the risk of malnutrition, while poor nutritional status continues to be a major driver of morbidity and mortality among older adults. The global aging of populations is leading to a rise in the number of individuals at high risk for degenerative diseases, such as Alzheimer's disease (AD), for which no cure is currently available. Mild cognitive impairment (MCI), the prodromal stage of AD, is also becoming more prevalent, with the risk of both conditions heightened by malnutrition (Littlejohns et al., 2014). The aging process is accompanied by physiological alterations that impair nutrient absorption, including diminished gastric acid secretion, shifts in gut microbiota composition, and reduced intrinsic factor activity, all increasing susceptibility to vitamin malabsorption syndromes. Such deficiencies contribute to heightened oxidative stress, neuroinflammation, and elevated homocysteine levels, pathophysiological mechanisms known to underlie both structural and functional brain alterations. As a result, suboptimal vitamin status has been consistently associated with cognitive decline, MCI, and a greater risk of dementia in longitudinal population-based studies (Porter et al., 2016).

This chapter seeks to integrate current epidemiological evidence on vitamin deficiencies, focusing particularly on age-related alterations in nutrient absorption and their impact on cognitive function. Exploring prevalence patterns, underlying biological mechanisms, and associated clinical outcomes underscores both the mechanistic insights and the translational challenges in mitigating micronutrient-related cognitive decline.

4.2 OVERVIEW OF ESSENTIAL VITAMINS AND THEIR ROLE IN COGNITIVE HEALTH

Vitamins and minerals are vital for human health, as they participate in numerous key metabolic pathways that sustain essential cellular functions. They play crucial roles in energy production, DNA synthesis, oxygen transport, and

DOI: 10.1201/9781003686675-4

nervous system activity, which are fundamental to optimal brain and muscle performance. These physiological effects, in turn, influence cognitive and psychological well-being, including mental and physical fatigue levels. This chapter focuses on B vitamins (B1, B2, B3, B5, B6, B8, B9, and B12), vitamin C, iron, magnesium, and zinc—nutrients well recognized for their contributions to these processes (Fekete et al., 2023).

1. **B Vitamins**: B vitamins act as cofactors in enzymatic reactions that drive energy production, neurotransmitter synthesis, and DNA repair—critical for brain health.
   Vitamin B1 (Thiamine): Supports glucose metabolism in the brain, essential for nerve conduction and neurotransmitter production (e.g., acetylcholine). Deficiency can lead to memory impairment, confusion, and even Wernicke–Korsakoff syndrome.
   Vitamin B2 (Riboflavin): Component of FAD and FMN, coenzymes in oxidative metabolism; helps reduce oxidative stress in neurons. Low riboflavin may contribute to fatigue and poor cognitive endurance.
   Vitamin B3 (Niacin): This vitamin is key in NAD/NADP formation, supporting ATP production and DNA repair. Severe deficiency causes pellagra, with dementia as a hallmark symptom.
   Vitamin B5 (Pantothenic Acid): It is a component of coenzyme A, vital for acetylcholine synthesis. It may influence memory and alertness by supporting neurotransmitter production.
   Vitamin B6 (Pyridoxine): Coenzyme in the synthesis of serotonin, dopamine, and GABA. Deficiency of this vitamin can cause irritability, confusion, and depression-like symptoms.
   Vitamin B8 (Inositol): Functions as a structural basis for secondary messengers in signal transduction. Linked to mood regulation and cognitive processing speed.
   Vitamin B9 (Folate): It is crucial for DNA synthesis and methylation; it regulates homocysteine levels. Low folate is associated with cognitive decline and depression.
   Vitamin B12 (Cobalamin) maintains myelin sheath integrity and assists in neurotransmitter synthesis. Deficiency can cause memory loss, confusion, and irreversible nerve damage.
2. **Vitamin C (Ascorbic Acid)** is a potent antioxidant that protects neurons from oxidative damage and supports dopamine production. Adequate vitamin C levels are associated with better memory performance and reduced cognitive fatigue (Travica et al., 2017).
3. **Iron** is essential for oxygen transport (hemoglobin) and neurotransmitter synthesis (dopamine, serotonin). Especially in early life, deficiency can impair attention, memory, and learning capacity.
4. **Magnesium** regulates NMDA receptor activity, which is involved in synaptic plasticity (critical for learning and memory). Low magnesium may contribute to brain fog, anxiety, and reduced cognitive flexibility.
5. **Zinc**: Supports synaptic transmission and neurogenesis; regulates brain-derived neurotrophic factor (BDNF). Deficiency is linked to impaired learning, depression, and slower cognitive processing.

4.3 GLOBAL EPIDEMIOLOGY OF VITAMIN DEFICIENCIES

Vitamin deficiencies are a significant global public health problem, affecting populations in developing and developed countries. While deficiencies are more prevalent in low- and middle-income nations due to food insecurity, inadequate dietary diversity, and poor health systems, high-income countries report significant micronutrient gaps due to poor nutritional habits, chronic illnesses, and aging populations (Muthayya et al., 2021).

The consequences include impaired immune function, poor cognitive and physical development, increased risk of chronic diseases, and higher mortality rates.

Micronutrients are crucial for sustaining life and supporting optimal physiological function. Yet, deficiencies in these essential nutrients remain a significant global health challenge, with pregnant women and children under 5 being the most at risk. The most prevalent MNDs are iron, iodine, folate, vitamin A, and zinc deficiencies, which are associated with stunted growth, cognitive impairment, pregnancy-related complications, and a heightened risk of disease and death. MNDs rarely occur in isolation; multiple deficiencies often coexist, compounding their harmful effects. Their impact extends beyond individual health, undermining economic productivity and hindering national development. The intergenerational cycle of MNDs is of particular concern, in which poor nutrition passes its effects from generation to generation, a phenomenon that remains not yet fully understood. Tackling MNDs is a global health priority, commonly addressed through supplementation, food fortification, and dietary diversification. The first 1,000 days of life are regarded as the most crucial window to break the cycle of malnutrition. Achieving lasting progress, however, requires a coordinated, long-term international commitment to expanding and sustaining effective nutrition programs. A clear understanding of the epidemiology of MNDs is vital for designing targeted interventions that meet the specific needs of different populations and deliver enduring results (Bailey et al., 2015).

4.3.1 Prevalence by Age Group and Region

A pooled analysis by the Lancet Global Health shows that children aged 6–59 months are ~56% globally affected by at least one deficiency (iron, zinc, vitamin A). By region, prevalence is 58% in East Asia and Pacific, 57% in South Asia, and 80% in sub-Saharan Africa. Among non-pregnant women aged 15–49, ~69% are globally deficient. Regional figures are 74% in South Asia and 80% in sub-Saharan Africa (Stevens et al., 2022). Globally, around one-third of children under 5 suffer from vitamin A deficiency, most commonly in preschoolers and pregnant women, particularly in sub-Saharan Africa (48%) and South Asia (44%). From 1990 to 2019, significant reductions in age-standardized prevalence were seen, but sub-Saharan Africa still leads in burden. Nearly half of all cases occur in children under 5 (Jiang et al., 2024). One systematic review and meta-analysis explored vitamin D deficiency patterns worldwide and across regions in people aged 1 year and older between 2000 and 2022. Findings show that although the prevalence dropped slightly from 2000–2010 to 2011–2022, it remains high. Key risk factors include latitude and season, with the Eastern Mediterranean region and lower-middle-income countries showing the most significant burden. Women were more likely to be affected, and data from low-income countries remain scarce (Cui et al., 2023). One study reported serum B12 and erythrocyte folic acid (FA) levels and their prevalence in a representative sample of Indian children and adolescents across national, state, and regional levels. Findings showed that B12 deficiency was most common in adolescents (31%), nearly double the rate seen in preschool children (13.8%) and school-age children (17.3%). A similar pattern was observed for FA deficiency, affecting 35.6% of adolescents compared with 22.8% of preschoolers and 27.6% of school-age children (Shalini et al., 2023). In Israel, nutritional deficiencies are a widespread concern, impacting people of all ages, diverse minority groups, and various regions. The most common include iron, calcium, zinc, folic acid, and vitamins B12, C, D, and E. Among children, iron and vitamin D deficiencies are especially prevalent, with higher rates seen in groups such as Bedouins, vegetarians, Ethiopians, overweight children, pregnant women and their newborns, those with gluten intolerance, children affected by *H. pylori* infection, and those with behavioral disorders, anorexia, or diabetes. Still, these deficiencies are not confined to these high-risk groups and occur throughout the general Israeli population (Haimi & Lerner, 2014).

4.3.2 Socioeconomic and Lifestyle Risk Factors

Cognitive decline is a multifactorial process influenced by genetic predisposition, neurodegenerative pathology, vascular health, and nutrition. Adequate vitamin intake and absorption are critical for neuroprotection, synaptic plasticity, and neurotransmitter synthesis. However, a complex interaction of socioeconomic status (SES), lifestyle practices, and age-related physiological changes often predisposes individuals to MNDs, thereby accelerating neurocognitive deterioration (Dominguez et al., 2021). Socioeconomic Determinants of Vitamin Deficiency: Socioeconomic disparities are critical to nutritional health. Low-income populations are disproportionately affected by food insecurity, limited dietary diversity, and reduced access to nutrient-rich foods—educational attainment further shapes nutritional awareness, influencing dietary practices and supplementation behaviors. Geographic and cultural factors, such as residence in food deserts, reliance on carbohydrate-based staple diets, or adherence to restrictive diets, can exacerbate vitamin insufficiency. Moreover, inequitable access to healthcare delays the diagnosis of vitamin-related deficiencies (e.g., vitamin B12 and D), allowing subclinical impairments to progress undetected (Zhao et al., 2025).

4.3.2.1 Lifestyle-Related Risk Factors

Lifestyle behaviors substantially influence micronutrient status. Excessive alcohol intake impairs thiamine and B vitamin absorption, while chronic smoking accelerates oxidative stress and depletes vitamin C and folate reserves. Sedentary lifestyles not only contribute to obesity and insulin resistance but also exacerbate metabolic inefficiencies that hinder vitamin utilization. Polypharmacy, common in older adults, presents an additional challenge; for example, metformin use is associated with vitamin B12 deficiency, while proton-pump inhibitors impair cobalamin absorption by reducing gastric acidity. Together, these behaviors and exposures compound the risk of MND (Bhatti et al., 2020).

4.4 AGE-RELATED CHANGES IN NUTRIENT ABSORPTION

Several examples of physical changes associated with aging include less acid in the gut, thinner skin, and muscle loss. While some modifications may impact your senses and standard of living, others may increase your susceptibility to nutritional deficiencies.

For instance, research indicates that 20% of the elderly suffer from atrophic gastritis, a disorder in which the cells responsible for producing stomach acid have been harmed by chronic inflammation.

Vitamin B12, as well as the minerals magnesium, calcium, and iron, can all be impacted by low stomach acid (The Lowdown on Low Stomach Acid, n.d.). A decreased requirement for calories is another issue associated with aging. Sadly, this leads to a dietary conundrum. While consuming fewer calories, older persons must obtain the same amount of specific nutrients, if not more. Thankfully, you can meet your nutrient requirements by taking a supplement and consuming a range of whole foods. A decrease in the body's capacity to detect basic signs like water retention and hunger is another problem that people may encounter as they age. You may become more susceptible to dehydration and inadvertent weight loss. Additionally, these repercussions may be more severe as you age (Raman, 2017).

Nutrition is broken down mechanically and combined with saliva when chewed. This is the first and most crucial phase in your digestive process. The simpler your bowels and digestive system accomplish their jobs, the more firmly we digest.

Enzyme molecules and acidic substances in the stomach aid in the breakdown of food nutrients. After that, the meal enters the small intestine, where most absorption occurs. The gut lining absorbs nutrients throughout the human system via the veins and arteries. Particular nutrients require particular circumstances to be effectively absorbed: Dietary fat is necessary for the lipid-soluble vitamins (A, D, E, and K).

Non-heme metals, or iron obtained from plants, are better absorbed when vitamin C is present. Calcium consumption has an impact on zinc. An excess of calcium might hinder absorption. For proper absorption, B12 requires bile and a protein known as intrinsic factor (Youness et al., 2022).

4.4.1 Gastrointestinal and Metabolic Alterations

The morphological and physiological dysfunctions of the GI system, which frequently occur in combination, are the hallmarks of digestive (GI) diseases, and their incidence has increased in recent decades. These disorders include altered microbiota in the gastrointestinal tract, movement disorders, abdominal hypersensitivity, and altered intestinal and immune function. The most significant GI illnesses include ulcerative colitis (UC), which exclusively affects the inner wall of the colon, and inflammatory bowel diseases (IBDs), which include Crohn's disease (CD), a long-term bowel disease that produces regions of inflammation throughout the GI tract (Figure 4.1). Irritable bowel syndrome (IBS), which is described as "abdominal distress linked to changed intestinal habits," and diverticular disease, which is a chronic illness of tiny pockets of intestine, are other prevalent, idiopathic, and chronic inflammatory diseases of the GI tract. Genetic predisposition, pharmaceutical treatments (such as antibiotics, not only those used for human use but also possibly those used in farm animals and crops and consumed with the resulting products), nonsteroidal anti-inflammatory drugs (NSAIDs) (such as aspirin, ibuprofen, and diclofenac), and unhealthy lifestyle choices (such as smoking, irregular eating, physical inactivity, and low-fiber diets) are the leading causes of gastrointestinal disorders.

Furthermore, the species *Helicobacter pylori*, a Gram-negative bacterial pathogen that may cause IBD and functional gastrointestinal disorders, affects roughly 50% of the world's population. Abdominal pain, constipation, diarrhea, distension of the abdomen, gastric acidity, GERD, bleeding in the GI tract, malnutrition, and intestinal tract obstruction are common symptoms of all GI disorders. Laxatives, opioids, anti-diarrheals, anti-emetics, movement boosts, and anti-acidity, anti-ulcer, and anti-inflammatory medicines are some of the medications now used to treat GI issues. Corticosteroids, immunosuppressants, and anti-tumor necrosis factor (TNF)-α antibodies are conventional treatments for IBD. These treatments are frequently associated with an increased risk of acquired infections and dysplasias, which can have costly effects on the management of the health system.

Evidence shows that the widespread prevalence of GERD results from the Western pattern diet (WPD). According to

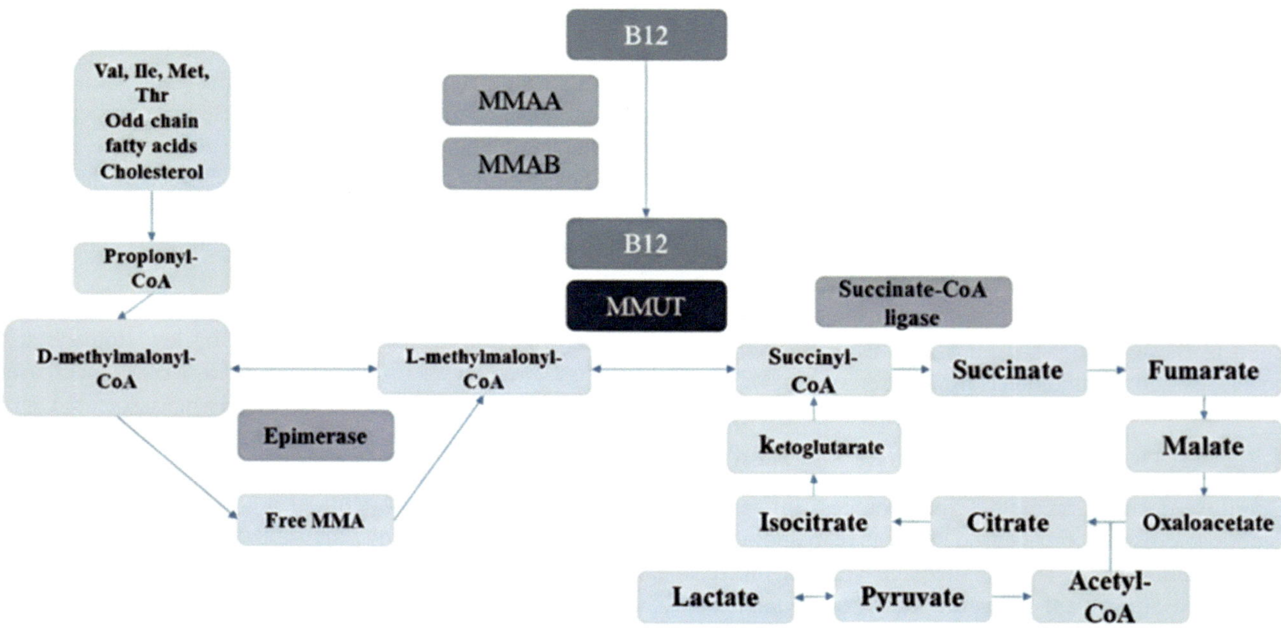

FIGURE 4.1 This figure shows how MMUT converts L-methylmalonyl-CoA into succinyl-CoA, linking amino acid metabolism with the TCA cycle.

studies, the incidence of GERD is 8.8%–25.9% in Europe, 11.6% in Australia, 23% in South America, 2.5%–7.8% in East Asia, 8.7%–33.1% in the Middle East, and 18.1%–27.8% in North America. Even though the prevalence of gastritis may vary by country, Ford et al. find a pooled prevalence of 21%. Diarrhea causes more than 128,000 inpatient stays and 3,000 deaths in the elderly population of America every year (Fujiwara & Arakawa, 2014).

On average, annual personal medical costs for certain GI conditions per person visiting a GI clinic were estimated to be $5049 for IBS, $6140 for diarrheal, $7522 for constipation, and $7646 for abdominal pain. Although our understanding of the pathophysiological mechanisms behind GI diseases has advanced recently, their etiopathogenesis remains unclear, and there is currently no marker that can be used to make a conclusive diagnosis. Visceral irritation, diseases, hereditary and genetic variables, stress, and gut microbiota alterations that result in symbiosis are some of the known etiological causes.

Environmental and genetic factors can contribute to metabolic syndrome (MS). This low-grade chronic inflammatory disorder includes several risk factors for major illnesses such as diabetes, belly fat, elevated cholesterol levels, and hypertension. When discussing the function of diabetes in human illness, Warren et al (1982) emphasized the clinical relevance of multiple sclerosis. Hepatic abnormalities such as hepatic statuses, cardiovascular illnesses (coronary heart disease [CHD], heart failure, and stroke), and type 2 diabetes mellitus (T2DM) are the most prevalent pathologies that might result from an untreated and prolonged condition of multiple sclerosis. According to Sadeghi, several metabolic disorders, such as gout, hypertension, and hyperglycemia, were already documented in the nineteenth century. Later, in 1947, Vague explained that metabolic problems, cardiovascular disease (CVD), and the development of T2DM all contributed to the obese phenotype. The European Group for the Research of Insulin Resistance and the World Health Organization (WHO) criteria recognize insulin resistance and glucose intolerance as key elements of multiple sclerosis. Although there are several definitions of this syndrome, the International Diabetes Federation (IDF) has had a single definition of MS since 2004. This definition emphasizes the critical role that obesity plays as a risk factor for the illnesses mentioned above.

Beneficial genealogy, smoking, drinking too much alcohol, aging, a low income, after menopause status, a lack of movement, unhealthy eating habits, and use of certain drugs (atypical antipsychotics) are the main risk variables for the development of multiple sclerosis. By depicting the pathogenic route of multiple sclerosis, Figure 4.2 illustrates the effects of an irregular biochemistry on the body's functioning. Unique pharmacologic entities are employed together for medical therapy because metabolic diseases share standard biological processes.

Examples of effective interventions include exercise, dietary changes, and pharmaceuticals like glucose sensitizers, renin–angiotensin–aldosterone system (RAAS) medications, statins, and fibrates. There has been a significant global rise in MS cases over the last 20 years, which has been linked to an obesity and type 2 diabetes epidemic. Globally, between 20% and 25% of individuals suffer from the neurological disorder. According to WHO estimates, 422 million people have diabetes, 1.13 billion have hypertension, and 650 million are obese (Habas et al., 2025).

There is mounting evidence that the pathophysiology of intestinal and extraintestinal disorders is linked to gut microbiota dysbiosis, which may be caused by various intrinsic and extrinsic factors, including genetic differences, nutrition, stress, and pharmaceutical treatment (Figure 4.3). Indeed, an emerging knowledge of the pathophysiology of a wide range of illnesses, including MS and GI disorders, is thought to depend on the interplay between the immunity of the host and microbiota. According to recent research, an adequate microbiota in the gut can influence the beginning of several gastrointestinal disorders, including IBS, carcinoma of the colon, celiac disease, and IBD. Furthermore, one of the leading causes of an irregular metabolism is dysbiosis. According to Hurt et al., the gut microbiota plays a significant role in the pathophysiology of type 2 diabetes by affecting insulin resistance, body weight, and pro-inflammatory activity. The breakdown of dietary carbohydrates that the body would otherwise be unable to digest is a crucial function of gut bacteria. The refractory portion of carbs in plant meals contains nutritional fiber, which can be either insoluble or soluble. Soluble fibers are degraded by enzymes generated from the gut microbes into short-chain fatty acids (SCFAs). Butyrate, acetate, and froth (SCFAs) are absorbed in the gastrointestinal tract and used as fuel by the host. Besides their function as energy materials, SCFAs regulate inflammation and dietary intake. Butyrate encourages intestinal cell regeneration and protection and mucin synthesis, lowers hypercholesterolemia, releases chemicals and/or neurotransmitters crucial for controlling stool movement, and is insulin-resistant. Because butyrate alters neurotransmitter release and reduces intestinal mechanoreceptor irritation, which may lead to decreased luminal pressure and/or peristalsis, it has been shown to alleviate stomach discomfort in IBS patients after consuming it (Powell, 2024).

Nowadays, it is acknowledged that nutrition and lifestyle are the main factors in preventing diseases, including MS, T2DM, GI problems, and CVDs. Furthermore, the most recent research suggests that the balance of gut microbiota can be restored by modulating intestinal dysbiosis using dietary supplements. From this chapter, we can provide an overview of the current understanding of the safeguarding roles of beneficial dietary factors in GI and MS and aid in developing an overview of these broad areas. This is because dysbiosis may be associated with digestive problems and MS, and a proper diet can restore the damaged gut microbiota and is thought to have been the initial approach for curing both GI problems and MS (De Filippis et al., 2020).

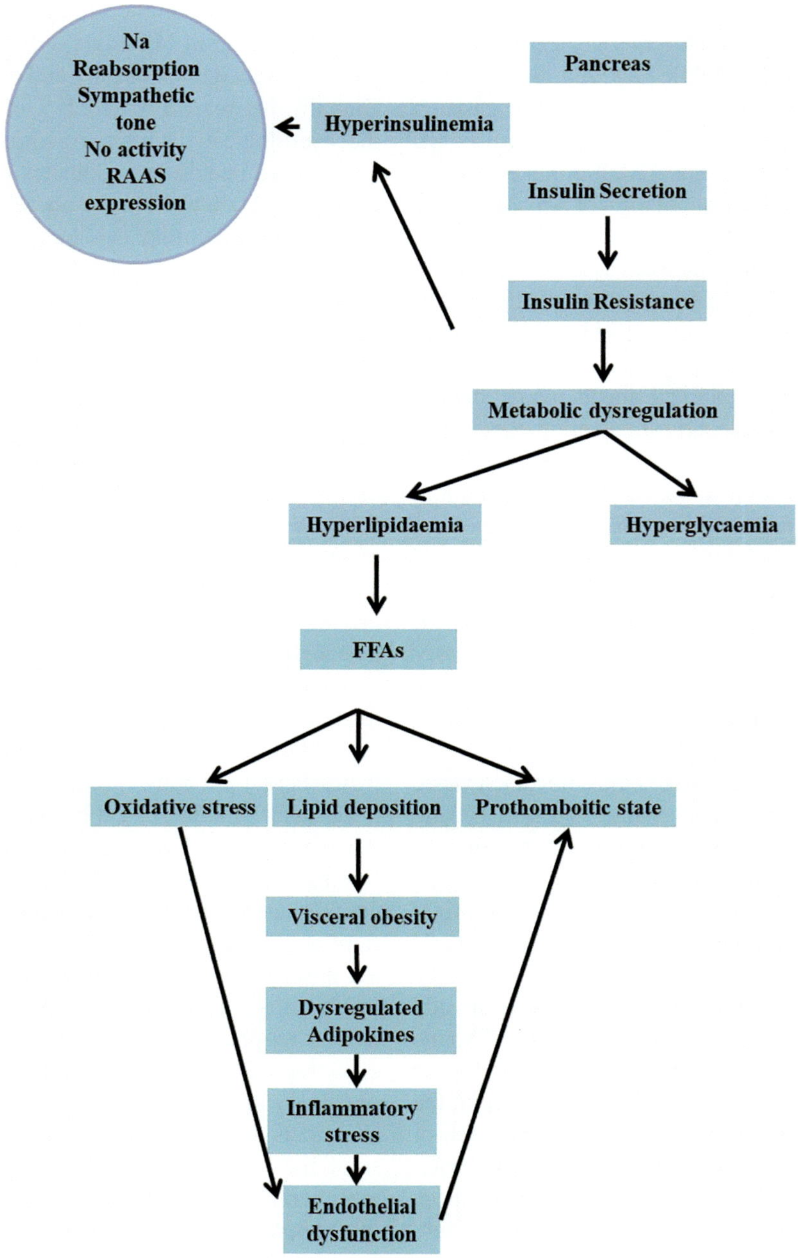

FIGURE 4.2 This figure illustrates how insulin resistance and obesity disrupt metabolism, hormones, and circulation, driving metabolic syndrome in humans.

4.4.2 Influence of Chronic Illness and Polypharmacy

Life expectancy has grown throughout the past century and continues due to notable advancements in medical research, socioeconomic situations, and access to healthcare. The number of senior individuals has significantly increased due to longer life expectancies. In 2021, 20.8% of the total population in each of the EU-27 member nations was aged (those 65 and above), which is predicted to increase significantly over the coming decades. Multiple illnesses and a higher frequency of chronic diseases are linked to aging. The literature has differing definitions of multimorbidity; some define it as having a minimum of two persistent diseases, while others define it as having three or more. Despite this variation, multimorbidity has been more common and now affects 82% of people over 85 and 65% of older individuals 65–84.

Moreover, multiple medical conditions are closely associated with polypharmacy, which is the simultaneous use of over five medications daily. With an overall frequency of 45%, polypharmacy is prevalent among older adults. Polypharmacy in older individuals has been recorded at rates as high as 65% in the United States and as low as 26.3% in Zürich and 39.9% in the nation of Portugal in Europe (Masnoon et al., 2017). When numerous prescription drugs are frequently required and influenced by various single-disease guidelines, polypharmacy may be suitable

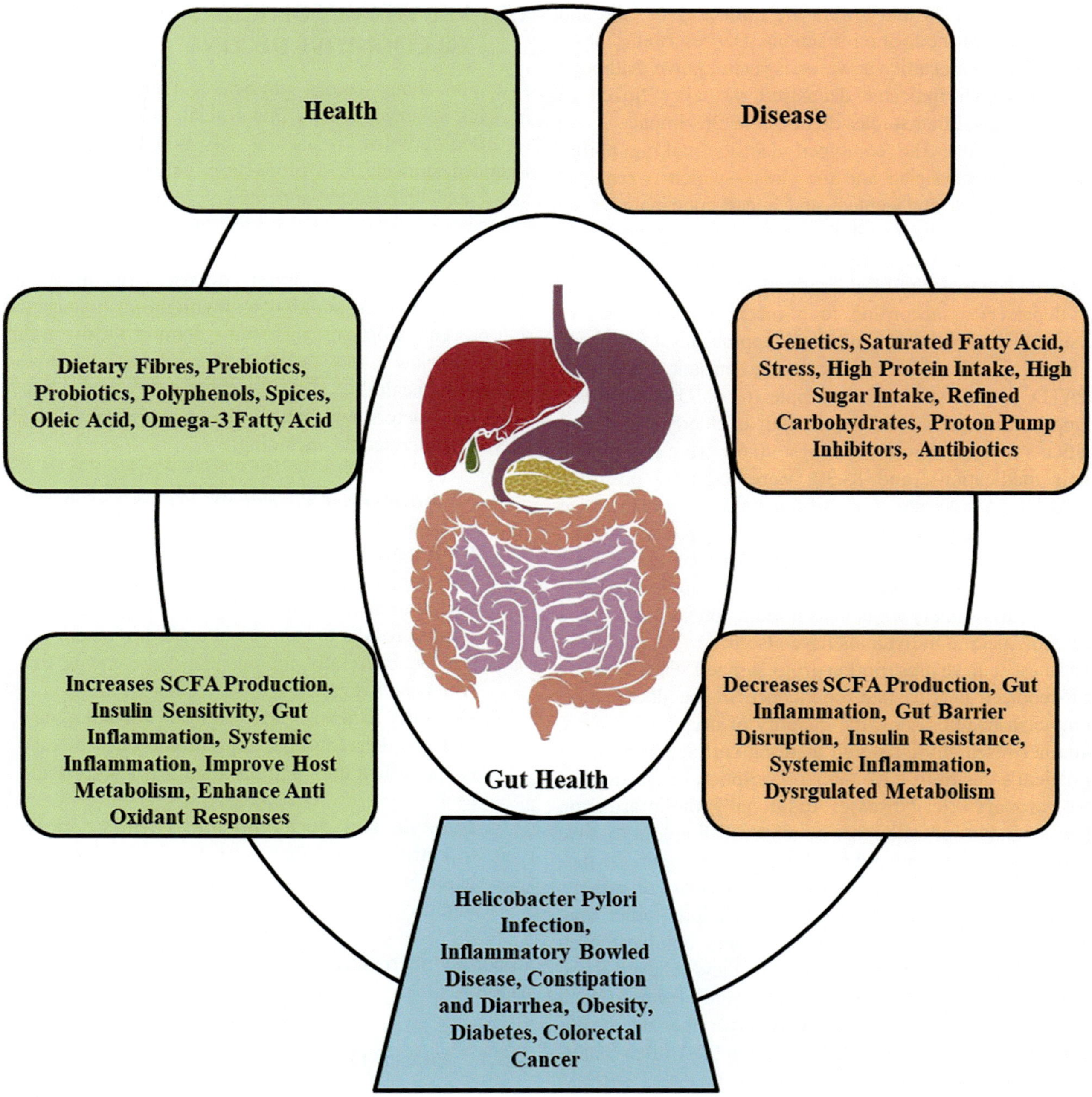

FIGURE 4.3 Diagram illustrating how internal and environmental factors contribute to gastrointestinal breakdown in wellness and illness

in situations of multiple illnesses, which is common in older persons. It is essential to recognize, nevertheless, that polypharmacy and advanced age may work in concert to increase the danger of adverse drug events (ADEs) and the use of potentially inappropriate medications (PIMs).

Age-related alterations in pharmacokinetics and pharmacodynamics, together with multiple drugs and multiple diagnoses, promote drug–drug and drug–disease interactions and raise reactions to drugs. The adverse effects of taking several medications are further compounded by the aging process, which is linked to deterioration in cognitive ability, lean body mass, hearing, eyesight, and balance, as well as a decrease in renal and hepatic functions. Adverse effects in older people, including a higher incidence of tumbles, vulnerability, memory loss, diminished functioning, being hospitalized, and mortality, are independently linked to polypharmacy. Additionally, several chronic medical disorders such as diabetes mellitus, hypertension, chronic renal disease, and cardiovascular illnesses have been significantly related to it.

DE prescribing, which involves stopping (or reducing) the dosage of an unsuitable medicine under a doctor's supervision, is a patient-centered approach to treating improper polypharmacy. This method is safe and effective in lowering the use of inappropriate medications and promotes shared decision-making between patients and doctors. It has been demonstrated that deprescribing lowers the prevalence of PIM and lowers the frequency of falls and severe adverse medication reactions. Deprescribing therapies' effect on mortality, however, is yet unknown. Although some research indicates decreased mortality linked to deprescribing, other studies do not show this impact.

To maximize the combined decision-making ability between the physician and the client—which is required to boost patient engagement and hence contribute to the effectiveness of deprescribing—consumers' opinions and ideas regarding their drugs must be recognized and considered. The majority of elderly patients (84%) were eager to deprescribe, according to a recent meta-analysis of research employing the patients' approaches toward the removal questionnaire and its updated variant (PATD and rPATD, respectively). Most people (67%–93%) remained happy with their medications. A study of elderly Portuguese patients revealed similar findings: 90.6% were happy with their medications, and 83.3% were eager to prescribe (NACDA Guidelines for Administration of Drugs to Human Subjects|National Institute on Drug Abuse, 2023).

Furthermore, two rPATD characteristics have been linked to patients' desire to prescribe: the appropriateness of drugs and anxiety about discontinuing prescriptions. The traits of specific individuals have also been linked to their readiness to have prescription drugs stopped. These include self-reported health state, confidence in the doctor, age, long-term health issues, the quantity of drugs, and so on. Another research discovered a strong correlation between a patient's desire to have their prescription drugs stopped and coronary artery disease. Additionally, the information now available indicates that patients' willingness to have their prescriptions prescribed varies depending on the particular class of drug. For instance, the majority of patients agreed to stop taking preventive medications rather than symptom-relieving ones, and they might not have been willing to stop a specific medication they believed to be necessary, according to Green et al.'s study from 2021 on the relationship between desire to prescribe and health results priorities among older patients in the United States. Furthermore, alternative cardiometabolic drugs such as insulins and antihypertensives have been thought to be more suitable than statins. Additionally, compared to other medications, older individuals were said to be more open to having a statin prescribed. Depending on the medication, various factors, such as the alleged seriousness of the underlying ailment, may contribute to these disparate patient attitudes and beliefs toward prescribing. Qualitative research of African American men found that they thought high blood pressure was worse than high cholesterol levels. According to another study, older individuals choose antihypertensive drugs over statins, which may be explained by their sense of the severity of their condition. Therefore, it is possible to speculate as to whether some medical issues have a distinct impact on people's desire to have a prescription drug prescribed. The likelihood of such a connection throughout a sample of elderly Portuguese individuals is investigated in this study (Pereira et al., 2024).

4.5 KEY VITAMIN DEFICIENCIES LINKED TO COGNITIVE DECLINE

Attention about seniors' physical and mental well-being is rising as societies age. One crucial, but sometimes disregarded, problem impacting this population is MND. Reduced stomach acid production, changed taste awareness, and decreased appetite are some physiological changes linked to aging that raise the risk of malnutrition. Poor nutritional intake and absorption are also exacerbated by conditions such as chronic disease, drug usage, and social isolation. In addition to highlighting management and preventive options, this study attempts to give a thorough overview of how MNDs affect older adults' physical and mental health.

Several factors contribute to the increased prevalence of MND in older people, including:

1. **Modifications in Physiology**: The body's capacity to absorb and digest nutrients declines with age. For instance, iron and vitamin B12 absorption may be hampered by reduced stomach acid production.
2. **Nutritional Elements**: Reduced food intake frequently results in changes in taste, fragrance, and hunger, especially for nutrient-dense meals like fruits and veggies.
3. **Chronic Illnesses and Drugs**: Polypharmacy, diabetes, heart failure, and chronic renal disease can all affect how well nutrients are absorbed and used.
4. **Lifestyle and Socioeconomic Factors**: Access to nutritious, well-balanced meals may be hampered by finances, social exclusion, and decreased mobility.

Important micronutrient inadequacies and their effects are as follows.

4.5.1 Vitamin D

The antioxidant vitamin D is crucial for bone health because it regulates the balance of phosphorus and calcium in the bones. Because of diminished skin production, less sun exposure, and low food intake, vitamin D insufficiency is frequent in older people. A deficiency raises the probability of falls by causing bone loss, cracks, and weakened muscles. Furthermore, low levels of vitamin D have been linked with an increased likelihood of dementia, including memory loss, suggesting that vitamin D has a role in mental health (Voulgaridou et al., 2023).

4.5.2 Vitamin B12

The B12 vitamin is essential for the manufacture of DNA, brain activity, and the development of red blood cells. Reduced stomach acid production in older persons frequently results in deficiencies because it hinders food absorption. Tiredness, hemorrhage, and neurological dysfunctions such as memory loss, cognitive decline, and neuropathy are signs of a B12 deficiency. Conditions and dementia might arise due to a prolonged deficit (Umair, 2025).

4.5.3 Folic acid (Vitamin B9)

Folate is also essential for methylation, DNA synthesis, and repair. Cognitive decline, mental issues, and an elevated risk of cardiovascular illness may be associated with a deficiency in the elderly. Elevated levels of a substance have been linked to an increased risk for cognitive impairment and stroke, which have been linked to low levels of vitamin folate.

4.5.4 Iron

The breakdown of energy and the flow of oxygen depend on iron. Even while it is less frequent in older persons than in younger populations, anemia due to iron deficiency can still happen, especially in those who have long-term medical issues, including renal or gastrointestinal diseases. Weakness, exhaustion, and weakened immunity are among the symptoms. Cognitive deficits, such as poor recall and focus loss, have also been connected to iron insufficiency.

4.5.5 Zinc

The production of DNA, wound reconstruction, and immune system functioning all depend on zinc. Insufficient zinc in older adults can slow wound healing, weaken the immune system, and make them more vulnerable to infections. Deficits in zinc have been connected to anxiety, diminished memory, and a higher risk of dementia-related diseases like dementia and AD. Zinc also has a part in brain health.

4.6 BIOMARKERS AND DIAGNOSTIC TOOLS FOR VITAMIN DEFICIENCY

Supplemental nutritional supplements are vital components required for optimal bodily function. They are regarded as micronutrients that the body needs in an emergency. Micronutrients perform a wide range of tasks, including blood glucose control, the breakdown of proteins, micronutrients, minerals, point-of-care diagnostics, GCMS, HPLC, longitudinal flow immunoassay, LCMS, biosensors, and vitamins for cell development and differentiation. The human organism needs micronutrients for a muscular physique in addition to carbohydrates and protein since it performs several intricate processes at once (Table 4.1).

Thirteen essential vitamins, nine water-soluble nutrients (including the antioxidant vitamin C and eight of the vitamin B complex vitamins), and four vitamins that dissolve in lipids (A, D, K, and E) are present. For this reason, nutrient deficiencies that are water-soluble are more common than those that are fat-soluble. Substantial and lesser minerals are distinct categories. Calcium is between 920 and 1,200 g or 1.5 times one's overall body weight. Nonetheless, additional microelements such as magnesium, potassium, salt, iron, and chlorine comprise approximately 0.85% of the body weight. Numerous neurodegenerative illnesses, including Parkinson's disease (PD) and AD, may benefit from vitamin therapy. Because it has anti-inflammatory and anti-oxidative activities, the absence of minerals makes neurons more likely to die or develop pathologies (Figure 4.4). These days, food's lack of healthy nutrition is caused by adulteration, increasing pesticide usage, and imbalanced diet consumption. This gradually leads to several inadequacies in the body. According to the 2018 Global Nutrition Report (GNR), one in three people suffers from undernutrition. People of all ages and genders are affected by these hidden epidemics of inadequacies, with specific risk groups being more affected than others, according to the WHO. While proper nutrition intake, tailoring, or a combination

TABLE 4.1
The Global Prevalence of Selected Micronutrient Deficiencies and Their Associated Health Disorders Highlights the Widespread Burden Across Populations and the Specific Risks Linked to Each Nutrient Deficiency

Micronutrients	Deficiency Prevalence	Disorders	References
Vitamin A	254 million	High risk of mortality in pregnant women and children	Udhani et al. (2023)
Vitamin B 1	3 billion	Korsakov syndrome	
Vitamin B 6	2.5 billion	Neurological disorder	
Vitamin B 12	4 billion	Megaloblastic anemia	
Vitamin D	Common in all age groups	Rickets, colorectal cancer	
Iodine	2 billion	Goiter, cognitive impairment	
Selenium	Common in Asia	Progression in tumor and cardiovascular risk	
Iron	2 billion	Anemia	
Zinc	The highly developed countries are estimated	Genetic disorders	

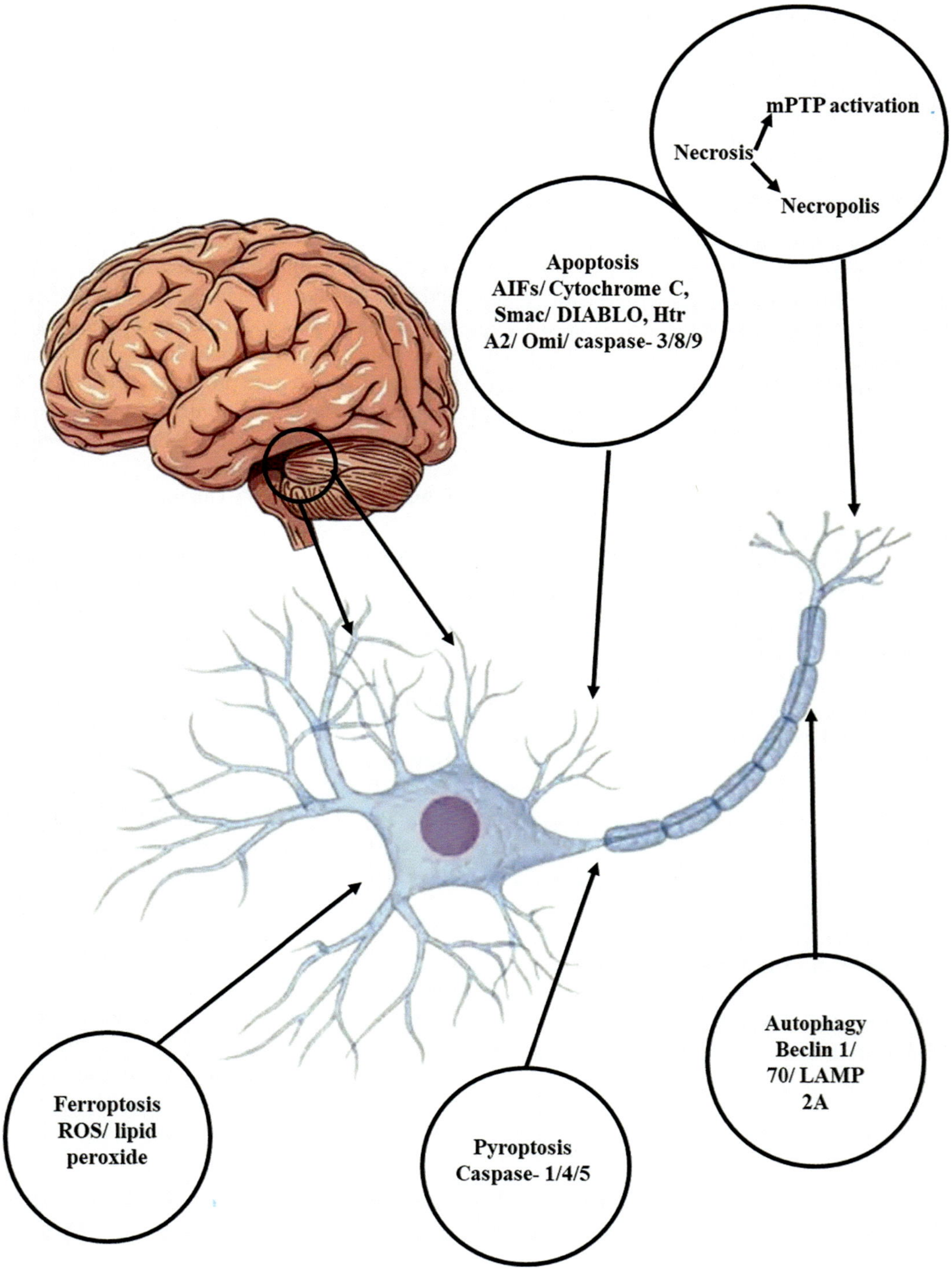

FIGURE 4.4 The pathophysiology of vitamins' preventive role in the death of neurons.

of particular nutrients can strengthen the body's functioning, inadequate nutrition can have significant, long-lasting effects on physical and mental health.

The body excretes water-soluble vitamins, whereas vitamins soluble in fat are stored in the liver and fatty tissues. Deficits in water-soluble nutrients are more prevalent than deficiencies in vitamins that dissolve in fat. Minerals can be classified as either minor or significant. Calcium is between 920 and 1,200 g or 1.5% of one's overall body weight. Nonetheless, additional macroelements such as magnesium, potassium, sodium, chlorine, and iron comprise around 0.85% of the body weight. 3,4. Numerous neurodegenerative illnesses, including PD and AD, may benefit from vitamin therapy because it has anti-inflammatory and anti-oxidative qualities; without nutrients, neurons are more likely to die or develop pathologies, as seen in Figure 4.4 (Johnson, 2024).

These days, food's lack of healthy nutrition is caused by adulteration, increasing pesticide usage, and imbalanced diet consumption. This gradually leads to several inadequacies in the body. According to the 2018 GNR, one in three people suffers from undernutrition. People of all ages and genders are affected by these hidden epidemics of inadequacies, with specific risk categories being more affected than others, according to the WHO. While proper nutritional intake, tailoring, or a combination of particular nutrients can strengthen the body's functioning, poor nutritional status may possess significant, long-lasting effects on one's mental and physical wellness. Given the distribution and the cause of malnutrition, it may contribute to the development of both acute and chronic conditions, such as thyroid gland (iodine deficiency), severe anemia (iron deficiency, vitamin B12), osteoporosis, and osteomalacia (calcium deficiency, vitamin D). Still, its deficiency also exacerbates heart disease, colon cancer, and other conditions. Micronutrient measurements in meals and drinks, dietary additives, and drugs are just as crucial for regulatory compliance and quality assurance as their detection in physiological fluids to determine the population's appropriate nutrient intake. Three stages are necessary to build an efficient and accurate approach for diagnosing deficiencies: choosing matrices, a biomarker, and a diagnostic tool.

To discover deficiencies, an advanced method that is both accurate and sensitive must be chosen. Micronutrient testing labs adhere to the guidelines established by the WHO, the European Society for Clinical Nutrition & Metabolism (ESPEN), the Clinical Laboratory Improvement Amendment, and the Centres for Disease Prevention and Control (CDC). ACCESS 2 Immunoassay System, ADVIA Centaur XP Immunoassay System, IMMULITE 2000 XPi Immunoassay System, etc., Alinity Immunoassayer for vitamin B12, VITROSV R 5600 for iron and calcium detection, vitamin D tests using DiaSorin, The LIAISONV R XL is a fully automated chemiluminescence analyzer that can conduct 120 assays simultaneously and handles all aspects of processing, including sample pre-dilutions, sample and reagent distribution, incubations, and wash procedures (Udhani et al., 2023).

4.7 EPIDEMIOLOGICAL EVIDENCE CONNECTING DEFICIENCY AND COGNITIVE IMPAIRMENT

The neuropathological cascade that results in AD and other prevalent types of dementia in later life has been linked to inflammation. Because of these findings, an observational epidemiological study has been conducted to determine how systemic and brain inflammatory indicators relate to dementia and cognitive impairment. One of the key mechanisms behind dementia and cognitive decline in the elderly may be inflammation. The neuropathological cascade that results in the formation of amyloid neurotic plaques, one of the defining features of AD, has been linked to inflammation (Figure 4.5). In AD, the immunological response in the brain is thought to be mediated by brain microglial cells. As a result, it has been suggested that lowering or eliminating inflammation may delay or even stop AD's growth. The goal of administering nonsteroidal anti-inflammatory medicines (NSAIDs) is to prevent or reduce AD. In this regard, blocking cyclo-oxygenase (such as COX-1 and COX-2) isoforms has been a reasonable line of inquiry (Weggen et al., 2001). Crucially, it has been demonstrated that NSAIDs alter beta-amyloid protein without affecting COX activity. Antiplatelet activity, vasomotor activity modulation, free radical scavenging, and peroxisome proliferator-activated receptor modification are other NSAID effects that may be significant in cognitively impairing processes.

Over time, we have learned that the strict categorization of dementias in later life into vascular forms, like vascular cognitive impairment (VCI), and neurodegenerative forms, like AD, may not apply to specific patients because the

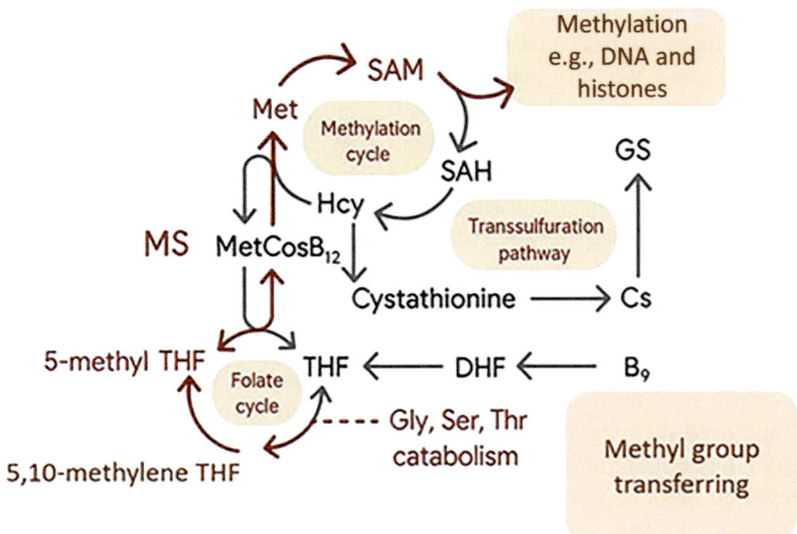

FIGURE 4.5 The relationship between cellular metabolism and the MS catalytic cycle. Red arrows show a methyl radical's route. MS stands for methionine synthase.

conditions may coexist or overlap in terms of pathophysiology and risk factors. Thus, similar pathways may trigger or intensify these prevalent types of cognitive impairment (Woodling et al., 2016). For instance, we now accept a vascular or angiogenesis theory for AD, and the β-amyloid protein may harm the brain parenchyma and the vascular endothelium by inducing inflammation. Moreover, several of the conventional vascular risk factors for AD, such as hypertension, diabetes, and atrial fibrillation, also function as risk factors for stroke or CHD (Miguel-Álvarez et al., 2015). Additionally, the components may be grouped as the MS or may be connected by T2DM, insulin resistance and inflammation, impaired glucose intolerance, obesity, dyslipidemia, hypertension, or increased blood pressure, which may increase the risk of stroke, CHD, and AD. It has previously been shown that AD and stroke coexist and interact during brain necropsy.

Therefore, diabetes mellitus may have detrimental consequences on the brain, which may be connected to cognitive impairment through a variety of routes. Insulin receptors in the brain's limbic system may be impacted by insulin resistance or decreased function, leading to reduced brain activity and cognitive impairment (Gebre et al., 2018). Furthermore, elevated tau phosphorylation, another characteristic of AD neuropathology, has been linked to disruption of insulin signaling, and elevated levels of beta-amyloid 42 in the cerebrospinal fluid may be linked to elevated insulin levels in the brain. Angiotensin II and the RAAS AT1 brain receptor illustrate the possible molecular connection between AD and brain vascular disease and atherosclerosis (Royea & Hamel, 2020). In particular, angiotensin II's activation of the AT1 receptor may contribute to the development of atherosclerosis, at least in part, through inflammatory processes (Loera-Valencia et al., 2021).

4.7.1 Cohort and Longitudinal Study Findings

Dementia and cognitive decline may emerge due to low vitamin B12 levels, although these studies' quality, methodology, and outcomes vary widely. We have conducted a thorough analysis of the data supporting the link between older persons' cognitive deterioration and their vitamin B12 status. Using keywords associated with vitamin B12 and cognition, we conducted a database search. All prospective cohort studies that evaluated the relationship between serum vitamin B12 and biomarkers were included. Two researchers conducted the data extraction and quality assessment. The quality assessment instrument assigns a good, neutral, or negative rating. Serum levels of vitamin B12 did not correlate with dementia or cognitive deterioration. However, four investigations that employed methylmalonic acid and holotranscobalamin (holoTC), two more recent biomarkers of vitamin B12 status, revealed links between low vitamin B12 status and a higher risk of cognitive decline or dementia diagnosis (O'Leary et al., 2012). To determine whether vitamin B12 status and cognitive decline are related, future research should be long enough (at least 6 years), recruit participants from the seventh decade, use standardized neurocognitive assessment tools rather than screening tests, and select markers of vitamin B12 status with sufficient specificity, such as methylmalonic acid and/or holoTC (Clarke et al., 2007). Research into modifiable risk factors is crucial since dementia and cognitive decline significantly affect the freedom and quality of life of those who suffer from them. AD is the most common type of dementia, accounting for up to 70% of cases. Other common types include vascular dementia, frontotemporal dementia, and dementia with Lewy bodies. Advanced age, heredity, low educational attainment, CVD, and its component vascular risk factors are risk factors for dementia. The apoE4 allele, which significantly raises the risk of AD by two to three times, is the most significant known genetic risk factor for the development of dementia (Gordon et al., 2025).

Although the exact role of vitamin B12 in this process is unclear, poor vitamin B12 status has been associated with cognitive decline for at least 50 years. According to some preliminary research, vitamin B12 deficiency can result in reversible dementia. It also causes neurological deterioration with demyelination of the spinal cord. It is theoretically conceivable that vitamin B12 deficiency and cognitive decline are related. In addition to being a cofactor for the methylation of total homocysteine (tHcy) to methionine and the conversion of methylmalonyl-CoA to succinyl-CoA, vitamin B12 is necessary for the production of DNA and myelin (Lauer et al., 2022). Dementia and brain alterations are linked to hypovitaminosis D, a disorder that is quite common in those 65 and older. Clear guidelines are required for researchers and clinicians due to the fast-growing and complex contribution of the vitamin D and cognition literature. An invitational conference on "Vitamin D and Cognition in Older Adults" brought together scientists worldwide. The task group concentrated on important issues regarding the function of vitamin D in AD and related conditions, drawing on prior research and professional judgment. A Delphi-style method was used to discuss and vote on each subject.

However, due to a lack of specificity and insufficient evidence, vitamin D levels should not currently be used as a diagnostic or prognostic biomarker of AD. The experts agreed that hypovitaminosis D increases the risk of cognitive decline and dementia in older adults and may change the clinical presentation as a result of related comorbidities. Due to the high frequency of hypovitaminosis D, this population should be checked for it and, if required, given supplements; however, this recommendation was not unique to cognition. The potential of "critical periods" when vitamin D may have the most impact on the brain was discussed during the argument, as was whether hypovitaminosis D affects cognition passively through a loss of neuroprotection or actively through harmful effects. The international task committee established five broad guidelines regarding vitamin D and older individuals' cognitive abilities. There are still several unknowns, and the suggestions must be updated as new information becomes available.

4.7.2 Gender and Ethnic Disparities

Serum 25-hydroxyvitamin D (25-OHD) levels below 12 ng/mL (multiply by 2.496 to convert to nanomoles per liter) and between 12 and 20 ng/mL, respectively, are indicative of vitamin D (VitD) deficiency and insufficiency, which are very common in the United States. The incidence of combined deficiency and insufficiency was 42% in the total adult population, according to the National Health and Nutrition Examination Survey, and it was significantly higher among African American (82%) and Hispanic (69%) people. Vitamin D is now known to affect many organ systems in addition to supporting bone health and calcium absorption. All human organs, including the brain, express VitD receptor (VDR) and the enzyme that changes 25-OHD into 1,25-dihydroxyvitamin D, the active form of the vitamin.

The about 50% frequency of hypovitaminosis in people over 65, which rises to 70%–90% among those with cognitive impairment, serves as motivation for studies looking at how vitamin D deficiency affects cognitive decline with age 5, 6. Furthermore, there is epidemiological evidence linking low blood levels of vitamin D to a higher risk of incident dementia and AD. In the Cardiovascular Health Study, for instance, vitamin D deficiency was linked to hazard ratios of roughly 2.2–2.3 for incident AD and all-cause dementia. Additionally, there is evidence that the baseline vitamin D level is proportionate to the total risk for incident dementia. Vitamin D levels are adversely correlated with cognitive performance, especially on executive function tests. They are also lower in people with moderate cognitive impairment (MCI), which is consistent with an increased risk of dementia 10–12. Additionally, vascular risk factors, including hypertension and CVD, both of which are linked to dementia, are linked to lower vitamin D levels 13–16. Although the precise mechanism is unknown, these studies provide evidence that hypovitaminosis D may increase the risk of late-life dementia and cognitive decline (Miller et al., 2015).

Since people of different races and ethnicities are more likely to have low vitamin D status, most of the research reviewed was carried out in predominantly white cohorts, so they do not accurately represent the effects of hypovitaminosis D on the general population 2. Thus, in an ethnically diverse population of older adults, we examined the relationships between hypovitaminosis D and cognitive deterioration. High variability in illness appearance, progression, and risk factors is a hallmark of AD (Ganguly et al., 2024). Since high phenotypic variability is currently seen to be one of the biggest obstacles to clinical trial design and early diagnosis, there is a lot of interest in figuring out what causes variability so that it can be utilized for patient stratification. The sex and gender of the individual are becoming important determinants of phenotypic variability, in addition to lifestyle and genetic factors (Bhattacharya et al., 2024). There is evidence of gender and sex differences in the rates of brain shrinkage and cognitive decline, as well as in the impact of risk factors and diagnostic biomarker patterns. Clinical practice and trials need to pay attention to such evidence, which may be highly relevant (Ferretti et al., 2020). The central fallacy that has dominated the traditional understanding of PD is that it is a neurodegenerative condition that primarily affects the substantia nigra's dopamine-producing neurons, leading to motor symptoms. We now know that PD is a multisystem disorder that is diverse and includes a broad spectrum of both motor and non-motor symptoms. The frequency of these non-motor symptoms increases as the disease progresses (Adhikary et al., 2024; Mukherjee et al., 2024).

One of the most prevalent non-motor signs is cognitive impairment. Cognitive impairment, especially dementia, is one of the most severe and incapacitating non-motor symptoms of PD and has been statistically linked to death (Adhikary et al., 2024; Mukherjee et al., 2024). Furthermore, cognitive impairment increases the stress on caregivers, hinders social function, and raises the expense of disease-related medical treatment (Adhikary et al., 2025). The complicated etiology of cognitive impairment in PD probably entails the gradual breakdown of several different brain networks. People with PD differ significantly in the rate, character, and timing of cognitive impairment. Therefore, it is critical to have a deeper understanding of the clinical characteristics of cognitive impairment in PD as well as the risk factors for these cognitive alterations. Clinically, this knowledge is essential for managing patients and communicating prognoses (Gonzalez-Latapi et al., 2021; Adhikary et al., 2024). The clinical profile linked to cognitive impairment in PD, including dementia (PDD) and moderate cognitive impairment (MCI), has been highlighted in this chapter. The most recent data on risk and protective variables linked to cognitive alterations in PDD will also be reviewed. Given the morbidity related to cognitive decline, understanding these characteristics may help identify preventive and modifying measures that might have a significant impact. Lastly, we point out the gaps in the literature and potential avenues for further study.

4.8 THE VICIOUS CYCLE: MALNUTRITION AND COGNITIVE DYSFUNCTION

The link between malnutrition and cognitive impairment creates a harmful cycle where each condition worsens the other (Adhikary et al., 2024; Mitra et al., 2025). This makes recovery more challenging. Malnutrition stops the brain from getting the nutrients required for growth, maintenance, and function. Vitamins, minerals, and macronutrients such as iron, zinc, B vitamins, omega-3 fatty acids, and proteins are all essential for neurotransmitter production, myelination, and brain integrity (Adhikary et al., 2025; Adhikary et al., 2025). Lacking these nutrients, the population, especially children and older people, experiences poor memory, attention, and problem-solving capabilities (Adhikary et al., 2024; Bandyopadhyay et al., 2025; Mohanty et al., 2025). Over time, persistent deficiencies of these nutrients result in impaired cognitive ability, slow learning, and decreased productivity.

On the other hand, cognitive impairment may worsen malnutrition. Those with memory loss, dementia, or impaired executive function may forget to eat, struggle to prepare well-nourishing meals, or overlook their nutritional needs (Yuan and Larsson, 2023). Children with learning disabilities tied to poor nutrition may have their understanding of healthy eating delayed, and such goes the cycle. This two-way interaction results in malnutrition leading to cognitive impairment and worsens malnutrition (Dibello et al., 2024).

The impact extends beyond the individual to families, caregivers, and communities, as individuals with cognitive impairment might depend on others for simple nutritional care (Tamura et al., 2020). Arresting the cycle requires specific actions that address nutritional gaps and support cognitive function. Providing fortified foods, ensuring access to balanced diets, teaching caregivers, and offering mentally stimulating activities can help reduce the impact of malnutrition on brain function (Green and Miller, 2022). Early detection is crucial. Interventions during essential stages of brain development, such as infancy, early childhood, and older age, can prevent lasting harm (Roberts et al., 2020). Ultimately, seeing malnutrition and cognitive dysfunction as related rather than separate problems highlights the need for a combined approach to healthcare and public policy. Improving nutrition and brain function can break this harmful cycle and promote healthier, more productive lives at every stage (Ward et al., 2023).

4.9 PUBLIC HEALTH STRATEGIES AND PREVENTIVE APPROACHES

Public health interventions and prevention interventions in dealing with nutrition-related cognitive impairment must be planned with a vision for wholeness, considering early detection, nutritional interventions, and policy-level interventions in older people (Barnes et al., 2023; Feinstein et al., 2023). Screening and early detection are the cornerstone of prevention because intervention is possible before irreversible damage is done by identifying high-risk individuals at the right time. Regular cognitive assessment, together with nutritional screening for deficiency in essential micronutrients such as vitamin B12, folate, vitamin D, iron, and omega-3 fatty acids, can help clinicians identify susceptibility among older people (Godos et al., 2024; Baker et al., 2024). Inclusion of such assessment in routine healthcare consultations makes clinicians less likely to overlook subtle memory, concentration, or diet alterations. Outside the clinical environment, community-based programs like health camps, mobile screening vehicles, and educational programs can reach underserved groups, allowing early detection of nutritional deficiency and cognitive impairment in their initial stages (Uchida et al., 2024). Once risk factors are detected, dietary and supplementation interventions are key resources to prevent further progression. A balanced diet of whole grains, fruits, vegetables, lean meats, nuts, and seeds forms the basis for maintaining brain health. Some eating patterns, such as the DASH or Mediterranean diets, focus on antioxidants, healthy fats, and anti-inflammatory foods (Baker et al., 2025). These diets have shown protective effects against age-related cognitive decline. Supplementation can help, especially when deficiencies or dietary intake is lacking. For example, vitamin D supplements are helpful for the bones and the brain. At the same time, B vitamin complexes help keep homocysteine levels in check, which is linked with vascular risks for cognitive impairment. Omega-3 fatty acids from fish or supplements help maintain neuronal integrity and synaptic function. It is worth mentioning that interventions should be individualized, taking into account culturally acceptable food items, food availability, and disease, and with education for older people and carers to ensure compliance (Huang et al., 2023).

National policy guidelines for older people are required to further such initiatives (Deckers et al., 2024). Governments and public health authorities must prioritize nutritional status as part of national strategies for healthy aging. Public policies can include subsidizing healthy foods, mandating food fortification programs, and making recommendations regarding long-term care facility geriatric nutrition (Kappe et al., 2024). Public health campaigns also play a vital role in increasing awareness of the connection between food and cognitive performance, allowing older individuals to make informed dietary decisions. In addition, providing nutrition and mental health education as part of healthcare provider training ensures that preventive care is delivered effectively at any level. At the community level, delivering supportive environments such as seniors' nutrition centers, meal delivery services, and caregiver networks can reduce barriers to accessing proper nutrition for target populations. By integrating these strategies, early detection, targeted dietetic interventions, and supportive policy action, public health systems can address both the problem of malnutrition and cognitive impairment among older persons. In addition to reducing the load on healthcare systems, this integrated approach enhances quality of life by allowing older persons to live independently, preserve dignity, and remain productive for as long as possible (Brodaty et al., 2025).

4.10 CONCLUSION

Evidence evaluated in this chapter highlights the critical interplay between vitamin status, age-related changes in nutrient absorbability, and cognitive health trajectory in older people. With a rising human population increasingly trending toward old age, deficiencies of major vitamins such as B12, folate, vitamin D, and antioxidant nutrients increasingly assume modifiable risk factors for cognitive decline and risk of neurodegenerative disease. Mechanisms of such correlation are multifaceted, involving suboptimum absorbability for physiological reasons, impacts of chronic disease and polypharmacology, and acceleration of socioeconomic and lifestyle determinants. Deficiencies aside from rendering older people vulnerable to frailty and functional decline sustain a vicious cycle of malnutrition, aggravating cognitive dysfunction, and further reducing nutritional adequacy. From the disease epidemiology of populations, massive gaps persist between regions, males and females, and socioeconomic groups, with a clear rationale for differential context-specific prevention intervention. Advances in diagnostic indices of biomarkers raise new hopes of early recognition of deficiencies, but uneven conversion of such knowledge to better clinical and public health tactics remains persistent. Preventive maneuvers from routine screening and focused supplementation to population-level fortification of foods and dietary education offer real promise of cognitive disorder burden reduction. Future investigations need to emphasize longitudinal studies of better biomarker measurement, exploration of gene–nutrient interactions, and cost efficacy of combined nutrition cognition intervention. Preventive focus on vitamin shortcomings among the elderly population is not just a biomedical priority but also a disease epidemiology priority, with hopes of maintaining independence, reducing healthcare costs, and improving quality of life for millions worldwide.

AUTHOR CONTRIBUTIONS

Riya Sarkar, Rajat Kanti Ghoshal, and Supriya Kumar Moulick were involved in the conceptualization, writing of the original draft, mechanistic insight, input on critical points, and editing; Md Abubakar and Abhishek Ambawatiya contributed to the manuscript writing and editing; Nitesh Kumar contributed to the manuscript editing and critical points; Krishnendu Adhikar was involved in the conceptualization, manuscript writing, reviewing, and conceptualization of figures and approved the version to be published.

FUNDING

None.

AVAILABILITY OF DATA

None.

CONFLICTS OF INTEREST

None.

ACKNOWLEDGEMENTS

We thankfully acknowledge the infrastructure and fellowship support to Md. Abubakar extended by the Department of Pharmaceuticals, Ministry of Chemicals and Fertilizers, Government of India, and NIPER HAJIPUR and Dr. Nitesh Kumar for the support and guidance in executing this work plan. The images were created using www.biorander.com and www.Grammarly.com to correct the grammar of this manuscript.

REFERENCES

Adhikary, K., Banerjee, A., Sarkar, R., Banerjee, R., Chowdhury, S. R., Ganguly, K., & Karak, P. (2025). HIV-associated neurocognitive disorders (HAND): Optimal diagnosis, antiviral therapy, pharmacological treatment, management, and future scopes. *Journal of the Neurological Sciences*, *470*, 123410. https://doi.org/10.1016/j.jns.2025.123410.

Adhikary, K., Banerjee, P., Barman, S., Bandyopadhyay, B., & Bagchi, D. (2024). Nutritional aspects, chemistry profile, extraction techniques of lemongrass essential oil and it's physiological benefits. *Journal of the American Nutrition Association*, *43*(2), 183–200. https://doi.org/10.1080/27697061.2023.2245435.

Adhikary, K., Kumari, S., Chatterjee, P., Dey, R., Maiti, R., Chakrabortty, S., Ahuja, D., & Karak, P. (2024). Unveiling bisphenol A toxicity: Human health impacts and sustainable treatment strategies. *Hormone Molecular Biology and Clinical Investigation*, *45*(4), 171–185. https://doi.org/10.1515/hmbci-2024-0034.

Adhikary, K., Mohanty, S., Bandyopadhyay, B., Maiti, R., Bhattacharya, K., & Karak, P. (2024). β-Amyloid peptide modulates peripheral immune responses and neuroinflammation in rats. *Biomolecular Concepts*, *15*(1). https://doi.org/10.1515/bmc-2022-0042.

Adhikary, K., Paul, A., Madan, A., Islam, A., Ashique, S., & Ramzan, M. (2025). Personalized precision: Revolutionizing cancer treatment with mRNA-based vaccines in melanoma therapy. *Advances in Immunology*, *166*, 137–167. https://doi.org/10.1016/bs.ai.2024.10.011.

Adhikary, K., Sarkar, R., Maity, S., Banerjee, I., Chatterjee, P., Bhattacharya, K., Ahuja, D., Sinha, N. K., & Maiti, R. (2024). The underlying causes, treatment options of gut microbiota and food habits in type 2 diabetes mellitus: A narrative review. *Journal of Basic and Clinical Physiology and Pharmacology*, *35*(3), 153–168. https://doi.org/10.1515/jbcpp-2024-0043.

Adhikary, K., Sarkar, R., Maity, S., Sadhukhan, I., Sarkar, R., Ganguly, K., Barman, S., Maiti, R., Chakraborty, S., Chakraborty, T. R., Bagchi, D., & Banerjee, P. (2024). Immunomodulation of macrophages in diabetic wound individuals by structurally diverse bioactive phytochemicals. *Pharmaceuticals*, *17*(10), 1294. https://doi.org/10.3390/ph17101294.

Bailey, R. L., West, K. P., & Black, R. E. (2015). The epidemiology of global micronutrient deficiencies. *Annals of Nutrition and Metabolism*, *66*(Suppl. 2), 22–33. https://doi.org/10.1159/000371618

Baker, L. D., Espeland, M. A., Whitmer, R. A., Snyder, H. M., Leng, X., Lovato, L., Papp, K. V., Yu, M., Kivipelto, M., Alexander, A. S., Antkowiak, S., Cleveland, M., Day, C., Elbein, R., Tomaszewski Farias, S., Felton, D., Garcia, K. R., Gitelman, D. R., Graef, S., Howard, M., … Carrillo, M. C. (2025). Structured vs self-guided multidomain lifestyle interventions for global cognitive function: The US pointer randomized clinical trial. *JAMA*, *334*(8), 681–691. https://doi.org/10.1001/jama.2025.12923.

Baker, L. D., Snyder, H. M., Espeland, M. A., Whitmer, R. A., Kivipelto, M., Woolard, N., Katula, J., Papp, K. V., Ventrelle, J., Graef, S., Hill, M. A., Rushing, S., Spell, J., Lovato, L., Felton, D., Williams, B. J., Ghadimi Nouran, M., Raman, R.,

Ngandu, T., Solomon, A., … U.S. POINTER Study Group (2024). Study design and methods: U.S. study to protect brain health through lifestyle intervention to reduce risk (U.S. POINTER). *Alzheimer's & Dementia: The Journal of the Alzheimer's Association*, *20*(2), 769–782. https://doi.org/10.1002/alz.13365.

Bandyopadhyay, U., Sen, D., Ahuja, D., Mahapatra, S. P., Biswas, D., Maiti, R., Chakraborty, S., Hazra, A., Parua, S., Basak, A. K., Das, A., Paul, N., Purkait, M. P., Syamal, A. K., Dey, R., Bhattacharya, K., Adhikary, K., & Bhattacharjee, A. (2025). Interplay of calcium, vitamin D, and parathormone in the milieu of infections and immunity: Reassessed in the context of COVID-19. *The Journal of Steroid Biochemistry and Molecular Biology*, *245*, 106624. https://doi.org/10.1016/j.jsbmb.2024.106624.

Barnes, L. L., Dhana, K., Liu, X., Carey, V. J., Ventrelle, J., Johnson, K., Hollings, C. S., Bishop, L., Laranjo, N., Stubbs, B. J., Reilly, X., Agarwal, P., Zhang, S., Grodstein, F., Tangney, C. C., Holland, T. M., Aggarwal, N. T., Arfanakis, K., Morris, M. C., & Sacks, F. M. (2023). Trial of the MIND diet for prevention of cognitive decline in older persons. *The New England Journal of Medicine*, *389*(7), 602–611. https://doi.org/10.1056/NEJMoa2302368.

Bhattacharya, K., Dey, R., Sen, D., Paul, N., Basak, A. K., Purkait, M. P., Shukla, N., Chaudhuri, G. R., Bhattacharya, A., Maiti, R., Adhikary, K., Chatterjee, P., Karak, P., & Syamal, A. K. (2024). Polycystic ovary syndrome and its management: In view of oxidative stress. *Biomolecular Concepts*, *15*(1). https://doi.org/10.1515/bmc-2022-0038.

Bhatti, G. K., Reddy, A. P., Reddy, P. H., & Bhatti, J. S. (2020). Lifestyle modifications and nutritional interventions in aging-associated cognitive decline and Alzheimer's disease. *Frontiers in Aging Neuroscience*, *11*, 369.

Brodaty, H., Chau, T., Heffernan, M., Ginige, J. A., Andrews, G., Millard, M., Sachdev, P. S., Anstey, K. J., Lautenschlager, N. T., McNeil, J. J., Jorm, L., Kochan, N. A., Maeder, A., Welberry, H., San Jose, J. C., Briggs, N. E., Popovic, G., Mavros, Y., Almendrales Rangel, C., Noble, Y., … Valenzuela, M. (2025). An online multidomain lifestyle intervention to prevent cognitive decline in at-risk older adults: A randomized controlled trial. *Nature Medicine*, *31*(2), 565–573. https://doi.org/10.1038/s41591-024-03351-6.

Clarke, R., Birks, J. S., Nexo, E., Ueland, P. M., Schneede, J., Scott, J., Molloy, A. M., & Evans, J. G. (2007). Low vitamin B-12 status and risk of cognitive decline in older adults. *American Journal of Clinical Nutrition*, *86*(5), 1384–1391. https://doi.org/10.1093/ajcn/86.5.1384

Corish, C.A., & Bardon, L.A. (2019). Malnutrition in older adults: Screening and determinants. *Proceedings of the Nutrition Society*, *78*, 372–379. https://doi.org/10.1017/S0029665118002628.

Cui, A., Zhang, T., Xiao, P., Fan, Z., Wang, H., & Zhuang, Y. (2023). Global and regional prevalence of vitamin D deficiency in population-based studies from 2000 to 2022: A pooled analysis of 7.9 million participants. *Frontiers in Nutrition*, *10*, 1070808.

DeFilippis, A., Ullah, H., Baldi, A., Dacrema, M., Esposito, C., Garzarella, E. U., Santarcangelo, C., Tantipongpiradet, A., & Daglia, M. (2020). Gastrointestinal disorders and metabolic syndrome: Dysbiosis as a key link and common bioactive dietary components useful for their treatment. *International Journal of Molecular Sciences*, *21*(14), 4929. https://doi.org/10.3390/ijms21144929.

Deckers, K., Zwan, M. D., Soons, L. M., Waterink, L., Beers, S., van Houdt, S., Stiensma, B., Kwant, J. Z., Wimmers, S. C. P. M., Heutz, R. A. M., Claassen, J. A. H. R., Oosterman, J. M., de Heus, R. A. A., van de Rest, O., Vermeiren, Y., Voshaar, R. C. O., Smidt, N., Broersen, L. M., Sikkes, S. A. M., Aarts, E., … van der Flier, W. M. (2024). A multidomain lifestyle intervention to maintain optimal cognitive functioning in Dutch older adults-study design and baseline characteristics of the FINGER-NL randomized controlled trial. *Alzheimer's Research & Therapy*, *16*(1), 126. https://doi.org/10.1186/s13195-024-01495-8.

Dibello, V., Custodero, C., Cavalcanti, R., Lafornara, D., Dibello, A., Lozupone, M., Daniele, A., Pilotto, A., Panza, F., & Solfrizzi, V. (2024). Impact of periodontal disease on cognitive disorders, dementia, and depression: A systematic review and meta-analysis. *GeroScience*, *46*(5), 5133–5169. https://doi.org/10.1007/s11357-024-01243-8.

Dominguez, L. J., Veronese, N., Vernuccio, L., Catanese, G., Inzerillo, F., Salemi, G., & Barbagallo, M. (2021). Nutrition, physical activity, and other lifestyle factors in the prevention of cognitive decline and dementia. *Nutrients*, *13*(11), 4080.

Dror, Y., Stern, F., & Gomori, M. J. (2014). Vitamins in the prevention or delay of cognitive disability of aging. *Current Aging Science*, *7*(3), 187–213.

Feinstein, A., Amato, M. P., Brichetto, G., Chataway, J., Chiaravalloti, N. D., Cutter, G., Dalgas, U., DeLuca, J., Farrell, R., Feys, P., Filippi, M., Freeman, J., Inglese, M., Meza, C., Motl, R. W., Rocca, M. A., Sandroff, B. M., Salter, A., & CogEx Research Team (2023). Cognitive rehabilitation and aerobic exercise for cognitive impairment in people with progressive multiple sclerosis (CogEx): A randomised, blinded, sham-controlled trial. *The Lancet, Neurology*, *22*(10), 912–924. https://doi.org/10.1016/S1474-4422(23)00280-6.

Fekete, M., Lehoczki, A., Tarantini, S., Fazekas-Pongor, V., Csípő, T., Csizmadia, Z., & Varga, J. T. (2023). Improving cognitive function with nutritional supplements in aging: A comprehensive narrative review of clinical studies investigating the effects of vitamins, minerals, antioxidants, and other dietary supplements. *Nutrients*, *15*(24), 5116. https://doi.org/10.3390/nu15245116

Ferretti, M. T., Martinkova, J., Biskup, E., Benke, T., Gialdini, G., Nedelska, Z., … & Schmidt, R. (2020). Sex and gender differences in Alzheimer's disease: Current challenges and implications for clinical practice: Position paper of the Dementia and Cognitive Disorders Panel of the European Academy of Neurology. *European Journal of Neurology*, *27*(6), 928–943.

Fujiwara, Y., & Arakawa, T. (2014). Overlap in patients with dyspepsia/functional dyspepsia. *Journal of Neurogastroenterology and Motility*, *20*(4), 447–457. https://doi.org/10.5056/jnm14080.

Ganguly, K., Adhikary, K., Acharjee, A., Acharjee, P., Trigun, S. K., Mutlaq, A. S., Ashique, S., Yasmin, S., Alshahrani, A. M., & Ansari, M. Y. (2024). Biological significance and pathophysiological role of matrix metalloproteinases in the central nervous system. *International Journal of Biological Macromolecules*, *280*(Pt 3), 135967. https://doi.org/10.1016/j.ijbiomac.2024.135967.

Gebre, A. K., Altaye, B. M., Atey, T. M., Tuem, K. B., & Berhe, D. F. (2018). Targeting renin–angiotensin system against Alzheimer's disease. *Frontiers in Pharmacology*, *9*, 440.

Godos, J., Micek, A., Currenti, W., Franchi, C., Poli, A., Battino, M., Dolci, A., Ricci, C., Ungvari, Z., & Grosso, G. (2024). Fish consumption, cognitive impairment and dementia: An updated dose-response meta-analysis of observational studies. *Aging Clinical and Experimental Research*, *36*(1), 171. https://doi.org/10.1007/s40520-024-02823-6.

Gonzalez-Latapi, P., Bayram, E., Litvan, I., & Marras, C. (2021). Cognitive impairment in Parkinson's disease: Epidemiology, clinical profile, protective and risk factors. *Behavioral Sciences*, *11*(5), 74.

Gordon, S., Hoey, L., McNulty, H., Keenan, J., Pangilinan, F., Brody, L. C., … & Hughes, C. F. (2025). Associations of one-carbon metabolism, related B-vitamins and ApoE genotype with cognitive function in older adults: Identification of a novel gene-nutrient interaction. *BMC Medicine*, *23*(1), 440.

Green, R., & Miller, J. W. (2022). Vitamin B12 deficiency. *Vitamins and Hormones*, *119*, 405–439. https://doi.org/10.1016/bs.vh.2022.02.003.

Habas, C., Modglin, L., & SingleCare. (2025, July 25). *Multiple Sclerosis Statistics 2025*. The Checkup. https://www.singlecare.com/blog/news/multiple-sclerosis-statistics.

Haimi, M., & Lerner, A. (2014). Nutritional deficiencies in the pediatric age group in a multicultural developed country, Israel. *World Journal of Clinical Cases*, *2*(5), 120.

Huang, L., Tao, Y., Chen, H., Chen, X., Shen, J., Zhao, C., Xu, X., He, M., Zhu, D., Zhang, R., Yang, M., Zheng, Y., & Yuan, C. (2023). Mediterranean-dietary approaches to stop hypertension intervention for neurodegenerative delay (MIND) diet and cognitive function and its decline: A prospective study and meta-analysis of cohort studies. *The American Journal of Clinical Nutrition*, *118*(1), 174–182. https://doi.org/10.1016/j.ajcnut.2023.04.025.

Jiang, W., Li, X., Wang, R., Du, Y., & Zhou, W. (2024). Cross-country health inequalities of four common nutritional deficiencies among children, 1990 to 2019: Data from the Global Burden of Disease Study 2019. *BMC Public Health*, *24*(1), 486.

Johnson, L. E. (2024, August 8). *Overview of Vitamins*. Merck Manual Consumer Version. https://www.merckmanuals.com/home/disorders-of-nutrition/vitamins/overview-of-vitamins.

Kappe, J., Wittmann, F., Luppa, M., Cardona, M. I., Weise, S., Fuchs, S., Kosilek, R. P., Sanftenberg, L., Brettschneider, C., Döhring, J., Escales, C., Czock, D., Wiese, B., Thyrian, J. R., Hoffmann, W., Frese, T., Gensichen, J., König, H. H., Kaduszkiewicz, H., & Riedel-Heller, S. G. (2024). General practitioners' perspectives on lifestyle interventions for cognitive preservation in dementia prevention. *BMC Primary Care*, *25*(1), 301. https://doi.org/10.1186/s12875-024-02566-3.

Lauer, A. A., Grimm, H. S., Apel, B., Golobrodska, N., Kruse, L., Ratanski, E., … & Grimm, M. O. W. (2022). Mechanistic link between vitamin B12 and Alzheimer's disease. *Biomolecules*, *12*(1), 129.

Littlejohns, T. J., Henley, W. E., Lang, I. A., Annweiler, C., Beauchet, O., Chaves, P. H., … & Llewellyn, D. J. (2014). Vitamin D and the risk of dementia and Alzheimer disease. *Neurology*, *83*(10), 920–928.

Loera-Valencia, R., Eroli, F., Garcia-Ptacek, S., & Maioli, S. (2021). Brain renin–angiotensin system as novel and potential therapeutic target for Alzheimer's disease. *International Journal of Molecular Sciences*, *22*(18), 10139.

Masnoon, N., Shakib, S., Kalisch-Ellett, L., & Caughey, G. E. (2017). What is polypharmacy? A systematic review of definitions. *BMC Geriatrics*, *17*(1). https://doi.org/10.1186/s12877-017-0621-2.

Miguel-Álvarez, M., Santos-Lozano, A., Sanchis-Gomar, F., Fiuza-Luces, C., Pareja-Galeano, H., Garatachea, N., & Lucia, A. (2015). Non-steroidal anti-inflammatory drugs as a treatment for Alzheimer's disease: A systematic review and meta-analysis of treatment effect. *Drugs & Aging*, *32*(2), 139–147.

Miller, J. W., Harvey, D. J., Beckett, L. A., Green, R., Farias, S. T., Reed, B. R., … & DeCarli, C. (2015). Vitamin D status and rates of cognitive decline in a multiethnic cohort of older adults. *JAMA Neurology*, *72*(11), 1295–1303.

Mitra, A., Shahid, A., Kumari, S., Mukherjee, T., Pramanick, S., Mohanty, S., Ansari, M. A., Adhikary, K., Prabhakar, P. K., & Kesari, K. K. (2025). Optimizing wound healing: Insights from phytochemicals and advanced therapies. *Inflammopharmacology*, *33*(7), 4009–4035. https://doi.org/10.1007/s10787-025-01806-x.

Mohanty, S., Khanna, M. K., Adhikary, K., Mukherjee, T., Sahu, A., Kahwa, I., Prabhakar, P. K., Swain, M. P., Sahoo, A., & Parida, M. K. (2025). Combating TB pathology: Advanced diagnostics, innovative therapies, and public health strategies: A review. *Current Pharmaceutical Biotechnology*. https://doi.org/10.2174/011389201035543025033 0074301.

Mukherjee, T., Das, T., Basak, S., Mohanty, S., Adhikary, K., Chatterjee, P., Maiti, R., & Karak, P. (2024). Mucormycosis during COVID-19 era: A retrospective assessment. *Infectious Medicine*, *3*(2), 100112. https://doi.org/10.1016/j.imj.2024.100112.

Muthayya, S., Rah, J. H., & Sugimoto, J. D. (2021). Micronutrient malnutrition across the life course: global prevalence and consequences. *Proceedings of the Nutrition Society*, *80*(3), 279–282. https://doi.org/10.1017/S0029665121001968

NationalInstitute on Drug Abuse. (2023, October 9). *NACDA Guidelines for Administration of Drugs to Human Subjects*. National Institute on Drug Abuse. https://nida.nih.gov/research/clinical-research/nacda-guidelines-administration-drugs-to-human-subjects.

Norman, K., Haß, U., & Pirlich, M. (2021). Malnutrition in older adults—recent advances and remaining challenges. *Nutrients*, *13*(8), 2764.

O'Leary, F., Allman-Farinelli, M., & Samman, S. (2012). Vitamin B12 status, cognitive decline and dementia: A systematic review of prospective cohort studies. *British Journal of Nutrition*, *108*(11), 1948–1961.

Pereira, A., Veríssimo, M., & Ribeiro, O. (2024). Influence of chronic medical conditions on older patients' willingness to deprescribe medications: A cross-sectional study. *BMC Geriatrics*, *24*(1). https://doi.org/10.1186/s12877-024-04891-9.

Porter, J., Smith, A. D., & Refsum, H. (2016). Causes, consequences and public health implications of low B-vitamin status in ageing. *Proceedings of the Nutrition Society*, *75*(3), 227–235. https://doi.org/10.1017/S0029665114001554

Powell, J. (2024, November 7). *Fiber: The Nutrition Source*. Harvard Chan School. https://nutritionsource.hsph.harvard.edu/carbohydrates/fiber/.

Raman, R. (2017, September 5). *How Your Nutritional Needs Change as You Age*. Healthline. https://www.healthline.com/nutrition/nutritional-needs-and-aging#How-Does-Aging-Affect-Your-Nutritional-Needs.

Roberts, S. B., Franceschini, M. A., Silver, R. E., Taylor, S. F., de Sa, A. B., Có, R., Sonco, A., Krauss, A., Taetzsch, A., Webb, P., Das, S. K., Chen, C. Y., Rogers, B. L., Saltzman, E., Lin, P. Y., Schlossman, N., Pruzensky, W., Balé, C., Chui, K. K. H., & Muentener, P. (2020). Effects of food supplementation on cognitive function, cerebral blood flow, and nutritional status in young children at risk of undernutrition: Randomized controlled trial. *BMJ (Clinical Research Edition), 370,* m2397. https://doi.org/10.1136/bmj.m2397.

Royea, J., & Hamel, E. (2020). Brain angiotensin II and angiotensin IV receptors as potential Alzheimer's disease therapeutic targets. *Geroscience, 42*(5), 1237–1256.

Shalini, T., Pullakhandam, R., Ghosh, S., Kulkarni, B., Rajkumar, H., Sachdev, H. S., … & Reddy, G. B. (2023). Prevalence of vitamin B12 and folate deficiencies in Indian children and adolescents. *Nutrients, 15*(13), 3026.

Stevens, G. A., Beal, T., Mbuya, M. N., Luo, H., Neufeld, L. M., Addo, O. Y., … & Young, M. F. (2022). Micronutrient deficiencies among preschool-aged children and women of reproductive age worldwide: A pooled analysis of individual-level data from population-representative surveys. *The Lancet Global Health, 10*(11), e1590–e1599.

Tamura, Y., Omura, T., Toyoshima, K., & Araki, A. (2020). Nutrition management in older adults with diabetes: A review on the importance of shifting prevention strategies from metabolic syndrome to frailty. *Nutrients, 12*(11), 3367. https://doi.org/10.3390/nu12113367.

The Lowdown on Low Stomach Acid. (n.d.). ION.ac.uk. https://ion.ac.uk/articles/low-stomach-acid.

Travica, N., Ried, K., Sali, A., Scholey, A., Hudson, I., & Pipingas, A. (2017). Vitamin C status and cognitive function: A systematic review. *Nutrients, 9*(9), 960. https://doi.org/10.3390/nu9090960

Uchida, K., Meno, K., Korenaga, T., Liu, S., Suzuki, H., Baba, Y., Tagata, C., Araki, Y., Tsunemi, S., Aso, K., Inagaki, S., Nakagawa, S., Kobayashi, M., Kakuma, T., Asada, T., Ota, M., Takihara, T., & Arai, T. (2024). Effect of matcha green tea on cognitive functions and sleep quality in older adults with cognitive decline: A randomized controlled study over 12 months. *PloS ONE, 19*(8), e0309287. https://doi.org/10.1371/journal.pone.0309287.

Udhani, R., Kothari, C., & Sarvaiya, J. (2023). A comprehensive study: Traditional and cutting-edge analytical techniques for the biomarker based detection of the micronutrients & POC sensing directions for next-generation diagnostic. *Critical Reviews in Analytical Chemistry, 54*(7), 2378–2397. https://doi.org/10.1080/10408347.2023.2169823.

Umair. (2025, April 29). *Best Vitamins for Energy: Boost Your Stamina Naturally.* Tech.Mex. https://tech.mex.com/muneeb/what-vitamins-is-best-for-energy.

Voulgaridou, G., Papadopoulou, S. K., Detopoulou, P., Tsoumana, D., Giaginis, C., Kondyli, F. S., Lymperaki, E., & Pritsa, A. (2023). Vitamin D and calcium in osteoporosis, and the role of bone turnover markers: A narrative review of recent data from RCTs. *Diseases, 11*(1), 29. https://doi.org/10.3390/diseases11010029.

Ward, N. A., Reid-McCann, R., Brennan, L., Cardwell, C. R., de Groot, C., Maggi, S., McCaffrey, N., McGuinness, B., McKinley, M. C., Noale, M., O'Neill, R. F., Prinelli, F., Sergi, G., Trevisan, C., Volkert, D., Woodside, J. V., & McEvoy, C. T. (2023). Effects of protein enriched mediterranean diet and exercise on nutritional status and cognition in adults at risk of undernutrition and cognitive decline: The PROMED-EX randomised controlled trial. *BMJ Open, 13*(10), e070689. https://doi.org/10.1136/bmjopen-2022-070689.

Warren, S. A., & Warren, K. G. (1982). Multiple sclerosis and diabetes mellitus: Further evidence of a relationship. *Canadian Journal of Neurological Sciences, 9*(4), 415–418. https://doi.org/10.1017/S0317167100022123

Weggen, S., Eriksen, J. L., Das, P., Sagi, S. A., Wang, R., Pietrzik, C. U., … & Koo, E. H. (2001). A subset of NSAIDs lower amyloidogenic Aβ42 independently of cyclooxygenase activity. *Nature, 414*(6860), 212–216.

Woodling, N. S., Colas, D., Wang, Q., Minhas, P., Panchal, M., Liang, X., … & Andreasson, K. I. (2016). Cyclooxygenase inhibition targets neurons to prevent early behavioural decline in Alzheimer's disease model mice. *Brain*, 139(7), 2063–2081.

Youness, R. A., Dawoud, A., ElTahtawy, O., & Farag, M. A. (2022). Fat-soluble vitamins: Updated review of their role and orchestration in human nutrition throughout life cycle with sex differences. *Nutrition & Metabolism, 19*(1). https://doi.org/10.1186/s12986-022-00696-y.

Yuan, S., & Larsson, S. C. (2023). Epidemiology of sarcopenia: Prevalence, risk factors, and consequences. *Metabolism: Clinical and Experimental, 144,* 155533. https://doi.org/10.1016/j.metabol.2023.155533.

Zhao, Y., Zheng, S., Wang, B., Xiao, W., He, P., & Bian, Y. (2025). Vitamin status and risk of age-related diseases among adult residents of the Pearl River delta region. *Nutrients, 17*(10), 1637.

5 Antioxidant Vitamins (A, C, and E) in Neuroprotection

Neha Prajapati[1] *and Deepanshi*[*] *Jaiswal*[1]

[1]Department of Botany, Dr. Harisingh Gour Vishwavidyalaya, Sagar, India

5.1 INTRODUCTION

Sies used the phrase "oxidative stress" to describe an imbalance between the body's antioxidant defenses and reactive oxygen species (ROS). ROS comprise both non-radical molecules like hydrogen peroxide and oxygen-centered free radicals like superoxide anion and hydroxyl radical. Since oxygen is necessary for aerobic creatures to survive, this leads to the "oxygen paradox": while oxygen is necessary for life, it also produces toxic ROS. Important biomolecules including proteins, lipid membranes, and DNA can be harmed by these reactive species [1,2]. The brain's large lipid content, rapid oxygen consumption, and weak antioxidant defenses make it extremely vulnerable to oxidative injury [3]. The brain is particularly vulnerable to oxidative damage because of its extensive membrane surface, longer axons, and non-dividing neuronal cells. ROS have the ability to damage mitochondria, cause lipid peroxidation, break down the blood–brain barrier (BBB), and encourage an excessive release of neurotoxic amino acids into the brain [4]. Glutamate, the principal excitatory amino acid in the brain, primarily operates through NMDA receptors. The link between ROS and excitatory amino acid release seems to be reciprocal, with each enhancing the effects of the other [5]. Neurodegenerative illnesses, including multiple sclerosis (MS), amyotrophic lateral sclerosis (ALS), Parkinson's disease (PD), and Alzheimer's disease (AD), are characterized by a progressive loss of neurons. About 25 million people worldwide are presently afflicted by these disorders, whose incidence is continuously rising, particularly those linked to dementia. It is anticipated that by 2050, this number will have surpassed 115 million due to growing life expectancy [6,7]. Antioxidants are substances that catalyze reduction processes or donate electrons to neutralize reactive species. They fall into one of two categories: exogenous (obtained via diet or supplements) or endogenous (made in the body). Enzymes such as catalase, superoxide dismutase (SOD), and GPx, together with substances like glutathione and melatonin, are important endogenous antioxidants. Vitamins C and E, carotenoids (which are precursors to vitamin A), polyphenols (which are effective against lipid oxidation), sulforaphanes (found in cruciferous foods), and anthocyanins (the pigments found in colored fruits and vegetables) are examples of exogenous antioxidants [8–10]. Certain microbes and plants create fat-soluble substances called carotenoids, which include beta-carotene, retinol, and vitamin A. They are vital to human growth, development, and general well-being [11]. There are two primary ways to get vitamin A through diet: preformed vitamin A (retinol, retinal, retinoic acid [RA], and retinyl esters) from meat and dairy products and pro-vitamin A carotenoids from plants (Table 5.1). Both forms of vitamin A are transformed into the active alcohol form, all-trans-retinol, in the intestine. By transferring energy, carotenoids help neutralize ROS, acting as antioxidants [12]. Research on adult rats and mice demonstrates that cognitive decline results from vitamin A deficiency, underscoring the critical role that retinoid signaling plays in preserving neurological health and normal brain function [13]. Carotenoids can be classified as either pro-vitamin A (like β-carotene and β-cryptoxanthin) or non-pro-vitamin A (like lutein and lycopene) depending on whether they convert to retinal. They interact with free radicals through electron transfer and hydrogen abstraction, trap peroxyl radicals, and quench singlet oxygen to produce antioxidant actions. Carotenoids have an impact on signaling pathways such as MAPK, NF-κB, and Nrf2. A retinol precursor, β-carotene, is more quickly used than α-tocopherol and neutralizes free radicals in lipid environments. When α-tocopherol and ascorbic acid are used together, their synergistic action against oxidative stress is the highest [14]. The fat-soluble vitamin A is essential for brain activity and is well-known for boosting immunity and eyesight. RA, in its active state, is necessary for gene regulation, neurotransmission, and plasticity of synaptic neurons and neural development. Learning, memories, and intelligence are supported by encouraging synaptic alterations. Furthermore, it provides neurological protection and could potentially help in slowing down neurological aging due to its anti-allergic and antioxidant qualities. In order for the cerebellum to form properly, RA is also essential during the embryonic phase, directing the growth and differentiation of axons and glial cells [15].

[*] Corresponding author

DOI: 10.1201/9781003686675-5

TABLE 5.1
Dietary Sources and Recommended Intakes of Antioxidant Vitamins (A, C, and E)

Vitamins	Major Dietary Sources	Recommended Dietary Allowance (RDA)—Adults (19+ years)	References
Vitamin A	Plant-based foods (green and leafy vegetables), animal-based products (such as flesh, liver, and egg yolks), colored fruits, milk, etc.	Men: 900 μg RAE/day Women: 700 μg RAE/day	[28–31]
Vitamin C	Star fruit, guava, black currant, kiwi, and strawberry, broccoli, kale, and peppers, citrus fruits, etc.	Men: 90 mg/day Women: 75 mg/day +35 mg/day for smokers	[32,33]
Vitamin E	Coconut, cottonseed, olive, palm, peanut, rapeseed/canola, soybean, and sunflower oils	Men and women: 15 mg α-tocopherol/day	[34,35]

Vitamin A functions as an antioxidant, shielding neurons from oxidative damage, which is connected to conditions including PD and AD. It may help decrease the progression of AD by lowering inflammation in the brain through immune response modulation. Recent studies revealed that vitamin A could improve cognitive function as well as lessen plaque accumulation in Alzheimer's sufferers. Therefore, increased beta-amyloid accumulation of plaque in Alzheimer's patients is linked to impaired retinal control [16]. In the case of PD, RA may preserve dopamine-producing neurons by regulating genes related to dopamine synthesis and neuron existence. Vitamin A's neuroprotective function is facilitated by its antioxidant activities and assistance in dopamine production [17]. Ascorbic acid, also known as vitamin C, is a vitamin which dissolves in water and is found in our bodies in two forms: ascorbate, which is the reduced form and is active at the physiological pH, and dehydroascorbic acid (DHA), which is the result of two electrons oxidizing ascorbate [18]. As an essential water-soluble antioxidant, vitamin C promotes brain health by supporting neural protection, neurotransmitter production, and oxidative damage regulation—all of which are critical for prevention as well as treatment of neurological disorders [19]. Vitamin C is a powerful antioxidant in the brain that combats ROS and free radicals, preventing oxidative damage to neurons and maybe delaying [20]. Furthermore, vitamin C supports the production of collagen fibers, which preserve the structure of the brain's blood vessels and connective tissue. Maintaining the integrity of the BBB is important because many neurological disorders cause it to break down [21]. Vitamin C assists in decreasing inflammation in the neurological system, which is a contributing cause to diseases including AD and MS. Greater amounts of vitamin C may improve consciousness and mental ability in AD by reducing neuronal inflammation and oxidative stress, according to studies [22]. The sodium-dependent transporter sodium-dependent vitamin C transporter 1 (SVCT1) facilitates the intestinal absorption of ascorbate, which then circulates in the blood. Through the choroid plexus, it gradually penetrates the brain and enters the cerebral spinal fluid (CSF). Through the GLUT1 transporter, blood-derived DHA enters the brain. Through the SVCT2 or GLUT1 transporters, neurons in the CSF absorb ascorbate or DHA [23]. The invention of vitamin E was first reported by American anatomist and endocrinologist Herbert M. Evans and his coworker Katherine S. Bishop [24]. An essential lipid-soluble antioxidant vitamin E, especially alpha-tocopherol, shields neuronal cells against damaging oxidants associated with neurological conditions by scavenging free radicals [25]. Vitamin E consumption may boost amyloid-β synthesis in Alzheimer's disorder, according to the latest research. This raises concerns about possible hazards and explains uneven results, emphasizing the requirement for cautious use [26,27]. This chapter focuses on the neural protective activities of curative vitamins C, A, and E, highlighting how they might lessen the oxidative injury and brain injury linked to neurological disorders such as MS, PD, and dementia. The BBB and nerve signaling are maintained by vitamin C, cellular membranes are shielded from peroxidation of lipids by vitamin E, and neural development is supported by vitamin A. Notwithstanding encouraging data, drawbacks such as poor absorption, erratic clinical results, and ambiguous associations with additional nourishment still exist. The purpose of this chapter is to discuss the curative value of such antioxidants in the prevention or treatment of neurological conditions, provide recent findings, and fill in the latest findings' gaps.

5.2 VITAMIN A AND NEUROPROTECTION

A lipid-soluble vital nutrient-rich antioxidant vitamin A (retinol) is derived either from plant-based products like carotene or from animal products as retinyl esters, which intestinal enzymes transform into retinol [36]. This is the exclusive forerunner of the RA, which adheres to nuclear receptors to control the regulation of genes [37]. There are various biologically active compounds known as retinoids, which are metabolic products of vitamin A. All-trans-RA serves as the main biologically active form, but alternative isomers, such as 9-cis-RA and 13-cis-RA, also have strong physiological effects [36]. The three subtypes of steroid-family nuclear receptors known as RA receptors (RARs)—RAR-α, RAR-β, and RAR-γ—are responsible for the transmission of RA [13]. Mainly, receptor subtypes vary in where they are found throughout the body, specifically in particular parts of the brain [38], and in order to adhere to the RA response elements (RAREs) at promoter regions of genes and control gene expression, RARs and retinoid X receptors (RXRs) form heterodimers (Figure 5.1). In order to modify the transcription of genes, RXRs, which come in forms including RXR-α, RXR-β, and RXR-γ variants, also collaborate with other nuclear receptors [39]. Dopaminergic signaling and brain plasticity are two of the many pathways and genes whose transcription is regulated by RA [40]. RA controls the lifelong processes of cell division, development, and apoptosis. Throughout development, it is indispensable to tissue organizing, especially in the central nervous system. In adulthood, it promotes immunological control, maintenance of cellular equilibrium, functioning of hormones, procreation, and visual perception [41,42]. The hepatic cell is where vitamin A and RA are kept until they are required to affect certain organs, such as the central nervous system. Retinaldehyde is produced by retinol dehydrogenase (RDH) enzymes from retinol, and retinaldehyde dehydrogenase (RALDH) enzymes catalyze it into the active compound RA [43]. Antioxidant activation is one way that RA has neuroprotective benefits. It lowers dangerous ROS into neuronal cells by upregulating antioxidant enzymes including catalase and SOD [44,45]. This compound affects the expression of genes related to the functioning of mitochondria, the longevity of cells, and responses to oxidative stress via stimulating nuclear receptors (RARs and RXRs) [46]. Since oxidative harm is linked to neurodegenerative illnesses, this type of antioxidant function aids in shielding neurons from it. To further maintain the health of neurons, RA also promotes neurogenesis and reduces neuroinflammation [17]. In general, their antioxidant response plays a major role in preserving the integrity of the brain and halting the progression of neurodegenerative diseases. Vitamin A provides neuroprotection against a number of neurodegenerative illnesses via an array of approaches, which mainly include its active form, called RA. Since neurogenesis is compromised for AD, RA plays a crucial role in promoting neurons to differentiate across the hippocampus [47]. Research indicates that administering RA to AD animal models can boost neurogenesis and memory function [48]. RA regulates the AMPA as well as NMDA receptor activity, which is necessary to regulate synaptic plasticity and learning [49]. RA could improve neurological communication in Huntington's condition animals, restoring synapse function and lowering motor impairments [40,50]. In conditions like PD as well as in people with MS, neurological inflammation remains the primary cause of progression of the disorder [51,52]. RA

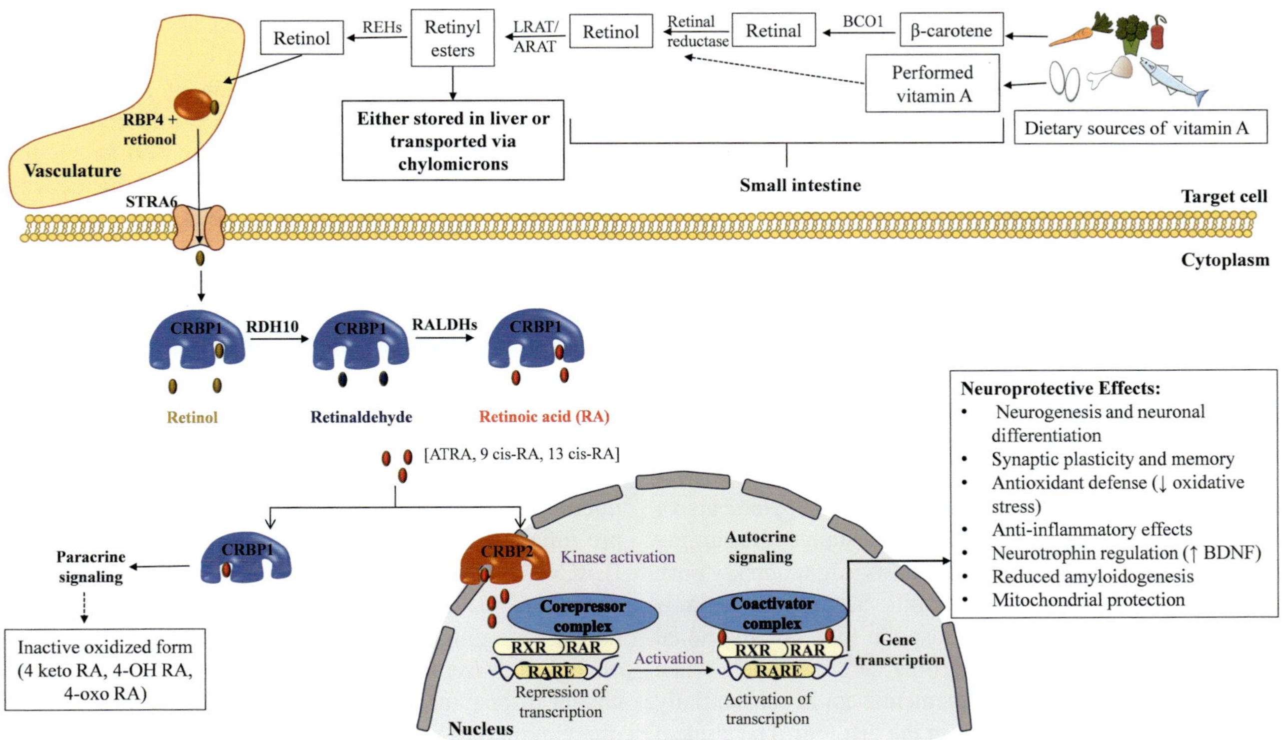

FIGURE 5.1 Vitamin A–mediated retinoic acid formation and related signaling pathways in the regulation of target gene transcription. The biochemical process by which vitamin A is converted to retinoic acid (RA) followed by its function in regulating gene transcription is depicted in this image. In order to ensure the bioavailability of retinoic acid, which is necessary to facilitate neuroprotection and promote the development and functioning of the nervous system due to its significance in the regulation of some specific genes, nutritional supplement sources that contain vitamin A—whether coming from plant or animal sources—will be crucial. **Abbreviations**: RBP4, retinol-binding protein 4; CRBP1, cellular retinol-binding protein 1; RA, retinoic acid; RAR, retinoic acid receptor; RXR, retinoid X receptor; RARE, retinoic acid response element; CBP/p300, CREB binding protein/E1A binding protein p300; N-CoR, nuclear receptor corepressor; RALDH, retinaldehyde dehydrogenase.

can hinder the stimulation of microglial cells and pro-inflammatory cytokines (such as TNF-α as well as IL-6) [53]. Administering RA into MS models decreases demyelination followed by the severity of the disease, suggesting strong immunomodulatory actions [54]. Apoptosis contributes to the death of dopaminergic neurons in PD. By inhibiting caspase processes and upregulating proteins that prevent apoptosis (Bcl-2), RA shields neurons against poisons like MPTP, which are employed in experiments to simulate PD pathology [55,56]. A major factor in conditions including AD, PD, and ALS is an increase in oxidative stress. The levels of ROS are decreased by RA's enhancement of antioxidant enzymes like SOD. Furthermore, RA can also maintain muscle strength and slow motor neuron deterioration in the ALS model [15]. When taken as a whole, these processes highlight the medicinal value of vitamin A toward decreasing the symptoms and the progression of a number of neurodegenerative diseases, providing possible directions for further therapies. Vitamin A serves as essential for neurological, immunological, and vision processes; however, it is important to take the right amount to prevent toxicity as well as deficiency [57]. The standard adult suggested dietary allowances or RDA tends to be 700 μg of retinol action equivalents (RAEs) per day for women and 900 μg of RAEs for men [58]. This encompasses primarily pro-vitamin A carotenoids coming from plants as well as preformed vitamin A via animal-based sources [59]. Meals including animal liver contain more than 6,000 μg RAE per 100 g, greatly above daily requirements; therefore, portion control is essential [60]. On the other hand, beta-carotene, which significantly transforms the body effectively into retinol, can be found in sources derived from plants. Notable plant-based sources of vitamin A include sweet potatoes, carrots, and spinach. Every meal contains more than 500 μg of RAE, which can be safely obtained via beta-carotene conversion, resulting in a lower risk of toxicity [61]. The hazardous condition known as hypervitaminosis A, which is characterized by symptoms including dizziness, headaches, weakness, skin peeling, discomfort with bones, and, in extreme cases, liver destruction or high intracranial pressure, could be brought on by a high level of vitamin A, particularly via supplements or regular animal liver eating [62]. For instance, consuming the oil from cod liver on a regular basis might quickly push consumption over acceptable limits because it contains about 4,000 μg RAE in each teaspoon [63]. An additional instance would be fortified morning cereals, which, when paired with various other supplements, could substantially boost the quantity of vitamin A in the diet [64]. The clinical recommendations stress that in order to prevent toxicity, total amounts taken via food and dietary supplements cannot exceed the average adult tolerated maximal consumption threshold of 3,000 μg RAE for every day. Because excessive intake of vitamin A could result in birth abnormalities, caution becomes particularly important during pregnancy [65]. Toxicological risks are avoided by balancing food sources and preventing needless high-dose supplementation, which guarantees vitamin A's nutritional and neuroprotective advantages. The two main dietary sources of vitamin A are sources derived 241 from animals including retinol (which are present in fish, flesh, dairy items, and egg 242 yolks) that can be accessible directly in metabolic processes conversion, while sources 243 derived from plants like beta-carotene (which are present in carrot, broccoli, and several other 244 vegetables) that can be transformed into retinol in the human body. RBP4 transports retinol 245 to specific cells in the circulation. Retinol oxidizes to retinaldehyde after interacting with 246 CRBP1 within the target cell. Retinaldehyde is further oxidized via enzyme RALDH, of which 247 then produces RA, an active metabolite. RA interacts with the RAR-RXR, 248 a heterodimer combination inside the nucleus. When RA binds to RAR, conformational 249 alterations occur that enable the recruitment associated with corepressors (like N-CoR) that 250 hinder transcription or coactivators (like CBP/p300) that stimulate transcription. This allows 251 the regulation of specific gene expression related to immune system responses, differentiation 252 of cells, and growth as given in Figure 5.1.

5.3 VITAMIN C AND NEUROPROTECTION

Ascorbic acid, also known as vitamin C, serves as an essential hydrophilic antioxidant that has important neuroprotective properties. This neutralizes ROS biochemically, avoiding oxidative degeneration of neurons' lipids and proteins, followed by DNA. This is important in neurodegenerative illnesses including PD as well as AD [66]. Additionally, it serves as a cofactor regulating enzymes that participate in the generation of neurotransmitters, like dopamine β-hydroxylase, responsible for converting dopamine into norepinephrine, all of which serve as crucial for mood as well as cognition [18]. Additionally, it alters serotonin processes, which affects emotional equilibrium and mental wellness [67]. In addition to neurotransmission, the antioxidant vitamin C has strong immunological modulatory effects, lowering neuroinflammation through controlling the activation of the brain's defense cells called microglia [68]. Additionally, it reduces the production of pro-inflammatory cytokines including TNF-α as well as IL-6, which are linked to neurological illnesses notably PD as well as AD. Moreover, dopamine levels must be maintained regarding PD, because vitamin C supports these biological processes [69]. Furthermore, by protecting mitochondrial enzymes and lowering oxidative damage, vitamin C promotes mitochondrial activity and ATP synthesis, guaranteeing that brain energy metabolism stays effective. Research indicates that taking vitamin C supplements increases the mitochondria's ability to cope with stress, which becomes crucial in disorders such as AD whenever malfunctioning mitochondria have become a major characteristic [70]. Vitamin C's entry into brain tissue is strictly controlled. Since passive diffusion is blocked by the BBB, the SVCT2 is the main pathway by which vitamin C

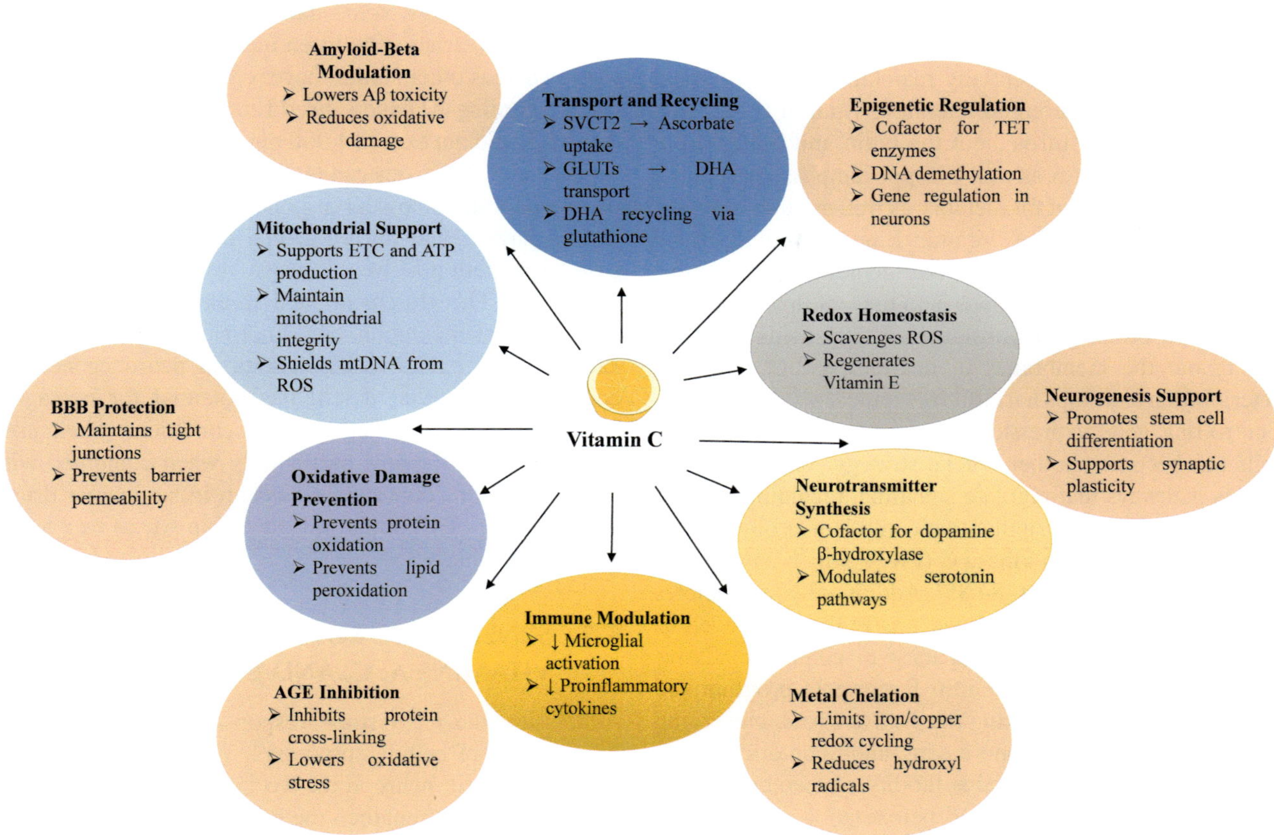

FIGURE 5.2 Multifaceted neuroprotective functions of vitamin C and its anti-neurodegenerative impact. This diagram highlights multiple neuroprotective roles of vitamin C (ascorbate) in brain health. Through SVCT2 and GLUT-mediated transport, along with glutathione recycling, vitamin C maintains antioxidant defenses by scavenging ROS, regenerating vitamin E, and preventing lipid/protein oxidation. It supports neurotransmitter synthesis (dopamine, norepinephrine, serotonin), modulates neuroinflammation, and preserves mitochondrial function for ATP production and mtDNA protection. Vitamin C also influences epigenetic regulation via TET enzymes and promotes neurogenesis, synaptic plasticity, and stem cell differentiation. Additionally, it inhibits AGEs, chelates metals, reduces Aβ aggregation, and maintains blood–brain barrier integrity, making it a multifunctional therapeutic agent against neurodegenerative diseases. **Abbreviations**: Vitamin C, ascorbic acid; Aβ, amyloid-beta; AGEs, advanced glycation end products; BBB, blood–brain barrier; DHA, dehydroascorbic acid; ETC, electron transport chain; GLUTs, glucose transporters; mtDNA, mitochondrial DNA; ROS, reactive oxygen species; SVCT2, sodium-dependent vitamin C transporter 2; TET enzymes, ten–eleven translocation enzymes.

reaches the human nervous system. This pathway ensures antioxidant defense right where the regulation is needed by concentrating ascorbic acid within neurons with levels as much as ten times greater than within plasma [71]. Utilizing glucose transporters (GLUT1), oxidative vitamin C (DHA) could cross the BBB and subsequently be converted to its original active ascorbate within brain cells. An effective supply can be assured under a variety of circumstances by this dual transit system [72]. Examples demonstrate the significance of vitamin C: by combating free radicals as well as lowering inflammation, vitamin C lowers the size of infarction and neurological impairments regarding ischemic stroke condition [21,73]. Supplementing reduces the amount of amyloid plaque and enhances memory regarding Alzheimer's models [74]. As shown in older populations and those suffering from depression as well as schizophrenia, individuals who possess reduced plasma antioxidant levels clinically exhibit significant cognitive impairment [75]. Vitamin C is abundant in foods like spicy peppers, cabbage, oranges, kiwis, raspberries, and other dietary sources that boost such protective effects [76,77]. Vitamin C represents a crucial component of neuroprotection because of its biochemistry and a unique way of entering the brain, which highlights its vital function in maintaining neuronal health, preventing oxidative stress, and regulating neurotransmission as shown in Figure 5.2.

5.4 VITAMIN E AND NEUROPROTECTION

As a lipophilic natural antioxidant, vitamin E is essential for preserving the integrity of neurons by shielding brain cells from oxidative stress. The two main isoform types regarding vitamin E, tocopherols and tocotrienols, are essential for neuroprotection for the CNS [78]. Although the chromanol ring has been shared by both isoforms, tocotrienols include unsaturated isoprenoid chain residues that improve membrane accessibility as well as boost their antioxidant properties [79]. Later in the activity of α-tocopherol transfer protein (α-TTP), which is especially generated by the liver, α-tocopherol has become the most physiologically active tocopherol and is preferentially maintained within human tissues [80]. As preferential retention, vitamin E gets distributed differently throughout the brain and builds up in regions like the cortex as well as hippocampus, a part of the brain that is important for memory as well as cognition [81]. In order to combat stress, vitamin E neutralizes ROS and halts lipid peroxidation across the membranes of neurons [82]. One of its main functions is to prevent lipid peroxidation, which protects polyunsaturated fatty acids (PUFAs) throughout the membranes of neurons against oxidative deterioration and stops harmful free radical production [83]. Controlling gene expression along with signaling among cells has become another crucial strategy. Vitamin E also affects pathways that regulate signaling, including NF-κB, a component of transcription that controls neuroinflammation, as well as protein kinase C (PKC), which participates in neuronal development [84,85]. Additionally, it promotes cellular survival mechanisms such as PI3K/Akt as well as MAPK, strengthening the capacity to combat oxidative stress [86]. In addition, vitamin E prevents programmed death of neurons by lowering nitrosative stress, preserving mitochondrial integrity, and having anti-apoptotic properties [87]. Vitamin E maintains fluidity of membranes and functioning by preventing lipid peroxidation, which guarantees effective neurotransmission and cellular communication [26]. For instance, α-tocopherol administration can be demonstrated to enhance memory function and decrease β-amyloid plaque deposition via animal models regarding AD [88]. While its therapeutic effectiveness in people is still up for debate, vitamin E has shown promise in PD models in preventing dopaminergic neuronal death by lowering damage caused by oxidation throughout the substantia nigra [89]. Furthermore, by reducing the synthesis of cytokines that promote inflammation notably TNF-α and IL-1β while altering the expression of enzymes that include cyclooxygenase-2 (COX-2), the antioxidant vitamin E promotes anti-inflammatory properties. The aforementioned anti-inflammatory properties play a crucial role in neurological conditions because long-term inflammation makes neuronal damage worse [90]. Vitamin E may decrease demyelination and enhance motor functioning in clinical models regarding MS, indicating a wider involvement in inflammation and autoimmune neurological disorders [91]. Vitamin E supplementation appears to have demonstrated the ability to preserve mitochondrial activity and lower ROS generation regarding stroke models, hence lowering infarct size as well as improving neurological prognosis [92]. Dietary sources of vitamin E include leafy vegetables that are green, nuts, seeds, and plant oils [93]. Additionally, the bioactivity of the synthesized (all-rac-α-tocopherol) versus natural (RRR-α-tocopherol) versions varies, resulting in the natural variety generally being more effective [94]. Research studies on supplementation have shown conflicting findings; some have shown minimal efficacy, emphasizing variation between individuals and dose problems, whereas other studies have reported cognitive advantages regarding AD and slowed functional decline [95,96]. Adults should not exceed 1,000 mg/day using α-tocopherol, because elevated doses can lead to hemorrhagic stroke and decreased immunological responses [97]. In addition, vitamin E plays a dual role regarding neurodegenerative diseases with both positive and negative aspects as shown in Figure 5.3. The aforementioned studies have shown vitamin E supplementation can improve behavioral results and decrease oxidative biomarkers in neurodegenerative conditions. However, it still has potential for managing and preventing neurodegenerative conditions when used in combination therapy, particularly when combined with additional antioxidants as well as anti-inflammatory drugs. The vital role of vitamins A, C, and E in AD as well as PD is given in Table 5.2.

5.5 SYNERGISTIC ROLES OF VITAMINS A, C, AND E

Addressing the CNS stress response, inflammation, and preservation of cellular processes, vitamins A, C, and E have synergistic neuroprotective benefits. The aforementioned vitamins strengthen the brain's resistance to neurodegeneration by acting in concert across several divisions of cells. Although vitamin C is an antioxidant that dissolves in water, eliminates ROS throughout the aqueous phase, and restores oxidized vitamin E to its vital state, vitamin E, a lipid-soluble antioxidant, shields the membranes of neurons from lipid peroxidation. Extended neural plasticity and resilience are facilitated by vitamin A, mainly in the form of RA, which controls genes related to neuronal development, differentiation, and immunological regulation [109–111]. In addition to its antioxidant actions, this synergy also involves the regulation of shared intracellular signaling processes. For instance, the NF-κB process, an essential controller of neuroinflammation, is influenced by each of these vitamins. Both vitamin C and vitamin E decrease ROS, which, in turn, suppresses the generation of pro-inflammatory cytokines including TNF-α as well as IL-6. Vitamin A, on the other hand, suppresses NF-κB activation via RARs [112–114]. The aforementioned vitamins also improve the MAPK/ERK and PI3K/Akt signaling systems, which support neurogenesis, longevity of cells, and synapse maintenance. Although RA improves ERK signaling in order to support neurodevelopment and plasticity, vitamin E has additionally been demonstrated to trigger PI3K/Akt, increasing neuronal tolerance against oxidative stress [115–117]. Such coordinated effects are also supported by research conducted in cell lines as well as animal models. Supplementing with all three vitamins together decreased the buildup of β-amyloid, increased the expression of antioxidant enzymes such as SOD, and boosted

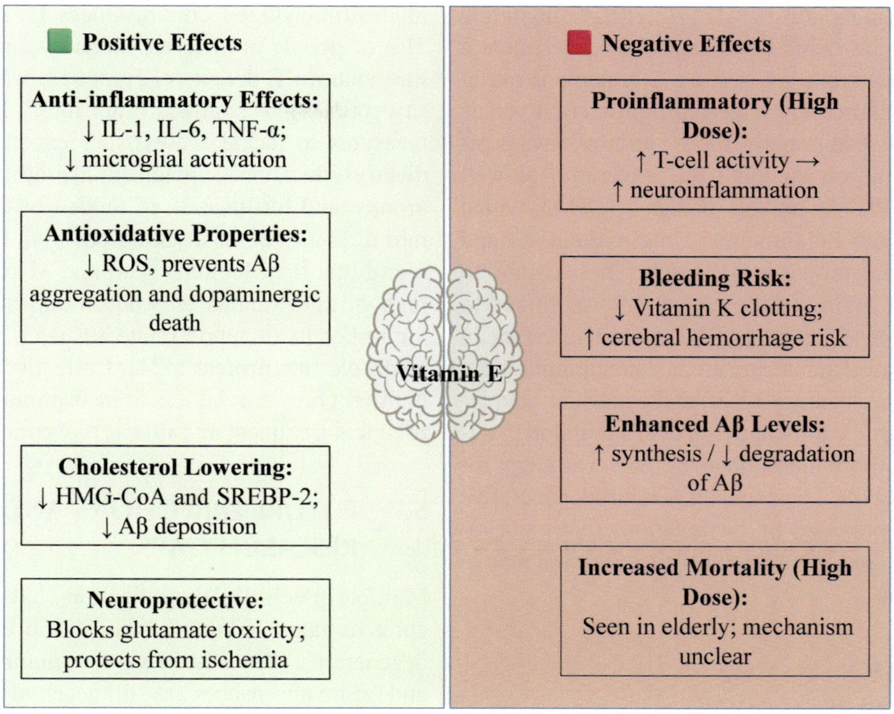

FIGURE 5.3 Dual role of vitamin E in neurodegenerative diseases: An overview of its positive and negative aspects. This diagram illustrates the dual roles of vitamin E (α-tocopherol) in neurodegeneration. Beneficially, it reduces pro-inflammatory cytokines (IL-1, IL-6, TNF-α), limits microglial activation, scavenges ROS, prevents Aβ aggregation in AD, protects dopaminergic neurons in PD, regulates cholesterol metabolism (via SREBP-2 and HMG-CoA reductase), and guards against ischemia and glutamate excitotoxicity. However, excessive vitamin E may enhance T-cell activation, trigger pro-inflammatory effects, interfere with vitamin K–dependent clotting (risking cerebral hemorrhage), and potentially worsen Aβ pathology. High doses are also linked to increased all-cause mortality in elderly populations. Thus, vitamin E demonstrates dose-dependent neuroprotective yet potentially harmful effects. **Abbreviations**: Aβ, amyloid-beta; IL, interleukin; TNF-α, tumor necrosis factor-alpha; ROS, reactive oxygen species; HMG-CoA, 3-hydroxy-3-methyl-glutaryl-coenzyme A; SREBP-2, sterol regulatory element-binding protein 2; AD, Alzheimer's disease; PD, Parkinson's disease.

TABLE 5.2
Roles of Vitamins A, C, and E in Alzheimer's and Parkinson's Diseases

Disease	Vitamin	Key Genes/ Proteins	Mechanism/Pathway	Neuroprotective Function	References
Alzheimer's disease (AD)	Vitamin A	RAR, RXR, ALDH1A1, BACE1	Downregulates BACE1 (reducing β-amyloid); modulates neuroinflammation via RAR/RXR; promotes synaptic plasticity	Decreases Aβ plaque formation; supports memory and learning	[48,98]
	Vitamin C	SVCT2, SOD1, GPX, BACE1	Antioxidant; reduces Aβ aggregation; enhances neurotransmitter production; stabilizes BBB	Reduces oxidative damage and neuroinflammation; improves cognition	[99,100]
	Vitamin E	TTPA, GPX4, APOE, NRF2	Inhibits lipid peroxidation; stabilizes cell membranes; modulates Nrf2-ARE antioxidant genes	Slows cognitive decline; reduces oxidative stress and inflammation	[101–104]
Parkinson's disease (PD)	Vitamin A	RARβ, RXRα, TH, GDNF	Regulates dopaminergic neuron survival; modulates glial cell activation	Supports dopaminergic pathways; reduces α-synuclein toxicity	[17,105]
	Vitamin C	SVCT2, DJ-1, SOD2	Scavenges ROS; maintains dopamine levels; protects mitochondria	Reduces oxidative stress in substantia nigra; preserves dopaminergic neurons	[18,106]
	Vitamin E	PINK1, PRKN, NRF2	Prevents mitochondrial dysfunction; modulates ferroptosis-related lipid peroxidation	Protects dopaminergic neurons; delays motor decline	[89,107,108]

memory function via models of AD [118,119]. Comparable synergistic effects are being documented in experimental models of ischemic stroke as well as AD, demonstrating that combination vitamins provide more protection versus monotherapy [120]. Additionally, these vitamins have positive interactions via polyphenols (like resveratrol as well as curcumin) and PUFAs notably omega-3 (DHA), which strengthen their anti-inflammatory, anti-oxidative, and membrane-stabilizing properties [121,122]. These combinations are typical of brain-healthy eating patterns linked to less cognitive loss, especially the MIND and Mediterranean diets [123,124]. Based on a nutritional standpoint, these results highlight the superiority of multi-nutrient supplements through holistic eating patterns over standalone vitamin medication. A more thorough and efficient strategy for avoiding or delaying the advancement of neurodegenerative illnesses could be provided by formulations that incorporate vitamins A, C, and E combined with additional neuroprotective chemicals.

5.6 TRANSLATIONAL AND CLINICAL PERSPECTIVES

The vitamins have been shown in preclinical research to have neuroprotective, anti-inflammatory properties, and antioxidant properties. For instance, using transgenic AD mice models, α-tocopherol (vitamin E) decreased β-amyloid toxicity as well as enhanced cognitive performance [125]. When compared with a placebo, 2,000 IU/day of vitamin E reduced the loss of functioning in moderately impacted patients with AD approximately 6–7 months [96]. Increased daily intake of vitamins C and E as in food material was also linked to a 13% reduced probability of AD, according to a 2017 meta-analysis [126]. Whereas RA controls genes essential for dopamine signals, taking vitamin C supplements has been demonstrated to shield dopaminergic neurons from PD via scavenging ROS [127,128]. Clinical implementation is difficult, even with encouraging preclinical outcomes. Results are complicated by variations in dose, bioavailability, and disease stage. For example, vitamin E and selenium failed to significantly decrease dementia as demonstrated by the PREADViSE trial (2015), which may have been caused by late intervention or inadequate baseline antioxidant intake [129]. This emphasizes the significance of nutrigenomics and customized therapy. Vitamin metabolism and response are influenced by genetic polymorphisms, for example, those influencing SOD2 as well as α-TTP, triggering the need for customized supplementation strategies [130]. Vitamin deficiencies are still common when viewed from a nutritional and overall health basis. The World Health Organization (WHO) estimates that more than 190 million preschool-aged children worldwide suffer from vitamin A insufficiency, which impairs neurodevelopment and raises the possibility of infection [131]. Approximately 30% of older adults have insufficient plasma vitamin C contents (<23 μmol/L), especially in institutionalized circumstances [132]. Similarly, 13%–21% of people in different locations suffer from a borderline vitamin E deficit (<12 μmol/L), which increases their susceptibility to oxidized brain injury [133]. Public health measures to tackle these issues encompass brain-healthy dietary education, supplemental programs for susceptible groups, and fortification of foods (for instance, vitamin A into milk as well as cooking oil) [64]. Morris et al. (2015) found that the Mediterranean and MIND diets, which are high in antioxidants as well as anti-inflammatory foods, are linked to an approximate 30%–53% lower incidence of cognitive impairment [134]. Collectively, these initiatives highlight how crucial it is to incorporate nutrition into neurological treatment and illness prevention on a global scale.

5.7 FUTURE DIRECTIONS AND RESEARCH GAPS

Multiple preclinical investigations have demonstrated the curative value of vitamins A, C, and E in treating neurodegenerative illnesses; however, a number of research gaps and future approaches are still necessary to achieve clinical translation [135]. The limited bioavailability and poor brain accessibility of numerous antioxidants, especially vitamins that are fat-soluble including A and E, are major drawbacks [136]. Innovative delivery methods including solid lipid carriers, liposomes, and nanoparticles have been developed to get around limitations [137]. For instance, when administered to a rat model regarding AD, α-tocopherol-loaded liposomes greatly enhanced BBB permeability and decreased neuronal death [138]. In preclinical settings, RA-conjugated nanoparticles additionally showed potential in improving retinoid signaling and lowering the burden of amyloid plaque [139]. At the same time, omics-based methods such as proteomics, metabolomics, and genomics can be creating new opportunities for customized vitamin treatment [140,141]. For example, person reactions to vitamin E are influenced by genetic variations associated with the TTPA gene, which codes for α-TTP, as well as SOD2, which is implicated in oxidative defense [142,143]. The significance of the APOE genotype for determining the efficacy of antioxidant therapies in AD underscores the necessity of customized diets driven by nutrigenomics. Such biomarkers could be used in future clinical investigations to customize antioxidant treatment depending on metabolic and genetic characteristics [144]. The absence of extensive, long-term randomized controlled trials (RCTs) assessing the effectiveness of these kinds of vitamins in avoiding or delaying neurodegeneration remains a significant research gap. To date, the majority of clinical investigations have either been small or lack solid evidence [145]. To illustrate the need for greater participant segmentation, earlier intervention, and superior trial design, known as the PREADViSE trial (2015), which assessed vitamin E and selenium for potential dementia mitigation, found no meaningful impact [129]. Furthermore, there is a lot of potential in investigating new interactions among antioxidant vitamins along with currently used pharmaceuticals or lifestyle modifications [146]. In individuals who had moderate cognitive decline, recent studies have shown that vitamin E and exercise regimens together showed more cognitive benefits than either approach alone [147]. Furthermore, interactions combining anti-inflammatory drugs, polyphenols, and omega-3 fatty acids continue to be studied for potential additive and synergistic effects [120,148]. To sum up, improving delivery technologies, incorporating omics for tailored treatment, carrying out reliable RCTs, and creating multimodal approaches that combine vitamin supplementation with pharmacological and lifestyle interventions are crucial steps forward in maximizing neuroprotection and cognitive performance.

5.8 CONCLUSION

The evidence suggests that vitamins A, C, and E play important neuroprotective roles, primarily via synergistic anti-oxidative, anti-inflammatory, and gene-regulatory actions in the brain and spinal cord. The aforementioned vitamins not only resist oxidative and nitrosative damage but also alter critical signaling processes like NF-κB, PI3K/Akt, and MAPK, which have been involved throughout neuron survival, inflammation, and synaptic plasticity. The way they interact with additional neuroprotective supplements, particularly polyphenols as well as omega-3 fatty acids, together with their appearance in brain-healthy eating practices, which include the Mediterranean and MIND diets, adds to their unique potential for slowing down or avoiding the onset of neurological conditions including AD, PD, and ALS. However, obstacles remain in converting preclinical discoveries into reliable clinical consequences. Bioavailability, interindividual fluctuation, disease stage, and a lack of reliable clinical trials all limit the capacity to draw accurate inferences about supplement effectiveness. Emerging techniques, such as nanotechnology-based methods of administration, nutrigenomics, and biomarker-driven tailored nutrition, show promise in addressing these limitations and improving therapeutic outcomes. Public health initiatives need to incorporate education, tailored supplementation, and fortification to treat the prevalent micronutrient deficiencies, especially in older and vulnerable populations. Future studies should focus on widespread, well-monitored trials and investigate integrated therapy approaches that combine lifestyle, medication, and diet. In conclusion, current research supports the integration of vitamins A, C, and E in complete approaches to enhancing mental health and alleviating age-associated neurological decline, even though their complete therapeutic potential regarding neuroprotection has not yet been discovered.

CONFLICT OF INTEREST

The authors declare there are no conflicts of interest.

REFERENCES

1. Y. K. Gupta, M. Gupta, and K. Kohli, "Neuroprotective role of melatonin in oxidative stress vulnerable brain," *Indian Journal of Physiology and Pharmacology,* vol. 47, pp. 373–386, 2003.
2. C. Prajapati, D. Kumar, V. Ambastha, and S. K. Singh, "Fungal endophytes as potential anticancer candidate over synthetic drugs: latest development and future prospects," in Singh, B. P., Abdel-Azeem, A. M., Gautam, V., Singh, G., and Singh, S. K. (eds.), *Endophytic Fungi: The Hidden Sustainable Jewels for the Pharmaceutical and Agricultural Industries.* Springer, pp. 27–56, 2024.
3. S. D. Skaper, M. Floreani, M. Ceccon, L. Facci, and P. Giusti, "Excitotoxicity, oxidative stress, and the neuroprotective potential of melatonin," *Annals of the New York Academy of Sciences,* vol. 890, no. 1, pp. 107–118, 1999.
4. S. C. Gilman, M. J. Bonner, and T. C. Pellmar, "Effect of oxidative stress on excitatory amino acid release by cerebral cortical synaptosomes," *Free Radical Biology and Medicine,* vol. 15, no. 6, pp. 671–675, 1993.
5. J. T. Coyle and P. Puttfarcken, "Oxidative stress, glutamate, and neurodegenerative disorders," *Science,* vol. 262, no. 5134, pp. 689–695, 1993.
6. I. Xu Lou, K. Ali, and Q. Chen, "Effect of nutrition in Alzheimer's disease: a systematic review," *Frontiers in Neuroscience,* vol. 17, p. 1147177, 2023.
7. S. N. Rai et al., "Autophagy modulating potential of mucuna pruriens in Parkinson's disease," *Current Traditional Medicine,* 2025. doi:10.2174/0122150838346643241211055438.
8. I. Korczowska-Łącka et al., "Disorders of endogenous and exogenous antioxidants in neurological diseases," *Antioxidants,* vol. 12, no. 10, p. 1811, 2023.
9. S. N. Rai, P. Singh, H. W. M. Steinbusch, E. Vamanu, G. Ashraf, and M. P. Singh, "The role of vitamins in neurodegenerative disease: an update," *Biomedicines,* vol. 9, no. 10, p. 1284, 2021.
10. C. Prajapati et al., "An update of fungal endophyte diversity and strategies for augmenting therapeutic potential of their potent metabolites: recent advancement," *Applied Biochemistry and Biotechnology,* vol. 197, no. 5, pp. 2799–2866, 2025.
11. S. R. Shannon, A. R. Moise, and P. A. Trainor, "New insights and changing paradigms in the regulation of vitamin A metabolism in development," *Wiley Interdisciplinary Reviews: Developmental Biology,* vol. 6, no. 3, p. e264, 2017.
12. L. Muller and V. Bohm, "Antioxidant activity of β-carotene compounds in different in vitro assays," *Molecules,* vol. 16, no. 2, pp. 1055–1069, 2011.
13. M. A. Lane and S. J. Bailey, "Role of retinoid signalling in the adult brain," *Progress in Neurobiology,* vol. 75, no. 4, pp. 275–293, 2005.
14. K. H. Lee, M. Cha, and B. H. Lee, "Neuroprotective effect of antioxidants in the brain," *International Journal of Molecular Sciences,* vol. 21, no. 19, p. 7152, 2020.
15. J. N. Clark, A. Whiting, and P. McCaffery, "Retinoic acid receptor-targeted drugs in neurodegenerative disease," *Expert Opinion on Drug Metabolism & Toxicology,* vol. 16, no. 11, pp. 1097–1108, 2020.
16. P. Joshi et al., "Combinations of vitamin A and vitamin E metabolites confer resilience against amyloid-β aggregation," *ACS Chemical Neuroscience,* vol. 14, no. 4, pp. 657–666, 2023.

17. A. Marie, M. Darricau, K. Touyarot, L. C. Parr-Brownlie, and C. Bosch-Bouju, "Role and mechanism of vitamin A metabolism in the pathophysiology of Parkinson's disease," *Journal of Parkinson's Disease,* vol. 11, no. 3, pp. 949–970, 2021.
18. J. Kocot, D. Luchowska-Kocot, M. Kiełczykowska, I. Musik, and J. Kurzepa, "Does vitamin C influence neurodegenerative diseases and psychiatric disorders?" *Nutrients,* vol. 9, no. 7, p. 659, 2017.
19. S. N. Hansen, P. Tveden-Nyborg, and J. Lykkesfeldt, "Does vitamin C deficiency affect cognitive development and function?" *Nutrients,* vol. 6, no. 9, pp. 3818–3846, 2014.
20. F. E. Harrison and J. M. May, "Vitamin C function in the brain: vital role of the ascorbate transporter SVCT2," *Free Radical Biology and Medicine,* vol. 46, no. 6, pp. 719–730, 2009.
21. C.-Y. Chang, J.-Y. Chen, M.-H. Wu, and M.-L. Hu, "Therapeutic treatment with vitamin C reduces focal cerebral ischemia-induced brain infarction in rats by attenuating disruptions of blood brain barrier and cerebral neuronal apoptosis," *Free Radical Biology and Medicine,* vol. 155, pp. 29–36, 2020.
22. Y. Feng and X. Wang, "Antioxidant therapies for Alzheimer′ s disease," *Oxidative Medicine and Cellular Longevity,* vol. 2012, no. 1, p. 472932, 2012.
23 L. Zylinska, M. Lisek, F. Guo, and T. Boczek, "Vitamin C modes of action in calcium-involved signaling in the brain," *Antioxidants,* vol. 12, no. 2, p. 231, 2023.
24. H. M. Evans and K. S. Bishop, "On the existence of a hitherto unrecognized dietary factor essential for reproduction," *Science,* vol. 56, no. 1458, pp. 650–651, 1922.
25. S. Q. Rodríguez-Lara et al., "Alternative interventions to prevent oxidative damage following ischemia/reperfusion," *Oxidative Medicine and Cellular Longevity,* vol. 2016, no. 1, p. 7190943, 2016.
26. L. Regner-Nelke et al., "Enjoy carefully: the multifaceted role of vitamin E in neuro-nutrition," *International Journal of Molecular Sciences,* vol. 22, no. 18, p. 10087, 2021.
27. C. Prajapati et al., "Intellectual property rights in neuroprotective biomaterials," in Kumar, G., Mukherjee, S., and Kumar, S. (eds.), *Biomaterials and Neurodegenerative Disorders.* Springer, pp. 251–269, 2025.
28. A. A. O. Xavier and A. Pérez-Gálvez, "Carotenoids as a source of antioxidants in the diet," in Herrero, R. G., and Britton, F. (eds.), *Carotenoids in Nature: Biosynthesis, Regulation and Function.* Springer, pp. 359–375, 2016.
29. L. Su et al., "Dietary total vitamin A, β-carotene, and retinol intake and the risk of diabetes in Chinese adults with plant-based diets," *Journal of Clinical Endocrinology & Metabolism,* vol. 107, no. 10, pp. e4106–e4114, 2022.
30. L. Englberger, I. Darnton-Hill, T. Coyne, M. H. Fitzgerald, and G. C. Marks, "Carotenoid-rich bananas: a potential food source for alleviating vitamin A deficiency," *Food and Nutrition Bulletin,* vol. 24, no. 4, pp. 303–318, 2003.
31. N. B. Tomlekova, P. J. White, J. A. Thompson, E. A. Penchev, and S. Nielen, "Mutation increasing β-carotene concentrations does not adversely affect concentrations of essential mineral elements in pepper fruit," *PloS ONE,* vol. 12, no. 2, p. e0172180, 2017.
32. M. Doseděl et al., "Vitamin C—sources, physiological role, kinetics, deficiency, use, toxicity, and determination," *Nutrients,* vol. 13, no. 2, p. 615, 2021.
33. S. A. Bsoul and G. T. Terezhalmy, "Vitamin C in health and disease," *Journal of Contemporary Dental Practice,* vol. 5, no. 2, pp. 1–13, 2006.
34. F. Shahidi, A. C. C. Pinaffi-Langley, J. Fuentes, H. Speisky, and A. C. de Camargo, "Vitamin E as an essential micronutrient for human health: common, novel, and unexplored dietary sources," *Free Radical Biology and Medicine,* vol. 176, pp. 312–321, 2021.
35. F. Visioli, "Effects of vitamin E on the endothelium: equivocal? α-tocopherol and endothelial dysfunction," *Cardiovascular Research,* vol. 51, no. 2, pp. 198–201, 2001.
36. M. A. Asson-Batres and C. Rochette-Egly, *Biochemistry of Retinoid Signaling,* pp. 9–10. Berlin: Springer, 2016.
37. M. Maden, "Retinoic acid in the development, regeneration and maintenance of the nervous system," *Nature Reviews Neuroscience,* vol. 8, no. 10, pp. 755–765, 2007.
38. W. Krezel, P. Kastner, and P. Chambon, "Differential expression of retinoid receptors in the adult mouse central nervous system," *Neuroscience,* vol. 89, no. 4, pp. 1291–1300, 1999.
39. W. Krężel, R. Rühl, and A. R. de Lera, "Alternative retinoid X receptor (RXR) ligands," *Molecular and Cellular Endocrinology,* vol. 491, p. 110436, 2019.
40. A. Niewiadomska-Cimicka et al., "Genome-wide analysis of RARβ transcriptional targets in mouse striatum links retinoic acid signaling with Huntington's disease and other neurodegenerative disorders," *Molecular Neurobiology,* vol. 54, no. 5, pp. 3859–3878, 2017.
41. K. D. Shearer, P. N. Stoney, P. J. Morgan, and P. J. McCaffery, "A vitamin for the brain," *Trends in Neurosciences,* vol. 35, no. 12, pp. 733–741, 2012.
42. R. Blomhoff and H. K. Blomhoff, "Overview of retinoid metabolism and function," *Journal of Neurobiology,* vol. 66, no. 7, pp. 606–630, 2006.
43. P. Sauvant, C. Feart, J. Sauvant, and V. Pallet, "Vitamin A and cognitive impairment," in Martin, C. R., and Preedy, V. R. (eds.), *Diet and Nutrition in Dementia and Cognitive Decline.* Elsevier, pp. 623–635, 2015.
44. B. Ahlemeyer et al., "Retinoic acid reduces apoptosis and oxidative stress by preservation of SOD protein level," *Free Radical Biology and Medicine,* vol. 30, no. 10, pp. 1067–1077, 2001.
45. C. C. Teixeira, I. M. Shapiro, M. Hatori, R. Rajpurohit, and C. Koch, "Retinoic acid modulation of glutathione and cysteine metabolism in chondrocytes," *Biochemical Journal,* vol. 314, no. 1, pp. 21–26, 1996.
46. S. Sharma et al., "Retinoid X receptor: cellular and biochemical roles of nuclear receptor with a focus on neuropathological involvement," *Molecular Neurobiology,* vol. 59, no. 4, pp. 2027–2050, 2022.
47. P. McCaffery, J. Zhang, and J. E. Crandall, "Retinoic acid signaling and function in the adult hippocampus," *Journal of Neurobiology,* vol. 66, no. 7, pp. 780–791, 2006.
48. Y. Ding et al., "Retinoic acid attenuates β-amyloid deposition and rescues memory deficits in an Alzheimer's disease transgenic mouse model," *Journal of Neuroscience,* vol. 28, no. 45, pp. 11622–11634, 2008.
49. K. L. Arendt, Y. Zhang, S. Jurado, R. C. Malenka, T. C. Südhof, and L. Chen, "Retinoic acid and LTP recruit postsynaptic AMPA receptors using distinct SNARE-dependent mechanisms," *Neuron,* vol. 86, no. 2, pp. 442–456, 2015.
50. M. Bosch et al., "Induction of GABAergic phenotype in a neural stem cell line for transplantation in an excitotoxic model of Huntington's disease," *Experimental Neurology,* vol. 190, no. 1, pp. 42–58, 2004.
51. C. Prajapati, S. N. Rai, S. S. Mishra, S. K. Singh, and A. K. Singh, "Natural sources of drugs for neurodegenerative

diseases," in Rai, S. N., Singh, S. K., and Singh, A. K. (eds.), *Neurodegenerative* Diseases. CRC Press, pp. 120–134, 2024.
52. H. Birla et al., "Tinospora cordifolia suppresses neuroinflammation in parkinsonian mouse model," *Neuromolecular Medicine,* vol. 21, no. 1, pp. 42–53, 2019.
53. J. Xu and P. D. Drew, "9-Cis-retinoic acid suppresses inflammatory responses of microglia and astrocytes," *Journal of Neuroimmunology,* vol. 171, no. 1–2, pp. 135–144, 2006.
54. S. M. Abtahi Froushani, N. Delirezh, R. Hobbenaghi, and G. Mosayebi, "Synergistic effects of atorvastatin and all-trans retinoic acid in ameliorating animal model of multiple sclerosis," *Immunological Investigations,* vol. 43, no. 1, pp. 54–68, 2014.
55. G. C. Kost, S. Selvaraj, Y. B. Lee, D. J. Kim, C.-H. Ahn, and B. B. Singh, "Clavulanic acid inhibits MPP+-induced ROS generation and subsequent loss of dopaminergic cells," *Brain Research,* vol. 1469, pp. 129–135, 2012.
56. G. K. Ulusoy, T. Celik, H. Kayir, M. Gürsoy, A. T. Isik, and T. I. Uzbay, "Effects of pioglitazone and retinoic acid in a rotenone model of Parkinson's disease," *Brain Research Bulletin,* vol. 85, no. 6, pp. 380–384, 2011.
57. P. Marzeda and J. J. Luszczki, "Role of vitamin A in health and illness," *Journal of Pre-Clinical and Clinical Research,* vol. 13, no. 3, 2019.
58. EFSA Panel on Dietetic Products, Nutrition and Allergies (NDA), "Scientific opinion on dietary reference values for vitamin A," *EFSA Journal,* vol. 13, no. 3, p. 4028, 2015.
59. T. Chacón-Ordóñez, R. Carle, and R. Schweiggert, "Bioaccessibility of carotenoids from plant and animal foods," *Journal of the Science of Food and Agriculture,* vol. 99, no. 7, pp. 3220–3239, 2019.
60. A. S. Green and A. J. Fascetti, "Meeting the vitamin A requirement: the efficacy and importance of β-carotene in animal species," *Scientific World Journal,* vol. 2016, no. 1, p. 7393620, 2016.
61. G. Tang, "Vitamin A value of plant food provitamin A—evaluated by the stable isotope technologies," *International Journal for Vitamin and Nutrition Research,* vol. 84, no. Suppl 1, pp. 25–29, 2014.
62. J. K. Roop, "Hypervitaminosis-an emerging pathological condition," *International Journal of Health Sciences and Research,* vol. 8, p. 280, 2018.
63. D. Wetzel, *Cod Liver Oil—Notes on the Manufacture of Our Most Important Dietary Supplement.* Weston A. Price Foundation, 2009.
64. O. Dary and J. O. Mora, "Food fortification to reduce vitamin A deficiency: international Vitamin A Consultative Group recommendations," *The Journal of Nutrition,* vol. 132, no. 9, pp. 2927S–2933S, 2002.
65. EFSA Panel on Dietetic Products, Nutrition and Allergies (NDA), "Scientific opinion on the tolerable upper intake level for preformed vitamin A and β-carotene," *EFSA Journal,* vol. 22, no. 6, p. e8814, 2024.
66. S. Parab et al., "Unraveling the mechanisms of hydrophilic vitamins in Alzheimer's and Parkinson's: preclinical and clinical evidence," in Shah, A. K., Tappia, P. S., and Dhalla, N. S. (eds.), *Hydrophilic Vitamins in Health and Disease.* Springer, pp. 181–217, 2024.
67. M. Masoudi et al., "Vitamin C improved anxiety and depression like behavior induced by chronic unpredictable mild stress in adolescent rats by influencing on oxidative stress balance, neurotransmitter systems, and inflammatory response," *Nutritional Neuroscience,* vol. 28, no. 4, pp. 503–512, 2025.
68. X.-Y. Zhang et al., "Vitamin C alleviates LPS-induced cognitive impairment in mice by suppressing neuroinflammation and oxidative stress," *International Immunopharmacology,* vol. 65, pp. 438–447, 2018.
69. A. Alberts, E.-T. Moldoveanu, A.-G. Niculescu, and A. M. Grumezescu, "Vitamin C: a comprehensive review of its role in health, disease prevention, and therapeutic potential," *Molecules,* vol. 30, no. 3, p. 748, 2025.
70. F. Monacelli, E. Acquarone, C. Giannotti, R. Borghi, and A. Nencioni, "Vitamin C, aging and Alzheimer's disease," *Nutrients,* vol. 9, no. 7, p. 670, 2017.
71. J. M. May, "Vitamin C transport and its role in the central nervous system," in Stanger, O. (ed.), *Water Soluble Vitamins: Clinical Research and Future Application.* Springer, pp. 85–103, 2011.
72. D. B. Agus et al., "Vitamin C crosses the blood-brain barrier in the oxidized form through the glucose transporters," *The Journal of Clinical Investigation,* vol. 100, no. 11, pp. 2842–2848, 1997.
73. X. Tang, H. Liu, Y. Xiao, L. Wu, and P. Shu, "Vitamin C intake and ischemic stroke," *Frontiers in Nutrition,* vol. 9, p. 935991, 2022.
74. S. Y. Kook et al., "High-dose of vitamin C supplementation reduces amyloid plaque burden and ameliorates pathological changes in the brain of 5XFAD mice," *Cell Death & Disease,* vol. 5, no. 2, pp. e1083, 2014.
75. K. Subotičanec, V. Folnegović-Šmalc, M. Korbar, B. Meštrović, and R. Buzina, "Vitamin C status in chronic schizophrenia," *Biological Psychiatry,* vol. 28, no. 11, pp. 959–966, 1990.
76. A. R. Proteggente et al., "The antioxidant activity of regularly consumed fruit and vegetables reflects their phenolic and vitamin C composition," *Free Radical Research,* vol. 36, no. 2, pp. 217–233, 2002.
77. S. Mulukutla, J. Riddle, and L. Heberling, "Vitamin C in fruits: does organic make a difference?," *Journal of Emerging Investigators,* vol. 21, pp. 1–5, 2015.
78. H.-M. Yap and K.-L. Lye, "An insight of vitamin E as neuroprotective agents," *Progress in Microbes & Molecular Biology,* vol. 3, no. 1, p. a0000071, 2020.
79. J. Atkinson, R. F. Epand, and R. M. Epand, "Tocopherols and tocotrienols in membranes: a critical review," *Free Radical Biology and Medicine,* vol. 44, no. 5, pp. 739–764, 2008.
80. D. Manor and S. Morley, "The α-tocopherol transfer protein," *Vitamins & Hormones,* vol. 76, pp. 45–65, 2007.
81. A. Gurel, O. Coskun, F. Armutcu, M. Kanter, and O. A. Ozen, "Vitamin E against oxidative damage caused by formaldehyde in frontal cortex and hippocampus: biochemical and histological studies," *Journal of Chemical Neuroanatomy,* vol. 29, no. 3, pp. 173–178, 2005.
82. N. Sharma and B. Nehru, "Beneficial effect of vitamin E in rotenone induced model of PD: behavioural, neurochemical and biochemical study," *Experimental Neurobiology,* vol. 22, no. 3, p. 214, 2013.
83. K. Stachowicz, "The role of polyunsaturated fatty acids in neuronal signaling in depression and cognitive processes," *Archives of Biochemistry and Biophysics,* vol. 737, p. 109555, 2023.
84. Y. Wang, N.-Y. Park, Y. Jang, A. Ma, and Q. Jiang, "Vitamin E γ-tocotrienol inhibits cytokine-stimulated NF-κB activation by induction of anti-inflammatory A20 via stress

adaptive response due to modulation of sphingolipids," *Journal of Immunology,* vol. 195, no. 1, pp. 126–133, 2015.
85. G. Fatima, A. Dzupina, A. A. Mahdi, J. Fedacko, A. Magomedova, and N. G. Yousif, "Role of vitamin-E as a vital nutrient in health and diseases," *Indian Journal of Clinical Biochemistry,* pp. 1–14, 2025. doi:10.1007/s12291-025-01309-0.
86. A. Ungurianu, A. Zanfirescu, G. Nițulescu, and D. Margină, "Vitamin E beyond its antioxidant label," *Antioxidants,* vol. 10, no. 5, p. 634, 2021.
87. M. Z. Sadikan et al., "Molecular mechanisms of vitamin E in ocular neurodegenerative disorders: an update on the emerging evidence and therapeutic implications," *Journal of Ocular Pharmacology and Therapeutics,* vol. 41, no. 3, pp. 89–100, 2025.
88. A. Gugliandolo, P. Bramanti, and E. Mazzon, "Role of vitamin E in the treatment of Alzheimer's disease: evidence from animal models," *International Journal of Molecular Sciences,* vol. 18, no. 12, p. 2504, 2017.
89. T. Schirinzi et al., "Dietary vitamin E as a protective factor for Parkinson's disease: clinical and experimental evidence," *Frontiers in Neurology,* vol. 10, p. 148, 2019.
90. U. Singh, S. Devaraj, and I. Jialal, "Vitamin E, oxidative stress, and inflammation," *Annual Review of Nutrition,* vol. 25, no. 1, pp. 151–174, 2005.
91. P. Theodosis-Nobelos and E. A. Rekka, "The multiple sclerosis modulatory potential of natural multi-targeting antioxidants," *Molecules,* vol. 27, no. 23, p. 8402, 2022.
92. O. Gonzalez-Perez et al., "Beneficial effects of α-lipoic acid plus vitamin E on neurological deficit, reactive gliosis and neuronal remodeling in the penumbra of the ischemic rat brain," *Neuroscience Letters,* vol. 321, no. 1–2, pp. 100–104, 2002.
93. F. Zaaboul and Y. Liu, "Vitamin E in foodstuff: nutritional, analytical, and food technology aspects," *Comprehensive Reviews in Food Science and Food Safety,* vol. 21, no. 2, pp. 964–998, 2022.
94. K. M. Ranard and J. W. Erdman Jr, "Effects of dietary RRR α-tocopherol vs all-racemic α-tocopherol on health outcomes," *Nutrition Reviews,* vol. 76, no. 3, pp. 141–153, 2018.
95. E. R. Miller Iii, R. Pastor-Barriuso, D. Dalal, R. A. Riemersma, L. J. Appel, and E. Guallar, "Meta-analysis: high-dosage vitamin E supplementation may increase all-cause mortality," *Annals of Internal Medicine,* vol. 142, no. 1, pp. 37–46, 2005.
96. G. La Fata, P. Weber, and M. H. Mohajeri, "Effects of vitamin E on cognitive performance during ageing and in Alzheimer's disease," *Nutrients,* vol. 6, no. 12, pp. 5453–5472, 2014.
97. A. D. Kaye et al., "Vitamin E (α-Tocopherol): emerging clinical role and adverse risks of supplementation in adults," *Cureus,* vol. 17, no. 2, 2025.
98. J. Almaguer, A. Hindle, and J. J. Lawrence, "The contribution of hippocampal all-trans retinoic acid (ATRA) deficiency to Alzheimer's disease: a narrative overview of ATRA-dependent gene expression in post-mortem hippocampal tissue," *Antioxidants,* vol. 12, no. 11, p. 1921, 2023.
99. A. E. Collins, T. M. Saleh, and B. E. Kalisch, "Naturally occurring antioxidant therapy in Alzheimer's disease," *Antioxidants,* vol. 11, no. 2, p. 213, 2022.
100. K. Murakami et al., "Vitamin C restores behavioral deficits and amyloid-β oligomerization without affecting plaque formation in a mouse model of Alzheimer's disease," *Journal of Alzheimer's Disease,* vol. 26, no. 1, pp. 7–18, 2011.
101. Y. Nishida et al., "Depletion of vitamin E increases amyloid β accumulation by decreasing its clearances from brain and blood in a mouse model of Alzheimer disease," *Journal of Biological Chemistry,* vol. 284, no. 48, pp. 33400–33408, 2009.
102. S. Dong et al., "Dietary vitamin E status dictates oxidative stress outcomes by modulating effects of fish oil supplementation in Alzheimer disease model APPswe/PS1dE9 mice," *Molecular Neurobiology,* vol. 55, no. 12, pp. 9204–9219, 2018.
103. M. Grundman, "Vitamin E and Alzheimer disease: the basis for additional clinical trials," *American Journal of Clinical Nutrition,* vol. 71, no. 2, pp. 630S–636S, 2000.
104. P. Baruah, H. Moorthy, M. Ramesh, D. Padhi, and T. Govindaraju, "A natural polyphenol activates and enhances GPX4 to mitigate amyloid-β induced ferroptosis in Alzheimer's disease," *Chemical Science,* vol. 14, no. 35, pp. 9427–9438, 2023.
105. H. Katsuki et al., "Retinoic acid receptor stimulation protects midbrain dopaminergic neurons from inflammatory degeneration via BDNF-mediated signaling," *Journal of Neurochemistry,* vol. 110, no. 2, pp. 707–718, 2009.
106. H. Liu et al., "Effect of plasma vitamin C levels on Parkinson's disease and age at onset: a Mendelian randomization study," *Journal of Translational Medicine,* vol. 19, no. 1, p. 221, 2021.
107. A. Atiq et al., "Vitamin E analog trolox attenuates MPTP-induced Parkinson's disease in mice, mitigating oxidative stress, neuroinflammation, and motor impairment," *International Journal of Molecular Sciences,* vol. 24, no. 12, p. 9942, 2023.
108. U. Tripathi et al., "Upregulated ECM genes and increased synaptic activity in Parkinson's human DA neurons with PINK1/PRKN mutations," *NPJ Parkinson's Disease,* vol. 10, no. 1, p. 103, 2024.
109. M. Navidhamidi, A. Nazari, S. Dehghan, A. Ebrahimpour, R. Nasrnezhad, and F. Pourabdolhossein, "Therapeutic potential of combined therapy of vitamin A and vitamin C in the experimental autoimmune encephalomyelitis (EAE) in Lewis rats," *Molecular Neurobiology,* vol. 59, no. 4, pp. 2328–2347, 2022.
110. C. Behl, "Vitamin E and other antioxidants in neuroprotection," *International Journal for Vitamin and Nutrition Research,* vol. 69, no. 3, pp. 213–219, 1999.
111. A. Martin, K. Youdim, A. Szprengiel, B. Shukitt-Hale, and J. Joseph, "Roles of vitamins E and C on neurodegenerative diseases and cognitive performance," *Nutrition Reviews,* vol. 60, no. 10, pp. 308–326, 2002.
112. Y. Peng et al., "Ascorbic acid inhibits ROS production, NF-κB activation and prevents ethanol-induced growth retardation and microencephaly," *Neuropharmacology,* vol. 48, no. 3, pp. 426–434, 2005.
113. C. Behl, "Vitamin E protects neurons against oxidative cell death in vitro more effectively than 17-β estradiol and induces the activity of the transcription factor NF-κB," *Journal of Neural Transmission,* vol. 107, no. 4, pp. 393–407, 2000.

114. M. A. de Bittencourt Pasquali et al., "Gene expression profile of NF-κB, Nrf2, glycolytic, and p53 pathways during the SH-SY5Y neuronal differentiation mediated by retinoic acid," *Molecular Neurobiology,* vol. 53, no. 1, pp. 423–435, 2016.
115. E. Cañón, J. M. Cosgaya, S. Scsucova, and A. Aranda, "Rapid effects of retinoic acid on CREB and ERK phosphorylation in neuronal cells," *Molecular Biology of the Cell,* vol. 15, no. 12, pp. 5583–5592, 2004.
116. L. Chen, A. G. Lau, and F. Sarti, "Synaptic retinoic acid signaling and homeostatic synaptic plasticity," *Neuropharmacology,* vol. 78, pp. 3–12, 2014.
117. A. S. Zaulkffali et al., "Vitamins D and E stimulate the PI3K-AKT signalling pathway in insulin-resistant SK-N-SH neuronal cells," *Nutrients,* vol. 11, no. 10, p. 2525, 2019.
118. L. A. Boothby and P. L. Doering, "Vitamin C and vitamin E for Alzheimer's disease," *Annals of Pharmacotherapy,* vol. 39, no. 12, pp. 2073–2080, 2005.
119. K. Ono and M. Yamada, "Vitamin A and Alzheimer's disease," *Geriatrics & Gerontology International,* vol. 12, no. 2, pp. 180–188, 2012.
120. O. F. Sultana, R. A. Hia, and P. H. Reddy, "A combinational therapy for preventing and delaying the onset of Alzheimer's disease: a focus on probiotic and vitamin co-supplementation," *Antioxidants,* vol. 13, no. 2, p. 202, 2024.
121. J.-M. Zingg and M. Meydani, "Interaction between vitamin E and polyunsaturated fatty acids," in Weber, P., Birringer, M., Blumberg, J. B., Eggersdorfer, M., and Frank, J. (eds.), *Vitamin E in Human Health.* Springer, pp. 141–159, 2019.
122. L. Pogačnik, A. Ota, and N. Poklar Ulrih, "An overview of crucial dietary substances and their modes of action for prevention of neurodegenerative diseases," *Cells,* vol. 9, no. 3, p. 576, 2020.
123. S. Kheirouri and M. Alizadeh, "MIND diet and cognitive performance in older adults: a systematic review," *Critical Reviews in Food Science and Nutrition,* vol. 62, no. 29, pp. 8059–8077, 2022.
124. J. A. Seabrook et al., "Dietary patterns and brain health in middle-aged and older adults: a narrative review," *Nutrients,* vol. 17, no. 9, p. 1436, 2025.
125. S. Sung et al., "Early vitamin E supplementation in young but not aged mice reduces Aβ levels and amyloid deposition in a transgenic model of Alzheimer's disease," *FASEB Journal,* vol. 18, no. 2, pp. 323–325, 2004.
126. X. Jiang, J. Huang, D. Song, R. Deng, J. Wei, and Z. Zhang, "Increased consumption of fruit and vegetables is related to a reduced risk of cognitive impairment and dementia: meta-analysis," *Frontiers in Aging Neuroscience,* vol. 9, p. 18, 2017.
127. T. A. Samad, W. Krezel, P. Chambon, and E. Borrelli, "Regulation of dopaminergic pathways by retinoids: activation of the D2 receptor promoter by members of the retinoic acid receptor–retinoid X receptor family," *Proceedings of the National Academy of Sciences,* vol. 94, no. 26, pp. 14349–14354, 1997.
128. Y. Sun, A. N. Pham, and T. D. Waite, "The effect of vitamin C and iron on dopamine-mediated free radical generation: implications to Parkinson's disease," *Dalton Transactions,* vol. 47, no. 12, pp. 4059–4069, 2018.
129. R. J. Kryscio et al., "Association of antioxidant supplement use and dementia in the prevention of Alzheimer's disease by vitamin E and selenium trial (PREADViSE)," *JAMA Neurology,* vol. 74, no. 5, pp. 567–573, 2017.
130. A. Niforou, V. Konstantinidou, and A. Naska, "Genetic variants shaping inter-individual differences in response to dietary intakes: a narrative review of the case of vitamins," *Frontiers in Nutrition,* vol. 7, p. 558598, 2020.
131. P. Song et al., "The prevalence of vitamin A deficiency and its public health significance in children in low-and middle-income countries: a systematic review and modelling analysis," *Journal of Global Health,* vol. 13, p. 04084, 2023.
132. J. F. Pearson et al., "Vitamin C status correlates with markers of metabolic and cognitive health in 50-year-olds: findings of the CHALICE cohort study," *Nutrients,* vol. 9, no. 8, p. 831, 2017.
133. J. Dobrovolny, M. Smrcka, and J. Bienertova-Vasku, "Therapeutic potential of vitamin E and its derivatives in traumatic brain injury-associated dementia," *Neurological Sciences,* vol. 39, no. 6, pp. 989–998, 2018.
134. M. C. Morris, C. C. Tangney, Y. Wang, F. M. Sacks, D. A. Bennett, and N. T. Aggarwal, "MIND diet associated with reduced incidence of Alzheimer's disease," *Alzheimer's & Dementia,* vol. 11, no. 9, pp. 1007–1014, 2015.
135. L. Changkakoti et al., "Exploration of the role of vitamins in preventing neurodegenerative diseases: comprehensive review on preclinical and clinical findings," *Current Neuropharmacology,* vol. 23, no. 5, pp. 547–563, 2024.
136. E. Andrès, N. Lorenzo-Villalba, J.-E. Terrade, and M. Méndez-Bailon, "Fat-soluble vitamins A, D, E, and K: review of the literature and points of interest for the clinician," *Journal of Clinical Medicine,* vol. 13, no. 13, p. 3641, 2024.
137. N. Itrat, B. Israr, R. Saroosh, D. Munawwar, F. Ameen, and S. Ajab, "Vitamins and minerals: nanotech solutions for enhanced bioavailability," in Chatterjee, A., Sarkar, T., and Smaoui, S. (eds.), *Nanofuel: The Future of Sports Nutrition: Boosting Performance with Nanotech Nutrients.* Springer, pp. 161–182, 2025.
138. L. Vasileva et al., "Mitochondria-targeted delivery strategy of dual-loaded liposomes for Alzheimer's disease therapy," *International Journal of Molecular Sciences,* vol. 24, no. 13, p. 10494, 2023.
139. R. Ferreira, J. Napoli, T. Enver, L. Bernardino, and L. Ferreira, "Advances and challenges in retinoid delivery systems in regenerative and therapeutic medicine," *Nature Communications,* vol. 11, no. 1, p. 4265, 2020.
140. G. Lagoumintzis and G. P. Patrinos, "Triangulating nutrigenomics, metabolomics and microbiomics toward personalized nutrition and healthy living," *Human Genomics,* vol. 17, no. 1, p. 109, 2023.
141. C. Prajapati, S. N. Rai, A. K. Singh, N. Rajak, N. Garg, and S. K. Singh, "Untargeted metabolomics and bioactivities assessment of *Xylaria ellisii,* an endophytic fungus isolated from the leaf of the plant *Acorus calamus* Linn," *Current Pharmaceutical Design,* vol. 31, no. 22, pp. 1781–1799, 2025.
142. L. Ulatowski, C. Dreussi, N. Noy, J. Barnholtz-Sloan, E. Klein, and D. Manor, "Expression of the α-tocopherol transfer protein gene is regulated by oxidative stress and

common single-nucleotide polymorphisms," *Free Radical Biology and Medicine,* vol. 53, no. 12, pp. 2318–2326, 2012.

143. T. K. Das et al., "Effect of vitamin E supplementation on mRNA expression of superoxide dismutase and interleukin-2 in arsenic exposed goat leukocytes," *Bulletin of Environmental Contamination and Toxicology,* vol. 89, no. 6, pp. 1133–1137, 2012.

144. E. Mocchegiani et al., "Vitamin E–gene interactions in aging and inflammatory age-related diseases: implications for treatment. A systematic review," *Ageing Research Reviews,* vol. 14, pp. 81–101, 2014.

145. N. M. D'Cunha et al., "Effect of long-term nutraceutical and dietary supplement use on cognition in the elderly: a 10-year systematic review of randomised controlled trials," *British Journal of Nutrition,* vol. 119, no. 3, pp. 280–298, 2018.

146. C. C. Danta and P. Piplani, "The discovery and development of new potential antioxidant agents for the treatment of neurodegenerative diseases," *Expert Opinion on Drug Discovery,* vol. 9, no. 10, pp. 1205–1222, 2014.

147. E. Cetin, E. C. Top, G. Şahin, Y. G. Özkaya, H. Aydin, and F. Toraman, "Effect of vitamin E supplementation with exercise on cognitive functions and total antioxidant capacity in older people," *Journal of Nutrition, Health & Aging,* vol. 14, no. 9, pp. 763–769, 2010.

148. L. d. C. Velasco-Rodríguez, H. S. García, and M. P. Rascón-Díaz, "Curcumin and omega-3 polyunsaturated fatty acids as bioactive food components with synergistic effects on Alzheimer's disease," *Psychogeriatrics,* vol. 24, no. 3, pp. 701–718, 2024.

6 The B Vitamins and Cognitive Health

Methylation, Mitochondrial Function, and Neuronal Integrity

Kanika Goyal[1], *Ritu Dahiya*[2], *Md Abubakar*[3], *Mamta Bishnoi*[4], *Saqib Hameed*[3], *and Nitesh Kumar*[3]

[1]Department of Pharmacy, School of Healthcare and allied sciences, Ganpati Group of Institutions, Bilaspur, India
[2]School of Pharmacy, Lingaya's Vidyapeeth (Deemed to be university), Faridabad, India
[3]Department of Pharmacology and Toxicology, National Institute of Pharmaceutical Education and Research, Hajipur, India
[4]Department of Pharmacy, School of Medical and Allied Science, Galgotias University, Greater Noida, India

6.1 INTRODUCTION TO B VITAMINS AND THE BRAIN

The human body requires micronutrients such as vitamins in minute amounts to regulate essential physiological processes, including immune function, metabolic activities, and overall health [1]. As indispensable micronutrients, vitamins contribute to overall health and play multifaceted roles in the brain, including protection against oxidative stress, regulation of neural communication, facilitation of energy production, and synthesis of neurotransmitters. Insufficient availability of these micronutrients precipitates abnormal brain function characterized by oxidative imbalance, mitochondrial failure, toxic protein aggregation (α-synuclein and amyloid-β (Aβ)), excitotoxic damage, and ultimately neurodegeneration [2]. Adequate consumption of these vitamins through diet or supplements is fundamental for sustaining brain function and mitigating the progression of neurological pathologies. Vitamin C and the B vitamin family (thiamine [B1], riboflavin [B2], niacin [B3], pantothenic acid [B5], vitamin B6, folate [B9], and vitamin B12) are among the nine water-soluble vitamins that are necessary for human health. The other four vitamins—A, D, E, and K—are fat-soluble. The B vitamin complex works as an essential cofactor in various metabolic pathways, supporting energy production, cellular replication, metabolic intermediate biosynthesis, and the synthesis of crucial neurotransmitters [3]. B vitamin availability must be sufficient to promote appropriate foetal and neonatal growth, maternal health during gestation, reproductive well-being, and cognitive function throughout life. In older populations, when impaired intake and decreased absorption can raise the risk of insufficiency, the need for specific nutrients increases [4].

Due to the fast ageing of the population and the rising incidence of neurodegenerative illnesses like Parkinson's disease (PD) and Alzheimer's disease (AD), cognitive decline has spread like wildfire throughout the world [5]. These disorders have a significant socioeconomic impact on patients, families, and healthcare systems worldwide, impairing memory, learning, and executive function. Age and genetic makeup are unchangeable risk factors, but new research shows that nutritional inadequacies are a significant—yet sometimes overlooked—factor in the acceleration of cognitive decline [6]. One-carbon metabolism is disturbed by B vitamin deficiency, which results in hyperhomocysteinemia, poor methylation, and the buildup of neurotoxic proteins such as α-synuclein in PD and Aβ in AD [7].

6.2 THE IMPACT OF VITAMINS ON AD PROGRESSION

Of all the neurodegenerative diseases, AD is the most common worldwide and the leading cause of dementia in the elderly. Progressive neuronal destruction, the extracellular buildup of Aβ plaques, and the intracellular buildup of neurofibrillary tangles (NFTs) made of hyperphosphorylated tau protein are the neural pathological cornerstones of AD [8,9]. Neuropathological evaluation of AD patients after death reveals a significant reduction in cerebral cortical volume accompanied by intracellular and extracellular accumulations of pathological deposits. Oxidative stress is recognized as one of the primary mechanisms underlying progressive neurodegeneration in AD, consistent with its role in other neurodegenerative diseases. Evidence supports the role of vitamins as antioxidants with potential neuroprotective effects, which has prompted their application as adjuvant approaches in AD

DOI: 10.1201/9781003686675-6

treatment [10]. Homocysteine (Hcy) metabolism depends on several B vitamins, notably B9, B12, B6, and B2, facilitating its methylation and transulphuration. Excess circulating Hcy can adversely affect cognitive performance and is implicated in the pathogenesis of dementia. Treatment with B vitamins decreases total Hcy concentrations, thereby contributing to the management of cognitive dysfunction. Vitamin B12 and folate at increased levels positively affected cognition and were linked to therapeutic improvements in AD. Vitamin B12 has been shown to inhibit tau fibrillization, thereby preventing the aggregation of tau protein and the formation of NFTs, which are key contributors to the progression of AD [11]. A multivitamin-based study highlighted the neuroprotective effects of vitamins B6, B12, B9, and choline under hypoxic conditions. Administration of this multivitamin regimen significantly mitigated hypoxia-induced memory impairments. Treatment with these vitamins was associated with reductions in tau hyperphosphorylation and decreased plasma Hcy levels, key contributors to neurodegeneration. Consequently, memory performance was markedly improved, demonstrating the potential of this multivitamin approach in preserving cognitive function under stress conditions [12].

6.3 THE IMPACT OF VITAMINS ON PD PROGRESSION

It is a neurodegenerative disorder marked by a range of motor and non-motor manifestations, resulting primarily from the loss of dopaminergic neurons in the midbrain [13]. Elevated Hcy, a sulphur-containing metabolite of methionine metabolism, has been implicated in multiple neurotoxic mechanisms across several neurodegenerative disorders, including PD [14]. In PD patients, higher Hcy levels strongly correlate with disease pathogenesis, suggesting that hyperhomocysteinemia may contribute directly to dopaminergic neuronal degeneration and symptom progression [15]. Enhanced plasma Hcy has been linked to dopaminergic neuronal loss within the nigrostriatal pathway in PD. Therefore, controlling Hcy concentrations may help slow the progression of PD [16]. B vitamins serve as essential cofactors in the methylation of Hcy to methionine, and studies have indicated a strong correlation between circulating Hcy levels and B vitamin status. B vitamin administration effectively reduces circulating Hcy levels in the blood [17]. Deficiency of thiamine promotes dopaminergic neuronal death, a hallmark feature of PD pathology.

6.4 WHY THE BRAIN IS UNIQUELY VULNERABLE TO NUTRIENT DEFICIENCIES

The human brain is uniquely vulnerable to nutrient deficiencies due to its exceptional metabolic demands, limited regenerative capacity, and selective nutrient transport mechanisms. Although it represents only about 2% of body weight, the brain consumes nearly 20% of the body's energy, primarily to sustain synaptic activity, neurotransmitter synthesis, and maintenance of ionic gradients critical for neuronal signalling [18]. This energy-intensive environment relies heavily on essential cofactors, particularly B vitamins, to support mitochondrial bioenergetics, oxidative phosphorylation, and one-carbon metabolism. Even subtle vitamin deficiencies such as B1, B6, B9, or B12 can disrupt adenosine triphosphate (ATP) production, impair methylation reactions, elevate Hcy levels, and trigger oxidative stress, compromising neuronal integrity. Unlike other cell types, neurons are largely post-mitotic, making them incapable of dividing to replace lost cells, so damage caused by nutrient insufficiency is often irreversible [19]. While protective, the blood–brain barrier (BBB) restricts nutrient entry, further amplifying vulnerability during periods of increased demand, such as growth, pregnancy, ageing, or metabolic stress [20]. Together, these factors emphasize the necessity of maintaining sufficient amounts of nutrients that support brain function, essential for maintaining the brain's extraordinary ability to function every day of life and preventing neurotoxicity and cognitive dysfunction [21].

6.5 CHAPTER OBJECTIVES AND SCOPE

The chapter "The B Vitamins and Cognitive Health: Methylation, Mitochondrial Function, and Neuronal Integrity" delves further into the crucial function of B vitamins in maintaining neuronal vitality and cognitive efficiency [22]. It investigates the molecular and physiological mechanisms by which B vitamins regulate one-carbon metabolism, Hcy transcription, and epigenetic activities, including deoxyribonucleic acid (DNA) and histone methylation, essential for synaptic restructuring and genetic diversity in neurons. The function of B vitamins in mitochondrial bioenergetics is also covered in this chapter, emphasizing how they serve as cofactors in enzymatic processes that enhance ATP generation, reduce oxidative stress, and maintain brain viability [23]. Through the integration of evidence from preclinical studies in living organisms and clinical studies, the chapter emphasizes the neuroprotective potential of B vitamin supplementation in lowering cognitive decline and advancing neurodegenerative disorders like PD and AD [24]. The objectives include elucidating the molecular pathways linking B vitamin biosynthesis to mental benefits, assessing translational and therapeutic implications, and providing a framework for pharmacological and nutritional strategies to preserve brain function. By linking biochemical pathways to operational cognitive impacts, the following part offers a comprehensive resource for understanding how B vitamins' synergistic role influences methylation, mitochondrial efficiency, neural capacity for endurance, and, overall, their cognitive abilities. It also points out areas that need more research in the discipline [25].

6.6 THE CORE B VITAMINS IN NEUROLOGICAL FUNCTION

Water-soluble components known as B vitamins are essential for maintaining mental health. By acting as cofactors in crucial cellular processes, they support DNA methylation, myelin formation, neurotransmitter generation, and energy production—all of which are necessary for normal brain functioning [26]. Optimum B vitamin supplements are required to preserve neuronal strength, cognitive performance, and synaptic plasticity, while deficiencies are linked to neurological deficits, cognitive decline, and neurodegenerative diseases. The critical B vitamins—thiamine (B1), riboflavin (B2), niacin (B3), pantothenic acid (B5), pyridoxine (B6), folate (B9), and cobalamin (B12)—and their molecular implications to brain functionality and cognitive function are examined in this chapter [27].

6.6.1 Thiamine (B1)

Vitamin B1, also known as B1, or thiamine, is primarily active as thiamine pyrophosphate and is a mediator in several enzymatic processes critical to the nervous system's functioning. Along with aiding in the production of complex dietary carbohydrates, hormones, cholesterol, chemical messengers, and genetic material, it also helps nerve messages travel more easily [28]. Vitamin B1, which is essential for optimal neural system performance, has been connected to major circulatory problems, including premature cardiac death, as well as mood swings, tension, and hopelessness. Axonal degradation, oxidative stress, formation of free radicals, disturbed myelin development, and excitotoxicity caused by glutamatergic excessive stimulation are some of the processes via which a lack of thiamine provision causes neuronal harm. Thiamine deficiency disrupts immune homeostasis in the brain, promoting neuroinflammation through T-cell infiltration, increased secretion of IL-1, TNF-α, and IL-6, and excessive activation of glial cells via the overexpression of CD40/CD40L signalling pathways [29]. In the bloodstream, thiamine predominantly binds to serum proteins, particularly albumin. It is absorbed in the small intestine via passive diffusion at low concentrations and through active transport mechanisms at higher concentrations, with excess thiamine excreted in the urine as free thiamine and acidic metabolites. As an essential cofactor in neuronal metabolism, thiamine supports cellular energy processes and is critical for the synthesis of neurotransmitters, particularly acetylcholine [30]. Deficiency of thiamine is linked to multiple neuropathological mechanisms such as impaired oxidative metabolism, elevated neuroinflammation, endoplasmic reticulum stress, and altered autophagy, all of which culminate in neurodegeneration. Collectively, these dysfunctions are central to the pathogenesis of age-associated neurological diseases like dementia, AD, PD, and Huntington's disease [31].

6.6.2 Riboflavin (B2)

The free radical theory of ageing suggests that continuous aerobic metabolism generates reactive oxygen species (ROS), which accumulate over time and inflict oxidative damage on cells and tissues. Despite its recognized role as an antioxidant, riboflavin has received limited attention in studies investigating antioxidant-mediated mechanisms of healthy ageing and lifespan extension [32]. The antioxidant action of riboflavin is mediated via dual pathways, namely the suppression of lipid peroxidation and the reduction in reperfusion-induced oxidative stress. Riboflavin is fundamental to numerous enzymatic reactions through its active coenzyme forms flavin mononucleotide (FMN) and flavin adenine dinucleotide (FAD) [33]. These coenzymes regulate haemoglobin synthesis, nitric oxide (NO) generation, cytochrome P450-mediated metabolism, and electron transfer and oxygen-handling pathways, enabling the synthesis and regeneration of key vitamins such as niacin, folate, and vitamin B6. Flavoproteins also facilitate the breakdown of essential brain lipids, regulate iron absorption and utilization, and play a crucial role in controlling thyroid hormone activity [34]. Riboflavin derivatives, essential cofactors in the glutathione (GSH) redox cycle, enhance endogenous antioxidant capacity while displaying intrinsic antioxidant properties.

6.6.3 Niacin (B3)

Niacin (vitamin B3), which includes nicotinic acid and nicotinamide, is a vital precursor for the biosynthesis of the coenzymes nicotinamide adenine dinucleotide $(NAD)^+$ and NADPH, both of which are central to cellular energy and redox metabolism [35]. Through phosphorylation by NAD^+ kinase, NAD^+ is transformed into NADPH. NAD^+ itself is a pivotal cofactor required for the oxidative metabolism of lipids, carbohydrates, proteins, and ethanol, in addition to its involvement in intracellular signalling pathways and genomic maintenance via DNA repair. As a key reducing agent, NADPH supports fatty acid and cholesterol synthesis [36]. Insufficient niacin intake leads to neurological and psychiatric disturbances, including anxiety, depression, and mood instability, and has been linked to conditions such as schizophrenia. Insufficient niacin levels result in pellagra, a multisystem disorder marked by dermatitis, gastrointestinal disturbances including diarrhoea, neurological impairment progressing to dementia, and oral mucosal ulcerations. Through its action on the immune system, niacin downregulates pro-inflammatory cytokines and alleviates inflammatory responses, as observed in studies on atherosclerotic disease [37]. After being absorbed in the stomach and small intestine, niacin is stored in the liver, and excess amounts are excreted in urine. The kynurenine process involves the metabolic conversion of tryptophan, which provides another source of NAD.

6.6.4 Pantothenic Acid (B5)

This vitamin supports brain function by playing crucial roles in the production of metabolic energy, the synthesis of neurotransmitters and myelin, neuroprotection, antioxidative defence, hormone synthesis, and the preservation of cognitive function [38]. Maintaining the integrity of the nervous system and preventing psychological illnesses depend on consuming enough of this vitamin through diet. Its importance in promoting general brain function and cognitive resilience is highlighted by its pivotal role in basic metabolic processes [39]. CoA is essential for the biosynthesis of sex hormones, red blood cell development, and the adrenal glands' generation of hormones that cause stress. In addition to its crucial function in oxidative biochemical processes, it facilitates the manufacture of fatty acids, saturated fat, amino acids, and phospholipids, all of which assist neuronal metabolism. Muscular mass: In mammals, this vitamin's central biological role is to help the production of CoA and the metabolism of macronutrients such as proteins, lipids, and carbs [40]. Additionally, it is necessary to produce signalling molecules that carry neural messages to muscle tissue through the action of CoA.

6.6.5 Pyridoxine (B6)

Among its many roles in energy metabolism, mood modulation, neurotransmitter biosynthesis, cognitive development, antioxidant resistance, and haem formation, this vitamin is essential for brain health. Ensuring sufficient levels of this vitamin through nutrition is vital for mental performance, psychological well-being, and neurological health [41]. It contributes to immune defence by promoting antibody synthesis and helps preserve normal neuronal function. This vitamin functions as a cofactor in folate metabolism and as a central mediator of amino acid metabolism, thereby influencing the biosynthesis of neurotransmitters, including dopamine, serotonin, gamma-aminobutyric acid (GABA), noradrenaline, and melatonin [42]. Mild insufficiency of this vitamin may preferentially reduce the production of GABA and serotonin, leading to sleep disturbances, behavioural changes, cardiovascular irregularities, and impaired hypothalamic–pituitary hormonal control. GABA suppresses excessive neuronal firing, while vitamin B6 plays a multifaceted role in regulating DNA expression, immune system activity, and brain glucose homeostasis. Pyridoxal-5′-phosphate concentrations often determine the status of functional parameters and inflammatory biomarkers, with deficiency contributing to increased severity of inflammation. Pyridoxal 5′-phosphate (PLP) is essential for both tryptophan catabolism and one-carbon metabolism, which is why this impact occurs. Inflammation has a part in the development of disorders that impact mental wellness and cognitive function, which emphasizes the importance of this role [43].

6.6.6 Folate (B9) and Cobalamin (B12)

Both folate (commonly known as vitamin B9) and cobalamin (also known as vitamin B12) are necessary for maintaining mental performance and cerebral health. Their functions in DNA synthesis, neurotransmitter generation, myelin growth, and Hcy synthesis underscore their significance in fostering cognitive performance, regulating mood, and preventing neurodegenerative disorders [44]. A sufficient quantity of these vitamins, whether via diet or supplementation, is crucial for maintaining brain function and preventing cognitive impairment. They are critical during rapid cellular and tissue growth periods such as young adulthood, adolescence, and pregnancy, since they facilitate the creation of DNA and ribonucleic acid (RNA). Cobalamin and folate also increase the synthesis of cells that carry oxygen and optimize iron utilization [45]. These nutrients have closely connected functions because they cooperate in the methionine and folate synthesis cycles. Functional folate shortage can result from folate being trapped as methyl tetrahydrofolate (THF) due to a vitamin B12 shortage. The destabilization of DNA caused by folate shortage affects transcription and repair mechanisms, which may impede activity potential production, promote hippocampus atrophy, disturb myelination, hinder neural plasticity, and degrade membrane lipid layer function. It also reduces purine and pyrimidine synthesis and alters DNA methylation in brain tissue [46]. Deficiencies of folate or B12 impair nucleotide and protein synthesis, affecting rapidly dividing tissues and contributing to megaloblastic anaemia, fatal developmental abnormalities, and specific neurological dysfunctions. The folate cycle also supports the production and regeneration of tetrahydrobiopterin, a cofactor required for enzymes that convert amino acids into NO and monoamine neurotransmitters. Deficiency of vitamin B6 can manifest neurologically as depression, cognitive impairment, psychiatric disturbances, and autonomic dysfunction, whereas B12 deficiency often presents with neurological symptoms before typical haematological changes [47]. Interestingly, a subset of individuals with folate or B12 deficiencies exhibit only anaemia. In contrast, others display exclusively neuropsychiatric symptoms, and nearly one-third of patients in psychiatric care show deficiencies of these vitamins. Key B vitamins, their coenzyme forms, and primary neurological roles are given in Table 6.1.

6.7 MECHANISM I: METHYLATION AND THE ONE-CARBON CYCLE

Methylation, the enzymatic transfer of methyl groups to DNA, RNA, proteins, and lipids, is a fundamental biochemical process that regulates gene expression, protein activity, and cellular signalling [48]. Central to this process is the one-carbon cycle, which integrates the folate and methionine cycles to generate methyl donors such as S-adenosylmethionine (SAM). Key B vitamins—including folate (B9), cobalamin (B12), and pyridoxine (B6)—act as essential cofactors, supporting the

TABLE 6.1
Key B Vitamins, Their Coenzyme Forms, and Primary Neurological Roles

Vitamin	Active Form	Sources	Primary Neurological Role
Vitamin B1 (thiamine)	Thiamine pyrophosphate (TPP)	Whole grains and legumes	Glucose metabolism, neurotransmitter synthesis, nerve impulse transmission, myelin maintenance, neuroprotection
Vitamin B2 (riboflavin)	Flavin mononucleotide (FMN), flavin adenine dinucleotide (FAD)	Dairy products, eggs, and leafy greens	Antioxidant defence, mitochondrial energy metabolism, neurotransmitter synthesis, red blood cell formation
Vitamin B3 (niacin)	Nicotinamide adenine dinucleotide (NAD^+), NAD phosphate (NADPH)	Meat, fish, nuts, and whole grains	Metabolism and DNA repair, neurotransmitter synthesis, neuroprotection, and antioxidative defence
Vitamin B5 (pantothenic acid)	Coenzyme A (CoA)	Meat, fish, whole grains, and legumes	Synthesis of coenzyme A, essential for metabolism, myelin formation, and cholesterol synthesis
Vitamin B6 (pyridoxine)	Pyridoxal-5′-phosphate (PLP)	Poultry, fish, bananas, and potatoes	Amino acid metabolism, neurotransmitter synthesis (GABA, serotonin, dopamine, norepinephrine), and haemoglobin production
Vitamin B9 (folate)	Tetrahydrofolate (THF), 5-methyl-THF	Leafy greens, legumes, and fortified grains	DNA and RNA synthesis and cell division are critical during pregnancy and homocysteine metabolism
Vitamin B12 (cobalamin)	Methyl cobalamin, adenosyl cobalamin	Meat, fish, dairy, and eggs	Function related to the nervous system, including the synthesis of DNA and RBCs

conversion of Hcy to methionine, nucleotide biosynthesis, and critical methylation reactions [49]. Proper functioning of the one-carbon cycle is vital for maintaining neuronal integrity, myelin synthesis, neurotransmitter production, and epigenetic regulation of cognitive function. Dysregulation of this pathway has been associated with cognitive decline, neurodegenerative disorders, and psychiatric conditions, underscoring the central importance of B-vitamin-dependent methylation in sustaining brain health [50].

6.8 THE FUNDAMENTALS OF ONE-CARBON METABOLISM

Folates (vitamin B9) are present in various foods, predominantly as polyglutamate conjugates, influencing their bioavailability and absorption. Inadequate dietary intake of folate can lead to megaloblastic anaemia and is a well-established risk factor for neural tube defects (NTDs) in developing embryos [51]. Several countries fortify foods with synthetic folic acid to hamper folate deficiency, an oxidized and more stable form than naturally occurring polyglutamated folates. Although folic acid is biologically inactive, it is converted within cells by dihydrofolate (DHF) reductase (DHFR) into DHF and subsequently into THF [52]. THF is then polyglutamated by the enzyme folyl-polyglutamate synthetase (FPGS), which enhances its cellular retention and facilitates its role as a one-carbon carrier in critical biosynthetic and methylation reactions. This polyglutamate step is crucial for retaining THF within cells and enhancing its activity as it enters the one-carbon cycle. Enzymes such as serine hydroxymethyltransferase (SHMT) and the glycine cleavage system (GCS) catalyse the transfer of a one-carbon unit from serine to THF, producing the key intermediate 5,10-methylene-THF (5,10-CH_2-THF), which serves as a central donor in nucleotide biosynthesis and methylation reactions. When this intermediate, DHF, and the precursor THF endure oxidative reactions at the C-9–N-10 link, three distinct intermediates are produced: a pteridine moiety, p-aminobenzoyl glutamate (pABG), and formaldehyde [53]. This fragility emphasizes the importance of folate metabolites being appropriately polyglutamated and preserved in cells to maintain their effectiveness in the one-carbon cycle. Consequently, the cellular flux of one-carbon units naturally includes a supply of reactive formaldehyde, which could damage the cells' survival (Figure 6.1). The unintentional release of formaldehyde can result in proteotoxic stress, genotoxic stress, and the generation of ROS, among other negative consequences. This emphasizes how tightly regulated folate metabolism is essential to maintaining the health of neurons and cells overall [54]. Furthermore, the breakdown of THF derivatives into formaldehyde has been shown to reduce the availability of one-carbon units for essential metabolic pathways. Particularly in rapidly dividing brain and glial cells, this blockage may compromise cellular function and genomic integrity by causing nucleotide disturbances, replication-induced stress, and DNA damage [55].

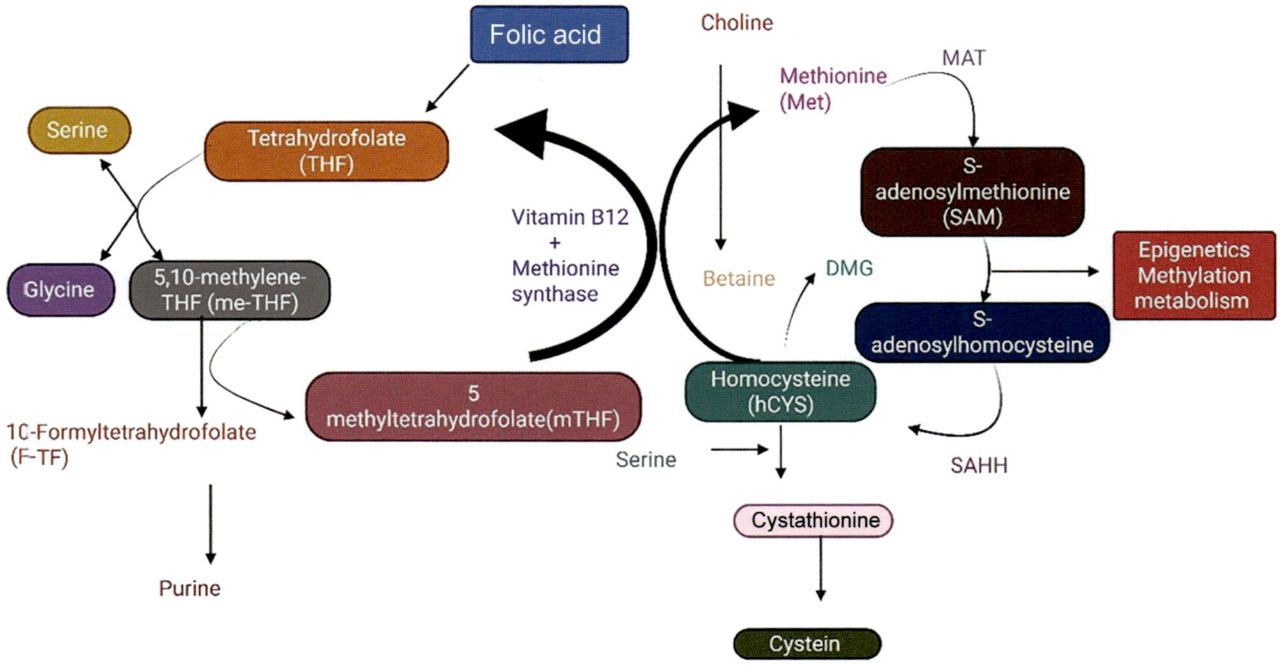

FIGURE 6.1 The one-carbon metabolism pathway.

6.9 THE CRITICAL ROLE OF FOLATE AND B12 IN THE METHYLATION CYCLE

The B vitamins cobalamin (B12), pyridoxine (B6), and folic acid (B9) are essential substrates in a one-carbon pathway. By moving a single carbon element from serine or glycine to the THF, this pathway produces methylene-THF, a vital molecule required for nucleotide metabolism and methylation, which are activities critical to neuronal function and genomic integrity [56]. The generated methylene-THF may follow several distinct pathways, contingent on the requirements of the cell. Purine production, essential for synthesizing DNA and RNA, can be oxidized to formyl-THF or used to generate the DNA nucleotide thymidylate (dTMP). On the other hand, the irreversible reduction of methylene-THF to 5-methyl-THF provides a methyl group for the methylation of Hcy to methionine. This process depends on vitamin B12 as a coenzyme [57]. The methylation of Hcy to methionine is catalysed by a methyltransferase enzyme that requires methylcobalamin (active B12) as a prosthetic group. An alternative methylation pathway involves betaine as the methyl donor, which does not require vitamin B12; however, this reaction occurs predominantly in the liver and minimizes Hcy metabolism in the central nervous system (CNS). Methionine is converted into SAM, the universal methyl donor that participates in many essential methylation reactions, including DNA, RNA, proteins, and neurotransmitters [58]. The methionine synthesis and its conversion to SAM represent a critical step in one-carbon metabolism. It is fundamental in maintaining neuronal integrity, myelin formation, and overall brain health. SAM-dependent methylation reactions are widespread in the brain, generating critical molecules such as neurotransmitters (including catecholamines and indoleamines), phospholipids, and myelin, all essential for proper neuronal function. Methylation plays a key role in modulating the expression of presenilin-1, the activity of γ-secretase, and the Aβ levels. It also regulates the expression of leucine carboxyl methyltransferase-1 and the B subunit of protein phosphatase 2A (PP2A), which is closely associated with the degree of tau protein phosphorylation. Folate and vitamin B12 deficiency markers, such as elevated levels of Hcy and methylmalonic acid (MMA), are commonly observed in older adults. Conditions such as atrophic gastritis and the frequent use of antacid medications can impair the intestinal absorption of dietary vitamin B12 [59]. The movement of folate and vitamin B12 into the brain, mediated by receptor-dependent endocytosis at the BBB, can become compromised with ageing. Neuropsychiatric disorders in older adults have multiple contributing factors, and the role of disrupted one-carbon metabolism in these conditions has primarily been evaluated through epidemiological research.

6.10 THE HOMOCYSTEINE HYPOTHESIS OF COGNITIVE DECLINE

6.10.1 Homocysteine as a Neurotoxin

Hcy, a sulphur-containing amino acid derived from methionine metabolism, has emerged as a critical factor in the pathophysiology of cognitive decline and neurodegenerative disorders. Elevated plasma levels of Hcy, known as hyperhomocysteinemia, are consistently associated with an increased risk of dementia, AD, and vascular cognitive impairment. The neurotoxic potential of Hcy arises from its multifaceted effects on neuronal structure and function, particularly through mechanisms involving excitotoxicity, oxidative stress, mitochondrial dysfunction, and impaired methylation pathways [60].

At the molecular level, Hcy exerts excitotoxic effects by overstimulating N-methyl-D-aspartate (NMDA) receptors, leading to excessive calcium influx, activation of calcium-dependent enzymes, and subsequent neuronal injury. This excitatory cascade disrupts synaptic signalling and promotes apoptosis and neuronal death [61]. At the same time, Hcy tilts the scales in favour of cellular oxidative stress by encouraging the production of ROS and reducing the activity of antioxidant enzymes. Brain degeneration and impaired synaptic function are further exacerbated by ROS-mediated damage to lipids, proteins, and nucleic acids [62]. Hcy is also linked to mitochondrial damage, where it inhibits the electron transport chain (ETC), lowers ATP synthesis, and causes a discharge of cytochrome c and other pro-apoptotic proteins. Hcy disrupts one-carbon metabolism by preventing methylation events that rely on SAM, a universal methyl donor. By affecting DNA and histone methylation, neurotransmitter production, and phospholipid synthesis, this methylation dysregulation ultimately leads to impairments in brain plasticity, cell structural integrity, and gene function. Additionally, Hcy contributes to vascular disorders by producing malfunctioning endothelial cells, accelerating smooth muscle expansion, and boosting the assembly of platelets [63]. These outcomes worsen its direct neurotoxic activity by reducing cerebral perfusion and BBB integrity. When considered collectively, these mechanisms demonstrate that Hcy is a pathogenic mediator that accelerates cognitive impairment by directly harming neurons and causing vascular injury.

6.10.2 Vascular Implications for Brain Health

Due to the significant impact that high Hcy levels have on cerebrovascular stability and function, hyperhomocysteinemia is a primary vascular indication for cognitive impairment. Hcy-mediated vascular toxicity contributes to both macrovascular and microvascular pathology, which, in turn, impairs cerebral perfusion, disrupts BBB homeostasis, and accelerates neurodegenerative processes [64]. The accumulative impact of these vascular assaults is significant in conditions like AD, dementia of the vascular system, and mixed dementia, when both brain neurons and vascular processes are involved [65].

Another significant consequence of high Hcy levels is the deterioration of the BBB. Increased Hcy also increases permeability and weakens rigid junction proteins, which facilitates the entry of neurotoxic agents and inflammatory messengers into the brain substrate [66].

- Dysfunction of the barrier that connects blood to brain exacerbates neuroinflammation and makes neurons more vulnerable to oxidative and metabolic stress. The unhealthy environment created by reduced oxygen distribution, microvascular injury, and interruption of the BBB eventually reduces the survival of neurons and episodic recall.
- The vasculature implications of hyperhomocysteinemia go beyond simple perfusion inadequacies and involve a complex interplay between angiogenic risk, vascular restructuring, damaged endothelial cells, and BBB degradation [67]. These channels provide a molecular link between the neurological modifications underlying dementia and systemic metabolic problems. A key therapy approach may concentrate on vascular channels affected by Hcy to preserve brain function and slow cognitive decline.
- By increasing oxidative stress and lowering NO accessibility, Hcy leads to dysfunction. In addition to rescuing NO, the formation of superoxide radicals results in the formation of peroxynitrite, a potent oxidant that damages endothelial cells and inhibits vascular dilation [68]. Abnormal endothelium–triggered stretching results in decreased cerebral blood flow and limited brain response to metabolic demands. By encouraging endothelial activation, which is characterized by the upregulation of adhesion molecules and inflammatory cytokines, Hcy also promotes the proliferation of leukocytes and inflammation in the arteries.
- Hcy fosters vascular remodelling and prothrombotic states. It stimulates vascular smooth muscle cell proliferation, enhances collagen deposition, and increases arterial stiffness, contributing to microvascular rarefaction and impaired haemodynamic regulation [69]. Furthermore, Hcy promotes platelet aggregation and enhances the coagulation cascade by interfering with natural anticoagulant systems such as protein C and antithrombin III. These changes increase the risk of cerebral small vessel disease, microinfarcts, and white matter lesions—hallmarks of cognitive decline associated with vascular pathology.

6.11 IMPACT ON NEUROTRANSMITTER SYNTHESIS AND EPIGENETIC REGULATION

One-carbon metabolism encompasses multiple interconnected pathways, including the one-carbon cycle, the methionine cycle, the transsulphuration pathway, and the formaldehyde cycle. In one-carbon metabolism, THF is a central carrier of one-carbon units, supplying them to various cellular compartments and biosynthetic processes. The cycle operates in a compartmentalized manner, spanning the mitochondria and cytosol, and during the S-phase, it also extends into the nucleus [70]. Epigenetics encompasses chromatin, DNA, and RNA modifications that regulate gene expression without altering the underlying nucleotide sequence. Central to these processes is the methionine cycle, which produces SAM (AdoMet), the universal methyl donor. AdoMet is utilized in both the nucleus and cytosol to support diverse methylation reactions critical for epigenetic regulation. Within DNA, the most common epigenetic modification is adding a methyl group at the C5 position of cytosine, forming 5-methylcytosine (5mC) [71]. This modification occurs primarily at CpG dinucleotides and is generally linked to transcriptional silencing. DNA methyltransferases (DNMTs) are the enzymes that transfer a methyl group from SAM (AdoMet) to cytosine residues in DNA. In addition to DNA, histone proteins in the nucleus undergo post-translational methylation, which is critical in regulating gene expression [72]. Multiple families of histone methyltransferases mediate this process, including protein arginine methyltransferases (PRMTs), enzymes containing a SET domain, and DOT1-like proteins. During methylation reactions, SAM (AdoMet) donates its methyl group and is subsequently converted to S-adenosyl Hcy (AdoHcy), which acts as a potent feedback inhibitor of most AdoMet-dependent methyltransferases. Hcy, derived from AdoHcy, can exit the nucleus and be methylated to methionine through the action of methionine synthase (MS), utilizing 5-methyl-THF (5-CH3-THF) as the methyl donor. In parallel, betaine-Hcy S-methyltransferase (BHMT), located in both cytosol and nucleus, can also regenerate methionine from Hcy by using betaine, a metabolite of choline, as an alternative methyl donor [73].

Apart from its methylation to methionine, Hcy can also be metabolized through the transsulphuration pathway to generate cysteine, which serves as a precursor for GSH, the primary cellular antioxidant [74]. Cystathionine (CTH) β-synthase (CBS), a cytosolic enzyme that can go to the nuclei after post-translational sumoylation and change Hcy into CTH, catalyses the initial step in this process. After that, CTH is hydrolysed by CTH γ-lyase to generate cysteine. Glutamate–cysteine ligase (GCL), which facilitates the condensation of cysteine and glutamate, the initial step in the synthesis of GSH, includes a combination of a catalytic (GCLC) component and a controlling (GCLM) component. Finally, glycine is added to this dipeptide structure by GSH synthase (GSS), which produces decreased GSH [75]. Cell-mediated GSH generation is crucial for maintaining redox balance, minimizing the adverse effects of formaldehyde, and modifying epigenetic regulation. The impact of differences in formaldehyde metabolism on epigenetic signalling and cell balance will require more investigation.

6.12 MECHANISM II: MITOCHONDRIAL FUNCTION AND BIOENERGETICS

6.12.1 The Neuron's High Demand for Energy (ATP)

The ROS are inevitable with mitochondrial energy consumption, especially in the ETC process. When the balance tips in favour of excessive ROS generation or antioxidant defences are compromised or depleted, oxidative stress results [76]. Widespread cellular damage, such as nuclear DNA lesions and changes to mitochondrial DNA (mtDNA), can result from this imbalance. Notably, mtDNA is highly susceptible to oxidative damage because it lacks protective histones and is close to the ETC, the central location where ROS are generated. One of the hallmarks of neurological illnesses is impaired or decreased mitophagy, which accelerates the death of neuronal cells. Mitochondrial transit disruption affects synaptic bioenergetics and may lead to neuronal death. A significant amount of ATP is needed to support neurotransmitter synthesis, release, and absorption during synaptic transmission, making it an extremely energy-demanding process. In addition to enabling the synthesis of neurotransmitters, mitochondria are essential for supplying these energy demands [77]. According to one theory, a decrease in mitochondrial density at synapses impairs signalling effectiveness and encourages the degeneration of neurons. Given this, it becomes clear that controlling mitophagy and mitochondrial biogenesis inside synapses is an important topic that needs more research. Transcellular mitophagy, a process by which neurons release mitochondria into the synaptic environment, where they are then taken up and broken down by glial cells, has been the subject of recent research [78]. This route may be a crucial connection between impaired mitophagy and synaptic dysfunction, beyond simply energy-related deficiencies, even though it is still little understood. Neurons have exceptionally high energy demands, particularly to sustain synaptic activity and the processes of neurotransmitter synthesis, release, and reuptake. Most of this ATP supply is derived from oxidative phosphorylation, and disruption of this pathway can result in profound functional deficits. Unlike other cell types, neurons possess a limited capacity to shift toward glycolysis or anaerobic ATP production due to the absence of key enzymatic components. Consequently, they depend heavily on astrocytes for metabolic support, primarily through lactate provision [79]. Under stress conditions, astrocytes enhance glycolytic activity to increase lactate availability for neuronal utilization. However, such metabolic adaptations may disturb lactate–glucose homeostasis, affecting mitochondrial activity and redox balance, thereby impairing feedback regulation of bioenergetic pathways. These perturbations contribute to mitochondrial dysfunction, excitotoxicity from glutamate accumulation, and progressive neurodegeneration.

6.12.2 B Vitamins as Essential Coenzymes in the Krebs Cycle and Oxidative Phosphorylation

B vitamins are indispensable coenzymes facilitating key enzymatic reactions in the Krebs cycle and oxidative phosphorylation, underpinning cellular bioenergetics [80]. Thiamine (B1), in the form of thiamine pyrophosphate, is critical for the function of pyruvate dehydrogenase and α-ketoglutarate dehydrogenase complexes, enabling the

TABLE 6.2
Role of B Vitamins in Various Metabolic Pathways

Vitamins	Major Role	Cofactor	Key Role in Metabolism
B1 (thiamine)	α-Keto acid decarboxylation and transketolase-mediated carbon transfer reactions	Thiamine pyrophosphate (TPP)	The pyruvate dehydrogenase complex is a crucial step linking mitochondrial breakdown of energy with cytosolic glycolysis through the Krebs cycle, which converts pyruvate into acetyl-CoA
B2 (riboflavin)	Redox reactions (electron transfer)	Flavin adenine dinucleotide (FAD), flavin mononucleotide (FMN)	The succinate dehydrogenase enzyme oxidizes succinate to fumarate in the TCA cycle while also transferring electrons to FAD to generate $FADH_2$, a component of the electron transport chain
B3 (niacin)	Redox reactions (electron transfer)	Nicotinamide adenine dinucleotide (NAD), nicotinamide adenine dinucleotide phosphate (NADP)	Glyceraldehyde-3-phosphate dehydrogenase catalyses the conversion of glyceraldehyde-3-phosphate to 1,3-bisphosphoglycerate, producing NADH and facilitating subsequent ATP generation in glycolysis
B5 (pantothenic acid)	Acyl group transfer	Coenzyme A (CoA)	Acetyl-CoA is converted into malonyl-CoA by acetyl-CoA carboxylase, supplying the activated two-carbon units required for fatty acid chain elongation
B9 (folate)	One-carbon metabolism (DNA synthesis)	Tetrahydrofolate (THF)	Thymidylate synthase catalyses the methylation of dUMP to dTMP, generating the thymidine nucleotide necessary for DNA synthesis and maintaining genomic integrity
B12 (cobalamin)	Methylation and rearrangement reactions	Methylcobalamin, 5′-deoxyadenosylcobalamin	Methionine synthase catalyses the methylation of homocysteine to methionine, using methylcobalamin and 5-methyltetrahydrofolate, supporting methyl group transfer and protein synthesis

conversion of glycolytic and amino acid–derived intermediates into acetyl-CoA and succinyl-CoA. Riboflavin (B2), as FMN and FAD, serves as a redox cofactor in succinate dehydrogenase and the ETC. At the same time, niacin (B3), through NAD^+/NADH, mediates electron transfer from multiple dehydrogenase reactions within the TCA cycle to complex I of the [81] mitochondrial respiratory chain. Pantothenic acid (B5) forms coenzyme A, essential for acetyl group transfer into the cycle, and pyridoxine (B6), as PLP, facilitates transamination reactions that replenish TCA intermediates. Biotin (B7) supports pyruvate carboxylase-mediated anaplerotic reactions, and cobalamin (B12) functions as a coenzyme for methylmalonyl-CoA mutase, generating succinyl-CoA to maintain cycle continuity. Collectively, these vitamins orchestrate the coupling of substrate oxidation to ATP production, maintain mitochondrial integrity, and are vital for tissues with high energetic demands, particularly the brain and skeletal muscle [82].

6.12.3 The Role of B1, B2, B3, B5, B6, B9, and B12

B vitamins comprise a group of eight water-soluble nutrients that are vital for numerous metabolic processes and overall human health. The following are included in this group: B1, also known as (thiamine), vitamin B2 (riboflavin), and vitamin B3 (niacin), vitamin B5 (pantothenic acid), B6 (pyridoxine), vitamin B9 (folic acid), and vitamin B12 (cobalamin) [83]. Each B vitamin is a cofactor in specific life processes to assist methylation, DNA creation, energy conversion, and regeneration. B vitamins are essential for preserving health and averting disease because they are pivotal in metabolic pathways and cellular function. Numerous clinical symptoms, such as neurological deficits, anaemia, and dermatological disorders, can arise from deficiencies in these vitamins [84]. Table 6.2 shows the role of B vitamins in various metabolic pathways.

6.13 LINKING BIOENERGETIC FAILURE TO COGNITIVE IMPAIRMENT

Maintaining optimal cellular energy metabolism is crucial for cognitive function, especially in neurons with high nutritional requirements. Since neurons' ability to produce energy through glycolysis is hindered, they mainly rely on oxidative phosphorylation in the mitochondria [85]. Therefore, a state of nutritional failure, marked by inadequate ATP production, elevated ROS generation, and disturbed calcium homeostasis, can result from any disturbance in mitochondrial function, whether brought on by diminished substrate availability, impaired ETC activity, or compromised mitochondrial dynamics. The brain is especially susceptible to energy shortages because a large amount of neuronal ATP is used for synaptic activity, neurotransmitter production, and ion gradient preservation. Neurotransmission, synaptic plasticity, and long-term potentiation—all essential for learning and memory processes—are hampered by decreased energy availability at synapses [86]. Furthermore, mitochondrial-mediated apoptotic pathways can be triggered by biological stress, which exacerbates neuronal death and adds to the cumulative loss of cognitive function seen in neurodegenerative illnesses like PD and AD.

Furthermore, the maintenance of neuronal energy balance depends on the metabolic support provided by glial cells, especially astrocytes [87]. Via the astrocyte–neuron lactate shuttle, astrocytes provide lactate to support the oxidative metabolism of neurons. Energy deficiencies, oxidative stress, and excessive excitotoxicity through glutamate buildup can all be exacerbated by disruptions in this molecular coupling. Bioenergetics and cognitive integrity are further compromised by nutrient shortages, especially in B vitamins and cofactors necessary for mitochondrial enzyme performance. All things considered, biological energy failure creates a mechanistic connection between synaptic deficiencies, mitochondrial malfunction, and cognitive impairment [88]. Understanding these pathways (Figure 6.2) highlights the importance of maintaining mitochondrial health and metabolic support in preventing and treating cognitive impairment. A schematic of a mitochondrion and its linkage to vitamin B is shown in Figure 6.2.

6.14 MECHANISM III: MAINTAINING NEURONAL INTEGRITY AND COMMUNICATION

6.14.1 Myelin Sheath Synthesis and Repair: The Role of Vitamin B12

Vitamin B12 (cobalamin) is a coenzyme in multiple metabolic pathways, including lipid, carbohydrate, and protein metabolism. Dietary cobalamin is typically obtained in deoxyadenosylcobalamin or methylcobalamin, which are biologically active. Research has demonstrated that methylcobalamin positively influences myelination by upregulating myelin basic protein expression in Schwann cells, thereby accelerating myelin formation [89]. Moreover, in experimental models of traumatic sciatic nerve injury in rats, co-administering vitamin B12 with dexamethasone—a glucocorticoid commonly used to reduce inflammation—enhances peripheral nerve regeneration [90]. This treatment increases Schwann cell proliferation, promotes the formation of myelinated nerve fibres, and enlarges the diameter of regenerating axons. Functional improvements were observed through elevated sciatic functional index scores, enhanced compound muscle action potentials, increased motor end plate numbers, and improved axonal and myelin morphology.

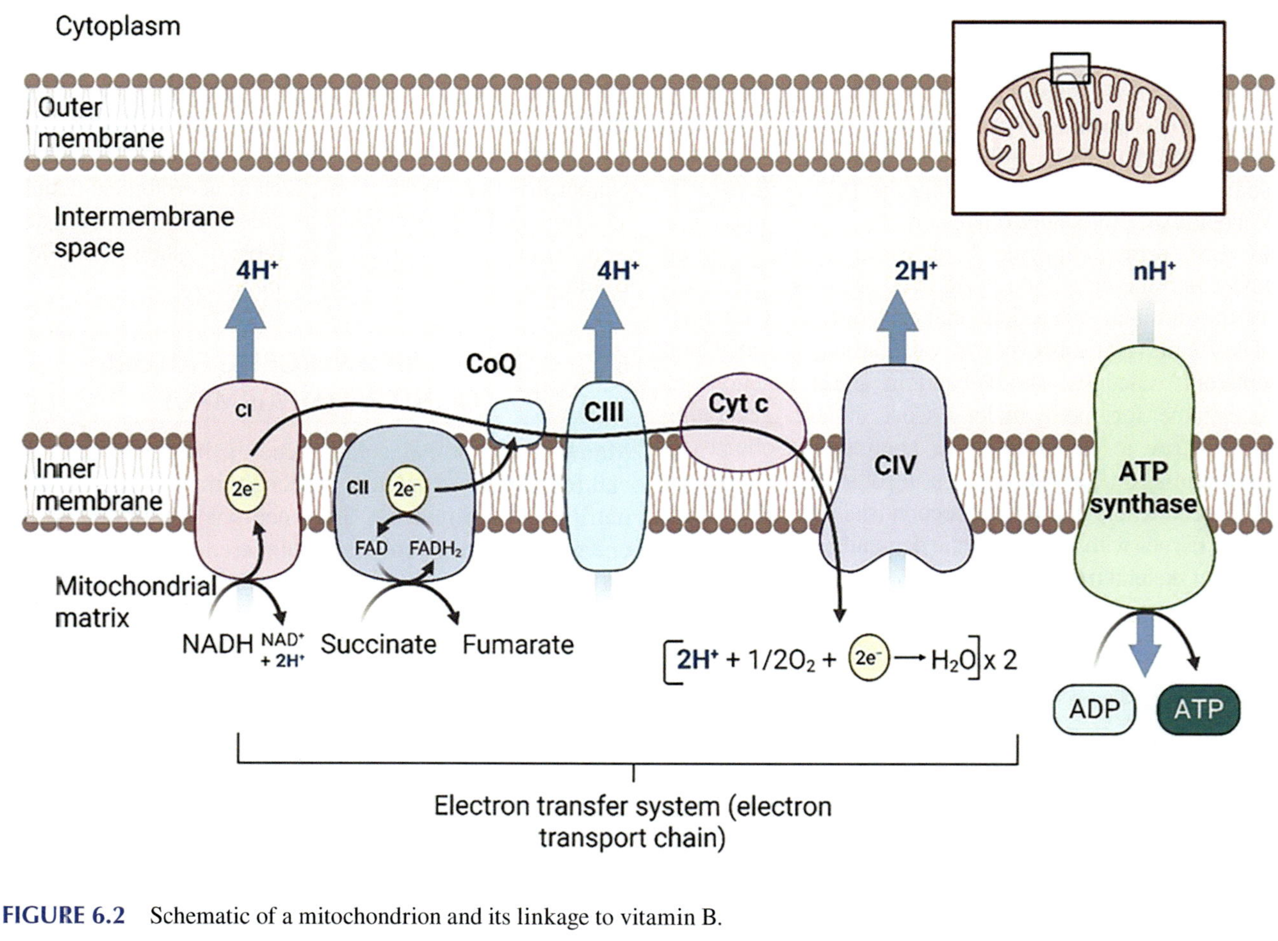

FIGURE 6.2 Schematic of a mitochondrion and its linkage to vitamin B.

Mechanistically, these effects of vitamin B12 are thought to be mediated, at least in part, by the upregulation of neurotrophic factors, including brain-derived neurotrophic factor (BDNF), which support neuronal survival, axonal growth, and myelination [91].

6.14.2 Synthesis of Key Neurotransmitters via Vitamin B6

Vitamin B6, predominantly in its active form, PLP, functions as an essential coenzyme in numerous enzymatic reactions involved in amino acid metabolism and neurotransmitter synthesis [92]. PLP-dependent enzymes catalyse decarboxylation, transamination, and racemization reactions that are critical for producing neurotransmitters necessary for proper CNS function. Deficiency in vitamin B6 leads to impaired neurotransmitter synthesis, contributing to neurological dysfunction and cognitive disturbances.

6.14.3 GABA (Gamma-Aminobutyric Acid) Synthesis

GABA, the primary inhibitory neurotransmitter in the CNS, is synthesized from glutamate through the action of glutamate decarboxylase (GAD). This enzyme requires PLP as a cofactor to catalyse the decarboxylation of glutamate to GABA. Adequate GABA levels are crucial for maintaining excitatory-inhibitory balance in neural circuits. Deficiency of vitamin B6 reduces GABA synthesis, potentially causing neuronal hyperexcitability, seizures, and related neurological disorders [93].

6.14.4 Dopamine, Norepinephrine, and Epinephrine Synthesis

Catecholamines are synthesized from the amino acid tyrosine. Tyrosine hydroxylase converts tyrosine to L-DOPA, which is decarboxylated to dopamine by aromatic L-amino acid decarboxylase (AADC), a PLP-dependent enzyme. Dopamine serves as a precursor for norepinephrine and epinephrine through the actions of dopamine β-hydroxylase and phenylethanolamine N-methyltransferase, respectively [94]. PLP is critical for dopamine formation; thus, B6 deficiency can impair catecholamine biosynthesis, affecting mood, motivation, and autonomic function.

6.14.5 Serotonin (5-Hydroxytryptamine) Synthesis

Serotonin is synthesized from tryptophan via two steps: hydroxylation by tryptophan hydroxylase to 5-hydroxytryptophan (5-HTP), followed by decarboxylation to serotonin by aromatic L-ADDC (PLP-dependent) [95]. Reduced PLP availability diminishes serotonin synthesis, which can contribute to mood disorders, irritability, and sleep disturbances.

6.14.6 Histamine Synthesis

Histamine, a neurotransmitter involved in arousal, immune modulation, and gastric acid secretion, is formed from histidine decarboxylase, another PLP-dependent enzyme. Insufficient vitamin B6 impairs histamine production, potentially affecting CNS signalling and immune responses [96].

6.15 NEUROINFLAMMATION AND THE MODULATORY ROLE OF B VITAMINS

Neuroinflammation is a hallmark of many neurological disorders, including PD, AD, MS, and traumatic brain injury. It is characterized by the activation of glial cells—particularly microglia and astrocytes—leading to the release of pro-inflammatory cytokines, chemokines, and reactive oxygen/nitrogen species (ROS/RNS) [97]. While acute neuroinflammatory responses can be protective, chronic or uncontrolled neuroinflammation contributes to neuronal dysfunction, synaptic loss, and progressive neurodegeneration. B vitamins play a pivotal role in modulating neuroinflammatory pathways through several mechanisms, including the regulation of Hcy metabolism, antioxidant defence, mitochondrial function, and neurotransmitter synthesis [98].

6.15.1 Regulation of Homocysteine and Neuroinflammation

Vitamins B6, B9 (folate), and B12 (cobalamin) are essential cofactors in Hcy metabolism, converting Hcy into methionine or cysteine via methylation and transsulphuration pathways [99]. Elevated Hcy is a known pro-inflammatory and pro-oxidative factor that can induce endothelial dysfunction, increase BBB permeability, and stimulate microglial activation. Adequate B vitamin status reduces Hcy accumulation, mitigating neuroinflammatory responses and oxidative stress in neural tissues.

6.15.2 Antioxidant and Mitochondrial Protective Effects

Several B vitamins support mitochondrial bioenergetics and antioxidant defence, including B2 (riboflavin), B3 (niacin), and B6. Riboflavin, as FAD/FMN, and niacin, as NAD^+/NAD phosphate $(NADP)^+$, are critical for redox reactions in the ETC, minimizing electron leakage and ROS generation [100]. PLP, the active form of B6, contributes to GSH synthesis by providing substrates for cysteine production, enhancing cellular antioxidant capacity. B vitamins indirectly suppress the pro-inflammatory activation of glial cells by maintaining mitochondrial function and limiting oxidative stress.

6.15.3 Modulation of Neurotransmitter Synthesis

Vitamin B6–dependent neurotransmitter synthesis can influence neuroinflammation. For example, GABA, synthesized via PLP-dependent GAD, exerts anti-inflammatory effects by inhibiting microglial activation. A balanced neuroinflammatory response can also be achieved by modulating neurological signalling through B6-mediated synthesis of dopaminergic and serotonergic receptors, respectively [101].

6.16 IMPACT ON NEUROTROPHIC AND ANTI-INFLAMMATORY SIGNALLING

Additionally, neurotrophic factors, including nerve growth factor, also known as NGF, and neurotrophic factor from the brain (BDNF), which support neuronal survival, repair, and anti-inflammatory signalling, are expressed more when B vitamins are taken. B vitamins mitigate the detrimental effects of chronic neuroinflammation on synaptic integrity and cognitive performance by promoting these pathways [102]. As a result, B vitamins modulate neuroinflammation through promoting the synthesis of neurotrophic factors and neurotransmitters, boosting antioxidant protection, controlling Hcy metabolism, and preserving mitochondrial function. These vitamin deficiencies can worsen glial activation, oxidative stress, and the release of pro-inflammatory cytokines, which can lead to neuronal damage and neurodegenerative diseases [100]. Therefore, it is essential to maintain neuroimmune balance and reduce inflammation-induced neurological dysfunction by consuming enough B vitamins. Table 6.3 shows the effects of specific B vitamin deficiencies on neuroinflammation.

6.17 CLINICAL PERSPECTIVES AND HUMAN STUDIES

6.17.1 Classic Deficiency Syndromes

The professional findings and signs of shortages of vitamins may vary greatly depending on the specific vitamin in issue, demographics, place of origin, dietary habits, inherited conditions that coexist, and socioeconomic status [103]. The kind and intensity of the deficit can significantly impact the appearance. However, pallor, exhaustion, disorientation, and widespread muscle weakness are some of the most commonly reported symptoms. Characteristic clinical signs and symptoms are frequently present in conjunction with vitamin deficiencies. For instance, a lack of vitamin B12 can cause megaloblastic anaemia and neurological deficits, while a lack of folate during gestation is associated with a higher risk of neural tube abnormalities in the growing foetus. Scurvy syndrome and night vision are linked to vitamin C and A deficits, respectively [104]. A thorough medical history, laboratory testing, and clinical evaluation are typically needed to diagnose vitamin deficits. The specific vitamin impacted and the degree of the deficiency determine the management approach. Therapy of underlying causes, supplements, and dietary changes are usually combined for an effective restoration. To avoid possibly irreversible harm, early intervention is crucial because severe deficits frequently show up as neurological symptoms [105]. However, some of these symptoms may be

TABLE 6.3
Effects of Specific B Vitamin Deficiencies on Neuroinflammation

Vitamins	Biochemical Role	Effects of Deficiency	Neuroinflammatory Consequences
B1 (thiamine)	Coenzyme for pyruvate dehydrogenase and α-ketoglutarate dehydrogenase; supports mitochondrial energy production	Reduced ATP, impaired Krebs cycle, increased ROS	Mitochondrial dysfunction, microglial/astrocyte activation, pro-inflammatory cytokine release, neuronal loss
B2 (riboflavin)	Forms FAD/FMN, essential for the electron transport chain and redox reactions	Impaired electron transport, NADH/$FADH_2$ deficiency, increased ROS	Oxidative stress-induced neuroinflammation, chronic glial activation
B3 (niacin)	NAD^+/$NADP^+$ precursor, critical for redox reactions and sirtuin signalling	Decreased NAD^+, impaired oxidative phosphorylation	Microglial activation, enhanced pro-inflammatory cytokine release, and neuronal injury
B5 (pantothenic acid)	Coenzyme A synthesis for energy and lipid metabolism	Reduced CoA, impaired mitochondrial flux	Bioenergetic stress, ROS accumulation, secondary neuroinflammation
B6 (pyridoxine)	PLP-dependent neurotransmitter synthesis (GABA, serotonin, dopamine) and glutathione production	Reduced GABA, dopamine, serotonin; impaired glutathione synthesis	Neuronal hyperexcitability, oxidative stress, and enhanced microglial activation
B9 (folate)	One-carbon metabolism, homocysteine methylation	Hyperhomocysteinemia, impaired DNA methylation	Endothelial dysfunction, BBB permeability, microglial/astrocyte activation, and increased cytokine release
B12 (cobalamin)	Coenzyme for methionine synthase; homocysteine methylation	Hyperhomocysteinemia, impaired methylation, neuronal injury	Microglial activation, oxidative stress, neuroinflammation, and cognitive decline

vague or poorly identified by medical professionals, making diagnosis and treatment more difficult.

6.17.2 Wernicke Encephalopathy (WE) of the Brain

A shortage of vitamin B1 (thiamine), also known as beriberi, can cause a wide range of symptoms that shift according to the degree and length of the shortage. Numerous organ systems, including the neurological, cardiovascular, and gastrointestinal, could be impacted. Especially in severe situations, such as those linked to long-term alcohol use, cardiovascular problems might manifest as arrhythmias, cardiomegaly, and acute or chronic heart failure. Peripheral numbness, tingling, paraesthesia, muscle weakness, loss of reflexes, confusion, memory impairment, delirium, seizures, behavioural changes (such as apathy, irritability, and depression), and visual disturbances (such as retinal detachment or nystagmus) are neurological symptoms of thiamine (vitamin B1) deficiency [106]. When the deficiency primarily presents as symmetrical peripheral neuropathy, it is referred to as "dry" beriberi, which is more commonly observed in children. In adults, a deficiency manifesting with cognitive impairment and dementia-like symptoms is often identified as WE, characterized by the classic triad of ataxia, confusion, and nystagmus. If untreated, WE can progress to Korsakoff syndrome or Wernicke–Korsakoff syndrome (WKS), which is marked by confabulation, altered mental status, and profound anterograde memory loss [107]. Early recognition and prompt treatment with thiamine supplementation can reverse initial neurological deficits. However, in severe cases or when treatment is delayed, the neurological damage may become permanent, particularly in patients who develop Korsakoff syndrome.

6.17.3 Pellagra

Niacin (vitamin B3) is a member of the vitamin B complex and functions as a coenzyme in numerous metabolic reactions, particularly in redox reactions involving NAD^+ and $NADP^+$. Niacin deficiency, which manifests as the condition known as pellagra, remains more prevalent in low-income regions, such as parts of Africa, where poverty limits access to diverse diets. In contrast, pellagra is uncommon in countries where diets are less reliant on niacin-poor staples like untreated corn, including Latin America, China, and India [108]. In developed countries such as the United States, pellagra is rare due to food fortification practices, such as flour enrichment. When it does occur, it is typically associated not with dietary insufficiency but with chronic alcoholism, malabsorption syndromes, or other conditions affecting nutrient utilization. The most frequent contributing reasons to secondary pellagra are drug addiction, gastrointestinal issues, and chronic alcoholism. Although it can cause severe neurological symptoms, pellagra is mainly characterized by gastrointestinal issues and dermatological problems [109]. Sleeplessness, anxiety, confusion, delusions, dementia, encephalopathy, and muscle weakness are some of these neurological symptoms. Laboratory tests of niacin, tryptophan, and the coenzymes NAD and NADP can confirm a suspected niacin shortage.

6.17.4 Carpal Tunnel Syndrome

Vitamin B6, or pyridoxine, is necessary in many metabolic reactions. Erythrocyte production, amino acid generation, and the formation of neurotransmitters all depend on it. Pyridoxine is found naturally in various foods, including fish, poultry, legumes, nuts, and bananas. It is also available for purchase as a dietary supplement [110]. Vitamin B6 pills are commonly used in therapeutic settings to address conditions such as morning sickness, wrist instability, and premenstrual disorder. The neurological system and brain need vitamin pyridoxine, vitamin B6, to function correctly. It also promotes healthy skin, eyes, and liver. Immunological regulation and hormone use are also significantly impacted. Pyridoxine deprivation can result in several clinical manifestations, including anaemia, dermatitis, seizures, depression, and cognitive impairments, all of which increase the possibility of heart attack and stroke, and numerous types of malignancies [111]. Severe malnutrition, prolonged alcohol use, liver disease, gestation, and the use of medications like isoniazid in excess can all result in pyridoxine shortages.

Cognitive deficiencies constitute some of the initial indications of insufficiency; minor examples include confusion, sorrow, disorientation, and deterioration of the legs' nerves, while severe cases can result in episodes. Moreover, pyridoxine-associated seizures and spasm, a genetic metabolic defect seen as refractory neonatal tremors, emphasize the critical function of vitamin B6 in the production of neurons and neural growth [112].

6.18 B VITAMINS IN AGE-RELATED COGNITIVE DECLINE AND MCI

Low-impact cognitive impairment (mild cognitive impairment [MCI]) and dementia caused by age (ARCD) are prevalent indicators in the aged population that often serve as precursors to more severe neurodegenerative diseases such as AD. Recent studies have shown that B vitamins, particularly folate, vitamin B6, and vitamin B12, are critical for maintaining mental capacity as people develop [113]. This chapter explores the relationship between B vitamin content and intellectual performance, with a focus on neuroprotection, the metabolism of Hcy, and treatment strategies. Elevated plasma Hcy levels have been identified as a significant risk factor for brain harm and dementia. B vitamins are necessary for the methylation and transsulphuration mechanisms that degrade Hcy. Deficits in folate, B6, or B12 may impair these processes, leading to hyperhomocysteinemia and subsequent neurological impairment [114]. Studies have shown that supplementing with these vitamins can lower Hcy levels and, in some cases, slow down the process of cognitive decline.

The B12 and folate vitamins aid in synthesizing SAM, a methyl donor in DNA methylation processes. Proper chemical methylation of the DNA is necessary for both neural flexibility and the regulation of gene transcription [115].

- B6 is a vitamin precursor to synthesizing neurotransmitters such as dopamine, serotonin, and GABA, which are critical for learning and psychological regulation.
- B vitamins aid in the production and upkeep of myelin, the sheath that surrounds neurons, and ensures the efficient communication of nerve signals.
- B vitamins have antioxidant qualities that aid in reducing stress caused by oxidation, which is linked to cognitive loss and neurological degeneration [116].

6.19 THE EVIDENCE FOR SUPPLEMENTATION IN ALZHEIMER'S DISEASE, PARKINSON'S DISEASE, AND DEMENTIA PREVENTION

With varying but promising results, more research has examined the potential of B vitamin supplementation as a therapeutic or preventive approach for dementias such as PD, AD, and others. Mechanistically, vitamins B6, B12, and folate are integral to one-carbon metabolism, a pathway essential for the regulation of Hcy levels, DNA synthesis and repair, methylation reactions, and neurotransmitter production [117]. Elevated Hcy has been consistently associated with accelerated brain atrophy, white matter hyperintensities, and increased risk of cognitive decline, making it a plausible therapeutic target. Randomized controlled trials have demonstrated that high-dose B vitamin supplementation can significantly reduce plasma Hcy concentrations and, in some cohorts, attenuate hippocampal and cortical atrophy, particularly in individuals with MCI who are at high risk of progressing to AD [118]. Moreover, studies have shown that supplementation with methylcobalamin and folic acid may enhance memory performance, slow the decline in executive function, and support the structural integrity of brain regions vulnerable to neurodegeneration. In PD, evidence is less robust but suggests that B vitamin supplementation, especially B12, may mitigate levodopa-induced hyperhomocysteinemia, thereby reducing peripheral neuropathy and potentially contributing to neuroprotection. However, despite these encouraging findings, several large-scale trials have failed to demonstrate consistent benefits on clinical endpoints such as global cognition, incidence of dementia, or disease progression. This discrepancy highlights the complexity of nutrient-based interventions, where baseline vitamin status, genetic polymorphisms (e.g. methylenetetrahydrofolate reductase [MTHFR] variants), lifestyle factors, and comorbidities strongly influence outcomes [119]. While the preventive role of B vitamin supplementation remains inconclusive, the available evidence underscores its potential importance in specific subpopulations, particularly those with deficiencies, elevated Hcy, or early cognitive impairment. Thus, future research should clarify optimal dosing strategies, treatment duration, and patient selection criteria to fully evaluate the role of B vitamins in neurodegenerative disease prevention and management. Table 6.4 shows the summary of major clinical trials on B vitamins.

TABLE 6.4
Summary of Major Clinical Trials on B Vitamins

Clinical Trial	Dosage	Duration	Primary Outcome	Summary
VITACOG	Folic acid 0.8 mg+vitamin B12 0.5 mg+vitamin B6 20 mg daily	24 months	Plasma homocysteine; rate of whole-brain atrophy by volumetric MRI; cognitive tests	Supplementation markedly lowered homocysteine and slowed whole-brain atrophy vs placebo. Cognitive improvement was seen in some memory/executive domains, and effects were most potent in participants with elevated baseline homocysteine; however, the results were limited by modest sample size and subgroup dependence
FACIT	Folic acid 0.8 mg/day (single agent)	3 years	Cognitive domains (processing speed, memory), vascular markers	Folic acid improved specific cognitive domains (processing speed, some memory measures) vs placebo. Strength: long duration; limitation: single-vitamin intervention and selection for elevated homocysteine
HOPE-2	Folic acid+vitamin B6+vitamin B12	5 years	Cardiovascular events (primary); cognitive endpoints secondary in some analyses	Vitamins lowered homocysteine by ~25%–30% but did not reduce major cardiovascular events and showed no clear population-level cognitive benefit—illustrating that biochemical change does not automatically translate to clinical cognitive outcomes in heterogeneous, advanced-disease cohorts.

(Continued)

TABLE 6.4 (*Continued*)
Summary of Major Clinical Trials on B Vitamins

Clinical Trial	Dosage	Duration	Primary Outcome	Summary
ADCS	High-dose B vitamin regimen (folate, B12, B6; homocysteine lowering)	18–24 months	Cognitive scales (ADAS-Cog, MMSE); homocysteine	Supplementation reduced homocysteine but did not slow global cognitive decline in established AD, suggesting limited efficacy once neurodegeneration is advanced
VITACOG	Same B vitamin regimen	2 years	Regional grey matter atrophy (hippocampus, temporal lobe)	B vitamin treatment reduced atrophy in AD-vulnerable regions (hippocampus/medial temporal cortex) in participants with elevated homocysteine—consistent with structural neuroprotection; power and generalizability limited by subset analyses

6.20 DIETARY SOURCES, BIOAVAILABILITY, AND SPECIAL POPULATIONS

B vitamins are widely distributed in plant- and animal-based foods, with leafy greens, legumes, whole grains, nuts, eggs, and dairy as primary sources. Bioavailability varies depending on the vitamin form, food matrix, cooking methods, and interactions with other nutrients or anti-nutritional factors, such as phytates. Certain B vitamins, particularly B12, require intrinsic factor and an intact gastrointestinal system for efficient absorption, highlighting the importance of digestive health [120]. Special populations, including vegetarians, vegans, pregnant women, and the elderly, are at increased risk of deficiency due to dietary restrictions, higher physiological demands, or age-related malabsorption. The effective use of folate and B12 may be further reduced by genetic polymorphisms, such as MTHFR variations, underscoring the necessity of focused nutritional measures. Customized strategies are often needed for the best neuropsychological and energy metabolism results, but micronutrients and fortified foods are effective treatments. It is necessary to comprehend these elements to create dietary suggestions that preserve proper vitamin statuses across a variety of demographics [121].

6.21 FACTORS AFFECTING ABSORPTION AND BIOAVAILABILITY

Vitamin B complex absorption and availability play a significant role in maintaining brain function and metabolism in cells. Despite their abundance in a wide variety of food sources, the body uses them in diverse ways based on a variety of external as well as internal variables. Comorbidities like age, gene variation, intestinal health, and disease of the liver or kidneys can all affect permeability [122]. Similarly, dietary constituents, alcohol use, smoking, and drug–nutrient combinations all further alter their biological effectiveness. While stabilization and supplementary methods have gained popularity as means of overcoming these challenges, they can also have drawbacks, such as differing individual responses or concealing other deficiencies [123]. Therefore, determining the factors that affect vitamin B absorption and its advantages and disadvantages is essential for developing targeted pharmacological and nutritional management solutions. Table 6.5 shows factors influencing the absorption and bioavailability of B vitamins.

6.22 CONSIDERATIONS FOR VEGETARIANS, VEGANS, AND THE ELDERLY

The needed amount of B vitamins is greatly impacted by dietary practices and age-related modifications in the body, necessitating special consideration for populations such as older people, vegetarians, and vegans. B12, or vitamin, insufficiency is the most significant because it is naturally present in almost all animal-based foods, including meat, poultry, eggs, fish, and cheese [124]. Vegans, who abstain from all animal products, are most susceptible to B12 deficiency, although vegetarians who eat small amounts of dairy or eggs may also be at risk. Chronic deficiency can manifest as megaloblastic anaemia, peripheral neuropathy, impaired methylation, and progressive cognitive decline. Although fortified foods and supplements can mitigate this risk, their effectiveness depends on consistency of use, formulation (e.g. cyanocobalamin vs. methylcobalamin), and individual absorption capacity [125]. Other B vitamins, such as folate, riboflavin, and niacin, can generally be obtained through plant-based diets. Still, their bioavailability may be reduced due to anti-nutritional factors like phytates and tannins in legumes, cereals, and certain vegetables. Vitamin B6, though abundant in plant sources, can also be affected by dietary patterns and interactions with medications, particularly in individuals with chronic conditions. For this reason, careful nutritional planning and, in many cases, supplementation are required to maintain optimal neurological health in vegetarian and vegan populations [126].

The elderly population faces additional challenges, independent of dietary choices. Age-related gastric atrophy and reduced intrinsic factor secretion impair vitamin B12

TABLE 6.5
Factors Influencing Absorption and Bioavailability of B Vitamins

Factor	Influence on Absorption	Advantages	Limitations	At-Risk Population
Dietary composition and food matrix	Naturally present in whole grains, legumes, fruits, vegetables, dairy, and animal foods; cooking and processing reduce bioavailability (especially thiamine, riboflavin, folate)	Whole foods provide synergistic nutrients; fortification enhances intake at the population level	Heat, light, and processing destroy labile vitamins; B12 in animal proteins requires gastric acid and intrinsic factor	Low-income populations, older people, and individuals on highly processed diets
Physiological states	Increased demand in pregnancy, lactation, and ageing reduces absorption efficiency	Enhanced transport of folate and B12 during pregnancy supports foetal neurodevelopment	Age-related gastric atrophy and low intrinsic factor impair B12 absorption; increased demands may outpace intake	Pregnant women, lactating mothers, and elderly individuals
Gastrointestinal health	Vitamins absorbed mainly in the small intestine; B12 in the ileum, folate in the jejunum, thiamine throughout the small intestine	A healthy gut ensures efficient digestion and absorption	GI disorders (celiac, Crohn's disease, tropical sprue, SIBO) impair absorption; bariatric surgery bypasses absorption sites	Patients with malabsorption syndromes, post-bariatric surgery, and chronic diarrhoea
Drug–nutrient interactions	Drugs interfere with the absorption or metabolism of specific vitamins	Awareness allows preventive supplementation (e.g. pyridoxine with isoniazid)	Metformin reduces B12; PPIs reduce B12 release; anticonvulsants lower folate; isoniazid interferes with B6 metabolism	Diabetic patients on metformin, long-term PPI users, TB patients, and epilepsy patients
Genetic polymorphisms	Variants in genes like *MTHFR* affect folate metabolism; transcobalamin polymorphisms alter B12 transport	Individuals without polymorphisms utilize vitamins efficiently	Mutations reduce enzyme efficiency, increasing homocysteine and risk of neurodegeneration despite adequate intake	individuals with the *MTHFR* C677T mutation, hereditary enzyme defects
Lifestyle and environmental factors	Diet, alcohol, smoking, and stress influence absorption and utilization	Diverse diets (animal+plant foods) promote optimal status; low alcohol intake supports metabolism	Alcohol inhibits thiamine, folate, and B6 absorption/ utilization; vegan diets lack B12; smoking increases oxidative stress and demand	Alcoholics, strict vegans/vegetarians, smokers, individuals with high stress

absorption, while chronic illnesses, polypharmacy, and decreased appetite compound the risk of multi-nutrient deficiencies. Absorption of vitamin B6 and folate may also be compromised, further contributing to elevated Hcy levels, oxidative stress, and neuroinflammation, which accelerate cognitive decline and increase the risk of dementia. Importantly, deficiencies in older adults are often underdiagnosed due to the overlap between clinical symptoms of vitamin deficiency (e.g. fatigue, confusion, neuropathy) and normal ageing processes. Therefore, targeted strategies are essential for these populations [127]. For vegetarians and vegans, regularly monitoring vitamin B12 status and consuming fortified foods or supplements is critical. For older adults, proactive screening, dietary fortification, and individualized supplementation regimens should be emphasized, alongside management of comorbidities and polypharmacy that may interfere with vitamin metabolism. Collectively, these tailored interventions can reduce the risk of neurological complications, preserve cognitive function, and improve overall quality of life in these vulnerable groups [128]. Microglial balance in neuronal integrity and cognitive function is shown in Figure 6.3.

6.23 CONCLUSION

B vitamins connect methylation dynamics, mitochondrial bioenergetics, and maintain neuronal integrity, forming the cornerstone of basic biochemical mechanisms that protect brain health. They are essential for cognitive resilience throughout life because they control Hcy metabolism, sustain energy generation, reduce oxidative stress, and promote neurotransmitter synthesis. According to new clinical and translational findings, improving B vitamin status may reduce the risk of cognitive decline and neurodegeneration, especially in susceptible groups like older people or people with genetic variants like MTHFR.

Although B vitamins alone cannot cure a disease, their incorporation into a comprehensive therapy approach that includes pharmaceutical, lifestyle, and nutritional methods has great potential to slow the progression of the disease and improve quality of life. Future studies focused on multi-omics integration, population-specific therapies, and precision supplementation will further elucidate their therapeutic potential. B vitamins are essential partners in the fight to maintain cognition, safeguard neural networks, encourage healthy brain ageing, and are necessary dietary nutrients.

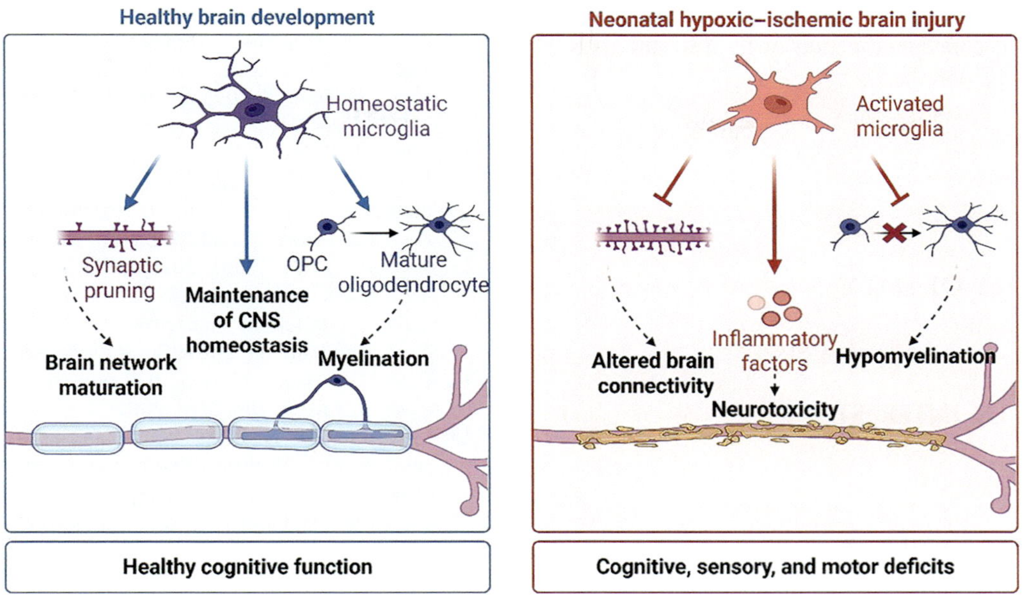

FIGURE 6.3 From homeostasis to neurotoxicity: microglial balance in neuronal integrity and cognitive function.

6.24 THE ROLE OF GENETIC POLYMORPHISMS (MTHFR)

The MTHFR genetic variations C677T and A1298C have been the subject of multiple investigations about their relationship with Hcy regulation, folate metabolic rate, and neurological risk. Notwithstanding the enormous progress, there are still many gaps that must be filled [129]. Future research needs to focus on customized nutrition approaches, which use DNA testing for MTHFR variants to identify individuals with impaired folate metabolism who may benefit from specific treatments, such as methyl folate or methylcobalamin dietary supplements [130]. With hyperhomocysteinemia playing a pathogenic role, such approaches may be particularly relevant in fighting the development of vascular dementia, AD, and mild-to-moderate cognitive decline (MCI). A further potential approach is the combination of epigenetics and nutrigenomics. In addition to affecting folate-dependent metabolism of one carbon, MTHFR mutations also have an impact on the activity of mitochondria and patterns of DNA methylation, both of which are essential for maintaining the integrity of neurons [131]. To modulate neuroinflammation and slow down cognitive loss, future research should examine how nutrient and gene relationships change the epigenetic control of neuroprotective and pro-inflammatory networks.

6.25 B VITAMINS AS ADJUNCTIVE THERAPY IN NEURODEGENERATIVE DISEASES

Because they target several pathogenic processes that underpin neuronal malfunction and cognitive decline, B vitamins are essential as an adjuvant therapy in neurodegenerative illnesses. These water-soluble vitamins enhance energy production and neuronal survival by functioning as vital cofactors in phosphorylation by oxidation, the Krebs cycle, and one-carbon biosynthesis [132]. Hcy metabolism depends on vitamins B6, B9 (folate), and B12. Supplementing these vitamins has decreased Hcy levels, reduced brain shrinkage, and prevented cognitive loss in diseases like AD. For mitochondrial bioenergetics, thiamine (B1), riboflavin (B2), niacin (B3), and pantothenic acid (B5) are essential because they guarantee practical ATP synthesis and reduce oxidative stress, a significant cause of brain degeneration [133]. Additionally, B vitamins influence neuroinflammatory systems; riboflavin promotes GSH recycling, which strengthens cellular defences against antioxidants, and niacin has anti-inflammatory effects on microglia. Furthermore, pyridoxine (B6) promotes the manufacture of neurotransmitters, including dopamine, serotonin, and GABA, which are frequently dysregulated in neurodegenerative diseases, whereas cobalamin (B12) is essential for myelin synthesis and axonal repair [134]. Growing data from clinical research support their positive benefits on peripheral neuropathies, ALS, AD, and PD, especially when combined with traditional therapy. Although B vitamins are unlikely to be used as stand-alone therapies, their incorporation into a multimodal therapeutic approach provides a clinically useful and biologically sound way to protect neuronal integrity, slow the growth of neurodegenerative illnesses, and enhance the quality of life for patients [135].

AUTHOR CONTRIBUTIONS

Kanika Goyal, Ritu Dahiya, and Md Abubakar were involved in the conceptualization, writing of the original draft, mechanistic insight, input on critical points, and editing; Mamta Bishnoi and Saqib Hameed contributed to the manuscript writing and editing; and Dr. Nitesh Kumar was involved in the conceptualization, manuscript writing, reviewing, and conceptualization of figures and approved the version to be published.

FUNDING

None.

AVAILABILITY OF DATA

None.

CONFLICT OF INTEREST

None.

ACKNOWLEDGEMENTS

We thankfully acknowledge Kanika Goyal from Ganpati Group of Institutions, Ritu Dahiya from School of Pharmacy, Lingaya's Vidyapeeth (deemed to be a university), Faridabad, Haryana, India, and Mamta Bishnoi from Galgotias University. We also thankfully acknowledge the infrastructure and fellowship support to Md Abubakar, which is supported by the Department of Pharmaceuticals, Ministry of Chemicals and Fertilizers, Government of India, and NIPER HAJIPUR, and Dr. Nitesh Kumar for the support and guidance in executing this work plan. The images were created using www.biorender.com and www.Grammarly.com to correct the grammar of this manuscript.

REFERENCES

1. Manasa K. Vitamin B12 role especially in DNA synthesis and its clinical manifestations. *Pharm Sci Anal Res J.* 2024;6(4):1–7.
2. Mikkelsen K, Apostolopoulos V. B vitamins and ageing. *Biochem Cell Biol Ageing Part I Biomed Sci.* 2019;451–70.
3. Kennedy DO. B vitamins and the brain: mechanisms, dose and efficacy—a review. *Nutrients.* 2016;8(2):68.
4. McGarel C, Pentieva K, Strain JJ, McNulty H. Emerging roles for folate and related B-vitamins in brain health across the lifecycle. *Proc Nutr Soc.* 2015;74(1):46–55.
5. Caffrey A, Lamers Y, Murphy MM, Letourneau N, Irwin RE, Pentieva K et al. Epigenetic effects of folate and related B vitamins on brain health throughout life: scientific substantiation and translation of the evidence for health improvement strategies. *Nutr Bull.* 2023;48(2):267–77.
6. Bowling A, Dieppe P. What is successful ageing and who should define it? *BMJ.* 2005;331(7531):1548–51.
7. Gareri P, Castagna A, Cotroneo AM, Putignano S, De Sarro G, Bruni AC. The role of citicoline in cognitive impairment: pharmacological characteristics, possible advantages, and doubts for an old drug with new perspectives. *Clin Interv Aging.* 2015;1421–9.
8. Swerdlow RH, Burns JM, Khan SM. The Alzheimer's disease mitochondrial cascade hypothesis: progress and perspectives. *Biochim Biophys Acta Mol Basis Dis.* 2014; 1842(8):1219–31.
9. Abubakar M, Giri A, Goel F, Khan M, Gupta J, Kumar D et al. Diabetes, Alzheimer's disease risk factors, and the cafeteria diet: a comprehensive review. *Curr Neuropharmacol.* 2025;23. doi:10.2174/011570159X384737250626094315.
10. Silva DF, Selfridge JE, Lu J, Lezi E, Roy N, Hutfles L et al. Bioenergetic flux, mitochondrial mass and mitochondrial morphology dynamics in AD and MCI cybrid cell lines. *Hum Mol Genet.* 2013;22(19):3931–46.
11. Yao J, Irwin RW, Zhao L, Nilsen J, Hamilton RT, Brinton RD. Mitochondrial bioenergetic deficit precedes Alzheimer's pathology in female mouse model of Alzheimer's disease. *Proc Natl Acad Sci.* 2009;106(34):14670–5.
12. Beal MF. Oxidatively modified proteins in aging and disease. *Free Radic Biol Med.* 2002;32(9):797–803.
13. Funayama M, Nishioka K, Li Y, Hattori N. Molecular genetics of Parkinson's disease: contributions and global trends. *J Hum Genet.* 2023;68(3):125–30.
14. Cassarino DS, Fall CP, Swerdlow RH, Smith TS, Halvorsen EM, Miller SW et al. Elevated reactive oxygen species and antioxidant enzyme activities in animal and cellular models of Parkinson's disease. *Biochim Biophys Acta Mol Basis Dis.* 1997;1362(1):77–86.
15. Höglinger GU, Carrard G, Michel PP, Medja F, Lombès A, Ruberg M et al. Dysfunction of mitochondrial complex I and the proteasome: interactions between two biochemical deficits in a cellular model of Parkinson's disease. *J Neurochem.* 2003;86(5):1297–307.
16. Tetrud JW, Langston JW. MPTP-induced Parkinsonism as a model for Parkinson's disease. *Acta Neurol Scand.* 1989;80:35–40.
17. Bao L, Avshalumov MV, Patel JC, Lee CR, Miller EW, Chang CJ et al. Mitochondria are the source of hydrogen peroxide for dynamic brain-cell signaling. *J Neurosci.* 2009;29(28):9002–10.
18. Wiley KD, Gupta M. *Vitamin B1 Thiamine Deficiency.* StatPearls Publishing, 2019.
19. Ropper AH. *Principios de Neurología.* McGraw-Hill, 2016.
20. Reynolds EH. The neurology of folic acid deficiency. *Handb Clin Neurol.* 2014;120:927–43.
21. Shulpekova Y, Nechaev V, Kardasheva S, Sedova A, Kurbatova A, Bueverova E et al. The concept of folic acid in health and disease. *Molecules.* 2021;26(12):3731.
22. Hrubša M, Siatka T, Nejmanová I, Vopršalová M, Kujovská Krčmová L, Matoušová K et al. Biological properties of vitamins of the B-complex, part 1: vitamins B1, B2, B3, and B5. *Nutrients.* 2022;14(3):484.
23. Bocobza SE, Aharoni A. Switching the light on plant riboswitches. *Trends Plant Sci.* 2008;13(10):526–33.
24. Jurgenson CT, Begley TP, Ealick SE. The structural and biochemical foundations of thiamin biosynthesis. *Annu Rev Biochem.* 2009;78(1):569–603.
25. Yoshii K, Hosomi K, Sawane K, Kunisawa J. Metabolism of dietary and microbial vitamin B family in the regulation of host immunity. *Front Nutr.* 2019;6:434971.
26. Mecocci P, MacGarvey U, Beal MF. Oxidative damage to mitochondrial DNA is increased in Alzheimer's disease. *Ann Neurol Off J Am Neurol Assoc Child Neurol Soc.* 1994;36(5):747–51.
27. Smith AG, Croft MT, Moulin M, Webb ME. Plants need their vitamins too. *Curr Opin Plant Biol.* 2007;10(3):266–75.
28. Serra-Majem L, Bes-Rastrollo M, Román-Vinas B, Pfrimer K, Sánchez-Villegas A, Martínez-González MA. Dietary patterns and nutritional adequacy in a Mediterranean country. *Br J Nutr.* 2009;101(S2):S21–8.

29. DI MATTEO A, Hancock RD, Frusciante L, Viola R. Characterisation of *Chlorella pyrenoidosa* L-ascorbic acid accumulating mutants: identification of an enhanced biosynthetic enzyme activity and cloning of the putative gene from *Arabidopsis thaliana*. *Comp Biochem Physiol A Mol Integr Physiol*. 2003;134(3, Supplement):S155.
30. Imai S ichiro, Guarente L. NAD+ and sirtuins in aging and disease. *Trends Cell Biol*. 2014;24(8):464–71.
31. Lopez-Gonzalez R, Lu Y, Gendron TF, Karydas A, Tran H, Yang D et al. Poly (GR) in C9ORF72-related ALS/FTD compromises mitochondrial function and increases oxidative stress and DNA damage in iPSC-derived motor neurons. *Neuron*. 2016;92(2):383–91.
32. Ashoori M, Saedisomeolia A. Riboflavin (vitamin B2) and oxidative stress: a review. *Br J Nutr*. 2014;111(11): 1985–91.
33. Balázs R, Lewis PD, Patel AJ. Nutritional deficiencies and brain development. *Hum Growth Neurobiol Nutr*. 1979;3:415–80.
34. Antonow-Schlorke I, Schwab M, Cox LA, Li C, Stuchlik K, Witte OW et al. Vulnerability of the fetal primate brain to moderate reduction in maternal global nutrient availability. *Proc Natl Acad Sci*. 2011;108(7):3011–6.
35. Gazzaniga F, Stebbins R, Chang SZ, McPeek MA, Brenner C. Microbial NAD metabolism: lessons from comparative genomics. *Microbiol Mol Biol Rev*. 2009;73(3):529–41.
36. Magnúsdóttir S, Ravcheev D, de Crécy-Lagard V, Thiele I. Systematic genome assessment of B-vitamin biosynthesis suggests co-operation among gut microbes. *Front Genet*. 2015;6:148.
37. Shibata K. Organ co-relationship in tryptophan metabolism and factors that govern the biosynthesis of nicotinamide from tryptophan. *J Nutr Sci Vitaminol (Tokyo)*. 2018;64(2):90–8.
38. Han A, Almeida L, Anand N, Salloum I, Kanaan S, Gadad BS et al. Exploring neuropsychiatric manifestations of vitamin B complex deficiencies. *Front Psychiatry*. 2025;16:1569826.
39. Tardy AL, Pouteau E, Marquez D, Yilmaz C, Scholey A. Vitamins and minerals for energy, fatigue and cognition: a narrative review of the biochemical and clinical evidence. *Nutrients*. 2020;12(1):228.
40. Orywal K, Socha K, Iwaniuk P, Kaczyński P, Farhan JA, Zoń W et al. Vitamins in the prevention and support therapy of neurodegenerative diseases. *Int J Mol Sci*. 2025;26(3):1333.
41. Mukhopadhyay M, Ghosh P, Chattopadhyay A, Bandyopadhyay D. An insight into the importance of B vitamins and melatonin in the prevention of diabetes through modulation of the brain energy metabolism-a comprehensive review. *Melatonin Res*. 2023;6(3): 397–430.
42. McGlynn C, Dermawan D, Canonizado GM. Neuron foundry Mach 8: a concept of multi-modal neuro-supportive energy therapy device for treating neurodegenerative disorders. *Brain Sci Adv*. 2024;10(1):5–37.
43. Hammond N, Wang Y, Dimachkie MM, Barohn RJ. Nutritional neuropathies. *Neurol Clin*. 2013;31(2):477–89.
44. Godonu SS. Role of vitamin B9 (folate) in neuronal plasticity. *Int J Sci Res Technol*. 2025;2(1):77–82.
45. Kumar S, Mishra DK, Usmani N, Kumar H. Vitamins and their role in brain function. In: R Yadav, S Kumar, and AD Dubey (Eds.), *Optimizing Brain Health and Performance through Essential Nutrients and Lifestyle Changes*. SAAR Publications, 2024, pp. 83–97.
46. Cunnane SC, Harbige LS, Crawford MA. The importance of energy and nutrient supply in human brain evolution. *Nutr Health*. 1993;9(3):219–35.
47. Briani C, Dalla Torre C, Citton V, Manara R, Pompanin S, Binotto G et al. Cobalamin deficiency: clinical picture and radiological findings. *Nutrients*. 2013;5(11):4521–39.
48. O'Keeffe S, Garcia L, Chen Y, Law RC, Liu C, Park JO. Bringing carbon to life via one-carbon metabolism. *Trends Biotechnol*. 2025;43(3):572–85.
49. Mazat JP. One-carbon metabolism in cancer cells: a critical review based on a core model of central metabolism. *Biochem Soc Trans*. 2021;49(1):1–15.
50. Qiu Y, Xie E, Xu H, Cheng H, Li G. One-carbon metabolism shapes T cell immunity in cancer. *Trends Endocrinol Metab*. 2024;35(11):967–80.
51. Daroch M, You D, Rasul F, Liu X, Jiang Y. C1 photochemotrophy–rethinking one-carbon metabolism in phototrophs. *Trends Biotechnol*. 2025;43(8):1836–1848.
52. Pham VN, Bruemmer KJ, Toh JDW, Ge EJ, Tenney L, Ward CC et al. Formaldehyde regulates S-adenosylmethionine biosynthesis and one-carbon metabolism. *Science*. 2023; 382(6670):eabp9201.
53. Ren X, Wang X, Zheng G, Wang S, Wang Q, Yuan M et al. Targeting one-carbon metabolism for cancer immunotherapy. *Clin Transl Med*. 2024;14(1):e1521.
54. Bernasocchi T, Mostoslavsky R. Subcellular one carbon metabolism in cancer, aging and epigenetics. *Front Epigenet Epigenomic*. 2024;2:1451971.
55. Cai S, Quan S, Yang G, Ye Q, Chen M, Yu H et al. One carbon metabolism and mammalian pregnancy outcomes. *Mol Nutr Food Res*. 2021;65(2):2000734.
56. Lim EW, Metallo CM. Tracing the diverse paths of one-carbon metabolism in cancer and beyond. *Cold Spring Harb Perspect Med*. 2024;14(9):a041533.
57. Kalecký K, Ashcraft P, Bottiglieri T. One-carbon metabolism in Alzheimer's disease and Parkinson's disease brain tissue. *Nutrients*. 2022;14(3):599.
58. Jiao Z, Zhang M, Ning J, Yao H, Yan X, Wu Z et al. The oncoprotein SET promotes serine-derived one-carbon metabolism by regulating SHMT2 enzymatic activity. *Proc Natl Acad Sci*. 2025;122(19):e2412854122.
59. Depeint F, Bruce WR, Shangari N, Mehta R, O'Brien PJ. Mitochondrial function and toxicity: role of B vitamins on the one-carbon transfer pathways. *Chem Biol Interact*. 2006;163(1–2):113–32.
60. Uchida Y, Ito K, Ohtsuki S, Kubo Y, Suzuki T, Terasaki T. Major involvement of Na+-dependent multivitamin transporter (SLC5A6/SMVT) in uptake of biotin and pantothenic acid by human brain capillary endothelial cells. *J Neurochem*. 2015;134(1):97–112.
61. Mentch SJ, Locasale JW. One-carbon metabolism and epigenetics: understanding the specificity. *Ann N Y Acad Sci*. 2016;1363(1):91–8.
62. Yang M, Vousden KH. Serine and one-carbon metabolism in cancer. *Nat Rev Cancer*. 2016;16(10):650–62.
63. Lucas S, Chen G, Aras S, Wang J. Serine catabolism is essential to maintain mitochondrial respiration in mammalian cells. *Life Sci Alliance*. 2018;1(2):e201800036.
64. Nikkanen J, Forsström S, Euro L, Paetau I, Kohnz RA, Wang L et al. Mitochondrial DNA replication defects disturb cellular dNTP pools and remodel one-carbon metabolism. *Cell Metab*. 2016;23(4):635–48.
65. Courtemanche C, Elson-Schwab I, Mashiyama ST, Kerry N, Ames BN. Folate deficiency inhibits the proliferation of primary human CD8+ T lymphocytes in

vitro. *J Immunol.* 2004;173(5):3186–3192. doi:10.4049/jimmunol.173.5.3186.
66. Bondi CO, Taha AY, Tock JL, Totah NKB, Cheon Y, Torres GE et al. Adolescent behavior and dopamine availability are uniquely sensitive to dietary omega-3 fatty acid deficiency. *Biol Psychiatry.* 2014;75(1):38–46.
67. Crawford MA. Long-chain polyunsaturated fatty acids in human brain evolution. In: SC Cunnane and K Stewart (Eds.), *Human Brain Evolution: The Influence of Freshwater and Marine Food Resources.* Wiley, 2010, pp. 13–31.
68. Takeda S, Sato N, Ikimura K, Nishino H, Rakugi H, Morishita R. Increased blood–brain barrier vulnerability to systemic inflammation in an Alzheimer disease mouse model. *Neurobiol Aging.* 2013;34(8):2064–70.
69. Waterland RA. Assessing the effects of high methionine intake on DNA methylation. *J Nutr.* 2006;136(6):1706S–1710S.
70. Lafourcade M, Larrieu T, Mato S, Duffaud A, Sepers M, Matias I et al. Nutritional omega-3 deficiency abolishes endocannabinoid-mediated neuronal functions. *Nat Neurosci.* 2011;14(3):345–50.
71. El Soury M, Fornasari BE, Carta G, Zen F, Haastert-Talini K, Ronchi G. The role of dietary nutrients in peripheral nerve regeneration. *Int J Mol Sci.* 2021;22(14):7417.
72. Strope TA, Birky CJ, Wilkins HM. The role of bioenergetics in neurodegeneration. *Int J Mol Sci.* 2022;23(16):9212.
73. Wilkins HM, Troutwine BR, Menta BW, Manley SJ, Strope TA, Lysaker CR et al. Mitochondrial membrane potential influences amyloid-β protein precursor localization and amyloid-β secretion. *J Alzheimer's Dis.* 2022;85(1):381–94.
74. Marcus C, Mena E, Subramaniam RM. Brain PET in the diagnosis of Alzheimer's disease. *Clin Nucl Med.* 2014;39(10):e413–26.
75. Parker Jr WD, Parks J, Filley CM, Kleinschmidt-DeMasters BK. Electron transport chain defects in Alzheimer's disease brain. *Neurology.* 1994;44(6):1090.
76. Pryde KR, Smith HL, Chau KY, Schapira AH V. PINK1 disables the anti-fission machinery to segregate damaged mitochondria for mitophagy. *J Cell Biol.* 2016;213(2):163–71.
77. Van Rheenen W, Van Der Spek RAA, Bakker MK, Van Vugt JJFA, Hop PJ, Zwamborn RAJ et al. Common and rare variant association analyses in amyotrophic lateral sclerosis identify 15 risk loci with distinct genetic architectures and neuron-specific biology. *Nat Genet.* 2021;53(12):1636–48.
78. Morellato AE, Umansky C, Pontel LB. The toxic side of one-carbon metabolism and epigenetics. *Redox Biol.* 2021;40:101850.
79. Sobczyńska-Malefora A, Delvin E, McCaddon A, Ahmadi KR, Harrington DJ. Vitamin B12 status in health and disease: a critical review. Diagnosis of deficiency and insufficiency–clinical and laboratory pitfalls. *Crit Rev Clin Lab Sci.* 2021;58(6):399–429.
80. Su X, Wellen KE, Rabinowitz JD. Metabolic control of methylation and acetylation. *Curr Opin Chem Biol.* 2016;30:52–60.
81. Škovierová H, Vidomanová E, Mahmood S, Sopková J, Drgová A, Červeňová T et al. The molecular and cellular effect of homocysteine metabolism imbalance on human health. *Int J Mol Sci.* 2016;17(10):1733.
82. Maddocks ODK, Athineos D, Cheung EC, Lee P, Zhang T, Van Den Broek NJF et al. Modulating the therapeutic response of tumours to dietary serine and glycine starvation. *Nature.* 2017;544(7650):372–6.
83. Amelio I, Melino G, Frezza C. Exploiting tumour addiction with a serine and glycine-free diet. *Cell Death Differ.* 2017;24(8):1311.
84. Yamaguchi T, Hirota K, Nagahama K, Ohkawa K, Takahashi T, Nomura T et al. Control of immune responses by antigen-specific regulatory T cells expressing the folate receptor. *Immunity.* 2007;27(1):145–59.
85. Nieto-Salazar MA, Ordóñez KNA, Carcamo ZDS, Cristina A, Ordóñez A, Saldana EA. Neurological dysfunction associated with vitamin deficiencies: a narrative review. *Open Access J Neurol Neurosurg.* 2023;18(1):555979.
86. Whitfield KC, Bourassa MW, Adamolekun B, Bergeron G, Bettendorff L, Brown KH et al. *Thiamine Deficiency Disorders: Diagnosis, Prevalence, and a Roadmap for Global Control Programs.* Wiley, 2018.
87. Whitney EN, Rolfes SR, Crowe T, Walsh A. *Understanding Nutrition*, 3rd ed. Cengage AU, 2019.
88. Lanska DJ. The discovery of niacin, biotin, and pantothenic acid. *Ann Nutr Metab.* 2012;61(3):246–53.
89. Nouri A, Patel K, Montejo J, Nasser R, Gimbel DA, Sciubba DM et al. The role of vitamin B12 in the management and optimization of treatment in patients with degenerative cervical myelopathy. *Glob Spine J.* 2019;9(3):331–7.
90. Scalabrino G. The multi-faceted basis of vitamin B12 (cobalamin) neurotrophism in adult central nervous system: lessons learned from its deficiency. *Prog Neurobiol.* 2009;88(3):203–20.
91. Singh NN, Thomas FP, Diamond AL, Diamond R. Vitamin B-12 associated neurological diseases. *E Med Neurol.* 2010;2010:1–22.
92. Dakshinamurti S, Dakshinamurti K, Zempleni J, Suttie JW, Gregory JF, Stover PJ. Vitamin B6. In: J Zempleni, RB Rucker, DB McCormick, and JW Suttie (Eds.), *Handbook of Vitamins.* CRC Press, 2014.
93. Felice JH. Effect of vitamin-B6 deficiency on certain neurotransmitters in two selected regions of the developing brain in the rat. Ph.D. Dissertation, Purdue University, 1978.
94. Footitt EJ. Vitamin B6 and serotonin metabolism in neurological disorders of childhood. Ph.D. Thesis, University College London, 2013.
95. Spinneker A, Sola R, Lemmen V, Castillo M, Pietrzik K, González-Gross M. Vitamin B6 status, deficiency and its consequences-an overview. *Nutr Hosp.* 2007;22(1):7–24.
96. Shabbir F, Patel A, Mattison C, Bose S, Krishnamohan R, Sweeney E et al. Effect of diet on serotonergic neurotransmission in depression. *Neurochem Int.* 2013;62(3):324–9.
97. Hernandez Garcia F. Moderate vitamin B6 deficiency and sulfur amino acid metabolism in male and female rats. Master of Science Thesis, Faculty of Graduate Studies, Food and Human Nutritional Sciences, University of Manitoba, 2024.
98. Wu Y, Zhang L, Li S, Zhang D. Associations of dietary vitamin B1, vitamin B2, vitamin B6, and vitamin B12 with the risk of depression: a systematic review and meta-analysis. *Nutr Rev.* 2022;80(3):351–66.
99. Rakić M, Lunić T, Bekić M, Tomić S, Mitić K, Graovac S et al. Vitamin B complex suppresses neuroinflammation in activated microglia: in vitro and in silico approach combined with dynamical modeling. *Int Immunopharmacol.* 2023;121:110525.
100. Ehmedah A. Effect of vitamin B complex therapy on process of neuroinflammation and regeneration of rat peripheral nerve after injury. Ph.D. Thesis, University of Belgrade, 2024.

101. Nguyen HD, Jo WH, Hoang NHM, Kim MS. Anti-inflammatory effects of B vitamins protect against tau hyperphosphorylation and cognitive impairment induced by 1, 2 diacetyl benzene: an in vitro and in silico study. *Int Immunopharmacol.* 2022;108:108736.
102. Babar Q, Ali A, Saeed A. Anti-inflammatory, antioxidant and immunostimulatory properties of the B vitamin complex. In: J LeBlanc (Ed.), *B-Complex Vitamins—Sources, Intakes and Novel Applications.* IntechOpen, 2022, p. 231.
103. Zdunek R, Strus K, Madera M, Dąbrowska N, Mazur S, Mazur A et al. Deficiency of thiamine and Wernicke-Korsakoff syndrome–review of pathophysiology, clinical presentation, diagnostics and treatment. *J Educ Heal Sport.* 2024;72:51248.
104. Mateos-Díaz AM, Marcos M, Chamorro AJ. Wernicke–Korsakoff syndrome and other diseases associated with thiamine deficiency. *Med Clínica Eng Ed.* 2022;158(9):431–6.
105. Benameur K, Pitter D, Clarke K. Nutritional peripheral neuropathies-review and practical approach. *Curr Treat Options Neurol.* 2025;27(1):1–8.
106. Dupuy G, Roux CJ, Barrois R, Imbard A, Pontoizeau C, Dangles MT et al. Vitamin deficiencies in children: lessons from clinical and neuroimaging findings. *Eur J Paediatr Neurol.* 2024;50:6–15.
107. Limketkai BN, Matarese LE, Mullin GE. Vitamins and minerals. In: T C Wang, M Camilleri, B Lebwohl, AS Lok, G . Wu, W J Sandborn, and K K Wang (Eds.), *Yamada's Textbook of Gastroenterology.* Wiley, 2022, pp. 426–56.
108. Ahire ED, Keservani RK, Surana KR, Singh S, Kesharwani RK. *Vitamins as Nutraceuticals: Recent Advances and Applications.* Wiley, 2023.
109. Badawy AAB. Pellagra and alcoholism: a biochemical perspective. *Alcohol Alcohol.* 2014;49(3):238–50.
110. Gasperi V, Sibilano M, Savini I, Catani MV. Niacin in the central nervous system: an update of biological aspects and clinical applications. *Int J Mol Sci.* 2019;20(4):974.
111. Senol MG, Sonmez G, Ozdag F, Saracoglu M. Reversible myelopathy with vitamin B12 deficiency. *Singapore Med J.* 2008;49(11):e330–2.
112. Caballero B, Finglas P, Toldrá F. *Encyclopedia of Food and Health.* Academic Press, 2015.
113. Refsum H, Smith AD. Low vitamin B-12 status in confirmed Alzheimer's disease as revealed by serum holotranscobalamin. *J Neurol Neurosurg Psychiatry.* 2003;74(7):959–61.
114. Morris MS. The role of B vitamins in preventing and treating cognitive impairment and decline. *Adv Nutr.* 2012;3(6):801–12.
115. Li S, Guo Y, Men J, Fu H, Xu T. The preventive efficacy of vitamin B supplements on the cognitive decline of elderly adults: a systematic review and meta-analysis. *BMC Geriatr.* 2021;21(1):367.
116. Li MM, Yu JT, Wang HF, Jiang T, Wang J, Meng XF et al. Efficacy of vitamins B supplementation on mild cognitive impairment and Alzheimer's disease: a systematic review and meta-analysis. *Curr Alzheimer Res.* 2014;11(9):844–52.
117. Behrens A, Graessel E, Pendergrass A, Donath C. Vitamin B—Can it prevent cognitive decline? A systematic review and meta-analysis. *Syst Rev.* 2020;9(1):111.
118. Hughes CF, Ward M, Tracey F, Hoey L, Molloy AM, Pentieva K et al. B-vitamin intake and biomarker status in relation to cognitive decline in healthy older adults in a 4-year follow-up study. *Nutrients.* 2017;9(1):53.
119. Chen Q, Huang J, Shi X, Peng Y, Chen A, Huang L et al. Associations between dietary B vitamin intakes and cognitive function among elderly individuals: an observational study. *Nutrition.* 2025;134:112716.
120. Jiang X, Guo Y, Cui L, Huang L, Guo Q, Huang G. Study of diet habits and cognitive function in the Chinese middle-aged and elderly population: the association between folic acid, B vitamins, vitamin D, coenzyme Q10 supplementation and cognitive ability. *Nutrients.* 2023;15(5):1243.
121. Fu Q, DeJager J, Gardner EM. Supplementation and mitigating cognitive decline in older adults with or without mild cognitive impairment or dementia: a systematic review. *Nutrients.* 2024;16(20):3567.
122. Oulhaj A, Jernerén F, Refsum H, Smith AD, De Jager CA. Omega-3 fatty acid status enhances the prevention of cognitive decline by B vitamins in mild cognitive impairment. *J Alzheimer's Dis.* 2016;50(2):547–57.
123. Chang J, Liu M, Liu C, Zhou S, Jiao Y, Sun H et al. Effects of vitamins and polyunsaturated fatty acids on cognitive function in older adults with mild cognitive impairment: a meta-analysis of randomized controlled trials. *Eur J Nutr.* 2024;63(4):1003–22.
124. Sheng LT, Jiang YW, Pan XF, Feng L, Yuan JM, Pan A et al. Association between dietary intakes of B vitamins in midlife and cognitive impairment in late-life: the Singapore Chinese Health Study. *Journals Gerontol Ser A.* 2020;75(6):1222–7.
125. O'Connor D, Molloy AM, Laird E, Kenny RA, O'Halloran AM. Sustaining an ageing population: the role of micronutrients in frailty and cognitive impairment. *Proc Nutr Soc.* 2023;82(3):315–28.
126. Rosenberg IH. Vitamin B12 and age-related cognitive decline—dementia and "Alzheimer's disease." *Food Nutr Bull.* 2024;45(1_suppl):S50–2.
127. Moretti R, Peinkhofer C. B vitamins and fatty acids: what do they share with small vessel disease-related dementia? *Int J Mol Sci.* 2019;20(22):5797.
128. PeaceHealth. *Age-Related Cognitive Decline (Holistic).* PeaceHealth Health Information Library.
129. Nachum-Biala Y, Troen AM. B-vitamins for neuroprotection: narrowing the evidence gap. *Biofactors.* 2012;38(2):145–50.
130. Gilbody S, Lewis S, Lightfoot T. Methylenetetrahydrofolate reductase (MTHFR) genetic polymorphisms and psychiatric disorders: a HuGE review. *Am J Epidemiol.* 2007;165(1):1–13.
131. Cortese C, Motti C. MTHFR gene polymorphism, homocysteine and cardiovascular disease. *Public Health Nutr.* 2001;4(2b):493–7.
132. Cai Y, Liu B, Zhang Y, Zhou Y. MTHFR gene polymorphisms in diabetes mellitus. *Clin Chim Acta.* 2024;561:119825.
133. Martin DN, Boersma BJ, Howe TM, Goodman JE, Mechanic LE, Chanock SJ et al. Association of MTHFR gene polymorphisms with breast cancer survival. *BMC Cancer.* 2006;6(1):257.
134. Jamil K. Clinical implications of MTHFR gene polymorphism in various diseases. *Biol Med.* 2014;6:e107.
135. Toffoli G, De Mattia E. Pharmacogenetic relevance of MTHFR polymorphisms. *Pharmacogenomics.* 2008;9(9): 1195–206.

7 Vitamins D and K
Immunomodulatory and Vascular Protective Roles in Brain Aging

Sumedha Gupta, Abhishek Kumar Gupta, Ritam Mukherjee, Sidharth Mehan, Pankaj Kumar Maurya, Gurudev Singh Raina, and Baljit Kaur

Division of Neuroscience, Department of Pharmacology, ISF College of Pharmacy, Moga, India (Affiliated to IK Gujral Punjab Technical University, Jalandhar, India)

ABBREVIATIONS

AD	Alzheimer's disease
AI	adequate intake
Akt	protein kinase B
ALS	amyotrophic lateral sclerosis
AMPK	AMP-activated protein kinase
ASD	autism spectrum disorder
BBB	blood–brain barrier
BDNF	brain-derived neurotrophic factor
BGLAP	bone γ-carboxyglutamate protein
BMD	bone mineral density
BMP-2	bone morphogenetic protein-2
BMPR	bone morphogenetic protein receptor
CerS	ceramide synthases
CNS	central nervous system
DEGS1	dihydroceramide desaturase 1
DRI	dietary reference intake
ERK	extracellular signal regulated kinase
Gas6	growth arrest-specific protein 6
GDNF	glial cell line-derived neurotrophic factor
ICH	International Council for Harmonisation
IFN-γ	interferon-γ
IKK	IκB kinase
IL	interleukin
IL-1β	interleukin-1β
IκBα	inhibitor of κB alpha
JAK	Janus kinase
LPS	lipopolysaccharide
MCI	mild cognitive impairment
MGP	matrix Gla protein
MK-4	menaquinone-4 (vitamin K2 isoform)
MS	multiple sclerosis
mTORC1	mechanistic target of rapamycin complex 1
MyD88	myeloid differentiation primary response 88
NADPH	nicotinamide adenine dinucleotide phosphate
NF-κB	nuclear factor kappa-B
NGF	nerve growth factor
NMDA	N-methyl-D-aspartate
PD	Parkinson's disease
PI3K	phosphoinositide 3-kinase
PIP2	phosphatidylinositol 4,5-bisphosphate
PIP3	phosphatidylinositol 3,4,5-trisphosphate
PROS1	protein S gene
ROS	reactive oxygen species
RRMS	relapsing remitting multiple sclerosis
RXR	retinoid X receptor
SGMS	sphingomyelin synthase
SMAD	mothers against decapentaplegic homolog
SOCS	suppressor of cytokine signaling
SOD	superoxide dismutase
SPT	serine palmitoyltransferase
STAT	signal transducer and activator of transcription
TAK1	transforming growth factor beta-activated kinase 1
TLR4	toll-like receptor 4
TNFR1	tumor necrosis factor receptor 1
TNF-α	tumor necrosis factor-alpha
TNF-α	tumor necrosis factor-α
TRADD	TNF receptor-associated death domain protein
TRAF2	TNF receptor-associated factor 2
TRAF6	TNF receptor-associated factor 6
UGT8	ceramide galactosyltransferase
UVB	ultraviolet-B radiation
VDR	vitamin D receptor
VDRE	vitamin D response element
VitD	vitamin D
Vit K1	phylloquinone (Vitamin K1)
VK	vitamin K
VK2	vitamin K2
VKDP	vitamin K-dependent protein
VKOR	vitamin K epoxide reductase

7.1 INTRODUCTION

Brain aging is a complex and gradual process that is characterized by structural, functional, and molecular changes,

DOI: 10.1201/9781003686675-7

which contribute to a decline in cognitive, emotional, and motor functions [1]. As individuals age, the brain experiences a reduction in volume, synaptic loss, decreased neurogenesis, mitochondrial dysfunction, oxidative stress, and dysregulation of neurotransmitter systems [2]. These changes collectively impair neural plasticity, reducing the brain's capacity to adapt and repair [3]. One of the significant issues related to brain aging is the heightened risk of neurodegenerative diseases like Alzheimer's disease (AD) [4], multiple sclerosis (MS) [5], Parkinson's disease (PD) [6], and other forms of dementia [7]. Additionally, age-related cognitive decline, memory impairment, and emotional instability create significant burdens on individuals and healthcare systems [8]. Immunomodulation and vascular protection are essential for maintaining brain health by safeguarding neuronal integrity and preventing neurodegenerative processes [9,10]. The brain's immune system, primarily regulated by microglia and interactions with the peripheral immune system, requires a suitable balance between pro-inflammatory and anti-inflammatory cytokine signaling to ensure optimal synaptic function and neuroplasticity [11]. Dysregulation of immune responses and chronic neuroinflammation are closely associated with neuronal damage, oxidative stress, and the progression of neurological disorders [12–15]. Likewise, vascular protection plays a crucial role in maintaining blood–brain barrier (BBB) integrity and cerebral perfusion, which ensure sufficient oxygen and nutrient supply. It is also responsible for preventing the infiltration of harmful molecules and immune cells [16]. Strategies that combine immunomodulatory and vasoprotective approaches not only mitigate inflammation and oxidative injury but also improve brain blood flow and cognitive function. Because of this, targeting these interconnected pathways has significant therapeutic potential for preserving brain health and preventing neurodegenerative and neuropsychiatric disorders [17–19].

Nowadays, nutritional factors, especially essential vitamins, are gaining more attention for their potential neuroprotective roles in maintaining brain health and preventing neurodegenerative disorders [20]. Vitamins are essential micronutrients that are required in very small quantities for various physiological processes, including metabolism, cellular signaling, and immune regulation [21]. They are generally classified into water-soluble (e.g., B-complex, C) and fat-soluble (e.g., A, D, E, K) groups and are mainly obtained through dietary sources such as fruits, vegetables, dairy products, whole grains, and fatty fish [22]. Among these, vitamins D and K have demonstrated emerging roles in supporting brain health [23,24]. Vitamin D is mainly synthesized through the skin upon exposure to sunlight, and it is present in foods like fatty fish, egg yolks, and fortified dairy products. It also functions as a neurosteroid. Vitamin D plays an essential role in neuroprotection by regulating calcium homeostasis, modulating neurotrophins such as brain-derived neurotrophic factor (BDNF), and suppressing inflammatory signaling pathways [25,26]. It has been shown that low serum levels of vitamin D are associated with cognitive decline and an increased risk of neurodegenerative diseases [27]. Similarly, vitamin K is abundant in green leafy vegetables, fermented foods, and plant oils [28]. Vitamin K supports central nervous system (CNS) function by aiding sphingolipid metabolism and activating vitamin K-dependent proteins (VKDPs) like Gas6 and protein S, which are key for neuronal survival, synaptic maintenance, and resistance to oxidative stress [29]. Together, the dual approach of modulating immune and vascular pathways along with nutritional support, especially through vitamins D and K, offers a promising strategy for enhancing neuroprotection and promoting healthy brain aging.

The main aim of this chapter is to explain the importance of vascular protection, immunomodulation, and the role of essential vitamins in maintaining brain health and preventing neurodegenerative disorders. It aims to combine current evidence highlighting how neurovascular integrity, immune regulation, and nutritional therapies collectively contribute to neuroprotection and cognitive resilience. In this chapter, particular focus is placed on vitamins D and K due to their emerging roles in modulating inflammation, oxidative stress, and neurovascular stability. The scope of this chapter is to provide a comprehensive discussion of therapeutic potentials, molecular mechanisms, and the translational relevance of these pathways in neurological and psychiatric conditions. By linking fundamental insights with clinical perspectives, this chapter aims to offer a complete understanding of how immunological, vascular, and nutritional factors converge to preserve brain health and inform novel therapeutic strategies.

7.2 BRAIN AGING: MECHANISMS AND CHALLENGES

Aging of the brain is a multifactorial process driven by complex cellular, molecular, vascular, and other changes that together comprise neuronal integrity and increase susceptibility to neurodegenerative and neuropsychiatric diseases [30]. Understanding this mechanism is essential for developing therapeutic strategies to preserve cognitive health throughout the lifespan. At the cellular and molecular levels, brain aging is driven by several key mechanisms, as shown in Figure 7.1 [31].

Oxidative stress is recognized as a key mechanism involved in brain aging. It occurs when there is an imbalance between the production of reactive oxygen species (ROS) and the brain's natural antioxidant defenses [32]. Under normal physiological conditions, ROS are continuously generated as natural by-products of cellular metabolism, especially within mitochondria during oxidative phosphorylation. At low to moderate levels, ROS serve as vital signaling molecules involved in maintaining cellular homeostasis, synaptic plasticity, and neurogenesis [33,34]. However, when their production exceeds the neutralizing capacity of endogenous antioxidant defenses such as

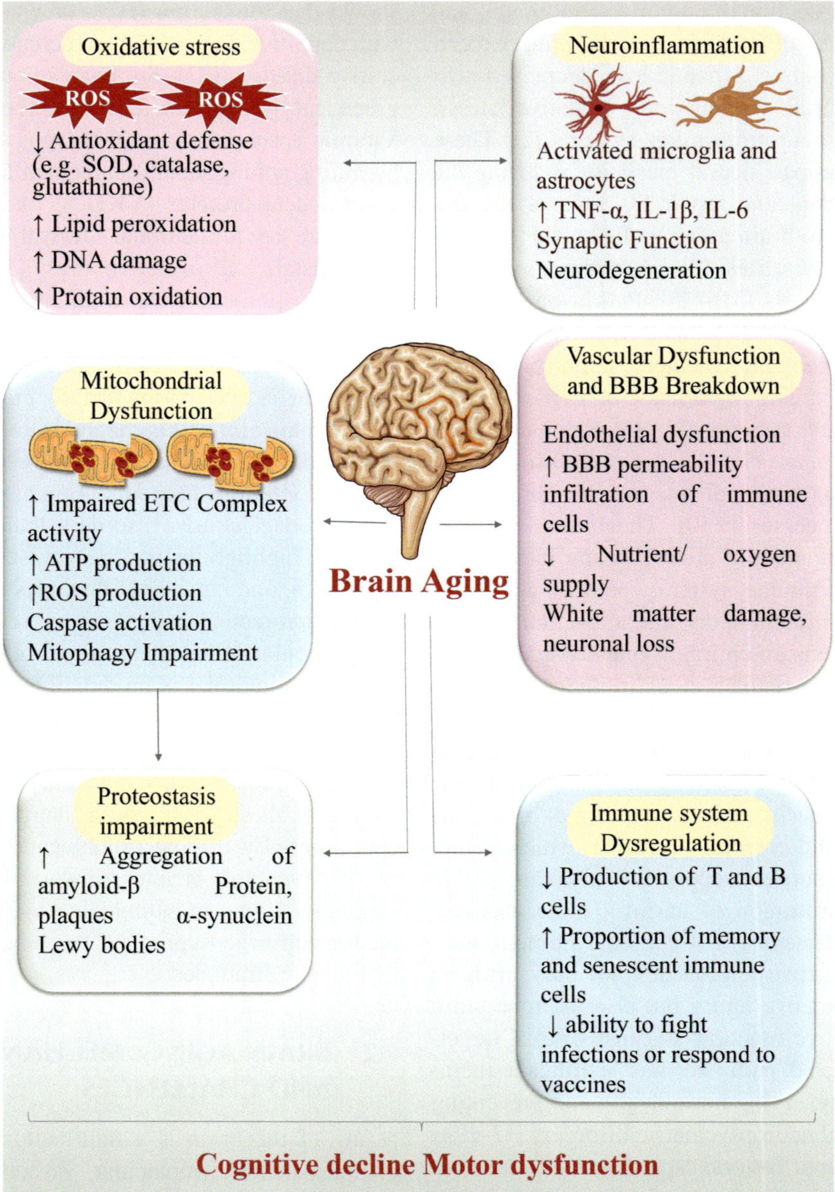

FIGURE 7.1 Integrated mechanisms underlying brain aging and neurodegeneration.

superoxide dismutase (SOD), catalase, and glutathione peroxidase (GPx), ROS accumulation leads to oxidative stress [35,36].

The brain is especially vulnerable to oxidative stress due to its high oxygen consumption, abundant lipid content, and comparatively limited antioxidant defenses compared to other organs. This renders neurons particularly prone to oxidative damage [37]. ROS attack polyunsaturated fatty acids in neuronal membranes, causing lipid peroxidation and compromising membrane integrity. Proteins undergo oxidative modifications that alter their structural shape and enzymatic functions, while oxidative DNA damage harms genomic stability and results in mutations or disrupted transcription. Together, these molecular damages contribute to progressive neuronal dysfunction [38–40].

A key factor in age-related oxidative stress is mitochondrial dysfunction. As we age, mitochondria become less efficient at producing Adenosine Triphosphate (ATP) and are more susceptible to electron leakage, which increases ROS production [41]. Dysfunctional mitochondria not only worsen oxidative stress but also trigger pro-apoptotic pathways, resulting in neuronal apoptosis [42]. This process is accompanied by synaptic loss, which is one of the hallmarks of cognitive decline during aging. Synaptic deterioration decreases the efficiency of neurotransmission and weakens neural network connectivity, ultimately leading to memory impairment and other age-related cognitive deficits [43].

Furthermore, oxidative stress works together with other pathological mechanisms of brain aging, including neuroinflammation, impaired autophagy, and dysregulation of calcium homeostasis [44]. ROS can activate microglia, leading to a chronic inflammatory state that speeds up neuronal damage. Simultaneously, oxidative damage to proteins involved in autophagy impairs the cell's ability to eliminate dysfunctional organelles and aggregated proteins, creating a vicious cycle that further heightens oxidative stress [45].

Thus, oxidative stress is a key link in the molecular process of brain aging, connecting mitochondrial dysfunction, neuroinflammation, synaptic degeneration, and neuronal death. Approaches that focus on reducing oxidative stress—such as boosting antioxidant defenses, maintaining mitochondrial health, or improving protein and organelle turnover—are actively being studied as possible treatments to slow down or lessen brain aging and related neurodegenerative diseases [46].

Neuroinflammation is also a major contributor to brain aging, characterized by a chronic, low-grade inflammatory state in the CNS. [47]. Overactivated microglia and astrocytes cause the release of pro-inflammatory cytokines such as IL-1β, IL-6, and tumor necrosis factor-alpha (TNF-α), which lead to disruption of synaptic function, impairment of neuroplasticity, and acceleration of neuronal degeneration [48,49]. Vascular dysfunction and the breakdown of the BBB are also key features of brain aging [50]. The age-related changes in cerebral vasculature, such as endothelial dysfunction and increased BBB permeability, impair nutrient and oxygen supply while allowing infiltration of peripheral immune cells and neurotoxins, leading to increased neuronal damage [51].

These interconnected processes result in observable cognitive and motor deficits [52]. Age-related cognitive decline usually appears as decreased memory retention, slower processing speed, and reduced executive function [53]. Upon progression of these deficits, they may progress into neurodegenerative diseases such as AD [54] and PD [55]. The AD is characterized by the buildup of β-amyloid plaques, hyperphosphorylation of tau protein, synaptic degeneration, and extensive neuronal loss [56]. PD involves dopaminergic neuronal degeneration in the substantia nigra and the formation of Lewy bodies made of α-synuclein, leading to motor dysfunction along with cognitive and behavioral symptoms. Other aging-related disorders, such as vascular dementia, Lewy body dementia, and frontotemporal dementia, share similar pathological mechanisms related to the brain aging [57,58].

The immune and vascular systems also have a crucial role in the development of brain aging [59]. The immune system helps clear cellular debris and pathogens; its dysregulation during aging contributes to chronic neuroinflammation and disrupts neuronal communication. Microglia and astrocytes, when overactivated, may hinder synaptic remodeling and worsen neurodegeneration. Similarly, aging of the vascular system, marked by arterial stiffening and endothelial dysfunction, leads to reduced cerebral perfusion and white matter damage [60]. The integrity of the BBB also decreases with age, permitting harmful substances to enter the brain and raising its vulnerability to injury and disease [61].

7.3 VITAMIN D: IMMUNOMODULATORY AND NEUROPROTECTIVE ROLES

Vitamin D is primarily known for maintaining calcium balance and supporting bone integrity, and it has attracted significant attention for its diverse biological effects on the CNS and immune systems [62]. It is a fat-soluble steroid-like hormone that exists in two forms: vitamin D2 and vitamin D3 [63]. Vitamin D2 is mainly produced from ergosterol, a sterol present in plants and fungi, upon exposure to ultraviolet B radiation [64]. Upon exposure to UV radiation, the skin synthesizes 7-hydroxycholesterol into vitamin D3, which is then obtained from dietary sources [65]. Once synthesized, vitamin D3 is transported to the liver, where it is activated through hydroxylation by the enzyme CYP2R1 into 25-hydroxyvitamin D3 [25(OH)D_3], the principal circulating form [66]. And this metabolite is further hydroxylated by the kidneys through CYP27B1 to produce the biologically active form, 1,25-dihydroxyvitamin D_3 [1,25$(OH)_2D_3$] [67]. The biologically active form, that is, calcitriol (1,25-dihydroxyvitamin D_3), exerts its effects by binding to the vitamin D receptor (VDR). VDRs are widely present in both the brain and immune cells, emphasizing the diverse roles of vitamin D beyond calcium metabolism, as shown in Figure 7.2. Within the CNS, VDRs are located in key regions such as the hippocampus, prefrontal cortex, hypothalamus, and substantia nigra, areas critically involved in cognition, memory, emotional regulation, and motor control. Activation of VDRs in neurons and glial cells influences the expression of neurotrophic factors, antioxidant enzymes, and neurotransmitter-synthesizing genes, thereby supporting neuroprotection and synaptic plasticity. In the immune system, VDRs are present on monocytes, macrophages, dendritic cells, and T and B lymphocytes, where they regulate cytokine production, immune cell differentiation, and inflammatory signaling. This dual distribution enables vitamin D to serve as a key immunomodulator while simultaneously maintaining neuronal integrity. Dysregulation of VDR signaling has been associated with increased neuroinflammation, impaired immune tolerance, and a higher risk of neurodegenerative and neuropsychiatric disorders [68].

Vitamin D exerts potent immunomodulatory effects through activation of the VDR, which regulates the balance between pro-inflammatory and anti-inflammatory cytokines that are essential for maintaining immune homeostasis and protecting against neuroinflammatory damage [69]. This leads to suppression of pro-inflammatory mediators such as IL-6, IL-17, and TNF-α, while it enhances the production of anti-inflammatory cytokines including IL-10 [70]. This cytokine modulation helps reduce chronic inflammation and lowers the risk of immune-mediated neuronal injury. Additionally, vitamin D affects adaptive immunity by promoting the differentiation and function of regulatory T cells (Tregs), which are essential for maintaining immune tolerance and preventing autoimmunity [71]. Within the CNS, vitamin D also modulates the activity

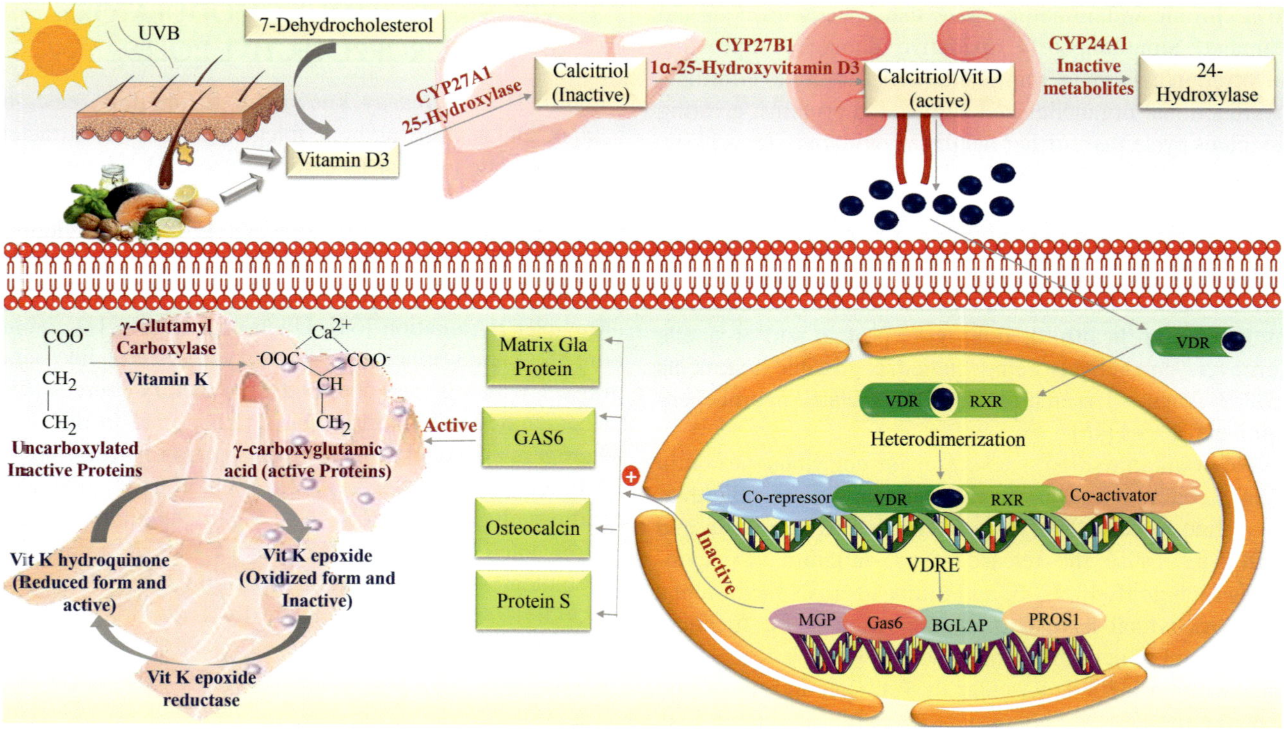

FIGURE 7.2 Interplay of vitamin D and vitamin K in the activation of VKDP.

of microglia, immune cells of the brain, by shifting them from a pro-inflammatory M1 to an anti-inflammatory M2 phenotype. These combined actions emphasize the central role of vitamin D as an immunoregulatory molecule that connects systemic and neuroimmune health, carrying significant implications for the prevention and management of neurodegenerative and neuropsychiatric disorders [72].

Beyond its immune function, vitamin D shows its neuroprotective actions through multiple interconnected mechanisms, particularly by regulating calcium homeostasis and demonstrating antioxidative and anti-apoptotic effects [73]. In the CNS, vitamin D regulates voltage-gated calcium channels and N-methyl-D-aspartate (NMDA) receptor activity, thereby preventing excessive calcium influx that can trigger excitotoxicity and neuronal injury [74]. By stabilizing intracellular calcium levels, it protects synaptic function and facilitates proper neuronal communication. Additionally, vitamin D enhances endogenous antioxidant defenses by upregulating enzymes such as GPx, SOD, and catalase, which reduce the accumulation of ROS and mitigate oxidative stress-induced cellular damage [75]. Beyond this, vitamin D protects against programmed cell death by downregulating pro-apoptotic proteins like Bax and caspases and upregulating anti-apoptotic mediators such as Bcl-2, thereby preserving mitochondrial integrity and preventing cytochrome c release. Collectively, these mechanisms work together to promote neuronal survival, reduce vulnerability to neurodegeneration, and emphasize the importance of vitamin D in maintaining brain health.

7.4 VITAMIN K: VASCULAR AND COGNITIVE PROTECTIVE MECHANISMS

7.4.1 Overview of Vitamin K and Its Metabolism

Vitamin K is a fat-soluble vitamin crucial for various physiological functions involving bone metabolism, homeostasis, and vascular health. The name "K" originates from the German word Koagulation, highlighting its early recognition and central role in blood coagulation [76]. Unlike many other vitamins that primarily serve as antioxidants or cofactors in metabolic reactions, Vitamin K has a unique role as a coenzyme. It functions as a coenzyme for the enzyme γ-glutamyl carboxylase, which carboxylates the glutamate residues on VKDPs to activate them [77]. It allows the VKDP to interact and bind with Ca2+ ions, which is a fundamental process for bone mineralization and blood coagulation [78]. The current adequate intakes (AIs) for vitamin K (as phylloquinone) range from 2.0 μg/day for infants (0–6 months) to 120 μg/day for men and 90 μg/day for women, with age- and gender-specific values set to support normal blood coagulation [79]. Vitamin K exists in three forms:

(a) Vitamin K1, also known as phylloquinone, is the primary dietary form of vitamin K. Being lipophilic, it is mainly found in green leafy vegetables such as spinach (250–1,139 Lg/100 g) and kale (240–1,220 Lg/100 g), as well as in oils like soybean oil. In the body, it plays a role in supporting the synthesis of clotting factors II, VII, IX, and X through hepatic carboxylation [80]. Absorbed in the small intestine along with dietary fats, it is rapidly cleared from the bloodstream [81]. (b) Vitamin K2, also called menaquinone, is present in meat, dairy, and fermented foods like natto. It is also naturally produced by gut bacteria. This form of vitamin K includes several types, from MK-4 to MK-13 [82]. Unlike Vitamin K1, K2 is extensively distributed to tissues outside the liver, such as bones and blood vessels, where it activates VKDPs like osteocalcin and matrix Gla protein (MGP) [83]. These proteins are crucial for bone mineralization and preventing vascular calcification, emphasizing K2's vital role in skeletal and cardiovascular health [84]. (c) Vitamin K3, also called menadione, is a synthetic precursor that can be converted into active forms of vitamin K2 in the body, supporting the activation of blood clotting factors. It is often used in veterinary care and included in animal feed because of its stability and affordability. However, its use in humans is strongly discouraged due to potential toxicity, including risks of oxidative stress and liver toxicity in adults [85].

Vitamin K deficiency is a clinically important condition that impairs the activity of VKDP involved in bone mineralization and vascular health. The most common sign is a bleeding disorder, usually shown by a prolonged prothrombin time, because the liver cannot produce functional clotting factors without enough vitamin K [86]. This causes easy bruising, mucosal bleeding, and, in severe cases, internal hemorrhage [87]. In infants, the deficiency may manifest as hemorrhagic disease of the newborn, a serious condition caused by low liver stores, inadequate placental transfer, and limited vitamin K in breast milk [88]. Several factors contribute to vitamin K deficiency. It can be caused by medications, especially with long-term use of anticoagulants like warfarin, which inhibit vitamin K epoxide reductase (VKOR) and prevent the regeneration of active vitamin K [89]. Liver diseases may also contribute by reducing the synthesis of vitamin K-dependent clotting factors [90]. In addition, fat malabsorption syndromes such as cystic fibrosis, inflammatory bowel disease, and biliary obstruction can hinder vitamin K absorption because it is fat-soluble. Besides its role in coagulation, deficiency also adversely affects skeletal health due to undercarboxylation of osteocalcin, leading to impaired bone mineralization and a higher risk of fractures [91]. Similarly, inadequate carboxylation of MGP predisposes to vascular issues, such as arterial calcification and atherosclerosis. Therefore, vitamin K deficiency is linked not only to bleeding but also to long-term effects on bone and cardiovascular health.

7.4.2 Vitamin K–Dependent Proteins and Their Biological Roles

VKDP represents a specialized group of proteins that require post-translational γ-carboxylation of specific glutamic acid residues to achieve functional calcium-binding capacity, as shown in Figure 7.2. This carboxylation, catalyzed by the enzyme γ-glutamyl carboxylase in the endoplasmic reticulum, is essential for the stability, structure, and biological activity of these proteins [83]. Representative VKDPs include osteocalcin, coagulation factors II (prothrombin), VII, IX, X, MGP, growth arrest-specific protein 6 (Gas6), and protein S [92]. The functional diversity of VKDPs enables vitamin K to affect various physiological systems, from blood coagulation and bone remodeling to vascular and neuroprotection [93].

7.4.2.1 Vascular and Skeletal Functions

MGP, a well-characterized VKDP, is primarily produced by vascular smooth muscle cells, osteoblasts, and chondrocytes. In its fully carboxylated form (cMGP), it strongly binds to calcium ions and hydroxyapatite crystals, preventing their nucleation and subsequent pathological deposition within arterial walls [94,95]. This anti-calcific activity is essential for preserving vascular elasticity and preventing the advancement of arterial stiffness and atherosclerosis [96].

In the skeletal system, MGP regulates bone morphogenetic protein-2 (BMP-2), a powerful osteogenic factor that promotes osteoblast differentiation [97]. By antagonizing BMP-2, MGP controls bone formation, prevents premature ossification, and maintains balanced bone turnover [98,99]. This balance is crucial for maintaining skeletal integrity and preventing abnormal mineralization patterns that could weaken bones.

7.4.2.2 Coagulation and Hemostatic Regulation

Vitamin K's role in hemostasis is mediated through its essential contribution to the maturation of coagulation factors II, VII, IX, and X, all of which are produced in the liver as inactive precursors [100]. The γ-carboxylation of glutamic acid residues within these proteins produces γ-carboxyglutamate (Gla) domains, which allow the proteins to bind calcium ions with high affinity [101]. This calcium-dependent binding helps anchor clotting factors to negatively charged phospholipid membranes, which is essential for assembling coagulation complexes, producing thrombin, and eventually forming a stable fibrin clot [102]. Without sufficient vitamin K, these proteins stay undercarboxylated and inactive, leading to impaired hemostasis, longer bleeding times, and greater risk of hemorrhage [103,104].

7.4.2.3 Vascular Integrity and Anti-Inflammatory Effect

Beyond its role in clotting, vitamin K is an important regulator of vascular health by activating MGP and Gas6 [105]. Gas6, another VKDP, interacts with the TAM (Tyro3, Axl, Mer) family of receptor tyrosine kinases to promote endothelial cell survival, inhibit apoptosis in vascular smooth muscle cells, and maintain endothelial barrier function [106,107]. These actions collectively reduce vascular injury and slow the progression of atherosclerotic lesions.

MK-4, the most biologically active form of vitamin K2, also exerts significant anti-inflammatory effects by activating Gas6-mediated signaling pathways that suppress nuclear factor kappa-B (NF-κB) activity [108,109]. NF-κB is a transcription factor that plays a central role in inducing pro-inflammatory cytokines such as TNF-α, IL-1β, and IL-6 [110,111]. Inhibiting NF-κB not only decreases inflammatory cytokine production but also reduces nicotinamide adenine dinucleotide phosphate (NADPH) oxidase activity, thereby lowering ROS generation [112,113]. This combined anti-inflammatory and antioxidant effect is crucial in preventing oxidative damage to vascular tissues.

7.4.2.4 Vitamin K-Mediated Neuroprotective Effect

Vitamin K also plays a prominent role in the CNS, where it helps maintain neuronal structure and function. One of its lesser-known yet essential functions is in regulating sphingolipid metabolism [114]. MK-4 functions as a cofactor for cerebroside sulfotransferase, an enzyme that catalyzes the sulfation of galactosylceramide, a crucial step in forming the myelin sheath [115]. Additionally, it promotes the activity of serine palmitoyltransferase (SPT), the rate-limiting enzyme in ceramide biosynthesis [116].

Through these actions, vitamin K helps maintain adequate sulfatide levels in the brain [117]. Sulfatides are essential for myelin integrity, axonal conduction, and synaptic efficiency. A deficiency in sulfatides can cause myelin breakdown, hinder signal transmission, and increase the risk of neurodegenerative diseases such as AD and other types of dementia [118,119]. Moreover, Gas6 and protein S in neural tissue offer neuroprotection by supporting neuronal survival, regulating neuroinflammation, and decreasing oxidative injury [120]. Clinical and epidemiological studies have shown that inadequate vitamin K intake is linked to faster cognitive decline, a higher risk of dementia, and worse performance in executive functions and verbal memory among older adults [121,122].

Collectively, the γ-carboxylation of VKDPs underpins essential biological processes across various organ systems. In the cardiovascular system, it prevents vascular calcification and supports endothelial stability. In the skeletal system, it helps maintain mineralization balance and bone strength. In the hemostatic system, it ensures the proper activation of coagulation factors. In the nervous system, it supports myelin integrity, neuronal survival, and cognitive function. By simultaneously regulating inflammation, oxidative stress, and lipid metabolism, vitamin K acts as a versatile regulator of both systemic and neurological health.

7.4.3 Evidence from Research Studies on Vitamin K's Role in Vascular Health and Brain Aging

See Table 7.1.

7.5 SYNERGISTIC ROLES OF VITAMINS D AND K

7.5.1 Vitamin D and K Interaction in Calcium Metabolism and Vascular Health

The interaction between the fat-soluble vitamins D and K is crucial for calcium metabolism, bone mineralization, and vascular health [141,142]. Vitamin D enhances intestinal absorption of dietary calcium and phosphate, increasing serum calcium levels, which are vital for many physiological functions like nerve conduction, muscle contraction, and blood clotting [143]. Elevated calcium levels stimulate the synthesis of VKDPs such as osteocalcin and MGP, but these are produced in an inactive, undercarboxylated form [105]. Vitamin K activates these proteins by γ-carboxylation, allowing them to bind calcium effectively [144]. Activated MGP inhibits abnormal calcium buildup in arterial walls, preventing vascular calcification, while activated osteocalcin binds calcium in the bone matrix, supporting mineralization and bone strength [145]. This coordinated action guarantees calcium is deposited in bones instead of soft tissues, thereby maintaining proper calcium distribution, supporting skeletal health, and reducing the risk of osteoporosis and cardiovascular issues [146].

7.5.2 Vitamin D and K Effect on Inflammation and Immune Regulation

Vitamin D and vitamin K work together as strong regulators of inflammation and immune response by the production and activation of VKDPs, such as MGP and Gas6 [147]. Vitamin D increases the expression of these proteins, while vitamin K activates them through γ-carboxylation, allowing their biological function [148]. Both vitamins help inhibit NF-κB activation, which reduces the production of pro-inflammatory cytokines like IL-6, IL-1β, TNF-α, and IFN-γ, while increasing the secretion of the anti-inflammatory cytokine IL-10, supporting immune tolerance [149,150]. Vitamin D also inhibits dendritic cell maturation and limits Th1 and Th17 cell differentiation, steering the immune response toward a more regulated state [151,152]. At the same time, vitamin K helps prevent excessive tissue damage by maintaining vascular and cellular integrity, thereby reducing chronic inflammatory responses and supporting overall immune balance [153]. Together, they play a vital role in managing inflammation and defending against immune-mediated damage.

7.5.3 Vitamin D and K Interaction in Preventing Neurodegeneration and Promoting Brain Health

Vitamin D and vitamin K work together to support brain health and healthy aging by promoting neuroprotection

TABLE 7.1
Preclinical and Clinical Insights into the Neuroprotective Role of Vitamin K

Sl.No.	Diseases	Dose and Duration	Study design	Final Outcome	References
1	Alzheimer's disease	Brain menaquinone-4 (MK-4) levels (postmortem tissue analysis)	Autopsy-based cohort study, rush memory, and aging project, $n \approx 500$	Higher brain MK-4 associated with 17%–20% lower odds of dementia/MCI and reduced AD neuropathology (tau tangles, Braak stage).	[123]
		Brain tissue quantification of K vitamers (HPLC/LC-MS)	Human brain samples, $n \approx 500$	MK-4 identified as the predominant form of vitamin K in brain regions; regional variations mapped.	[124]
		Higher dietary phylloquinone (Vit K1) intake	Cross-sectional CLIP study in older adults	Higher dietary K_1 intake correlated with **better cognition and behavior scores**.	[125]
		Dietary K1 intake (~70 µg/week vs ~22 µg/week)	Cross-sectional observational study.	Adequate vitamin K_1 intake linked to **better episodic memory consolidation**.	[126]
		PC12 cell	Cell line	The protective effects of vitamin K2 against H_2O_2 and Aβ (1-42)-induced apoptosis in PC12 cells may be through engaging both path ways. Vitamin K2 can be considered as a potential drug for the prevention of Alzheimer's disease progress or considered as a pretreatment to slow the progression of the disease.	[127]
		Diet rich in vitamin K	Prospective study of 960 participants 58–99 years	1 serving per day of green leafy vegetables and foods rich in phylloquinone, lutein, nitrate, folate, α-tocopherol, and kaempferol may help to slow cognitive decline with aging.	[128]
2.	Parkinson's disease	Vit K2 (2.5–100 µM)	PC12 cell lines,	↓ROS production, caspase-3. ↑pro-caspase-3, glutathione, cell viability.	[129]
		Vit K2	Case–control study 93 patients with PD with the age of 69.18±10.429 years and the gender ratio (male/female) of 42/51. Ninety-five healthy controls were recruited with the age of 66.42±9.971 years and the gender ratio (male/female) of 52/43.	VK2 level in PD patients was significantly lower than that in healthy controls. The decreased VK2 level may be related to the occurrence and development of PD. Anti-inflammatory effects, we speculate that PD patients may be caused by the lack of VK2.	[130]
		MK-4 (0.5–20 µmol/L) MK-4 (5–20 µmol/L)	Microglial cell line (BV2)	↑ iNOS, COX-2, TNF-α and IL-1β. ↓ ROS production, and caspase-1.	[131]
		Vit K1	Cross-sectional study Sixty-two women with Parkinson's disease (mean age 70.7 years) and 62 age-matched controls were studied. Patients were classified into two groups: Group A (independent, Hoehn & Yahr stages I–II, $n=26$) and Group B (dependent, stages III–V, $n=36$).	Group B (advanced PD) showed significantly lower bone mineral density, vitamin K1, and 25-OHD levels, with higher undercarboxylated osteocalcin compared to Group A. Vitamin K1 correlated positively with 25-OHD and negatively with undercarboxylated osteocalcin and disease severity. Regression analysis identified disease stage, vitamin K1, 25-OHD, and undercarboxylated osteocalcin as independent predictors of BMD.	[132]

(Continued)

TABLE 7.1 (*Continued*)
Preclinical and Clinical Insights into the Neuroprotective Role of Vitamin K

Sl.No.	Diseases	Dose and Duration	Study design	Final Outcome	References
		5–100 lM of 1,4-NQ, menadione, menaquinone, or phylloquinone	Cell line study	Vitamins K and many other bioactive molecules might represent a potential scaffold for designing drugs targeting a-syn fibrillization.	[133]
		Dietary intake 122.420±274.367 μg/d	Cross-sectional study aged 40–80 years	Increased total dietary intake of vitamins C and vitamins K from food, along with zinc intake, is linked to a reduced risk of PD.	[134]
3.	Multiple sclerosis	Vitamin K2 supplementation	A cross-sectional study overall, 45 MS (31 females and 14 males) patients with 39 RRMS, 5 secondary progressive MS (SPMS) and 1 PPMS patient and 29 healthy controls (19 females and 10 males) were included in the analyses	MS patients had much lower vitamin K2 levels than healthy controls. VK2 levels declined with age, disease duration, and relapse frequency. Female MS patients had especially low VK2. Results suggest VK2 deficiency may be linked to disease progression and neuroinflammation.	[135]
		Vitamin K (*i.p.* phylloquinone, 2 mg, three times per week)	C57Bl6/J mice	Vitamin K2 improved insulin sensitivity through modulation of osteocalcin carboxylation, suggesting a potential role in preventing metabolic disorder.	[136]
		Dietary intervention increasing vitamin K levels	Randomized control trial human pilot: modified paleolithic diet intervention	Diet increased serum vitamin K; associated with improved fatigue, motor/cognitive function, and quality of life.	[137]
4.	Amyotrophic lateral sclerosis	In vitro: VK_1 and MK-4 (EC_{50} ~2–30 nM) in vivo: dietary VK deficiency for weeks–months	Primary cell culture and rodent preclinical studies (oligodendrocyte precursors, cortical neurons; dietary models)	**Vitamin K1 and MK-4 prevent oxidative cell death** in oligodendrocyte precursors and neurons by inhibiting 12-lipoxygenase activation; dietary VK deficiency impairs neurogenesis provides mechanistic rationale relevant to neurodegenerative processes in ALS but not ALS-specific efficacy data.	[138]
5.	Autism	Menaquinone-4 (MK-4), a form of vitamin K2	Cross-sectional clinical study comparing children with ASD (n large cohort) vs neurotypical controls 731 children with ASD	**Children with ASD had significantly lower serum MK-4** than controls; MK-4 concentrations correlated with developmental quotient subscales (stronger in males). Suggests lower MK-4 may relate to developmental outcomes.	[115]
		Oral multivitamin/mineral (includes K sometimes)	Randomized open trials of multivitamin/mineral supplements in children with ASD The study involved 141 children and adults with autism	Multinutrient formulas improved some metabolic/nutritional parameters and minor symptom measures; studies are not vitamin-K specific but show feasibility of nutritional interventions in ASD.	[139]
		Dietary supplement	Pilot study A sample of 101 children with autism	At baseline, children with autism had overall adequate macronutrient and micronutrient intake compared to Dietary Reference Intakes (DRIs), but there was **large individual variability**. **Deficiencies** (<80% DRI) were noted in calcium (9/20 participants), pantothenic acid (6/20), vitamin D (5/20), and **vitamin K (8/20)**. During 2 months of risperidone treatment, participants showed **weight gain** and a **significant increase in vitamin K intake**, compared to placebo. However, risperidone treatment did **not significantly alter overall nutritional balance** after 6 months. While average diets met DRIs, many children with autism had **nutrient deficiencies (notably calcium, vitamin D, and vitamin K)**, and risperidone mainly contributed to weight gain without major changes in diet quality.	[140]

and managing neuroinflammation and oxidative stress [154,155]. Both vitamins are involved in sphingolipid metabolism, especially in the production of sulfatides, which are crucial for preserving myelin integrity, synaptic plasticity, and effective axonal conduction [156]. They also boost the expression of neurotrophic factors such as BDNF, nerve growth factor (NGF), and glial cell line-derived neurotrophic factor (GDNF), which are essential for neuronal repair, regeneration, and survival [157].

In addition, these vitamins help prevent cerebral vascular calcification, reduce lipid peroxidation, and decrease ROS production, thereby maintaining cerebral blood flow [158]. They downregulate the production of pro-inflammatory cytokines like IL-1β, IL-6, and TNF-α, modulate calcium channels and pumps to prevent excitotoxicity, and inhibit excessive microglial activation [159,160]. Through these combined actions, vitamins D and K help preserve neuronal integrity and cognitive function as we age [161].

7.6 POTENTIAL MECHANISMS OF ACTION IN BRAIN AGING AND OTHER CONDITIONS

7.6.1 Anti-Inflammatory Signaling Pathways and Oxidative Defense

The NF-κB signaling pathway is a key regulator of inflammation and is activated by various endogenous stimuli such as IL-1β, TNF-α, and lipopolysaccharide (LPS) [162]. These ligands interact with their respective receptors, TNFR1 and TLR4, triggering the recruitment and phosphorylation of adaptor protein complexes, specifically TRADD–TRAF2 and MyD88–TRAF6 [163,164]. TRAF6 (TNF receptor-associated factor 6) normally functions as a signaling adaptor that connects Toll-like receptor and interleukin-1 receptor pathways to downstream cascades, aiding in the initiation of immune responses against pathogens [165,166]. When overactivated, TRAF6 stimulates pro-inflammatory NF-κB signaling [167]. Activated TRAF6 further stimulates TAK1 (transforming growth factor beta-activated kinase 1), a serine/threonine kinase that normally functions in innate immunity and stress responses by activating both NF-κB and MAPK pathways [168]. TAK1 subsequently phosphorylates the IKK complex (IKKα, IKKβ, and IKKγ), which then phosphorylates IκBα, leading to its degradation and the release of NF-κB into the cytoplasm [169]. The free NF-κB moves into the nucleus, where it binds to κB sites and promotes the transcription of pro-inflammatory cytokines, including IL-6, TNF-α, and IL-1β, leading to inflammation [170]. Interestingly, the synergistic effect of vitamins K and D has been shown to reduce this inflammatory cascade through several overlapping mechanisms.

MGP is a VKDP that plays a crucial role in maintaining vascular health [171]. In vascular smooth muscle cells, carboxylated MGP binds to the BMPR1/2 receptor, preventing BMP-2/BMP-4 from accessing these receptors [172]. BMPs are multifunctional growth factors belonging to the TGF-β superfamily that normally regulate embryonic development, osteoblast differentiation, and bone repair [173]. However, in vascular tissues, excessive BMP signaling encourages calcification and inflammation [174]. BMPs typically activate SMAD1/5/8 (mothers against decapentaplegic homologs 1, 5, and 8), a family of transcription factors that, under normal physiological conditions, transmit BMP/TGF-β signals to regulate cell growth, differentiation, and extracellular matrix production [175,176]. In vascular pathology, ongoing BMP-SMAD signaling causes abnormal calcification and promotes inflammatory gene expression [177,178]. By inhibiting BMP-SMAD signaling, MGP prevents TAK1 activation, which results in lower NF-κB activity, decreased release of pro-inflammatory cytokines, and maintenance of vascular integrity [179,180].

Protein S is another VKDP with anti-inflammatory and anticoagulant properties. Under normal physiological conditions, protein S functions as a cofactor for activated protein C, thereby regulating blood coagulation and preventing thrombosis [181,182]. In endothelial cells, protein S also binds to TAM receptors (Tyro3, Axl, Mer), triggering the activation of SOCS1/3 (suppressor of cytokine signaling 1 and 3) [183,184]. SOCS proteins are part of a negative feedback loop in cytokine signaling; they normally regulate immune responses by inhibiting Janus kinase (JAK)/signal transducer and activator of transcription (STAT) activity and preventing excessive cytokine release [185,186]. This action not only decreases tissue factor expression and NADPH oxidase activity but also suppresses NF-κB-driven inflammation.

Osteocalcin, another VKDP, performs dual functions depending on its carboxylation state. The uncarboxylated form of osteocalcin typically binds to hydroxyapatite crystals in bone, reinforcing the matrix and promoting mineralization [187,188]. Conversely, the carboxylated form binds to the GPRC6A receptor, where it activates CaMKKβ and then the PI3K/Akt pathway [189,190]. This pathway typically supports cellular survival, energy metabolism, and mitochondrial health. By inhibiting mTORC1 downstream, carboxylated osteocalcin decreases IκBα degradation, which limits NF-κB activation and reduces ROS production while preserving mitochondrial function [191,192].

Gas6 is another ligand for TAM receptors with vital roles in vascular and immune regulation [193]. Under normal conditions, Gas6 controls cell survival, the clearance of apoptotic cells (efferocytosis), and vascular repair [194]. Binding of Gas6 to TAM receptors induces SOCS1/3 expression, which inhibits JAK/STAT signaling and downregulates TRAF6-TAK1 [195,196]. This produces a strong anti-inflammatory effect by suppressing NF-κB activity. Beyond its role in controlling inflammation, Gas6 signaling supports mitochondrial health, reduces oxidative stress, and promotes endothelial repair [197,198]. This integrative regulation emphasizes their therapeutic potential in fighting chronic inflammation, oxidative stress, and pathological vascular remodeling, as shown in Figure 7.3.

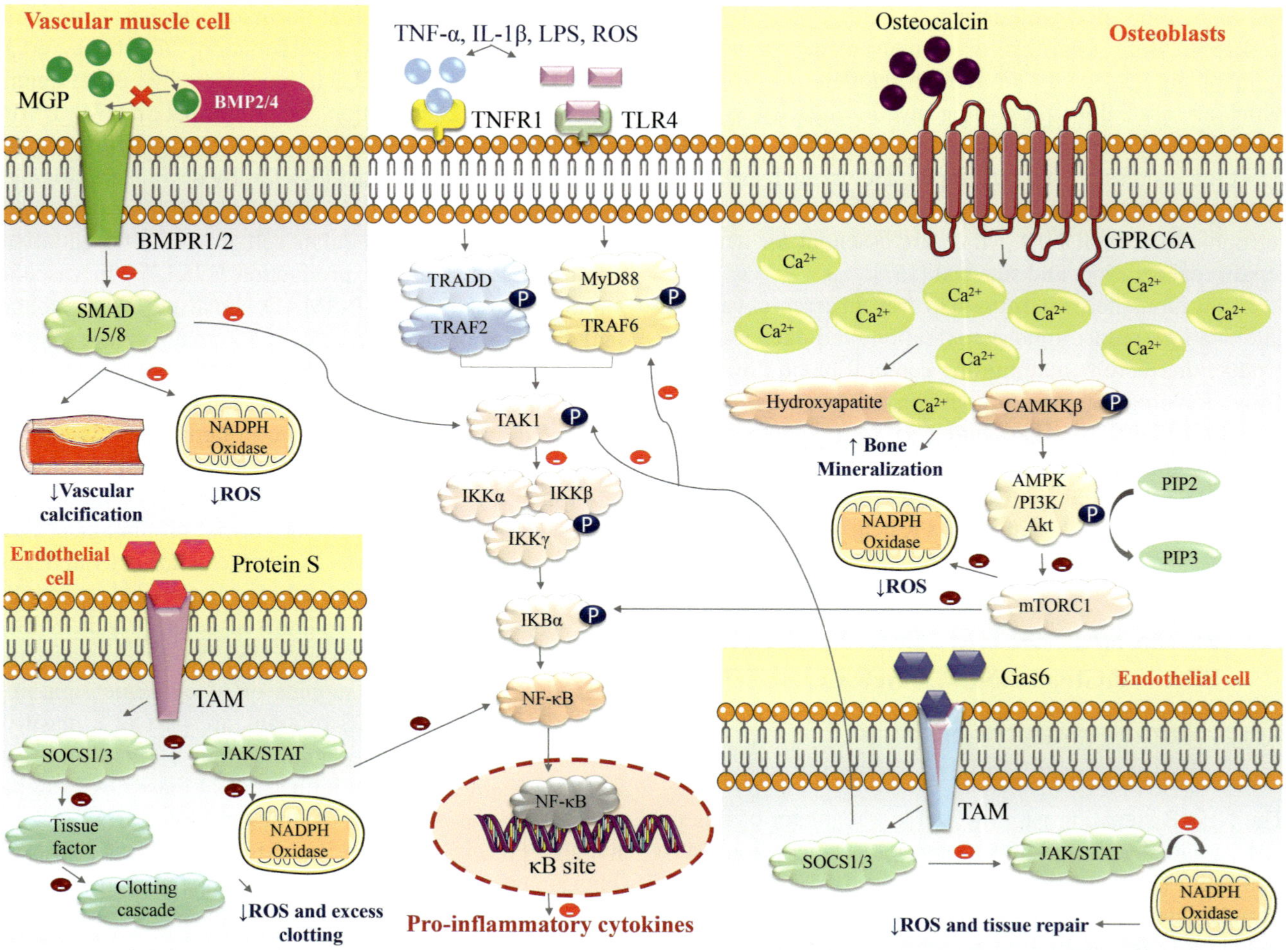

FIGURE 7.3 Schematic representation of VKDP-mediated anti-inflammatory and antioxidant action.

7.7 MECHANISM OF NEURONAL SURVIVAL AND SYNAPTIC PLASTICITY

Vitamin D metabolism is crucial for neuronal health and myelin formation. When activated, vitamin D is transformed into its active form, calcitriol, which attaches to the VDR in the cytoplasm of neuronal cells [199,200]. This VDR–calcitriol complex translocates into the nucleus, where it heterodimerizes with the retinoid X receptor (RXR) as shown in Figure 7.2 [201,202]. The heterodimer binds to vitamin D response elements (VDREs) on DNA, initiating the transcription and translation of genes such as DEGS1, UGT8, SGMS1/2, and VKDPs [203,204]. These genes encode enzymes vital for sphingolipid metabolism and myelin sheath formation in oligodendrocytes, thus supporting neuronal survival and functional integrity.

The lipid metabolic pathway involved in myelin biosynthesis begins with palmitoyl CoA and serine, which combine to form 3-ketosphinganine through the action of SPT [205]. This intermediate is then reduced by 3-ketodihydrosphingosine reductase to form sphinganine, which is subsequently converted into dihydroceramide via ceramide synthases (CerS) [206]. Dihydroceramide is desaturated by dihydroceramide desaturase (DEGS1) to produce ceramide, a key molecule in sphingolipid metabolism. Ceramide can then be converted into sphingomyelin by sphingomyelin synthase (SGMS1/2) or into galactosylceramide by ceramide galactosyltransferase (UGT8) [207]. With the involvement of vitamin K and the enzyme cerebroside sulfotransferase, galactosylceramide is further modified into sulfatide, a key lipid component of the myelin sheath [208,209]. Thus, vitamin D and vitamin K work together to support oligodendrocyte function and promote myelination.

Additionally, several proteins support neuronal survival. Gas6, when binding to TAM receptors (Tyro3, Axl, Mer) in neuronal cells, activates the pro-survival kinase ERK (extracellular signal-regulated kinase), which promotes oligodendrocyte survival and prevents degeneration [210–212]. Similarly, protein S, another ligand for TAM receptors, inhibits the JAK/STAT signaling pathway, thereby reducing ROS production and preventing neuroinflammation [213,214]. Osteocalcin, acting through the GPRC6A receptor, upregulates the expression of BDNF, a key neurotrophin that regulates neurotransmitter balance, including dopamine, glutamate, and serotonin [215,216]. This action helps prevent excitotoxicity, supports synaptic plasticity, and maintains cognitive function.

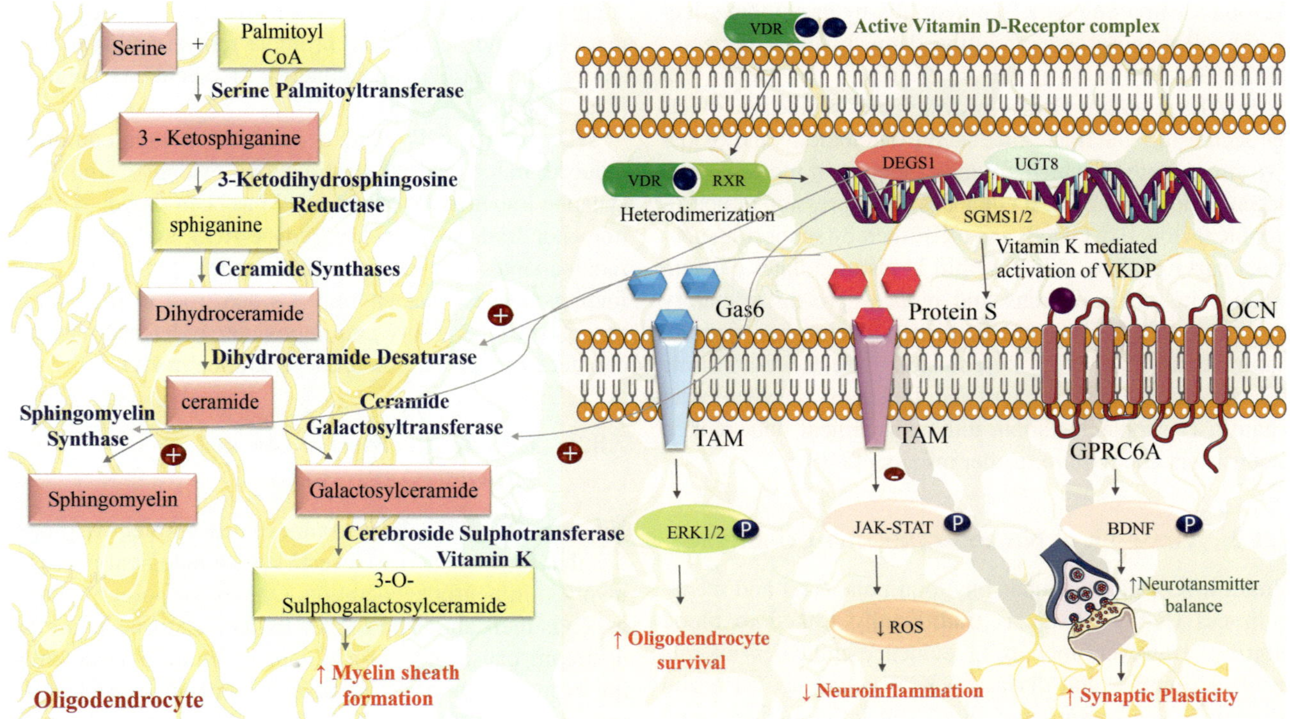

FIGURE 7.4 Schematic representation of vitamin K and D-mediated regulation of neuroprotection and brain health.

Collectively, these pathways show how vitamins D and K, along with their dependent proteins MGP, protein S, Gas6, and osteocalcin, work together to regulate myelin biosynthesis, neuronal survival, synaptic balance, and mitochondrial function. Through their combined actions, these molecules play a major role in maintaining brain health, preventing neuroinflammation, and slowing the progression of aging-related neurodegeneration as shown in Figure 7.4.

7.8 VITAMINS D AND K IN NEURODEGENERATIVE DISEASES

Neurodegenerative disorders, such as AD, PD, Huntington's disease (HD), and amyotrophic lateral sclerosis (ALS), pose an increasing public health challenge in aging societies. These disorders involve progressive neuronal loss, synaptic dysfunction, and chronic neuroinflammation. Recent progress in nutritional neuroscience has emphasized the role of vitamins D and K in influencing key pathological processes linked to these conditions, including oxidative stress, neuroimmune dysregulation, and cerebrovascular issues.

7.8.1 Vitamin D in Neurodegenerative Pathogenesis

Vitamin D plays a versatile role in the CNS, extending beyond its traditional function of regulating calcium and phosphate balance. The presence of VDRs and the enzyme 1α-hydroxylase in neuronal and glial cells highlights its autocrine and paracrine actions in the brain [217]. Vitamin D influences neuronal differentiation, neurotransmission, and neurotrophic signaling, including increasing NGF and BDNF [218].

In AD, vitamin D decreases amyloidogenic processing by downregulating β-secretase (BACE1) and enhances Aβ clearance through activating macrophages and microglia for phagocytosis [219,220]. It also prevents tau hyperphosphorylation and synaptic loss by reducing GSK-3β activity [221]. Furthermore, its immunomodulatory effects include suppressing pro-inflammatory cytokines such as IL-1β, IL-6, and TNF-α, while promoting anti-inflammatory cytokines like IL-10. These actions may help reduce chronic neuroinflammation, a hallmark of AD progression [222,223].

In PD, vitamin D acts as a neuroprotective agent by stabilizing intracellular calcium in dopaminergic neurons and preventing oxidative damage in the substantia nigra pars compacta [224,225]. Experimental models further confirm its role in maintaining dopaminergic neuron viability by reducing ROS and nitric oxide production [226,227].

Vitamin D exerts immunoregulatory effects that are crucial in the pathophysiology of MS, an autoimmune demyelinating disorder. It modulates both innate and adaptive immune responses by inhibiting Th1 and Th17 cell differentiation while promoting the development of Tregs, thereby reducing CNS-directed autoimmunity [228,229]. Vitamin D suppresses pro-inflammatory cytokines like IL-17 and IFN-γ and increases IL-10 expression, resulting in less microglial activation and decreased BBB permeability [230,231]. Preclinical and clinical studies also suggest that vitamin D supplementation may delay disease onset, reduce relapse rates, and limit lesion load on Magnetic

Resonance Imaging (MRI), potentially by modulating dendritic cell activity and inhibiting demyelination processes [232,233].

In ALS, vitamin D may provide neuroprotective effects by decreasing glutamate-mediated excitotoxicity and oxidative stress, both of which are major factors in motor neuron degeneration [234,235]. It enhances the expression of neurotrophic factors like GDNF and NGF, thereby supporting neuronal survival [236,237]. Vitamin D also regulates calcium balance in motor neurons and reduces pro-inflammatory mediators like TNF-α and IL-6, thus decreasing neuroinflammation [238,239]. Experimental models demonstrate delayed disease progression and increased survival after vitamin D supplementation, although clinical data remain inconsistent and require further research [240].

7.8.2 Vitamin K in Neurodegenerative Conditions

Vitamin K, which includes phylloquinone (K1) and menaquinones (K2 variants), is traditionally linked to blood clotting. However, increasing evidence indicates its role in CNS function and disease. VKDPs, such as Gas6 and protein S, have anti-apoptotic, anti-inflammatory, and myelin-protective effects [241]. Gas6, in particular, binds to the TAM family of tyrosine kinase receptors (Tyro3, Axl, and MerTK), activating downstream survival pathways such as the PI3K/Akt pathway and inhibiting inflammatory responses in microglia [242].

Reduced vitamin K levels in AD have been associated with disrupted sphingolipid metabolism and increased oxidative stress [243,244]. Since sphingolipids are essential for synaptic integrity and neuronal membrane structure, their dysregulation speeds up neurodegeneration [245]. Clinical evidence also links long-term use of vitamin K antagonists to cognitive decline, highlighting a protective role of vitamin K sufficiency in supporting cognitive functions [246].

Research on vitamin K in PD is still emerging, but preclinical findings suggest neuroprotective potential [247,248]. Vitamin K2 (menaquinone-4) restores mitochondrial membrane potential and ATP production, thereby reducing neuronal apoptosis in models of mitochondrial dysfunction—a central mechanism in PD pathogenesis. Its antioxidative action further supports dopaminergic neuron survival [249].

ALS involves progressive motor neuron degeneration caused by oxidative stress, excitotoxicity, and mitochondrial failure. Vitamin K2 has been shown to stabilize mitochondrial membrane potential, enhance antioxidant defenses, and reduce apoptosis in neuronal models. These mechanisms closely align with ALS pathology, making vitamin K a potential neuroprotective agent, although clinical validation is still required [250].

MS involves autoimmune demyelination and neuroinflammation. Vitamin K aids sphingolipid synthesis essential for myelin health, and through Gas6–TAM receptor signaling, it has anti-inflammatory and pro-survival effects [209,251]. Preclinical studies suggest vitamin K supplementation may decrease microglial activation, protect oligodendrocytes, and promote remyelination, emphasizing its potential as an adjunctive therapeutic target in MS [252,253].

HD pathology is characterized by mutant huntingtin-induced mitochondrial dysfunction, oxidative stress, and synaptic impairment. Vitamin K2 has been shown to restore mitochondrial bioenergetics, reduce ROS, and prevent neuronal apoptosis in experimental models [254,255]. These findings suggest that vitamin K's dual role in maintaining mitochondrial stability and antioxidative defense may help counteract progressive neurodegeneration in HD.

7.8.3 Convergent Mechanisms in Neurodegeneration

Vitamins D and K exhibit a range of neuroprotective properties that intersect across several key pathological processes involved in neurodegeneration. One major point of convergence is their ability to reduce oxidative stress, a common upstream cause of neuronal injury. Vitamin D enhances the expression of antioxidant enzymes such as SOD, GPx, and catalase [256]. While vitamin K—particularly in its menaquinone-4 (MK-4) form—serves as a powerful electron carrier in mitochondria, it reduces ROS production and maintains mitochondrial membrane potential. These antioxidant effects help prevent lipid peroxidation, protein oxidation, and DNA damage seen in neurodegenerative conditions [257]. Both vitamins also aid in regulating calcium homeostasis, which is vital for neuronal signaling and survival. Disrupted intracellular calcium levels can initiate excitotoxic cascades, mitochondrial dysfunction, and activate apoptotic pathways [258]. Vitamin D controls calcium influx by regulating L-type calcium channels and calcium-binding proteins like calbindin, while vitamin K2 supports calcium balance indirectly through activation of Gla-proteins, which prevent harmful calcium buildup in neuronal and vascular tissues [259,260].

Another important aspect of convergence is their role in regulating neuroimmune responses. Chronic neuroinflammation, marked by microglial activation and the release of pro-inflammatory cytokines (e.g., IL-1β, TNF-α, IL-6), is a key feature of many neurodegenerative diseases, including AD, PD, and MS. Vitamin D inhibits the activation of pro-inflammatory transcription factors like NF-κB and encourages microglia to adopt an anti-inflammatory M2 phenotype [222,261]. Vitamin K, especially K2, has been shown to reduce cytokine production and suppress the expression of inflammatory mediators through inhibition of NF-κB and activation of the steroid and xenobiotic receptor (SXR), indicating an immunoregulatory synergy with vitamin D [262,263].

Furthermore, both vitamins help preserve cerebrovascular integrity, which is essential for maintaining cerebral perfusion and reducing BBB disruption—a known factor in neurodegenerative progression [264]. Vitamin D boosts endothelial nitric oxide production, enhances vascular reactivity, and decreases vascular inflammation [265]. Vitamin K is necessary for activating MGP, which inhibits vascular calcification and may help prevent microvascular damage in the brain [266,267]. Taken together, these interconnected roles in oxidative balance, calcium signaling, immune modulation, and vascular protection highlight vitamins D and K as potential complementary agents in preventing or reducing neurodegenerative disease processes. Their combined use may produce additive or synergistic effects, making further investigation through both mechanistic studies and well-designed clinical trials necessary [268,269].

7.9 TRANSLATIONAL IMPLICATIONS AND THERAPEUTIC POTENTIAL

The biological plausibility of vitamins D and K as neuroprotective agents supports their potential use in clinical practice. Given their favorable safety profiles and the widespread prevalence of deficiency, especially in elderly populations, optimizing vitamin levels could serve as an accessible and cost-effective adjunct strategy for slowing brain aging and delaying neurodegenerative disease progression.

Age-related decline in skin production of vitamin D, lower dietary intake, and poor absorption all contribute to common vitamin D deficiency in older adults [270]. Similarly, insufficient vitamin K intake, especially in Western diets that are low in green leafy vegetables and fermented foods, is common. Both deficiencies have been independently linked to poorer cognitive outcomes, higher inflammatory markers, and increased risk of cerebrovascular events [271,272]. Maintaining serum 25(OH)D levels between 30 and 50 ng/mL is generally recommended for neuroprotective benefits, although the optimal levels for cognitive outcomes are still to be determined. Regarding vitamin K, while no standardized plasma biomarker is routinely used in clinical practice, undercarboxylated osteocalcin and MGP can act as surrogate markers of functional deficiency [273,274].

Recent interventional studies have examined the co-supplementation of vitamins D and K, hypothesizing synergistic effects on calcium regulation, inhibition of vascular calcification, and modulation of systemic inflammation [275–278]. Randomized controlled trials in non-neurological populations have shown improved vascular elasticity and decreased arterial stiffness, indicating potential benefits for maintaining cerebrovascular health—a key factor in cognitive function [279,280]. However, translating these findings into neurodegenerative disease contexts requires carefully designed trials with rigorous cognitive, functional, and imaging-based endpoints. Additionally, consideration must be given to confounding variables such as baseline nutritional status, comorbidities, medication use (e.g., anticoagulants), and genetic polymorphisms affecting vitamin metabolism. Despite their promise, several obstacles hinder the clinical implementation of vitamin-based neuroprotective strategies. These include limited awareness among clinicians, lack of consensus on dosage regimens, insufficient inclusion in clinical guidelines for neurodegenerative diseases, and the paucity of high-quality longitudinal data. Moreover, regulatory challenges exist in distinguishing nutritional supplements from pharmacological interventions, complicating reimbursement and access in some healthcare systems (Table 7.2).

7.10 FUTURE DIRECTIONS AND AREAS FOR RESEARCH

To fully understand the therapeutic potential of vitamins D and K in brain aging and neurodegeneration, collaborative research efforts are necessary across various fields. Long-term, sufficiently powered randomized controlled trials are crucial to establishing a causal relationship between vitamin supplementation and clinical outcomes in cognitive aging and neurodegenerative diseases. These trials should include stratification based on age, sex, baseline vitamin status, and genetic predispositions to enhance generalizability. Advances in genomics and metabolomics now allow for personalized assessments of nutritional needs and metabolic responses. Variants in genes like VDR, CYP27B1, or VKORC1 may influence the response to vitamin supplementation and could guide targeted strategies for vulnerable groups. Personalized nutrition approaches that use AI-driven decision tools and biomarkers of brain health may improve the timing and effectiveness of interventions.

Further investigation into the molecular targets of vitamins D and K within the CNS is necessary to clarify their full therapeutic potential. Innovative platforms such as human brain organoids, induced pluripotent stem cells (iPSC)-derived neurons, and microfluidic BBB models can support detailed studies of vitamin-driven effects on neural and vascular cells. Moreover, examining interactions between vitamins and other neuroprotective agents, including omega-3 fatty acids, flavonoids, and anti-inflammatory drugs, may reveal synergistic combinations appropriate for comprehensive interventions in brain aging.

7.11 CONCLUSION

Vitamins D and K are increasingly recognized as key modulators of neurobiological health, with a growing body of evidence linking them to immunoregulation, redox balance, vascular integrity, and neuronal survival. Their involvement in essential pathways of neurodegenerative disorders underscores their importance in preventing and managing age-related cognitive decline.

This chapter highlights the multifactorial roles of vitamins D and K, including their potential synergistic effects in regulating calcium metabolism, reducing neuroinflammation, and influencing neurovascular coupling. In addition to their individual functions, these vitamins may work

TABLE 7.2
Preclinical and Clinical Insights into the Neuroprotective Role of Vitamin D

Sl.No.	Diseases	Dose and Duration	Study design	Final Outcome	References
1	Alzheimer's disease	800 IU/day of vitamin D for 12 months once daily by oral gavage	Randomized, double-blind, placebo-controlled trial with 210 AD patients	**Significantly improved cognitive performance** (including memory, attention, vocabulary, visuospatial skills, and IQ) **Reduced amyloid beta-related biomarkers** (plasma Aβ42, APP, BACE1, and their mRNA levels) compared with placebo.	[281]
		100,000 IU drinking vial (2 mL)/day for 24 weeks, oral intake of vitamin D3	Unicentric, double-blind, randomized, placebo-controlled trial with 120 AD patients	**Baseline**: All groups similar MMSE. After **6 months**: • **Memantine + VitD:** ↑ MMSE **+4.0** • **Memantine alone:** no change • **VitD alone:** slight decline **Analysis**: Combo treatment significantly better than either alone	[282]
		Participants received 1,000 IU/day oral vitamin D2 for 8 weeks, then 6,000 IU/day (2 caps TID with food, adjusted to keep serum 25OHD 130–175 nM) for 8 weeks.	Double-blind, randomized, placebo-controlled, single-center trial with 63 recruited, 32 randomized, 31 completed high-dose/placebo phase	High-dose vitamin D2 on top of low-dose supplementation for 8 weeks did not improve cognition, memory, disability, mood, or pain in mild–moderate Alzheimer's disease.	[283]
		Recommended: ≥35 µg/week (~5 µg/day, ≥400 IU/day) Inadequate: <35 µg/week (~<5 µg/day, <400 IU/day) Mean: Recommended group ~69.5 µg/week; Inadequate group ~21.7 µg/week	Cross-sectional observational study. A total of 5,596 community-dwelling women	Older women with inadequate vitamin D intake (<35 µg/week) had significantly lower cognitive scores and higher odds of impairment (adjusted OR = 1.30, p = 0.024) compared to those with adequate intake (AI), independent of confounders	[284]
		Patients with a new diagnosis of AD, who had not taken anti-dementia drugs or vitamin D supplements. They prescribed memantine alone (n = 18), vitamin D alone (n = 17), or memantine plus vitamin D (n = 8) for an average of 6 months.	A Pre–Post pilot study of 43 white outpatients (mean 84.7 ± 6.3 years; 65.1% women)	Baseline MMSE is similar across groups. And Memantine + vitamin D: +4.0 points (P = 0.034). Memantine alone showed no change (P = 0.891) where vitamin D alone showed slight decline (P = 0.504). So, it can be concluded that the combo of memantine and VitD is significantly better (P < 0.01).	[285]
		500 IU/kg of chow for 5 months	Animal model of PPswe/PS1dE9 transgenic mice of 4 months	High vitamin D given early (M1–M6) improved working memory and increased hippocampal neurogenesis in AD mice. Normal vitamin D (NVD) given later (M4–M9) improved neurogenesis at an advanced age, but high doses had no effect then. Vitamin D deficiency (0VD) early → more amyloid plaques; late → impaired neurogenesis in both WT and AD mice. Gender effect: improvements varied between males and females (males responded better in presymptomatic phase, females in symptomatic phase). In vitro: calcitriol (active vitamin D) increased progenitor cell proliferation but did not enhance neuronal differentiation.	[286]

(Continued)

TABLE 7.2 (*Continued*)
Preclinical and Clinical Insights into the Neuroprotective Role of Vitamin D

Sl.No.	Diseases	Dose and Duration	Study design	Final Outcome	References
		4,000 IU/kg of chow for 2 months	Animal Studies of 8-week-old BALB/c male mice (22–25 g)	Vitamin D3 supplementation (4,000 IU/kg for 2 months) improved locomotor activity and reduced anxiety-like behavior in BALB/c mice. Treated mice showed better memory performance in the Y-maze and passive avoidance tests. **Neurochemical effects**: increased hippocampal serotonin and dopamine levels. **Oxidative stress**: reduced lipid peroxidation and increased antioxidant enzyme activity (SOD, catalase, glutathione). **Histology**: vitamin D reduced neuronal damage in the hippocampus.	[264]
2.	Parkinson's disease	BMI-based: <25 → 4,000 IU/day; 25–30 → 5,000 IU/day; >30 → 6,000 IU/day for 12 weeks	Randomized controlled trial, total 50 patients in the study with the average age of 65.6	Increased in VitD concentration Better functional performance Trend to reduced inflammation	[287]
		VitD3 (10,000 IU/d) plus calcium carbonate, placebo plus 1,000 mg calcium carbonate for 16 weeks	Pilot Randomized double-blind intervention trial, total 51 participants, average age of 66.6	No overall balanced improved Potentially improves balance in younger PD patients	[288]
		VitD3 (1,000 IU/d) for 3 months	Randomized, double-blind, placebo-controlled trial, total participants 120, average age of 49.9	VitD has no effects on improvement of levodopa induced dyskinesia in PD	[289]
		VitD3 (1,200 IU/d) for 12 months	Randomized, double-blind, placebo-controlled trial, total participants 114, average age of 71.9	VitD3 supplementation stabilizes PD symptoms FokI TT or CT genotypes have positive effects	[290]
		400 μM of vitamin D3 for 8 days	Rotenone-induced Parkinson's model of *C. elegans*	VitD3 rescued dopaminergic neuron loss in rotenone-exposed worms. Reduced α-syn accumulation in transgenic strain NL5901. Improved locomotion and chemotaxis. Lowered ROS, restored mitochondrial function, upregulated gst-4 and daf-16. Extended survival under rotenone stress.	[291]
		Vitamin D3 3 μg/kg/day, *i.p.* for 29 days	A total of 124 male C57BL/6J mice, 6–7 weeks old, weight **18–22 g**	Calcitriol improved rotarod and pole test performance in 6-OHDA mice. Preserved TH+ dopaminergic neurons in substantia nigra. Restored striatal dopamine, DOPAC, and HVA levels. Reduced microglial M1 polarization and pro-inflammatory cytokines. Lowered T-cell infiltration in midbrain, boosted splenic Tregs. Treg depletion abolished calcitriol's protective effects.	[226]
3.	Multiple sclerosis	Vitamin D3 dose—100,000 IU every 14 days for 24 months	Parallel, double-blind, randomized placebo-controlled clinical trial Total 316 participants enrolled with the average weight of and randomized into 163 in vitamin D group, 153 in placebo group	**Primary outcome:** Disease activity occurred in **60.3% (VitD) vs 74.1% (placebo)** **Time to disease activity:** Longer with VitD. **MRI outcomes: 57.1%** of patients in the vitamin D group vs **65.3%** in the placebo group showed MRI activity. **Relapse rate:** No significant difference (**17.9% VitD vs 21.8% placebo**	[292]

(*Continued*)

TABLE 7.2 (*Continued*)
Preclinical and Clinical Insights into the Neuroprotective Role of Vitamin D

Sl.No.	Diseases	Dose and Duration	Study design	Final Outcome	References
		High dose: 5,000 IU/day vs **low dose: 600 IU/day given daily for 96 weeks**	Randomized, phase 3, double-blind, multi-center, placebo-controlled clinical trial 172 randomized (89 high-dose, 83 low-dose)	**Relapse risk:** No significant difference between high dose and low dose; 32% (low-dose) vs 34% (high-dose) **Primary endpoint:** High-dose vitamin D did not reduce clinical relapse over 96 weeks. **Safety:** No hypercalcemia; 3 cases of nephrolithiasis/ureterolithiasis (1 low-dose, 2 high-dose).	[293]
		Oral high-dose cholecalciferol (vitamin D3) **14,007 IU/day for 48 weeks**	Phase II, randomized, double-blind, placebo-controlled, add-on clinical trial **229 randomized** (113 vitamin D, 116 placebo) **229 randomized** (113 vitamin D, 116 placebo)	**Primary endpoint (NEDA-3 at 48 weeks):** No difference (VitD 36.3% vs placebo 35.3%) **MRI secondary endpoints:** Fewer **combined unique active lesions** (IRR 0.68; $p=0.0045$). Smaller **increase in T2 lesion volume** **Clinical relapses & disability progression:** No significant differences reported. **Safety:** High-dose vitamin D was well tolerated, no major safety issues noted.	[294]
		Oral **cholecalciferol 100,000 IU every other week for 96 weeks**	Double-blind, placebo-controlled, parallel-group, 2-year clinical trial **181 randomized** (1:1 vitamin D vs placebo)	High-dose cholecalciferol in RRMS patients on interferon β-1a did **not meet the primary endpoint** of reducing annualized relapse rate (ARR). However, among trial completers, vitamin D was associated with **lower ARR, fewer and smaller T1 hypointense lesions, and less EDSS progression**. Treatment was **safe and well tolerated**, with adverse events similar to placebo.	[295]
		Oral cholecalciferol (vitamin D3) **High dose:** ~10,200 IU/day, **Low dose:** ~200 IU/day for 18 months	Multicentric, randomized, double-blind, controlled, **phase II pilot trial** **A total of 80** patients participated in the study	Primary endpoint (new T2-weighted lesions at 18 months): No difference between high- and low-dose groups. Other clinical/MRI outcomes: No significant differences observed in relapse rates, disability progression (EDSS), T2 lesion volume, gadolinium-enhancing lesions, or brain atrophy. Safety and tolerability**:** High daily dose (~10,200 IU) was well tolerated, with no serious adverse effects or hypercalcemia, and comparable adverse event rates in both arms.	[296]
		Weekly dosing of vitamin D3 at 20,000 IU for 1 year	Double-blind, randomized, placebo-controlled clinical trial 66 MS patients (34 in the VitD group and 32 in the placebo group)	Vitamin D3 added to interferon β-1b significantly increased serum vitamin D levels (from 54 to 110 nmol/L). Results showed a significant reduction in T1 gadolinium-enhancing lesions and trends toward lower T2 lesion burden and disability in the vitamin D group. No significant difference in relapse rate or adverse events was observed. Vitamin D3 add-on therapy was safe and reduced MRI disease activity in MS.	[297]
		Vitamin D supplementation with 400 IU (International Units) per week orally for 45 days	Experimental animal study using 45 male dark agouti rats	**Myelin:** VD+ preserved myelin at day 30; **Microglia:** Reduced activation at days 30 & 45; **Apoptosis:** Fewer apoptotic cells at days 15, 30 & 45; **Neurons:** Better survival at days 15 & 30; **NfL:** Lower serum levels → less axonal injury; **Oxidative stress:** ↓ Cu++-oxLDL & HOCl-ox LDL at all time points; **Antioxidant defense:** ↑ TAC and serum polyphenols; **Astrocytes:** Protective activation pattern with VD+	[298]

(*Continued*)

TABLE 7.2 (*Continued*)
Preclinical and Clinical Insights into the Neuroprotective Role of Vitamin D

Sl.No.	Diseases	Dose and Duration	Study design	Final Outcome	References
4.	Amyotrophic lateral sclerosis	Cholecalciferol, 50,000/75,000/100,000 IU per month via oral route, for 6 months	Randomized clinical trial Total 48 ALS patients were included in this study	Serum vitamin D: Significant dose-dependent rise in 25(OH)D levels; higher with 75,000 & 100,000 IU/month vs. 50,000 IU/month; ALSFRS-R (motor function): No significant improvement across groups; UMN score: No significant change among groups; Dose comparison: No difference between 75,000 IU and 100,000 IU groups; Overall: Vitamin D corrected deficiency but did not slow ALS progression over 6 months.	[299]
		Vitamin D given at 2,000 IU/day via oral route for 9 months	Observational cohort (9-month follow-up). A total 37 ALS patients were observed with an average weight of 55	Baseline deficiency: 81% had 25(OH)D <30 ng/mL; 43% <20 ng/mL; Serum levels: in supplemented group, median vitamin D rose from 18.5 → 31 ng/mL at 6 months; ALSFRS-R decline: slower decline at 9 months in vitamin D group; no significant difference at 3 or 6 months; Safety: No adverse effects reported; overall: 2,000 IU/day vitamin D was safe and showed potential functional benefit at 9 months.	[300]
		AI: 1 IU D_3/g feed Deficient (DEF): 0.025 IU D_3/g feed (2.5% of AI)	Preclinical, **G93A ALS transgenic mouse model, A total** 42 mice (22 males, 20 females), average weight of 25 days	Oxidative Stress: DEF mice had ↑ 4-HNE Antioxidant Response: DEF mice had ↑ GPx1 Apoptosis: DEF mice showed ↑ Bax/Bcl2 ratio Females (DEF vs AI): ↑ GPx1, ↑ IL-6 Males (DEF vs AI): ↑ 4-HNE, ↓ SOD2 Sex Differences (DEF males vs DEF females): ↓ SOD2, ↓ GPx1, ↓ IL-6, ↓ ChAT	[301]
		AI group: 1 IU D_3/g feed **High Dose group:** 50 IU D_3/g feed (50×AI)	Preclinical, **ALS G93A transgenic mouse model, A total** 100 mice (**55 males, 45 females**), 25 days old.	HiD females ate 9% less food → possible toxicity. HiD mice had better motor endurance (+12%) and higher clinical scores (+37%). HiD females showed +20% endurance AUC but worse pre-onset severity.	[302]
		AI (1 IU/g feed) High dose (HiD; 10 IU/g feed)	Preclinical (ALS transgenic G93A mouse model), A total of 32 mice (21 males, 11 females)	Paw grip endurance: HiD mice had 7% higher scores Motor performance: HiD mice had 22% higher scores Sex effect: Effect mainly in males—HiD males had 33% higher motor performance vs. AI males	[240]
5.	Autism	Vitamin D dose: 300 IU/kg/day for 15 days	Randomized, parallel, double-blind, clinical trial A total patient: 43 children with ASD (22 intervention, 21 control)	Serum 25(OH)D: Significantly increased after supplementation TNF-α: Significantly decreased; IL-37: Significantly decreased RLRS total score: Significantly improved in the vitamin D group RLRS subcategories: No significant changes (small improvements only). Between-group RLRS comparison: not significant	[303]
		2,000 IU of vitamin D3 daily for 12 months	Randomized, double-blind, placebo-controlled, 2×2 factorial clinical trial A total of 73 children with ASD	**Omega-3 (OM)** and **vitamin D + Omega-3 (VIDOM)** improved **SRS-social awareness** Final observation: • VIDOM → **better social communicative functioning (SRS)** • OM → **improved total SRS score** • VIDOM → **improved taste/smell sensory processing (SPM)** • OM → **better balance/motion sensory processing (SPM)** **Vitamin D alone** did **not show significant improvements** compared to placebo.	[304]

(Continued)

TABLE 7.2 (*Continued*)
Preclinical and Clinical Insights into the Neuroprotective Role of Vitamin D

Sl.No.	Diseases	Dose and Duration	Study design	Final Outcome	References
		Standard diet group: 1,000 IU/kg diet. Vitamin D-deficient group: 0 IU/kg diet. Supplementation group: 100,000 IU/kg intramuscular injection	Preclinical animal study Sprague Dawley rats, 8–10 weeks old at study start (200–250 g body weight)	Male VDD and ASD rats showed **reduced social interaction, impaired olfactory recognition, and increased stereotypic behaviors**, along with **GI dysfunction** They also had **lower brain and intestinal serotonin (5-HT), and reduced VDR/TPH1/2 expression**. **High-dose vitamin D (2 injections of 100,000 IU/kg)** significantly **improved behavior, GI function, and restored 5-HT and VDR/TPH expression** in ASD males.	[305]
		Vitamin D-deficient diet: contained 0 IU/kg VD3 and Control diet: contained 1,000 IU/kg VD3, starting from **4 weeks before mating,** continued **throughout pregnancy and lactation.**	Preclinical animal study **C57BL/6J mice,** age **6–8 weeks old** at the start	Maternal VD deficiency led to autism-like behaviors in offspring (↑ repetitive & anxious behaviors, ↓ social interaction). Offspring showed gut microbiota disturbances, ↓ short-chain fatty acids, and impaired intestinal barrier proteins (claudin-1, occludin). FMT treatment corrected microbiota imbalance, ↑ SCFAs, improved gut barrier, and rescued autistic-like behaviors.	[306]

together to strengthen brain resilience against metabolic, vascular, and inflammatory challenges. Such insights open new pathways for therapeutic development, especially in personalized nutrition and precision medicine approaches to brain aging.

While emerging findings are compelling, the field remains relatively early-stage, highlighting the need for carefully designed clinical trials, mechanistic studies at the cellular and molecular levels, and integrative omics-based research to establish causality and dose–response relationships. Moreover, individual variability caused by genetic polymorphisms, lifestyle choices, and comorbidities calls for personalized strategies to optimize vitamin D and K levels. Addressing both immune modulation and cerebrovascular protection through nutritional approaches signifies a paradigm shift in tackling neurodegeneration. In this context, maintaining adequate and consistent levels of vitamins D and K may serve as a key strategy to preserve synaptic health, support cognitive functions, and promote healthy brain aging throughout life.

ACKNOWLEDGMENTS

The authors thank Chairman Mr. Parveen Garg, ISF College of Pharmacy, Moga (Punjab), India, for his excellent vision and support.

AUTHOR'S CONTRIBUTIONS

Sumedha Gupta (S.G.), Abhishek Kumar Gupta (A.K.G.), Ritam Mukherjee (R.M.) wrote the original draft of the review and contributed to writing and review. Sidharth Mehan (S.M) was responsible for conceptualization and supervision. Pankaj Kumar Maurya (P.K.M.), Gurudev Singh Raina (G.S.R.), Baljit Kaur (B.K.) performed the formal analysis and editing. All authors agree to be accountable for all aspects of work, ensuring integrity and accuracy.

FUNDING

None.

ETHICS APPROVAL

Not applicable.

CONSENT TO PUBLISH

Authors approve the submission of this publication.

CONSENT TO PARTICIPATE

Not applicable.

DATA AVAILABILITY STATEMENTS

Data sharing is not applicable to this article as no datasets were generated or analyzed during the current study.

REFERENCES

1. J. Lee and H. J. Kim, "Normal Aging induces changes in the brain and neurodegeneration progress: review of the structural, biochemical, metabolic, cellular, and molecular changes," *Frontiers in Aging Neuroscience*, vol. 14, p. 931536, 2022. doi:10.3389/fnagi.2022.931536.
2. V. I. Lushchak, "Interplay between bioenergetics and oxidative stress at normal brain aging. Aging as a result of increasing disbalance in the system oxidative stress–energy provision," *Pflugers Archiv European Journal of Physiology*, vol. 473, no. 5, pp. 713–722, 2021. doi:10.1007/s00424-021-02531-4.
3. N. Aderinto, M. O. AbdulBasit, G. Olatunji, and T. Adejumo, "Exploring the transformative influence of neuroplasticity on stroke rehabilitation: a narrative review of current evidence," *Annals of Medicine & Surgery*, vol. 85, no. 9, pp. 4425–4432, 2023. doi:10.1097/ms9.0000000000001137.
4. H. M. Wilkins and R. H. Swerdlow, "Mitochondrial links between brain aging and Alzheimer's disease," *Translational Neurodegeneration*, vol. 10, no. 1, pp. 1–11, 2021. doi:10.1186/s40035-021-00261-2.
5. E. Ghione et al., "Aging and brain atrophy in multiple sclerosis," *Journal of Neuroimaging*, vol. 29, no. 4, pp. 527–535, 2019. doi:10.1111/jon.12625.
6. H. Kumar et al., "The role of free radicals in the aging brain and Parkinson's disease: convergence and parallelism," *International Journal of Molecular Sciences*, vol. 13, no. 8, pp. 10478–10504, 2012. doi:10.3390/ijms130810478.
7. T. T. Xu et al., "Spermidine and spermine delay brain aging by inducing autophagy in SAMP8 mice," *Aging*, vol. 12, no. 7, pp. 6401–6414, 2020. doi:10.18632/aging.103035.
8. C. F. Reynolds, D. V. Jeste, P. S. Sachdev, and D. G. Blazer, "Mental health care for older adults: recent advances and new directions in clinical practice and research," *World Psychiatry*, vol. 21, no. 3, pp. 336–363, 2022. doi:10.1002/wps.20996.
9. C. Ising and M. T. Heneka, "Functional and structural damage of neurons by innate immune mechanisms during neurodegeneration review-article," *Cell Death and Disease*, vol. 9, no. 2, 2018. doi:10.1038/s41419-017-0153-x.
10. S. Xu, J. Lu, A. Shao, J. H. Zhang, and J. Zhang, "Glial cells: role of the immune response in ischemic stroke," *Frontiers in Immunology*, vol. 11, no. February, pp. 1–16, 2020. doi:10.3389/fimmu.2020.00294.
11. L. C. Freeman and J. P. Y. Ting, "The pathogenic role of the inflammasome in neurodegenerative diseases," *Journal of Neurochemistry*, vol. 136, pp. 29–38, 2016. doi:10.1111/jnc.13217.
12. C. Greene et al., "Blood–brain barrier disruption and sustained systemic inflammation in individuals with long COVID-associated cognitive impairment," *Nature Neuroscience*, vol. 27, no. 3, pp. 421–432, 2024. doi:10.1038/s41593-024-01576-9.
13. V. Rajeev et al., "Pathophysiology of blood brain barrier dysfunction during chronic cerebral hypoperfusion in vascular cognitive impairment," *Theranostics*, vol. 12, no. 4, pp. 1639–1658, 2022. doi:10.7150/thno.68304.

14. Y. C. Fang, Y. C. Hsieh, C. J. Hu, and Y. K. Tu, “Endothelial dysfunction in neurodegenerative diseases,” *International Journal of Molecular Sciences*, vol. 24, no. 3, pp. 1–27, 2023. doi:10.3390/ijms24032909.
15. R. Zhan et al., “NAD+ rescues aging-induced blood-brain barrier damage via the CX43-PARP1 axis,” *Neuron*, vol. 111, no. 22, pp. 3634-3649.e7, 2023. doi:10.1016/j.neuron.2023.08.010.
16. Y. Lu et al., “Endothelial TFEB signaling-mediated autophagic disturbance initiates microglial activation and cognitive dysfunction,” *Autophagy*, vol. 19, no. 6, pp. 1803–1820, 2023. doi:10.1080/15548627.2022.2162244.
17. G. Bocheva, D. Bakalov, P. Iliev, and R. Tafradjiiska-Hadjiolova, “The vital role of melatonin and its metabolites in the neuroprotection and retardation of brain aging,” *International Journal of Molecular Sciences*, vol. 25, no. 10, 2024. doi:10.3390/ijms25105122.
18. L. Wei et al., “H3K18 lactylation of senescent microglia potentiates brain aging and Alzheimer’s disease through the NFκB signaling pathway,” *Journal of Neuroinflammation*, vol. 20, no. 1, pp. 1–18, 2023. doi:10.1186/s12974-023-02879-7.
19. H. M. Abdolmaleky and J. R. Zhou, “Underlying mechanisms of brain aging and neurodegenerative diseases as potential targets for preventive or therapeutic strategies using phytochemicals,” *Nutrients*, vol. 15, no. 15, 2023. doi:10.3390/nu15153456.
20. K. Orywal et al., “Vitamins in the prevention and support therapy of neurodegenerative diseases,” *International Journal of Molecular Sciences*, vol. 26, no. 3, pp. 1–21, 2025. doi:10.3390/ijms26031333.
21. N. Neufingerl and A. Eilander, “Nutrient intake and status in children and adolescents consuming plant-based diets compared to meat-eaters: a systematic review,” *Nutrients*, vol. 15, no. 20, 2023. doi:10.3390/nu15204341.
22. C. E. Ofoedu et al., “Revisiting food-sourced vitamins for consumer diet and health needs: a perspective review, from vitamin classification, metabolic functions, absorption, utilization, to balancing nutritional requirements,” *PeerJ*, vol. 9, p. e11940, 2021. doi:10.7717/peerj.11940.
23. M. Mohammadi-Kordkhayli et al., “Vitamins A and D enhance the expression of Ror-γ-targeting miRNAs in a mouse model of multiple sclerosis,” *Molecular Neurobiology*, 2023. doi:10.1007/s12035-023-03427-3.
24. R. Ribeiro, J. R. Nicoli, G. Santos, and J. Lima-Santos, “Impact of vitamin deficiency on microbiota composition and immunomodulation: relevance to autistic spectrum disorders,” *Nutritional Neuroscience*, vol. 24, no. 8, pp. 601–613, 2021. doi:10.1080/1028415X.2019.1660485.
25. J. Janoušek et al., “Vitamin D: sources, physiological role, biokinetics, deficiency, therapeutic use, toxicity, and overview of analytical methods for detection of vitamin D and its metabolites,” *Critical Reviews in Clinical Laboratory Sciences*, vol. 59, no. 8, pp. 517–554, 2022. doi:10.1080/10408363.2022.2070595.
26. E. Y. Khairy and M. M. Attia, “Protective effects of vitamin D on neurophysiologic alterations in brain aging: role of brain-derived neurotrophic factor (BDNF),” *Nutritional Neuroscience*, 2021. doi:10.1080/1028415X.2019.1665854.
27. A. Janse, O. Van De Rest, L. C. P. G. M. De Groot, and R. F. Witkamp, “The association of vitamin D status with mild cognitive impairment and dementia subtypes: a cross-sectional analysis in Dutch geriatric outpatients,” *Journal of Alzheimer’s Disease*, 2023. doi:10.3233/JAD-220732.
28. H. Bai et al., “Effects of dietary vitamin K3 supplementation on vitamin K1 and K2 (menaquinone) dynamics in dairy cows,” *Animal Science Journal*, 2022. doi:10.1111/asj.13680.
29. E. Ramazani, M. Fereidoni, and Z. Tayarani-Najaran, “Protective effects of vitamin K2 on 6-OHDA-induced apoptosis in PC12 cells through modulation bax and caspase-3 activation,” *Molecular Biology Reports*, 2019. doi:10.1007/s11033-019-05011-2.
30. F. Gaspar-Silva, D. Trigo, and J. Magalhaes, “Ageing in the brain: mechanisms and rejuvenating strategies,” *Cellular and Molecular Life Sciences*, vol. 80, p. 190, 2023. doi:10.1007/s00018-023-04832-6.
31. C. A. Schurman, J. B. Burton, J. Rose, L. M. Ellerby, T. Alliston, and B. Schilling, “Molecular and cellular crosstalk between bone and brain: accessing bidirectional neural and musculoskeletal signaling during aging and disease,” *Journal of Bone Metabolism*, vol. 30, no. 1, pp. 1–29, 2023. doi:10.11005/jbm.2023.30.1.1.
32. M. T. Islam, “Oxidative stress and mitochondrial dysfunction-linked neurodegenerative disorders,” *Neurological Research*, vol. 39, pp. 73–82, 2017. doi:10.1080/01616412.2016.1251711.
33. N. Esteras et al., “Mitochondrial ROS control neuronal excitability and cell fate in frontotemporal dementia,” *Alzheimer’s and Dementia*, vol. 18, no. 2, pp. 318–338, 2022. doi:10.1002/alz.12394.
34. B. Zhang et al., “Role of mitochondrial reactive oxygen species in homeostasis regulation,” *Redox Report,* vol. 27, pp. 45–52, 2022. doi:10.1080/13510002.2022.2046423.
35. G. Paradies, G. Petrosillo, V. Paradies, and F. M. Ruggiero, “Mitochondrial dysfunction in brain aging: role of oxidative stress and cardiolipin,” *Neurochemistry International,* vol. 58, pp. 447–457, 2011. doi:10.1016/j.neuint.2010.12.016.
36. P. Chaudhary et al., “Oxidative stress, free radicals and antioxidants: potential crosstalk in the pathophysiology of human diseases,” *Frontiers in Chemistry*, vol. 11, p. 1158198, 2023. doi:10.3389/fchem.2023.1158198.
37. K. H. Lee, M. Cha, and B. H. Lee, “Neuroprotective effect of antioxidants in the brain,” *Journal of Molecular Sciences*, vol. 21, p. 7152, 2020. doi:10.3390/ijms21197152.
38. N. Chandimali et al., “Free radicals and their impact on health and antioxidant defenses: a review,” *Cell Death Discovery*, vol. 11, no. 1, 2025. doi:10.1038/s41420-024-02278-8.
39. L. J. Su et al., “Reactive oxygen species-induced lipid peroxidation in apoptosis, autophagy, and ferroptosis,” *Oxidative Medicine and Cellular Longevity*, vol. 2019, p. 5080843, 2019. doi:10.1155/2019/5080843.
40. Y. Zheng, J. Sun, Z. Luo, Y. Li, and Y. Huang, “Emerging mechanisms of lipid peroxidation in regulated cell death and its physiological implications,” *Cell Death and Disease*, vol. 15, no. 11, pp. 1–19, 2024. doi:10.1038/s41419-024-07244-x.
41. X. Wang, A. A. Memon, A. Hedelius, A. Grundberg, J. Sundquist, and K. Sundquist, “Circulating mitochondrial long non-coding 7S RNA in primary health care patients with depression/anxiety,” *Journal of Affective Disorders*, 2024. doi:10.1016/j.jad.2023.12.053.
42. S. Waldbaum and M. Patel, “Mitochondria, oxidative stress, and temporal lobe epilepsy,” *Epilepsy Research*, vol. 88, pp. 23–45, 2010. doi:10.1016/j.eplepsyres.2009.09.020.
43. S. Navakkode and B. K. Kennedy, “Neural ageing and synaptic plasticity: prioritizing brain health in healthy longevity,” *Frontiers in Aging Neuroscience*, vol. 16, no. August, pp. 1–16, 2024. doi:10.3389/fnagi.2024.1428244.

44. C. Gemma, J. Vila, A. Bachstetter, and P. C. Bickford, "Oxidative stress and the aging brain: from theory to prevention," in D. R. Riddle (Ed.), *Brain Aging*. CRC Press, 2019. doi:10.1201/9781420005523-15.
45. R. von Bernhardi, L. Eugenín-von Bernhardi, and J. Eugenín, "Microglial cell dysregulation in brain aging and neurodegeneration," *Frontiers in Aging Neuroscience*, vol. 7, p. 124, 2015. doi:10.3389/fnagi.2015.00124.
46. A. Ionescu-Tucker and C. W. Cotman, "Emerging roles of oxidative stress in brain aging and Alzheimer's disease," *Neurobiology of Aging*, vol. 107, pp. 86–95, 2021. doi:10.1016/j.neurobiolaging.2021.07.014.
47. T. Ozben and S. Ozben, "Neuro-inflammation and anti-inflammatory treatment options for Alzheimer's disease," *Clinical Biochemistry*, vol. 72, pp. 87–89, 2019. doi:10.1016/j.clinbiochem.2019.04.001.
48. N. Shibata et al., "Implications for soluble iron accumulation, oxidative stress, and glial glutamate release in motor neuron death associated with sporadic amyotrophic lateral sclerosis," *Neuropathology*, vol. 45, no. 3, pp. 177–201, 2025. doi:10.1111/neup.13033.
49. N. Hamadi et al., "Increased pro-inflammatory cytokines, glial activation and oxidative stress in the hippocampus after short-term bilateral adrenalectomy," *BMC Neuroscience*, 2016. doi:10.1186/s12868-016-0296-1.
50. N. E. Propson, E. R. Roy, A. Litvinchuk, J. Köhl, and H. Zheng, "Endothelial C3a receptor mediates vascular inflammation and blood-brain barrier permeability during aging," *Journal of Clinical Investigation*, 2021. doi:10.1172/JCI140966.
51. A. C. Yang et al., "Physiological blood–brain transport is impaired with age by a shift in transcytosis," *Nature*, 2020. doi:10.1038/s41586-020-2453-z.
52. P. Singh, B. Barman, and M. K. Thakur, "Oxidative stress-mediated memory impairment during aging and its therapeutic intervention by natural bioactive compounds," *Frontiers in Aging Neuroscience*, vol. 14, no. July, pp. 1–14, 2022. doi:10.3389/fnagi.2022.944697.
53. L. J. Dominguez et al., "Nutrition, physical activity, and other lifestyle factors in the prevention of cognitive decline and dementia," *Nutrients*, vol. 13, p. 4080, 2021. doi:10.3390/nu13114080.
54. M. M. Gonzales et al., "Biological aging processes underlying cognitive decline and neurodegenerative disease," *Journal of Clinical Investigation*, vol. 132, p. e158453, 2022. doi:10.1172/JCI158453.
55. C. Oedekoven, L. Egeri, F. Jessen, M. Wagner, and R. Dodel, "Subjective cognitive decline in idiopathic Parkinson´s disease: a systematic review," *Ageing Research Reviews*, vol. 76, p. 101508, 2022. doi:10.1016/j.arr.2021.101508.
56. R. C. Knopp, K. K. Baumann, M. L. Wilson, W. A. Banks, and M. A. Erickson, "Amyloid beta pathology exacerbates weight loss and brain cytokine responses following low-dose lipopolysaccharide in aged female Tg2576 mice," *International Journal of Molecular Sciences*, 2022. doi:10.3390/ijms23042377.
57. B. Kalyanaraman, G. Cheng, and M. Hardy, "Gut microbiome, short-chain fatty acids, alpha-synuclein, neuroinflammation, and ROS/RNS: relevance to Parkinson's disease and therapeutic implications," *Redox Biology*, vol. 71, p. 103092, 2024. doi:10.1016/j.redox.2024.103092.
58. S. Prasad et al., "Recent advances in Lewy body dementia: a comprehensive review," *Disease-a-Month*, vol. 69, p. 1014412023. doi:10.1016/j.disamonth.2022.101441.
59. J. A. Trejo-Lopez, A. T. Yachnis, and S. Prokop, "Neuropathology of Alzheimer's disease," *Neurotherapeutics*, vol. 19, no. 1, pp. 173–185, 2022. doi:10.1007/s13311-021-01146-y.
60. C. Gao, J. Jiang, Y. Tan, and S. Chen, "Microglia in neurodegenerative diseases: mechanism and potential therapeutic targets," vol. 8, no. 1, 2023. doi:10.1038/s41392-023-01588-0.
61. E. G. Knox, M. R. Aburto, G. Clarke, J. F. Cryan, and C. M. O'Driscoll, "The blood-brain barrier in aging and neurodegeneration," *Molecular Psychiatry*, vol. 27, no. 6, pp. 2659–2673, 2022. doi:10.1038/s41380-022-01511-z.
62. N. Charoenngam and M. F. Holick, "Immunologic effects of vitamin d on human health and disease," *Nutrients*, vol. 12, no. 7, pp. 1–28, 2020. doi:10.3390/nu12072097.
63. T. L. Pop, C. Sîrbe, G. Benţa, A. Mititelu, and A. Grama, "The role of vitamin D and vitamin D binding protein in chronic liver diseases," *International Journal of Molecular Sciences,* vol. 23, p. 10705, 2022. doi:10.3390/ijms231810705.
64. F. Kawagoe, S. Mototani, and A. Kittaka, "Design and synthesis of fluoro analogues of vitamin D," *International Journal of Molecular Sciences*, vol. 22, no. 15, 2021. doi:10.3390/ijms22158191.
65. D. A. Glencross, C. Cheadle, and C. M. Hawrylowicz, "Vitamin D and adaptive immunity in health and disease," in M. Hewison, R. Bouillon, E. Giovannucci, D. Goltzman, M. B. Meyer, and J. Welsh (Eds.), *Feldman and Pike's Vitamin D: Volume Two: Disease and Therapeutics*. Elsevier, pp. 1035–1056, 2023. doi:10.1016/B978-0-323-91338-6.00046-X.
66. L. Kattner, "Recent developments towards the synthesis of Vitamin D metabolites," *Anticancer Research*, 2020. doi:10.21873/anticanres.13980.
67. F. Kawagoe, S. Mototani, and A. Kittaka, "The synthesis and biological evaluation of D-ring-modified vitamin D analogues," *Biomolecules*, vol. 11, no. 11, 2021. doi:10.3390/biom11111639.
68. S. Christakos, P. Dhawan, A. Verstuyf, L. Verlinden, and G. Carmeliet, "Vitamin D: metabolism, molecular mechanism of action, and pleiotropic effects," *Physiological Reviews*, 2015. doi:10.1152/physrev.00014.2015.
69. A. Siddiq, A. K. Naveed, N. Ghaffar, M. Aamir, and N. Ahmed, "Association of pro-inflammatory cytokines with vitamin D in Hashimoto's thyroid autoimmune disease," *Medicina (Lithuania)*, vol. 59, no. 5, 2023. doi:10.3390/medicina59050853.
70. V. Rai and D. K. Agrawal, "Immunomodulation of IL-33 and IL-37 with vitamin D in the neointima of coronary artery: a comparative study between balloon angioplasty and stent in hyperlipidemic microswine," *International Journal of Molecular Sciences*, vol. 22, no. 16, 2021. doi:10.3390/ijms22168824.
71. V. Rai, M. M. Radwan, and D. K. Agrawal, "Il-33, il-37, and vitamin D interaction mediate immunomodulation of inflammation in degenerating cartilage," *Antibodies*, 2021. doi:10.3390/antib10040041.
72. F. K. Alswailmi, S. I. A. Shah, H. Nawaz, and G. M. Al-Mazaideh, "Molecular mechanisms of vitamin D-mediated immunomodulation," *Galen Medical Journal*, vol. 10, p. e2097, 2021. doi: 10.31661/gmj.v10i0.2097
73. A. Banerjee et al., "Role of pro-inflammatory cytokines and vitamin D in probable Alzheimer's disease with depression," *Aging and Disease*, 2017. doi:10.14336/AD.2016.1017.

74. T. J. Littlejohns et al., "Vitamin D and the risk of dementia and Alzheimer disease," *Neurology*, 2014. doi:10.1212/WNL.0000000000000755.
75. A. Ali et al., "Vitamin D exerts neuroprotection via SIRT1/nrf-2/NF-kB signaling pathways against D-galactose-induced memory impairment in adult mice," *Neurochemistry International*, 2021. doi:10.1016/j.neuint.2020.104893.
76 K. Azuma, Y. Osuka, N. Kojima, H. Sasai, H. Kim, and S. Inoue, "Association of vitamin K insufficiency with cognitive dysfunction in community-dwelling older adults," *Frontiers in Nutrition*, vol. 8, no. January, pp. 1–6, 2022. doi:10.3389/fnut.2021.811831.
77. K. L. Berkner and K. W. Runge, "Vitamin K-dependent protein activation: normal gamma-glutamyl carboxylation and disruption in disease," *International Journal of Molecular Sciences*, vol. 23, no. 10, 2022. doi:10.3390/ijms23105759.
78. S. S. Ahmad, S. Karim, I. M. Ibrahim, H. M. Alkreathy, M. Alsieni, and M. A. Khan, "Effect of vitamin K on bone mineral density and fracture risk in adults : systematic review and meta-analysis," *Biomedicines*, vol. 10, no. 1084, pp. 1–13, 2022.
79. M. K. Shea and S. L. Booth, "Vitamin K," *Advances in Nutrition*, vol. 13, no. 1, pp. 350–351, 2022. doi:10.1093/advances/nmab133.
80. N. S. S. Ambarwati, N. T. Sukma, A. A. Yesi Desmiaty, M. A. Setia Budi, and I. Ahmad, "Peel extract : secondary metabolite composition , antioxidant , and elastase inhibitory activity evaluation," *Journal of Advanced Pharmaceutical Technology & Research*, vol. 15, no. 1, pp. 8–12, 2024. doi:10.4103/japtr.japtr.
81. P. Mladěnka et al., "Vitamin K—sources, physiological role, kinetics, deficiency, detection, therapeutic use, and toxicity," *Nutrition Reviews*, vol. 80, no. 4, pp. 677–698, 2022. doi:10.1093/nutrit/nuab061.
82. J. L. Ellis et al., "Dietary vitamin K is remodeled by gut microbiota and influences community composition," *Gut Microbes*, vol. 13, no. 1, pp. 1–16, 2021. doi:10.1080/19490976.2021.1887721.
83. A. Kieronska-Rudek, A. Kij, A. Bar, A. Kurpinska, and T. Mohaissen, "Phylloquinone improves endothelial function, inhibits cellular senescence, and vascular inflammation," *GeroScience*, vol. 46, pp. 4909–4935, 2024.
84. J. K. Nowak et al., "Fat-soluble vitamins in standard vs. liposomal form enriched with vitamin K2 in cystic fibrosis: a randomized multi-center trial," *Journal of Clinical Medicine*, vol. 11, p. 462,2022.
85. A. Despotović et al., "Combination of ascorbic acid and menadione induces cytotoxic autophagy in human glioblastoma cells," *Oxidative Medicine and Cellular Longevity*, vol. 2022, 2022. doi:10.1155/2022/2998132.
86. S. Fiesack et al., "Belgian consensus recommendations to prevent vitamin k deficiency bleeding in the term and preterm infant," *Nutrients*, vol. 13, no. 11, pp. 1–16, 2021. doi:10.3390/nu13114109.
87. A. Annadurai, C. G. Delhi Kumar, P. S. M., and N. Parameswaran, "Late-onset vitamin K deficiency bleeding: a preventable yet prevailing cause of intracranial hemorrhage in young infants—an ambispective descriptive study," *Intensive Care Medicine—Paediatric and Neonatal*, vol. 2, no. 1, 2024. doi:10.1007/s44253-024-00028-3.
88. P. Clarke et al., "Exclusively breastmilk-fed preterm infants are at high risk of developing subclinical vitamin K deficiency despite intramuscular prophylaxis at birth," *Journal of Thrombosis and Haemostasis*, vol. 20, no. 12, pp. 2773–2785, 2022. doi:10.1111/jth.15874.
89. N. Mathews and C. P. M. Hayward, "Vitamin K deficiency: diagnosis and management," *Annals of Laboratory Medicine*, vol. 45, no. 4, pp. 358–366, 2025. doi:10.3343/alm.2024.0590.
90. A. Sakwit et al., "Vitamin K deficiency bleeding in children with cholestatic liver disease: a systematic review and meta-analysis," *Research and Practice in Thrombosis and Haemostasis,* vol. 9, no. 3, p. 102847, 2025. doi:10.1016/j.rpth.2025.102847.
91. N. Alonso, A. Meinitzer, E. Fritz-Petrin, D. Enko, and M. Herrmann, "Role of vitamin K in bone and muscle metabolism," *Calcified Tissue International*, vol. 112, no. 2, pp. 178–196, 2023. doi:10.1007/s00223-022-00955-3.
92. S. Aggarwal et al., "Vitamin K2: an emerging essential nutraceutical and its market potential," *Journal of Applied Biology & Biotechnology*, vol. 10, no. 2, pp. 173–184, 2022. doi:10.7324/JABB.2022.100221.
93. D. Mehta and V. Badmaev, "Preventing inflammaging with the multitasking vitamin K," *EC Neurology,* vol. 14, pp. 33–35, 2022.
94. M. R. Kulikauskas, X. Shaka, and V. L. Bautch, "The versatility and paradox of BMP signaling in endothelial cell behaviors and blood vessel function," *Cellular and Molecular Life Sciences*, vol. 79, no. 2, pp. 1–28, 2022. doi:10.1007/s00018-021-04033-z.
95. M. Xiong, P. Cheng, B. Liu, Y. Zhao, T. Gao, and Z. Li, "The GgcxK325Q mutation does not affect the calcium homeostasis of the epididymis and male fertility in mice," *Current Issues in Molecular Biology*, vol. 46, no. 6, pp. 5052–5065, 2024. doi:10.3390/cimb46060303.
96. R. Malhotra et al., "Matrix GLA protein levels are associated with arterial stiffness and incident heart failure with preserved ejection fraction," *Arteriosclerosis, Thrombosis, and Vascular Biology*, vol. 42, no. 2, pp. E61–E73, 2022. doi:10.1161/ATVBAHA.121.316664.
97. M. Cozzolino, M. Fusaro, P. Ciceri, L. Gasperoni, and G. Cianciolo, "The role of vitamin K in vascular calcification," *Advances in Chronic Kidney Disease*, vol. 26, no. 6, pp. 437–444, 2019. doi:10.1053/j.ackd.2019.10.005.
98. X. Wang et al., "MGP regulates the adipogenic differentiation of mesenchymal stem cells in osteoporosis via the Ca2+/CaMKII/RIP140/FABP3 axis," *Cell Death Discovery*, vol. 11, no. 1, 2025. doi:10.1038/s41420-025-02472-2.
99. L. Zhang, X. Wu, X. Cai, X. Qiao, Y. Yao, and K. I. Bostrom, "Abstract 486: role of matrix Gla protein in vascular regulation in brown adipose tissue," *Arteriosclerosis, Thrombosis, and Vascular Biology*, vol. 42, p. 48, 2022. doi:10.1161/atvb.42.suppl_1.486.
100. S. Tang, Z. Ruan, A. Ma, D. Wang, and J. Kou, "Effect of vitamin K on wound healing: a systematic review and meta-analysis based on preclinical studies," *Frontiers in Pharmacology*, vol. 13, p. 1063349, 2022.
101. J. Aaberg Lauridsen et al., "Investigating the associations between uncarboxylated matrix gla protein as a proxy for vitamin K status and cardiovascular disease risk factors in a general adult population," *European Journal of Nutrition,* vol. 64, p. 17, 2024.
102. J. K. Tie and D. W. Stafford, "Structural and functional insights into enzymes of the vitamin K cycle," *Journal of Thrombosis and Haemostasis,* vol. 14, pp. 27–40, 2016. doi:10.1111/jth.13217.
103. J. M. Azar, R. Lambert, F. A. Maffei, and T. A. Thomas, "Late-onset vitamin K deficiency presenting as haemorrhagic shock and severe multi-system organ failure," *BMJ*

Case Reports, vol. 17, no. 8, pp. 17–20, 2024. doi:10.1136/bcr-2024-261126.

104. K. Yasuda et al., "Intrauterine fetal death due to fetal intracranial hemorrhage associated with maternal vitamin K deficiency: a case report," *Case Reports in Women's Health*, vol. 40, no. November, pp. 19–21, 2023. doi:10.1016/j.crwh.2023.e00566.
105. M. Zhang, Q. Zhang, P. Du, X. Chen, and Y. Zhang, "Roles of vitamin K-dependent protein in biomineralization," *International Journal of Molecular Medicine,* vol. 53, Article 6, 2023. doi:10.3892/ijmm.2023.5330.
106. T. Li, Y. Wang, and W.-P. Tu, "Vitamin K supplementation and vascular calcification: a systematic review and meta-analysis of randomized controlled trials," *Frontiers in Nutrition,* vol. 10, p. 1115069, 2023.
107. T.-H. Kim, H. Kim, H. H. Lee, and J. H. Sang, "Vitamin K: calcium metabolism modulator for menopausal women," *Journal of Menopausal Medicine*, vol. 30, no. 3, p. 152, 2024. doi:10.6118/jmm.24023.
108. H. Jung et al., "Anti-inflammatory clearance of amyloid-β by a chimeric Gas6 fusion protein," *Nature Medicine*, 2022. doi:10.1038/s41591-022-01926-9.
109. X. Li, D. Guo, Y. Hu, and Y. Chen, "Pro-oxidative stress, TLR4/NF-κB/MyD88 and GAS6 pathways, and atherosclerotic susceptibility: development of tibial artery occlusions in elderly patients with chronic total coronary occlusions," *SSRN Electronic Journal*, 2021. doi:10.2139/ssrn.3830356.
110. A. Shah, M. Varma, and R. Bhandari, "Exploring sulforaphane as neurotherapeutic: targeting Nrf2-Keap & Nf-Kb pathway crosstalk in ASD," *Metabolic Brain Disease,* vol. 38, pp. 373–385, 2024. doi:10.1007/s11011-023-01224-4.
111. L. Andrei blaj, A. I. Cucu, B. I. Tamba, and M. D. Turliuc, "The role of the NF-KB pathway in intracranial aneurysms," *Brain Sciences,* vol. 13, p. 1660, 2023.
112. P. Mohammadzadeh and C. Amberg, "AXL/Gas 6 signaling mechanisms in the hypothalmic-pituitary-gonadal axis," *Frontiers in Endocrinology,* vol. 14, p. 1212104, 2023.
113. M. Fukatsu et al., "The suppressive effects of Mer inhibition on inflammatory responses in the pathogenesis of LPS-induced ALI/ARDS," *Science Signaling*, 2022. doi:10.1126/scisignal.abd2533.
114. S. AlBlooshi, "Vitamin K and women's health: a review," *Frontiers in Global Women's Health*, vol. 6, no. July, 2025. doi:10.3389/fgwh.2025.1590414.
115. H. Dong, B. Wang, J. Feng, X. Yue, and F. Jia, "Correlation between serum concentrations of menaquinone-4 and developmental quotients in children with autism spectrum disorder," *Frontiers in Nutrition*, vol. 8, no. September, pp. 1–7, 2021. doi:10.3389/fnut.2021.748513.
116. A. I. Diachenko, I. A. Rodin, T. N. Krasnova, O. I. Klychnikov, and L. N. Nefedova, "The role of vitamin K in the development of neurodegenerative diseases," *Biochemistry (Moscow),* vol. 89, pp. S57–S70, 2024. doi:10.1134/S0006297924140049.
117. M. H. Hashim et al., "Exploring the link between vitamin K and depression: a systematic review," *Medicina (Lithuania)*, vol. 61, no. 5, pp. 1–15, 2025. doi:10.3390/medicina61050861.
118. E. Emekli-Alturfan and A. A. Alturfan, "The emerging relationship between vitamin K and neurodegenerative diseases: a review of current evidence," *Molecular Biology Reports*, vol. 50, no. 1, pp. 815–828, 2023. doi:10.1007/s11033-022-07925-w.
119. K. Chatterjee, P. M. Mazumder, and S. Banerjee, "Vitamin K: a potential neuroprotective agent," *Revista Brasileira de Farmacognosia,* vol. 33, pp. 676–687, 2023. doi:10.1007/s43450-023-00378-7.
120. C. Qi et al., "Platycodon grandiflorus polysaccharide with anti-apoptosis, anti-oxidant and anti-inflammatory activity against LPS/D-GalN induced acute liver injury in mice," *Journal of Polymers and the Environment*, vol. 29, no. 12, pp. 4088–4097, 2021. doi:10.1007/s10924-021-02179-2.
121. A. Wang, M. Zhao, J. Luo, T. Zhang, and D. Zhang, "Association of dietary vitamin K intake with cognition in the elderly," *Frontiers in Nutrition,* vol. 9, p. 900887, 2022.
122. H. A. Elkattawy et al., "Vitamin K2 (menaquinone-7) reverses age-related structural and cognitive deterioration in naturally aging rats," *Antioxidants*, vol. 11, no. 3, 2022. doi:10.3390/antiox11030514.
123. S. L. Booth et al., "Association of vitamin K with cognitive decline and neuropathology in community-dwelling older persons," *Alzheimer's and Dementia: Translational Research and Clinical Interventions*, vol. 8, no. 1, pp. 1–9, 2022. doi:10.1002/trc2.12255.
124. X. Fu et al., "Vitamin D and vitamin K concentrations in human brain tissue are influenced by freezer storage time: the memory and aging project," *Journal of Nutrition*, 2021. doi:10.1093/jn/nxaa336.
125. J. Chouet et al., "Dietary vitamin K intake is associated with cognition and behaviour among geriatric patients: the CLIP study," *Nutrients*, vol. 7, no. 8, pp. 6739–6750, 2015. doi:10.3390/nu7085306.
126. N. Presse et al., "Vitamin K status and cognitive function in healthy older adults," *Neurobiology of Aging*, 2013. doi:10.1016/j.neurobiolaging.2013.05.031.
127. E. Hadipour, Z. Tayarani-Najaran, and M. Fereidoni, "Vitamin K2 protects PC12 cells against Aβ (1-42) and H_2O_2-induced apoptosis via p38 MAP kinase pathway," *Nutritional Neuroscience*, vol. 23, no. 5, pp. 343–352, 2020. doi:10.1080/1028415X.2018.1504428.
128. M. C. Morris, Y. Wang, L. L. Barnes, D. A. Bennett, B. Dawson-Hughes, and S. L. Booth, "Nutrients and bioactives in green leafy vegetables and cognitive decline: prospective study," *Neurology*, vol. 90, no. 3, pp. E214–E222, 2018. doi:10.1212/WNL.0000000000004815.
129. E. Ramazani, M. Fereidoni, and Z. Tayarani-Najaran, "Protective effects of vitamin K2 on 6-OHDA-induced apoptosis in PC12 cells through modulation bax and caspase-3 activation," *Molecular Biology Reports*, 2019. doi:10.1007/s11033-019-05011-2.
130. Y. Yu, X. Yu, Q. Cheng, L. Tang, and M. Shen, "The association of serum vitamin K_2 levels with Parkinson's disease: from basic case-control study to big data mining analysis," Aging (Albany NY), vol. 12, no. 16, pp. 16410–16419, 2020.
131. Y. X. Yu et al., "Vitamin K2 suppresses rotenone-induced microglial activation in vitro," *Acta Pharmacologica Sinica*, vol. 37, no. 9, pp. 1178–1189, 2016. doi:10.1038/aps.2016.68.
132. Y. Sato, M. Kaji, T. Tsuru, K. Satoh, and I. Kondo, "Vitamin K deficiency and osteopenia in vitamin D-deficient elderly women with Parkinson's disease," *Archives of Physical Medicine and Rehabilitation*, 2002. doi:10.1053/apmr.2002.27376.
133. F. L. Da Silva, E. Coelho Cerqueira, M. S. De Freitas, D. L. Gonçalves, L. T. Costa, and C. Follmer, "Vitamins K interact with N-terminus α-synuclein and modulate the protein fibrillization in vitro: exploring the interaction between

quinones and α-synuclein," *Neurochemistry International*, vol. 62, no. 1, pp. 103–112, 2013. doi:10.1016/j.neuint.2012.10.001.

134. L. Liu et al., "Association between dietary intake and risk of Parkinson's disease: cross-sectional analysis of survey data from NHANES 2007–2016," *Frontiers in Nutrition*, vol. 10, no. December, pp. 1–9, 2023. doi:10.3389/fnut.2023.1278128.
135. R. Lasemi, M. Kundi, N. B. Moghadam, H. Moshammer, and J. A. Hainfellner, "Vitamin K2 in multiple sclerosis patients," *Wiener Klinische Wochenschrift*, vol. 130, no. 9–10, pp. 307–313, 2018. doi:10.1007/s00508-018-1328-x.
136. D. C. Popescu et al., "Vitamin K enhances the production of brain sulfatides during remyelination," *PLoS ONE*, vol. 13, no. 8, pp. 1–19, 2018. doi:10.1371/journal.pone.0203057.
137. A. Irish, C. Erickson, T. Wahls, L. Snetselaar, and W. Darling, "Randomized control trial evaluation of a modified Paleolithic dietary intervention in the treatment of relapsing-remitting multiple sclerosis: a pilot study," *Degenerative Neurological and Neuromuscular Disease*, vol. 7, pp. 1–18, 2017. doi:10.2147/dnnd.s116949.
138. N. Mabelele, "Genetic alterations in bone," *Bone*, vol. 23, no. 1, pp. 1–7, 2008..
139. S. Mcdonough-means et al., "Effect of a vitamin/mineral supplement on children and adults with autism," *BMC Pediatrics*, vol. 11, no. 1, p. 111, 2011.
140. R. L. Lindsay et al., "Dietary status and impact of risperidone on nutritional balance in children with autism: a pilot study," *Journal of Intellectual and Developmental Disability*, vol. 31, no. 4, pp. 204–209, 2006. doi:10.1080/13668250601006924.
141. F. G. Rodrigues, M. S. Ormanji, I. P. Heilberg, S. J. L. Bakker, and M. H. de Borst, "Interplay between gut microbiota, bone health and vascular calcification in chronic kidney disease," *European Journal of Clinical Investigation*, vol. 51, no. 9, pp. 1–14, 2021. doi:10.1111/eci.13588.
142. J. C. Fleet, "Vitamin D-mediated regulation of intestinal calcium absorption," *Nutrients*, vol. 14, no. 16, 2022. doi:10.3390/nu14163351.
143. Z. H. Al-Oanzi et al., "The role of vitamin D in reducing the risk of metabolic disturbances that cause cardiovascular diseases," *Journal of Cardiovascular Development and Disease*, vol. 10, no. 5, 2023. doi:10.3390/jcdd10050209.
144. J. Lacombe et al., "Vitamin K-dependent carboxylation regulates Ca2+ flux and adaptation to metabolic stress in β cells," *Cell Reports*, vol. 42, no. 5, p. 112500, 2023. doi:10.1016/j.celrep.2023.112500.
145. M. Stock and G. Schett, "Vitamin k-dependent proteins in skeletal development and disease," *International Journal of Molecular Sciences*, vol. 22, no. 17, 2021. doi:10.3390/ijms22179328.
146. R. Bouillon, D. Manousaki, C. Rosen, K. Trajanoska, F. Rivadeneira, and J. B. Richards, "The health effects of vitamin D supplementation: evidence from human studies," *Nature Reviews Endocrinology*, vol. 18, no. 2, pp. 96–110, 2022. doi:10.1038/s41574-021-00593-z.
147. N. Aydin, B. Ouliass, G. Ferland, and S. Hafizi, "Modification of Gas6 protein in the brain by a functional endogenous tissue vitamin K cycle," *Cells*, vol. 13, no. 10, 2024. doi:10.3390/cells13100873.
148. J. Kaźmierczak-Barańska and B. T. Karwowski, "The protective role of vitamin K in aging and age-related diseases," *Nutrients*, vol. 16, no. 24, 2024. doi:10.3390/nu16244341.
149. Y. J. Bai et al., "Vitamin D supplementation reduced blood inflammatory cytokines expression and improved graft function in kidney transplant recipients," *Front Immunology*, vol. 14, no. July, pp. 1–11, 2023. doi:10.3389/fimmu.2023.1152295.
150. M. Agar, K. Güngör, N. D. Güngör, M. Kavrut, and A. A. Madenli, "Vitamin D supplementation inhibits NF-κB pathway in lean and obese women with PCOS," European Review for Medical and Pharmacological Sciences, vol. 26, pp. 3973–3977, 2022.
151. D. Chauss et al., "Autocrine vitamin D signaling switches off pro-inflammatory programs of TH1 cells," *Nat Immunol*, vol. 23, no. 1, pp. 62–74, 2022. doi:10.1038/s41590-021-01080-3.
152. R. Zhao et al., "Immunomodulatory function of vitamin D and its role in autoimmune thyroid disease," *Front Immunology*, vol. 12, no. February, pp. 1–6, 2021. doi:10.3389/fimmu.2021.574967.
153. H. V. T. Vo, Y. T. Nguyen, N. Kim, and H. J. Lee, "Vitamin A, D, E, and K as Matrix Metalloproteinase-2/9 regulators that affect expression and enzymatic activity," *International Journal of Molecular Sciences*, vol. 24, no. 23, 2023. doi:10.3390/ijms242317038.
154. C. S. Latimer et al., "Vitamin D prevents cognitive decline and enhances hippocampal synaptic function in aging rats," *Proceedings of the National Academy of Sciences of the United States of America,* vol. 111, no. 41, pp. E4359–E4366, 2014. doi:10.1073/pnas.1404477111.
155. D. C. Simes, C. S. B. Viegas, N. Araújo, and C. Marreiros, "Vitamin K as a diet supplement with impact in human health: current evidence in age-related diseases," *Nutrients*, vol. 12, no. 1, pp. 1–24, 2020. doi:10.3390/nu12010138.
156. M. Garcia-Gil, F. Pierucci, A. Vestri, and E. Meacci, "Crosstalk between sphingolipids and vitamin D3: potential role in the nervous system," *British Journal of Pharmacology*, vol. 174, no. 8, pp. 605–627, 2017. doi:10.1111/bph.13726.
157. E. Y. Khairy and M. M. Attia, "Protective effects of vitamin D on neurophysiologic alterations in brain aging: role of brain-derived neurotrophic factor (BDNF)," *Nutritional Neuroscience*, vol. 24, no. 8, pp. 650–659, 2021. doi:10.1080/1028415X.2019.1665854.
158. W. Lasoń, D. Jantas, M. Leśkiewicz, M. Regulska, and A. Basta-Kaim, "The vitamin D receptor as a potential target for the treatment of age-related neurodegenerative diseases such as Alzheimer's and Parkinson's diseases: a narrative review," *Cells*, vol. 12, no. 4, 2023. doi:10.3390/cells12040660.
159. E. L Bishop, A. Ismailova, S. Dimeloe, M. Hewison, and J. H. White, "Vitamin D and immune regulation: antibacterial, antiviral, anti-inflammatory," *Journal of Bone and Mineral Research Plus*, vol. 5, no. 1, pp. 1–23, 2021. doi:10.1002/jbm4.10405.
160. E. Dogan-Sander et al., "Inflammation and the association of vitamin d and depressive symptomatology," *Nutrients*, vol. 13, no. 6, pp. 1–15, 2021. doi:10.3390/nu13061972.
161. T. Kim, D. Kim, Y. Kim, J. Kim, S. Kang, and J. Cho, "Combined effects of exercise and vitamin D on neuroinflammation, blood-brain barrier integrity, oxidative stress, and cognitive function in nonpathological mice," *International Neurourology Journal*, vol. 29, no. Suppl 1, pp. S22–S34, 2025. doi:10.5213/inj.2550140.070.

162. M. Song et al., "DNMT1-mediated DNA methylation in toll-like receptor 4 (TLR4) inactivates NF-κB signal pathway-triggered pyroptotic cell death and cellular inflammation to ameliorate lipopolysaccharides (LPS)-induced osteomyelitis," *Molecular and Cellular Probes*, vol. 71, no. April, p. 101922, 2023. doi:10.1016/j.mcp.2023.101922.
163. P. G. Sangaran, Z. A. Ibrahim, Z. Chik, Z. Mohamed, and A. Ahmadiani, "LPS preconditioning attenuates apoptosis mechanism by inhibiting NF-κB and caspase-3 activity: TLR4 pre-activation in the signaling pathway of LPS-induced neuroprotection," *Molecular Neurobiology*, vol. 58, no. 5, pp. 2407–2422, 2021. doi:10.1007/s12035-020-02227-3.
164. B. Zhang, J. Shen, Z. Zhong, and L. Zhang, "PKM2 aggravates cerebral ischemia reperfusion-induced neuroinflammation via TLR4/MyD88/TRAF6 signaling pathway," *Neuroimmunomodulation*, vol. 28, no. 1, pp. 29–37, 2021. doi:10.1159/000509710.
165. N. H. Eisa, S. A. Helmy, D. H. El-Kashef, M. El-Sherbiny, and N. M. Elsherbiny, "Pramipexole protects against diabetic neuropathy: effect on oxidative stress, TLR4/IRAK-1/TRAF-6/NF-κB and downstream inflammatory mediators," *International Immunopharmacology*, 2024. doi:10.1016/j.intimp.2024.111514.
166. J. Yang, H. J. Kim, K. H. I. N. M. Herath, and Y. Jee, "Cellucalst enzyme-assisted extraction of Sargassum horneri enhances the immunomodulation by regulating TLR4/MyD88/NF-kB pathway in murine splenocytes with or without Concanavalin A," *Biomedicine and Pharmacotherapy*, vol. 181, p. 117673, 2024. doi:10.1016/j.biopha.2024.117673.
167. T. Zhang, C. Ma, Z. Zhang, H. Zhang, and H. Hu, "NF-κB signaling in inflammation and cancer," *MedComm (Beijing)*, vol. 2, no. 4, pp. 618–653, 2021. doi:10.1002/mco2.104.
168. L. Zhao, Y. Hao, Z. Song, Y. Fan, and S. Li, "TRIM37 negatively regulates inflammatory responses induced by virus infection via controlling TRAF6 ubiquitination," *Biochemical and Biophysical Research Communications*, vol. 556, pp. 87–92, 2021. doi:10.1016/j.bbrc.2021.03.147.
169. M. R. Zinatizadeh, B. Schock, G. M. Chalbatani, P. K. Zarandi, S. A. Jalali, and S. R. Miri, "The nuclear factor kappa B (NF-kB) signaling in cancer development and immune diseases," *Genes & Diseases*, vol. 8, no. 3, pp. 287–297, 2021. doi:10.1016/j.gendis.2020.06.005.
170. J. Tang, L. Xu, Y. Zeng, and F. Gong, "Effect of gut microbiota on LPS-induced acute lung injury by regulating the TLR4/NF-kB signaling pathway," *International Immunopharmacology*, vol. 91, no. September 2020, p. 107272, 2021. doi:10.1016/j.intimp.2020.107272.
171. M. A. Ortega et al., "Vascular calcification: molecular networking, pathological implications and translational opportunities," *Biomolecules*, vol. 14, no. 3, pp. 1–21, 2024. doi:10.3390/biom14030275.
172. Y. C. Cao et al., "Histone lysine methylation modification and its role in vascular calcification," *Front Endocrinol (Lausanne)*, vol. 13, no. June, pp. 1–17, 2022. doi:10.3389/fendo.2022.863708.
173. Z. Okur et al., "Control of neuronal excitation–inhibition balance by BMP–SMAD1 signalling," *Nature*, vol. 629, no. 8011, pp. 402–409, 2024. doi:10.1038/s41586-024-07317-z.
174. A. Manzari-Tavakoli, A. Babajani, M. H. Farjoo, M. Hajinasrollah, S. Bahrami, and H. Niknejad, "The cross-talks among bone morphogenetic protein (BMP) signaling and other prominent pathways involved in neural differentiation," *Frontiers in Molecular Neuroscience*, vol. 15, no. March, pp. 1–15, 2022. doi:10.3389/fnmol.2022.827275.
175. P. V. Iyengar et al., "TRAF4 inhibits bladder cancer progression by promoting BMP/SMAD signaling," *Molecular Cancer Research*, vol. 20, no. 10, pp. 1516–1531, 2022. doi:10.1158/1541-7786.MCR-20-1029.
176. E. Varga, R. Pap, G. Jánosa, K. Sipos, and E. Pandur, "IL-6 regulates hepcidin expression via the BMP/SMAD pathway by altering BMP6, TMPRSS6 and TfR2 expressions at normal and inflammatory conditions in BV2 microglia," *Neurochemical Research*, vol. 46, no. 5, pp. 1224–1238, 2021. doi:10.1007/s11064-021-03322-0.
177. J. Huang et al., "KLF2 mediates the suppressive effect of laminar flow on vascular calcification by inhibiting endothelial BMP/SMAD1/5 signaling," *Circulation Research*, vol. 129, no. 4, pp. E87–E100, 2021. doi:10.1161/CIRCRESAHA.120.318690.
178. A. M. Valentin Cabrera, S. K. Ashbrook, and J. D. Hutcheson, "Considerations on the development of therapeutics in vascular calcification," *Journal of Cardiovascular Development and Disease*, vol. 12, no. 6, 2025. doi:10.3390/jcdd12060206.
179. J. B. Kaushal, V. Shukla, P. Sankhwar, R. K. Jha, and A. Dwivedi, "Targeted inhibition of TAK1 abrogates TGFβ1 non-canonical signaling axis, NFκB/Smad7 inhibiting human endometriotic cells proliferation and inducing cell death involving autophagy," *Cytokine*, 2021. doi:10.1016/j.cyto.2021.155700.
180. H. S. Cui, D. H. Kim, S. Y. Joo, Y. S. Cho, J. B. Kim, and C. H. Seo, "Exosomes derived from human hypertrophic scar fibroblasts induces smad and TAK1 signaling in normal dermal fibroblasts," *Archives of Biochemistry and Biophysics*, 2022. doi:10.1016/j.abb.2022.109215.
181. T. Kobayashi, K. Sugiura, T. Ojima, M. Serizawa, K. Hirai, and E. Morishita, "Thrombosis-related characteristics of pregnant women with antithrombin deficiency, protein C deficiency and protein S deficiency in Japan," *Thrombosis Journal*, vol. 22, no. 1, pp. 1–12, 2024. doi:10.1186/s12959-024-00581-z.
182. A. Usman, A. Jibbran, U. Ahmad, F. Tariq, M. M. Saleem, and M. Mudassar, "Portal vein thrombosis due to inherited combined deficiency of protein C and S in a young adult: a case report," *Radiology Case Reports*, vol. 19, no. 12, pp. 6235–6240, 2024. doi:10.1016/j.radcr.2024.09.052.
183. T. Prouse, S. Majumder, and R. Majumder, "Functions of TAM receptors and ligands protein S and Gas6 in atherosclerosis and cardiovascular disease," *International Journal of Molecular Sciences*, vol. 25, no. 23, pp. 1–19, 2024. doi:10.3390/ijms252312736.
184. D. DeRyckere, J. M. Huelse, H. S. Earp, and D. K. Graham, "TAM family kinases as therapeutic targets at the interface of cancer and immunity," Nature Reviews Clinical Oncology, vol. 20, pp. 755–77, 2023. doi:10.1038/s41571-023-00813-7.
185. D. Zhang, T. Ning, and H. Wang, "Vitexin alleviates inflammation and enhances apoptosis through the regulation of the JAK/STAT/SOCS signaling pathway in the arthritis rat model," *Journal of Biochemical and Molecular Toxicology*, vol. 36, no. 12, pp. 1–9, 2022. doi:10.1002/jbt.23201.

186. J. B. Ruganzu et al., "TREM2 overexpression rescues cognitive deficits in APP/PS1 transgenic mice by reducing neuroinflammation via the JAK/STAT/SOCS signaling pathway," *Experimental Neurology*, vol. 336, no. June, p. 113506, 2021. doi:10.1016/j.expneurol.2020.113506.
187. J. K. Nowicki and E. Jakubowska-Pietkiewicz, "Osteocalcin: beyond bones," *Endocrinology and Metabolism*, vol. 39, no. 3, pp. 399–406, 2024. doi:10.3803/EnM.2023.1895.
188. M. Tavakol, J. Liu, S. E. Hoff, C. Zhu, and H. Heinz, "Osteocalcin: promoter or inhibitor of hydroxyapatite growth?" *Langmuir*, 2024. doi:10.1021/acs.langmuir.3c02948.
189. M. C. Diaz-Franco, R. Franco-Diaz de Leon, and J. R. Villafan-Bernal, "Osteocalcin-GPRC6A: an update of its clinical and biological multi-organic interactions (Review)," *Molecular Medicine Reports*, vol. 19, no. 1, pp. 15–22, 2019. doi:10.3892/mmr.2018.9627.
190. L. Ma, F. Gong, J. Xu, and J. Yang, "Uncarboxylated osteocalcin reverses the high glucose-induced inhibition of the osteogenic differentiation of MC3T3E1 cells via the GPRC6A/cAMP/PKA/AMPK signaling pathway," *International Journal of Molecular Medicine*, vol. 47, no. 5, pp. 1–11, 2021. doi:10.3892/IJMM.2021.4924.
191. Y. Zhang et al., "mTORC1 Inhibits NF-κB/NFATc1 signaling and prevents osteoclast precursor differentiation, in vitro and in mice," *Journal of Bone and Mineral Research*, vol. 32, no. 9, pp. 1829–1840, 2017. doi:10.1002/jbmr.3172.
192. Q. Jin et al., "CAMK1D inhibits glioma through the PI3K/AKT/mTOR signaling pathway," *Frontiers in Oncology*, vol. 12, no. April, pp. 1–12, 2022. doi:10.3389/fonc.2022.845036.
193. I. Malikova, A. Worth, D. Aliyeva, M. Khassenova, M. V. Kriajevska, and E. Tulchinsky, "Proteolysis of TAM receptors in autoimmune diseases and cancer: what does it say to us?," *Cell Death & Disease*, vol. 16, no. 1, pp. 1–13, 2025. doi:10.1038/s41419-025-07480-9.
194. R. Bassyouni, S. Gaber, A. Gomaa, and E. S. Ali, "Gas6 / TAM system: physiological insights and diseases," *Fayoum University Medical Journal*, vol. 1, no. 1, pp. 5–23, 2018. doi:10.21608/fumj.2018.5502.
195. S. Huang et al., "SOCS proteins participate in the regulation of innate immune response caused by viruses," *Frontiers in Immunology*, vol. 11, no. September, pp. 1–15, 2020. doi:10.3389/fimmu.2020.558341.
196. D. O. Moon, "Curcumin in cancer and inflammation: an in-depth exploration of molecular interactions, therapeutic potentials, and the role in disease management," *International Journal of Molecular Sciences*, vol. 25, no. 5, 2024. doi:10.3390/ijms25052911.
197. J. Engelmann, D. Ragipoglu, I. Ben-Batalla, and S. Loges, "The role of TAM receptors in bone," *International Journal of Molecular Sciences*, vol. 25, no. 1, pp. 1–22, 2024. doi:10.3390/ijms25010233.
198. S Zhang, Y. Liu, X. Wang, N. An, and X. Ouyang, "STAT1/SOCS1/3 are involved in the inflammation-regulating effect of GAS6/AXL in periodontal ligament cells induced by porphyromonas gingivalis lipopolysaccharide in vitro," *Journal of Immunology Research*, vol. 2021, 2021. doi:10.1155/2021/9577695.
199. A. G. De La Fuente et al., "Vitamin D receptor-retinoid X receptor heterodimer signaling regulates oligodendrocyte progenitor cell differentiation," *Journal of Cell Biology*, vol. 211, no. 5, pp. 975–985, 2015. doi:10.1083/jcb.201505119.
200. F. Kerlikowsky, J. P. Schuchardt, and A. Hahn, "Folate, vitamin B12 and vitamin D status in healthy and active home-dwelling people over 70 years," *BMC Geriatrics*, vol. 23, no. 1, pp. 1–11, 2023. doi:10.1186/s12877-023-04391-2.
201. M. E. Doost et al., "Synergistic activation of VDR-RXR heterodimers by vitamin D and rexinoids in human kidney and brain cells," *Cells*, vol. 13, no. 22, 2024. doi:10.3390/cells13221878.
202. Y. Li et al., "Modulation of the vitamin D/vitamin D receptor system in osteoporosis pathogenesis: insights and therapeutic approaches," *Journal of Orthopaedic Surgery and Research*, vol. 18, no. 1, pp. 1–12, 2023. doi:10.1186/s13018-023-04320-4.
203. M. Fusaro et al., "Calcimimetic and vitamin D analog use in hemodialyzed patients is associated with increased levels of vitamin K dependent proteins," *Endocrine*, vol. 51, no. 2, pp. 333–341, 2016. doi:10.1007/s12020-015-0673-z.
204. U. Sivagurunathan et al., "Effects of the interaction between dietary vitamin D3 and vitamin K3 on growth, skeletal anomalies, and expression of bone and calcium metabolism-related genes in juvenile gilthead seabream (*Sparus aurata*)," *Animals*, vol. 14, no. 19, 2024. doi:10.3390/ani14192808.
205. Amarjeet et al., "Understanding the cross-talk between the ceramide biosynthetic pathway and mitochondrial function in multiple sclerosis and demyelinating disorders," *Acta Neurologica Scandinavica*, vol. 2023, 2023. doi:10.1155/2023/7398037.
206. R. A. Sadler et al., "Beyond the coagulation cascade: vitamin K and its multifaceted impact on human and domesticated animal health," *Current Issues in Molecular Biology*, vol. 46, no. 7, pp. 7001–7031, 2024. doi:10.3390/cimb46070418.
207. D. L. Davis, U. Mahawar, V. S. Pope, J. Allegood, C. Sato-Bigbee, and B. W. Wattenberg, "Dynamics of sphingolipids and the serine palmitoyltransferase complex in rat oligodendrocytes during myelination," *Journal of Lipid Research*, vol. 61, no. 4, pp. 505–522, 2020. doi:10.1194/jlr.RA120000627.
208. J. A. Barnes-Vélez, F. B. Aksoy Yasar, and J. Hu, "Myelin lipid metabolism and its role in myelination and myelin maintenance," *Innovation*, vol. 4, no. 1, p. 100360, 2023. doi:10.1016/j.xinn.2022.100360.
209. D. C. Popescu et al., "Vitamin K enhances the production of brain sulfatides during remyelination," *PLoS One*, vol. 13, no. 8, pp. 1–19, 2018. doi:10.1371/journal.pone.0203057.
210. A. Kosta et al., "GAS6/TAM signaling pathway controls MICA expression in multiple myeloma cells," *Frontiers in Immunology*, vol. 13, no. July, pp. 1–13, 2022. doi:10.3389/fimmu.2022.942640.
211. S. E. Gilchrist, G. M. Pennelli, and S. Hafizi, "Gas6/tam signalling negatively regulates inflammatory induction of gm-csf in mouse brain microglia," *Cells*, vol. 10, no. 12, 2021. doi:10.3390/cells10123281.
212. D. Apostolo et al., "Could Gas6/TAM axis provide valuable insights into the pathogenesis of systemic sclerosis?," *Current Issues in Molecular Biology*, vol. 46, no. 7, pp. 7486–7504, 2024. doi:10.3390/cimb46070444.
213. M. Jain et al., "Role of JAK/STAT in the neuroinflammation and its association with neurological disorders," *Annals of Neurosciences*, vol. 28, no. 3–4, pp. 191–200, 2021. doi:10.1177/09727531211070532.

214. M. Yan et al., "SOCS modulates JAK-STAT pathway as a novel target to mediate the occurrence of neuroinflammation: molecular details and treatment options," *Brain Research Bulletin*, vol. 213, no. January, p. 110988, 2024. doi:10.1016/j.brainresbull.2024.110988.
215. J. Li, S. Lou, and X. Bian, "Osteocalcin and GPR158: linking bone and brain function," *Frontiers in Cell and Developmental Biology*, vol. 13, no. April, pp. 1–14, 2025. doi:10.3389/fcell.2025.1564751.
216. M. Nakamura, M. Imaoka, and M. Takeda, "Interaction of bone and brain: osteocalcin and cognition," *International Journal of Neuroscience*, vol. 131, no. 11, pp. 1115–1123, 2021. doi:10.1080/00207454.2020.1770247.
217. D. W. Eyles, S. Smith, R. Kinobe, M. Hewison, and J. J. McGrath, "Distribution of the Vitamin D receptor and 1α-hydroxylase in human brain," *Journal of Chemical Neuroanatomy*, vol. 29, no. 1, pp. 21–30, 2005. doi:10.1016/j.jchemneu.2004.08.006.
218. H. A. Shirazi, J. Rasouli, B. Ciric, A. Rostami, and G. X. Zhang, "1,25-Dihydroxyvitamin D3 enhances neural stem cell proliferation and oligodendrocyte differentiation," *Experimental and Molecular Pathology*, vol. 98, no. 2, 2015. doi:10.1016/j.yexmp.2015.02.004.
219. M. O. W. Grimm et al., "Vitamin D and its analogues decrease amyloid-β (Aβ) formation and increase Aβ-degradation," *International Journal of Molecular Sciences*, vol. 18, no. 12, 2017. doi:10.3390/ijms18122764.
220. M. R. Durk et al., "1α, 25-dihydroxyvitamin D3 reduces cerebral Amyloid-β accumulation and improves cognition in mouse models of Alzheimer's disease," *Journal of Neuroscience*, vol. 34, no. 21, 2014. doi:10.1523/JNEUROSCI.2711-13.2014.
221. T. Y. Wu, L. X. Zhao, Y. H. Zhang, and Y. G. Fan, "Activation of vitamin D receptor inhibits Tau phosphorylation is associated with reduction of iron accumulation in APP/PS1 transgenic mice," *Neurochemistry International*, vol. 153, 2022. doi:10.1016/j.neuint.2021.105260.
222. V. T. A. Thu, T. X. Hoang, and J. Y. Kim, "1,25-dihydroxy vitamin D3 facilitates the M2 polarization and β-amyloid uptake by human microglia in a TREM2-dependent manner," *BioMed Research International*, vol. 2023, 2023. doi:10.1155/2023/3483411.
223. T. Bellettini-Santos et al., "Vitamin D3 improves spatial memory and modulates cytokine levels in aged rats," *Metabolic Brain Disease*, vol. 38, no. 4, 2023. doi:10.1007/s11011-022-01152-9.
224. A. Pignolo et al., "Vitamin D and Parkinson's disease," *Nutrients*, vol. 14, p. 1220, 2022. doi:10.3390/nu14061220.
225. T. Behl et al., "Understanding the role of 'sunshine vitamin D' in Parkinson's disease: a review," *Frontiers in Pharmacology,* vol. 13, p. 993033, 2022. doi:10.3389/fphar.2022.993033.
226. Y. Xie, L. Chen, J. Chen, and Y. Chen, "Calcitriol restrains microglial M1 polarization and alleviates dopaminergic degeneration in hemiparkinsonian mice by boosting regulatory T-cell expansion," *Brain and Behavior*, vol. 14, no. 2, 2024. doi:10.1002/brb3.3373.
227. J. Hu et al., "Calcitriol Alleviates MPP+- and MPTP-induced parthanatos through the VDR/PARP1 pathway in the model of Parkinson's disease," *Frontiers in Aging Neuroscience*, vol. 13, 2021. doi:10.3389/fnagi.2021.657095.
228. J. Correale, M. C. Ysrraelit, and M. I. Gaitán, "Vitamin D-mediated immune regulation in multiple sclerosis," *Journal of the Neurological Sciences*, vol. 311, no. 1–2, 2011. doi:10.1016/j.jns.2011.06.027.
229. Z. Xie et al., "1,25-dihydroxyvitamin D3-induced dendritic cells suppress experimental autoimmune encephalomyelitis by increasing proportions of the regulatory lymphocytes and reducing T helper type 1 and type 17 cells," *Immunology*, vol. 152, no. 3, 2017. doi:10.1111/imm.12776.
230. M. Galoppin et al., "Full spectrum of vitamin D immunomodulation in multiple sclerosis: mechanisms and therapeutic implications," *Brain Communications*, vol. 4, no. 4, 2022. doi:10.1093/braincomms/fcac171.
231. R. Hashemi, S. S. Hosseini-Asl, S. R. Arefhosseini, and M. Morshedi, "The impact of Vitamin D3 intake on inflammatory markers in multiple sclerosis patients and their first-degree relatives," *PLoS ONE*, vol. 15, no. 4, 2020. doi:10.1371/journal.pone.0231145.
232. L. A. N. Mimura et al., "Preclinical therapy with vitamin D3 in experimental encephalomyelitis: efficacy and comparison with paricalcitol," *International Journal of Molecular Sciences*, vol. 22, no. 4, 2021. doi:10.3390/ijms22041914.
233. J. H. Laursen, H. B. Søndergaard, P. S. Sørensen, F. Sellebjerg, and A. B. Oturai, "Vitamin D supplementation reduces relapse rate in relapsing-remitting multiple sclerosis patients treated with natalizumab," *Multiple Sclerosis and Related Disorders*, vol. 10, 2016. doi:10.1016/j.msard.2016.10.005.
234. Z. Savran, S. B. Baltaci, T. Aladag, R. Mogulkoc, and A. K. Baltaci, "Vitamin D and neurodegenerative diseases such as multiple sclerosis (MS), Parkinson's disease (PD), Alzheimer's disease (AD), and amyotrophic lateral sclerosis (ALS): a review of current literature," *Current Nutrition Reports*, vol. 14, pp. 1–15, 2025. doi:10.1007/s13668-025-00663-y.
235. A. Gianforcaro and M. J. Hamadeh, "Vitamin D as a potential therapy in amyotrophic lateral sclerosis," *C*NS Neuroscience & Therapeutics, vol. 20, pp. 101–111, 2014. doi:10.1111/cns.12204.
236. R. A. N. Pertile, X. Cui, L. Hammond, and D. W. Eyles, "Vitamin D regulation of GDNF/Ret signaling in dopaminergic neurons," *FASEB Journal*, vol. 32, no. 2, 2018. doi:10.1096/fj.201700713R.
237. D. Gezen-Ak, E. Dursun, and S. Yilmazer, "The effect of vitamin D treatment on nerve growth factor (NGF) release from hippocampal neurons," *Archives of Neuropsychiatry*, vol. 51, no. 2, 2014. doi:10.4274/npa.y7076.
238. K. Sedaghat, R. Naderian, R. Pakdel, A. R. Bandegi, and Z. Ghods, "Regulatory effect of vitamin D on pro-inflammatory cytokines and anti-oxidative enzymes dysregulations due to chronic mild stress in the rat hippocampus and prefrontal cortical area," *Molecular Biology Reports*, vol. 48, no. 12, 2021. doi:10.1007/s11033-021-06810-2.
239. S. G. Menéndez and W. Manucha, "Vitamin d as a modulator of neuroinflammation: implications for brain health," *Current Pharmaceutical Design*, vol. 30, no. 5, 2024. doi:10.2174/0113816128281314231219113942.
240. A. Gianforcaro and M. J. Hamadeh, "Dietary vitamin D3 supplementation at 10× the adequate intake improves functional capacity in the G93A transgenic mouse model of ALS, a pilot study," *CNS Neuroscience & Therapeutics*, vol. 18, no. 7, pp. 547–557, 2012. doi:10.1111/j.1755-5949.2012.00316.x.
241. G. Ferland, "Vitamin K and the nervous system: an overview of its actions," *Advances in Nutrition,* vol. 3, pp. 204–212, 2012. doi:10.3945/an.111.001784.

242. T. Prouse, S. Majumder, and R. Majumder, "Functions of TAM receptors and ligands protein S and Gas6 in atherosclerosis and cardiovascular disease," *Multidisciplinary Digital Publishing Institute (MDPI)*, 2024. doi:10.3390/ijms252312736.
243. L. Alisi et al., "The relationships between vitamin k and cognition: a review of current evidence," *Frontiers in Neurology*, vol. 10, p. 239, 2019. doi:10.3389/fneur.2019.00239.
244. N. J. Haughey, V. V. R. Bandaru, M. Bae, and M. P. Mattson, "Roles for dysfunctional sphingolipid metabolism in Alzheimer's disease neuropathogenesis," *Biochimica et Biophysica Acta (BBA) — Molecular and Cell Biology of Lipids*, vol. 1801, pp. 878–886, 2010. doi:10.1016/j.bbalip.2010.05.003.
245. G. Ferland, "Vitamin K, an emerging nutrient in brain function," *BioFactors*, vol. 38, pp. 151–157, 2012. doi:10.1002/biof.1004.
246. S. L. Booth et al., "Association of vitamin K with cognitive decline and neuropathology in community-dwelling older persons," *Alzheimer's and Dementia: Translational Research and Clinical Interventions*, vol. 8, no. 1, 2022. doi:10.1002/trc2.12255.
247. Y. X. Yu et al., "Vitamin K2 suppresses rotenone-induced microglial activation in vitro," *Acta Pharmacologica Sinica*, vol. 37, no. 9, 2016. doi:10.1038/aps.2016.68.
248. Y. X. Yu, X. D. Yu, Q. Z. Cheng, L. Tang, and M. Q. Shen, "The association of serum vitamin K2 levels with Parkinson's disease: from basic case-control study to big data mining analysis," *Aging*, vol. 12, no. 16, 2020. doi:10.18632/aging.103691.
249. H. Tang, Z. Zheng, H. Wang, L. Wang, G. Zhao, and P. Wang, "Vitamin K2 modulates mitochondrial dysfunction induced by 6-hydroxydopamine in SH-SY5Y cells via mitochondrial quality-control loop," *Nutrients*, vol. 14, no. 7, 2022. doi:10.3390/nu14071504.
250. J. Prasuhn et al., "The use of vitamin K2 in patients with Parkinson's disease and mitochondrial dysfunction (PD-K2): a theranostic pilot study in a placebo-controlled parallel group design," *Frontiers in Neurology*, vol. 11, 2021. doi:10.3389/fneur.2020.592104.
251. D. D'Onghia et al., "Gas6/TAM system as potential biomarker for multiple sclerosis prognosis," *Frontiers in Neurology*, vol. 15, 2024. doi:10.3389/fimmu.2024.1362960.
252. M. Moriya, Y. Nakatsuji, T. Okuno, T. Hamasaki, M. Sawada, and S. Sakoda, "Vitamin K2 ameliorates experimental autoimmune encephalomyelitis in Lewis rats," *Journal of Neuroimmunology*, vol. 170, no. 1–2, 2005. doi:10.1016/j.jneuroim.2005.08.001.
253. R. Lasemi, M. Kundi, N. B. Moghadam, H. Moshammer, and J. A. Hainfellner, "Vitamin K2 in multiple sclerosis patients," *Wiener klinische Wochenschrift*, vol. 130, no. 9–10, 2018. doi:10.1007/s00508-018-1328-x.
254. A. McGarry et al., "A randomized, double-blind, placebo-controlled trial of coenzyme Q10 in Huntington disease," *Neurology*, vol. 88, no. 2, 2017. doi:10.1212/WNL.0000000000003478.
255. H. A. Elkattawy et al., "Vitamin K2 (menaquinone-7) reverses age-related structural and cognitive deterioration in naturally aging rats," *Antioxidants*, vol. 11, no. 3, 2022. doi:10.3390/antiox11030514.
256. J. Lin et al., "Chronic vitamin D3 supplementation alleviates cognition impairment via inhibition of oxidative stress regulated by PI3K/AKT/Nrf2 in APP/PS1 transgenic mice," *Neuroscience Letters*, vol. 783, 2022. doi:10.1016/j.neulet.2022.136725.
257. J. Li, H. Wang, and P. A. Rosenberg, "Vitamin K prevents oxidative cell death by inhibiting activation of 12-lipoxygenase in developing oligodendrocytes," *Journal of Neuroscience Research*, vol. 87, no. 9, 2009. doi:10.1002/jnr.22029.
258. X. Cui and D. W. Eyles, "Vitamin D and the central nervous system: causative and preventative mechanisms in brain disorders," *Nutrients*, vol. 14, p. 4353, 2022. doi:10.3390/nu14204353.
259. Y. Liu, Z. Zhong, J. Xie, B. Ni, and Y. Wu, "Neuroprotective roles of vitamin D: bridging the gap between mechanisms and clinical applications in cognitive decline," *International Journal of Molecular Sciences*, vol. 26, no. 15, p. 7146, 2025. doi:10.3390/ijms26157146.
260. L. D. Brewer, V. Thibault, K. C. Chen, M. C. Langub, P. W. Landfield, and N. M. Porter, "Vitamin D hormone confers neuroprotection in parallel with downregulation of L-type calcium channel expression in hippocampal neurons," *Journal of Neuroscience*, vol. 21, no. 1, 2001. doi:10.1523/jneurosci.21-01-00098.2001.
261. M. Cohen-Lahav, S. Shany, D. Tobvin, C. Chaimovitz, and A. Douvdevani, "Vitamin D decreases NFκB activity by increasing IκBα levels," *Nephrology Dialysis Transplantation*, vol. 21, no. 4, pp. 889–897, 2006. doi:10.1093/ndt/gfi254.
262. Y. Ohsaki et al., "Vitamin K suppresses the lipopolysaccharide-induced expression of inflammatory cytokines in cultured macrophage-like cells via the inhibition of the activation of nuclear factor αB through the repression of IKKα/β phosphorylation," *Journal of Nutritional Biochemistry*, vol. 21, no. 11, 2010. doi:10.1016/j.jnutbio.2009.09.011.
263. W. Luo, D. Ye, K. Zhao, L. Zhou, Y. Wu, and Q. Ge, "Associations between vitamin K and systemic immune and inflammation biomarkers: a population-based study from the NHANES (2007–2020)," *Frontiers in Nutrition*, vol. 12, 2025. doi:10.3389/fnut.2025.1625209.
264. S. Ito, S. Ohtsuki, Y. Nezu, Y. Koitabashi, S. Murata, and T. Terasaki, "1α,25-Dihydroxyvitamin D3 enhances cerebral clearance of human amyloid-β peptide(1-40) from mouse brain across the blood-brain barrier," *Fluids Barriers CNS*, vol. 8, no. 1, 2011. doi:10.1186/2045-8118-8-20.
265. D. H. Kim, C. A. Meza, H. Clarke, J. S. Kim, and R. C. Hickner, "Vitamin D and endothelial function," *Nutrients*, vol. 12, p. 575, 2020. doi:10.3390/nu12020575.
266. A. M. G. Jaminon et al., "Matrix Gla protein is an independent predictor of both intimal and medial vascular calcification in chronic kidney disease," *Scientific Reports*, vol. 10, no. 1, 2020. doi:10.1038/s41598-020-63013-8.
267. S. Roumeliotis, E. Dounousi, M. Salmas, T. Eleftheriadis, and V. Liakopoulos, "Vascular calcification in chronic kidney disease: the role of vitamin K-dependent matrix gla protein," *Frontiers in Medicine*, vol. 7, p. 154. doi:10.3389/fmed.2020.00154.
268. P. M. Kidd, "Vitamins D and K as pleiotropic nutrients: clinical importance to the skeletal and cardiovascular systems and preliminary evidence for synergy," *Alternative Medicine Review*, vol. 15, pp. 199–222, 2010.
269. A. J. Van Ballegooijen, S. Pilz, A. Tomaschitz, M. R. Grübler, and N. Verheyen, "The synergistic interplay between vitamins D and K for bone and cardiovascular health: a narrative review," *International Journal of Endocrinology*, vol. 2017, p. 7454376, 2017. doi:10.1155/2017/7454376.

270. F. Zhang and W. Li, "Vitamin D and sarcopenia in the senior people: a review of mechanisms and comprehensive prevention and treatment strategies," *Therapeutics and Clinical Risk Management*, vol. 20, pp. 577–595, 2024. doi:10.2147/TCRM.S471191.
271. K. Azuma, Y. Osuka, N. Kojima, H. Sasai, H. Kim, and S. Inoue, "Association of vitamin K insufficiency with cognitive dysfunction in community-dwelling older adults," *Frontiers in Nutrition*, vol. 8, 2022. doi:10.3389/fnut.2021.811831.
272. C. Wessinger, C. Hafer-Macko, and A. S. Ryan, "Vitamin K intake in chronic stroke: implications for dietary recommendations," *Nutrients*, vol. 12, no. 10, 2020. doi:10.3390/nu12103059.
273. A. C. Carvalho et al., "25-OH vitamin D levels and cognitive performance: longitudinal assessment in a healthy aging cohort," *Frontiers in Aging Neuroscience*, vol. 11, 2019. doi:10.3389/fnagi.2019.00330.
274. B. Galunska, Y. Yotov, M. Nikolova, and A. Angelov, "Extrahepatic vitamin K-dependent Gla-proteins–potential cardiometabolic biomarkers," *Multidisciplinary Digital Publishing Institute (MDPI)*, 2024. doi:10.3390/ijms25063517.
275. M. Razavi, M. Jamilian, M. Karamali, F. Bahmani, E. Aghadavod, and Z. Asemi, "The effects of vitamin D-K-calcium co-supplementation on endocrine, inflammation, and oxidative stress biomarkers in vitamin D-deficient women with polycystic ovary syndrome: a randomized, double-blind, placebo-controlled trial," *Hormone and Metabolic Research*, vol. 48, no. 7, 2016. doi:10.1055/s-0042-104060.
276. Z. Asemi et al., "The effects of Vitamin D, K and calcium co-supplementation on carotid intima-media thickness and metabolic status in overweight type 2 diabetic patients with CHD," *British Journal of Nutrition*, vol. 116, no. 2, 2016. doi:10.1017/S0007114516001847.
277. A. Florea, M. E. Kooi, W. Mess, L. J. Schurgers, J. Bucerius, and F. M. Mottaghy, "Effects of combined vitamin k2 and vitamin d3 supplementation on na18 f.f pet/mri in patients with carotid artery disease: the intricate rationale and trial design," *Nutrients*, vol. 13, no. 3, pp. 1–12, 2021. doi:10.3390/nu13030994.
278. O. Atieh et al., "Vitamins K2 and D3 improve long COVID, fungal translocation, and inflammation: randomized controlled trial," *Nutrients*, vol. 17, no. 2, 2025. doi:10.3390/nu17020304.
279. C. McGreevy et al., "The effect of vitamin D supplementation on arterial stiffness in an elderly community-based population," *Journal of the American Society of Hypertension*, vol. 9, no. 3, 2015. doi:10.1016/j.jash.2014.12.019.
280. M. H. J. Knapen, L. A. J. L. M. Braam, N. E. Drummen, O. Bekers, A. P. G. Hoeks, and C. Vermeer, "Menaquinone-7 supplementation improves arterial stiffness in healthy postmenopausal women: a double-blind randomised clinical trial," *Thrombosis and Haemostasis*, vol. 113, no. 5, 2015. doi:10.1160/TH14-08-0675.
281. J. Jia, J. Hu, X. Huo, R. Miao, Y. Zhang, and F. Ma, "Effects of Vitamin D supplementation on cognitive function and blood Aβ-related biomarkers in older adults with Alzheimer's disease: a randomised, double-blind, placebo-controlled trial," *Journal of Neurology, Neurosurgery, and Psychiatry*, vol. 90, no. 12, 2019. doi:10.1136/jnnp-2018-320199.
282. C. Annweiler, B. Fantino, E. Parot-Schinkel, S. Thiery, J. Gautier, and O. Beauchet, "Alzheimer's disease-input of vitamin D with mEmantine assay (AD-IDEA trial): study protocol for a randomized controlled trial," *Trials*, vol. 12, 2011. doi:10.1186/1745-6215-12-230.
283. M. S. Stein, S. C. Scherer, K. S. Ladd, and L. C. Harrison, "A randomized controlled trial of high-dose vitamin D2 followed by intranasal insulin in Alzheimer's disease," *Journal of Alzheimer's Disease*, vol. 26, no. 3, 2011. doi:10.3233/JAD-2011-110149.
284. C. Annweiler, A. M. Schott, Y. Rolland, H. Blain, F. R. Herrmann, and O. Beauchet, "Dietary intake of vitamin D and cognition in older women: a large population-based study," *Neurology*, vol. 75, no. 20, 2010. doi:10.1212/WNL.0b013e3181fd6352.
285. C. Annweiler, F. R. Herrmann, B. Fantino, B. Brugg, and O. Beauchet, "Effectiveness of the combination of memantine plus vitamin d on cognition in patients with Alzheimer disease: a pre-post pilot study," *Cognitive and Behavioral Neurology*, vol. 25, no. 3, 2012. doi:10.1097/WNN.0b013e31826df647.
286. M. Morello et al., "Vitamin D improves neurogenesis and cognition in a mouse model of Alzheimer's Disease," *Molecular Neurobiology*, vol. 55, no. 8, 2018. doi:10.1007/s12035-017-0839-1.
287. Z. K. Bytowska et al., "Effect of 12-week BMI-based vitamin D3 supplementation in Parkinson's disease with deep brain stimulation on physical performance, inflammation, and vitamin D metabolites," *International Journal of Molecular Sciences*, vol. 24, no. 12, 2023. doi:10.3390/ijms241210200.
288. A. L. Hiller, C. F. Murchison, B. M. Lobb, S. O'Connor, M. O'Connor, and J. F. Quinn, "A randomized, controlled pilot study of the effects of vitamin D supplementation on balance in Parkinson's disease: does age matter?" *PLoS One*, vol. 13, no. 9, 2018. doi:10.1371/journal.pone.0203637.
289. A. H. Habibi, A. Anamoradi, G. A. Shahidi, S. Razmeh, E. Alizadeh, and K. M. Kokhedan, "Treatment of levodopainduced dyskinesia with vitamin D: a randomized, double-blind, placebo-controlled trial," *Neurology International*, vol. 10, no. 3, 2018. doi:10.4081/ni.2018.7737.
290. M. Suzuki et al., "Randomized, double-blind, placebo-controlled trial of vitamin D supplementation in Parkinson disease," *American Journal of Clinical Nutrition*, vol. 97, no. 5, 2013. doi:10.3945/ajcn.112.051664.
291. Z. Zheng et al., "VDR and deubiquitination control neuronal oxidative stress and microglial inflammation in Parkinson's disease," *Cell Death Discovery*, vol. 10, no. 1, 2024. doi:10.1038/s41420-024-01912-9.
292. Thouvenot Eric et al., "High-dose vitamin D in clinically isolated syndrome typical of multiple sclerosis: the D-Lay MS randomized clinical trial," *JAMA*, vol. 333, pp. 1413–1422, 2025.
293. S. D. Cassard et al., "High-dose vitamin D3 supplementation in relapsing-remitting multiple sclerosis: a randomised clinical trial," *EClinicalMedicine*, vol. 59, 2023. doi:10.1016/j.eclinm.2023.101957.
294. R. Hupperts et al., "Randomized trial of daily high-dose Vitamin D3 in patients with RRMS receiving subcutaneous interferon β-1a," *Neurology*, vol. 93, no. 20, pp. E1906–E1916, 2019. doi:10.1212/WNL.0000000000008445.
295. W. Camu et al., "Cholecalciferol in relapsing-remitting MS: a randomized clinical trial (CHOLINE)," *Neurology: Neuroimmunology & Neuroinflammation*, vol. 6, no. 5, 2019. doi:10.1212/NXI.0000000000000597.

296. J. Dörr, S. Ohlraun, H. Skarabis, and F. Paul, "Efficacy of vitamin D supplementation in multiple sclerosis (EVIDIMS trial): study protocol for a randomized controlled trial," *Trials*, vol. 13, 2012. doi:10.1186/1745-6215-13-15.
297. M. Soilu-Hänninen et al., "A randomised, double blind, placebo controlled trial with vitamin D 3 as an add on treatment to interferon β-1b in patients with multiple sclerosis," Journal of Neurology, Neurosurgery, and Psychiatry, vol. 83, no. 5, 2012. doi:10.1136/jnnp-2011-301876.
298. M. T. Haindl et al., "Vitamin D: an effective antioxidant in an animal model of progressive multiple sclerosis," *Nutrients*, vol. 15, no. 15, 2023. doi:10.3390/nu15153309.
299. F. Trojsi et al., "Vitamin D supplementation has no effects on progression of motor dysfunction in amyotrophic lateral sclerosis (ALS)," *European Journal of Clinical Nutrition*, vol. 74, no. 1, 2020. doi:10.1038/s41430-019-0448-3.
300. C. Karam, M. J. Barrett, T. Imperato, D. J. L. Macgowan, and S. Scelsa, "Vitamin D deficiency and its supplementation in patients with amyotrophic lateral sclerosis," *Journal of Clinical Neuroscience*, vol. 20, no. 11, 2013. doi:10.1016/j.jocn.2013.01.011.
301. E. Moghimi, J. A. Solomon, A. Gianforcaro, and M. J. Hamadeh, "Dietary vitamin D3 restriction exacerbates disease pathophysiology in the spinal cord of the G93A mouse model of amyotrophic lateral sclerosis," *PLoS One*, vol. 10, no. 5, 2015. doi:10.1371/journal.pone.0126355.
302. A. Gianforcaro, J. A. Solomon, and M. J. Hamadeh, "Vitamin D3 at 50x AI attenuates the decline in paw grip endurance, but not disease outcomes, in the G93A mouse model of ALS, and Is toxic in females," *PLoS One*, vol. 8, no. 2, 2013. doi:10.1371/journal.pone.0030243.
303. Z. Javadfar et al., "Effect of vitamin D supplementation on inflammatory status and behavioral symptoms in children with autism spectrum disorders: a double-blind randomized clinical trial," *BMC Pediatrics*, vol. 25, no. 1, p. 615, 2025. doi:10.1186/s12887-025-05985-y.
304. H. Mazahery et al., "A randomised-controlled trial of vitamin d and omega-3 long chain polyunsaturated fatty acids in the treatment of core symptoms of autism spectrum disorder in children," *Journal of Autism and Developmental Disorders*, vol. 49, no. 5, 2019. doi:10.1007/s10803-018-3860-y.
305. B. Wang et al., "Sex-Specific effects of vitamin D on autistic behavior and gastrointestinal symptoms in rats via the regulation of serotonin metabolism," *Scientific Reports*, vol. 15, no. 1, 2025. doi:10.1038/s41598-025-05845-w.
306. J. Cui et al., "Induction of autism-related behavior in male mice by early-life vitamin D deficiency: association with disruption of the gut microbial composition and homeostasis," *Food & Function*, vol. 15, no. 8, 2024. doi:10.1039/d4fo00279b.

8 Nutritional Interactions
Cofactor Minerals, PUFA, Polyphenols, and Vitamin Synergy in Brain Health

Divya Choudhary, MD Nasiruddin Khan, Ravi Rana, Sidharth Mehan, Zuber Khan, Arun Kumar Sharma, and Manjeet Kumar

Division of Neuroscience, Department of Pharmacology, ISF College of Pharmacy, Moga, Punjab, India (Affiliated to IK Gujral Punjab Technical University, Jalandhar, Punjab, India)

8.1 INTRODUCTION

The brain is a vital organ that functions as a command center for the body, controlling essential processes such as memory, cognition, motor, and sensory functions, as well as maintaining homeostasis. Brain health is therefore a crucial factor for sustaining overall physical and mental well-being. The World Health Organization (WHO) defines brain health as the ability to operate across sensory, cognitive, socioemotional, motor, and behavioral domains, allowing a person to reach their full potential regardless of the presence or absence of disease (WHO). Brain health could be influenced by various genetic, environmental, and specific nutritional factors. Although the brain accounts for approximately 2% of body weight in humans, it has a very high metabolic demand and consumes around 20% of energy [1]. The various neuronal and non-neuronal cells require a constant supply of substrates such as oxygen, cofactor minerals, glucose, polyunsaturated fatty acid (PUFA), polyphenols, and vitamins to sustain their function [2]. The crucial role of various nutritional components in neural development and function starts in early fetal stages; adequate nutritional supply is essential for neurogenesis, brain growth, memory, and cognition [3]. It has also been reported that poor dietary patterns during fetal, neonatal, and early childhood phases negatively affect early brain development [4,5]. Studies suggest that a high-quality diet, including nutrients like minerals, PUFAs, vitamins, polyphenols, and antioxidants, can modulate various cellular pathways and enhance cognitive function, synaptic plasticity, neurogenesis, and reduce inflammation and oxidative stress [6].

8.2 NUTRITIONAL FOUNDATIONS OF BRAIN HEALTH

Dietary nutrients are essential for optimal brain function and health. Nutrients like PUFA are embedded in neuronal cell membranes and serve as their building blocks, helping maintain fluidity [7]. Minerals such as magnesium, zinc, iron, copper, and selenium function as cofactors for various enzymes involved in myelin formation, redox balance, neurotransmitter synthesis, immune response regulation, and synaptic plasticity [8]. Whereas polyphenols and antioxidant vitamins protect the brain against oxidative stress, aging, cognitive decline, and inflammation [9].

All these dietary nutrients can work synergistically, for example, PUFA as an anti-inflammatory agent, minerals as cofactors, vitamin C as a free radical stabilizer, and polyphenols as scavengers that augment the antioxidant defense system, mitigate inflammation, and provide increased neuroprotective activity [10]. Similarly, dietary nutrients can optimize energy metabolism, regulate neurogenesis, and control neurotransmitter synthesis and release in a synergistic manner.

8.3 COFACTOR MINERALS AND BRAIN HEALTH

Minerals are inorganic elements that, despite their minor requirements, are essential for regulating and maintaining a variety of physiological functions in the body [11]. Minerals play a crucial biological role as secondary messengers, regulators of acid–base balance, redox reactions, electrical gradients, and cellular pH [12]. Minerals such as magnesium, zinc, iron, copper, and selenium also play a vital role as enzyme cofactors and are involved in various cellular and physiological functions.

8.3.1 MAGNESIUM

Magnesium (Mg^{2+}) is the second most common intracellular cation that catalyzes over 300 enzymatic reactions in the body [13]. It is vital for the normal functioning and maturation of the Central Nervous System (CNS), Adenosine Triphosphate (ATP) production, memory, cognition, homeostasis, and the regulation of cellular electrophysiology [14,15]. The daily dietary intake of magnesium is 310 mg for adult females and 400 mg for males. The

DOI: 10.1201/9781003686675-8

serum Mg^{2+} levels between 0.75–0.95 mmol/L are considered normal in adults, while a serum Mg^{2+} level below 7.75 mmol/L indicates hypomagnesemia [16]. Various studies have reported that chronic hypomagnesemia is associated with cardiovascular disorders, inflammation, diabetes, osteoporosis, neuropsychiatric, and neurodegenerative disorders [17–19]. Mg^{2+} has been shown to directly block N-methyl-D-aspartate (NMDA) receptors and prevent excessive Ca^{2+} influx during the resting state, while allowing Ca^{2+} entry during depolarization. It is responsible for the voltage-dependent opening of N methyl D aspartate receptor (NMDAR), which further regulates long-term potentiation and long-term depression involved in synaptic plasticity. Experimental studies have reported that magnesium supplementation prevents synaptic loss and enhances cognition and memory in rodents [20,21].

Mg^{2+} also plays a pivotal role in neurotransmission, as it acts as a cofactor for enzymes involved in the synthesis, storage, and release of monoaminergic neurotransmitters [22]. It is reported to enhance synaptic function by activating the PI3K/Akt pathway and inhibiting GSK-3β signaling [23].

Reduced Mg^{2+} levels often cause excitotoxicity and heightened oxidative stress due to decreased NMDAR blockage by Mg^{2+} [24,25]. Treatment with dietary Mg^{2+} shows positive modulation of GABAA receptors, boosting inhibitory neurotransmission, which counterbalances the excitatory glutaminergic transmission [26]. Studies have reported that dietary Mg^{2+} supplementation results in neural calming, anxiolytic, and anti-depressant effects [27–29].

8.3.2 Zinc

Zinc is an essential trace metal ion present in the body; it acts as a cofactor mineral involved in various cellular and biological functions like neurogenesis, cell division and proliferation, immune modulation, memory and cognition, nucleic acid synthesis, and redox action [30]. The recommended daily dietary intake of zinc in adults ranges from 9 to 13 g/day. In adults, zinc content normally ranges between 1.4 and 2.3 g, of which almost 85% is stored in skeletal muscle and bones, while the remaining amount is distributed across the rest of the body [31]. Zinc also functions as a neuromodulator, since high concentrations of free zinc ions (Zn^{2+}) are present in brain regions such as the olfactory bulb, amygdala, hippocampus, and cortex [32]. A zinc-deficient diet when given to pregnant rodents results in embryos born with neural tube defects [33,34]. Similarly, low maternal serum zinc levels in humans are also linked to babies with neural tube defects, while dietary zinc intake during pregnancy decreases the risk of neural tube defects [35,36]. Multiple studies have demonstrated the extensive role of zinc in neurogenesis, and its deficiency leads to reduced expression of genes involved in neuronal proliferation and differentiation [37,38]. Zinc supplementation in human subjects leads to increased plasma levels of insulin-like growth factor-1, growth hormone, and Insulin Like Growth Factor Binding Protein 3 (IGFBP3), which are essential for neuronal growth, proliferation, and synaptic plasticity [39,40].

Many studies have demonstrated the negative impact of zinc deficiency on the maturation of T and B cells, Th1/Th2 balance, neutrophil function, and NK cell cytotoxicity [41–43]. During developmental phases in both humans and rodents, zinc deficiency is associated with impaired memory, social behavior, learning, and cognition, as well as an increased risk of autism [44,45]. Additionally, adequate zinc levels are crucial for brain-derived neurotrophic factor (BDNF) expression, learning, and memory formation in the hippocampus [46].

8.3.3 Iron

Iron is an essential trace element necessary for all life forms, and it is the most widely distributed transition metal ion in the human body [47]. Iron plays an essential role in many biological processes such as oxygen transport, neurogenesis, myelination, energy metabolism, and DNA replication [48]. The iron content in adults is approximately 4–5 g, so a daily dietary intake of 8 mg for adult males and 18 mg for females is essential to maintain iron homeostasis in the body. Dietary iron exists in two forms: heme-iron (absorbed via heme carrier protein) and non-heme-iron (absorbed via divalent metal transporter 1) [49]. About 70%–80% of total iron is bound to heme in red blood cells, which is responsible for the RBCs' oxygen-carrying capacity, and 5% is present as myoglobin in muscle cells [50]. The remaining smaller fraction of iron is either bound to plasma protein transferrin, stored intracellularly within ferritin, or as heme-containing proteins like cytochromes and iron–sulfur (Fe-S) clusters [51]. Iron plays an essential role in the transport of molecular O_2 via hemoglobin and myoglobin, where it is bound within the protoporphyrin IX complex [52]. Iron is significantly involved in regulating energy homeostasis and metabolism. Fe-S clusters are crucial for ATP generation in the mitochondria; they act as electron transfer groups during the electron transport chain [53,54]. It is also reported that iron deficiency potentially leads to reduced mitochondrial phosphorylation due to decreased expression of mitochondrial complex I, complex II, and complex III [55]. Fe-S clusters also play an important role as cofactors for enzymes such as aconitase, succinate dehydrogenase, and fumarase involved in the citric acid cycle [56]. However, iron supplementation leads to the upregulation of genes that encode heme, Fe-S clusters, and mitochondrial respiratory chain complexes [57].

Maintaining iron homeostasis in the CNS is also very crucial for normal physiological functioning. Iron is a vital mineral for various biological processes such as neurogenesis, neurotransmitter synthesis, DNA synthesis, mitochondrial respiration, and myelin production in the CNS [58]. Oligodendrocytes are the neural cells with the highest iron levels in the brain, where iron acts as a substrate for myelin production [59]. Multiple studies have demonstrated the essential role of iron in Oligodendrocyte Precursor Cells (OPC) proliferation, differentiation, and myelination [60]. Research studies also demonstrated that knockdown of genes expressing iron transporters, divalent metal transporter-1 and transferrin receptors, resulted in hypomyelination, decreased intracellular iron, and impaired OPC maturation [61,62]. Iron stores within OPC in the form of ferritin are essential for OPC maturation during the first postnatal weeks [63,64]. Iron is also essential as a cofactor for myelin production in oligodendrocytes, so the remyelination process is also impaired if OPCs and oligodendrocytes are iron deficient [65].

8.3.4 Copper

Copper is a crucial micronutrient and trace element that is essential for physiological cellular functions such as antioxidant activity, neurotransmitter synthesis, mitochondrial oxidative phosphorylation, and cellular transport [66]. The recommended daily dietary intake of copper is 900 μg/d for adults. Inside the body, copper actively participates in redox reactions because it can switch between Cu^+ and Cu^{2+} oxidation states [67]. Because it can easily accept or donate an electron, it functions as a cofactor for important enzymes such as superoxide dismutase, dopamine β hydroxylase, cytochrome C oxidase, and peptidylglycine α-amidating monooxygenase [68]. To maintain copper homeostasis in the body, adults need a daily dietary intake of 900 μg of copper [69]. Copper is transported in blood as ceruloplasmin, a plasma protein, throughout the body. Research studies reported that the transfer of copper ions from the copper chaperone for Speroxide Dismutase (SOD) to SOD1 is crucial for its activation and antioxidant activity [70,71]. Several studies also reported the role of copper-dependent enzymes (Peptidylglycine Alpha Amidating Monooxygenase (PAM) and Dopamine Beta Hydroxylase (DBH)) in the synthesis of peptide hormones and catecholamines [72,73].

Copper also plays a significant role as a cofactor for cytochrome C oxidase (Complex IV) of the Electron Transport Chain (ETC), where it transfers electrons from cytochrome C to oxygen and reduces it to water [74]. Copper deficiency is linked to decreased SOD expression and increased oxidative stress, mitochondrial dysfunction, and inflammation [75]. Research studies indicate that treating APP/PS1 mice with CuII (gtsm) enhances cognition and memory [74]. Similarly, CuII (atsm) also exhibits significant neuroprotective effects in rodent models of motor neuron disease, cerebral ischemia, and Parkinson's disease (PD) [76–78].

8.3.5 Selenium

Selenium is an essential trace element that acts as an antioxidant and is important for overall health, especially brain health. Selenium plays a key role in thyroid hormone metabolism, fertility, antioxidant defense, and supporting immune function [79]. The recommended daily dose of selenium ranges from 55 to 80 μg/day to maintain an adequate serum selenium concentration of 8 μg/dL [80]. In the human body, selenium is used as a precursor for the generation of amino acid selenocysteine (Sec), which is a constituent of selenoproteins. There are 25 selenoproteins, among which glutathione peroxidase, thioredoxin reductase, iodothyronine deiodinases, and selenoprotein P are the most studied and well categorized [81]. Selenoprotein P acts as the main selenium reservoir and transporter because it contains 10 Sec residues in its structure. It travels through the blood and delivers selenium to the brain by interacting with tissue-specific SELENOP receptors known as apolipoprotein E receptor 2 (ApoER2) [82]. Rodent models with genetic selenoprotein P deficiency show decreased selenium levels in the brain, leading to cerebello-cerebral atrophy, movement disorders, seizures, and neurodegeneration [83,84]. Low selenium levels are also linked to a higher risk of major depressive disorder, autism, and neurodegenerative disorders [85–87]. Several studies have reported that selenium has a protective effect against stress-induced depression and glucocorticoid-induced depression [88,89].

The physiological levels of selenium are vital for the regulation of redox balance, immune response, and inflammation. The major selenoenzymes include Glutathione Peroxidase (GPx), which detoxifies reactive oxygen species (ROS) (H_2O_2); Thioredoxin Reductase (TXNRD), which generates reduced thioredoxin that regulates redox functions in immune cells; and deiodinases, which regulate thyroid hormones [90]. It is reported that increased dietary selenium raises the expression of these selenoproteins, leading to reduced ROS production and improved oxidative capacity [91]. Selenium can regulate the differentiation of macrophages toward the M2 (anti-inflammatory) phenotype instead of the M1 (inflammatory) phenotype and reduce the levels of inflammatory cytokines IL-1β, iNOS, and NF-κB [92]. Some preclinical and human studies also suggest that selenium inhibits inflammation by suppressing NLRP3 inflammasome and NF-κB pathways, further inhibiting the production of inflammatory mediators [93,94]. Mechanism of minerals in brain health has been given in Figure 8.1. Table 8.1 shows the deficiency states of various minerals and their neurological consequences.

8.4 PUFAS AND COGNITIVE FUNCTION

PUFAs are essential fats with two or more double bonds in their hydrocarbon structure. PUFAs are fundamental components of cell membranes, responsible for maintaining their fluidity and permeability, and are also vital for proper neural and retinal function. PUFAs can be further classified into omega-3 and omega-6 fatty acids based on the position of the first double bond from the methyl end of the molecule. Omega-3 fatty acids include α-linolenic acid, docosahexaenoic acid (DHA), and eicosapentaenoic acid (EPA), while omega-6 fatty acids include linoleic acid and arachidonic acid. In humans, α-linolenic acid and linoleic acid must be obtained from dietary sources because

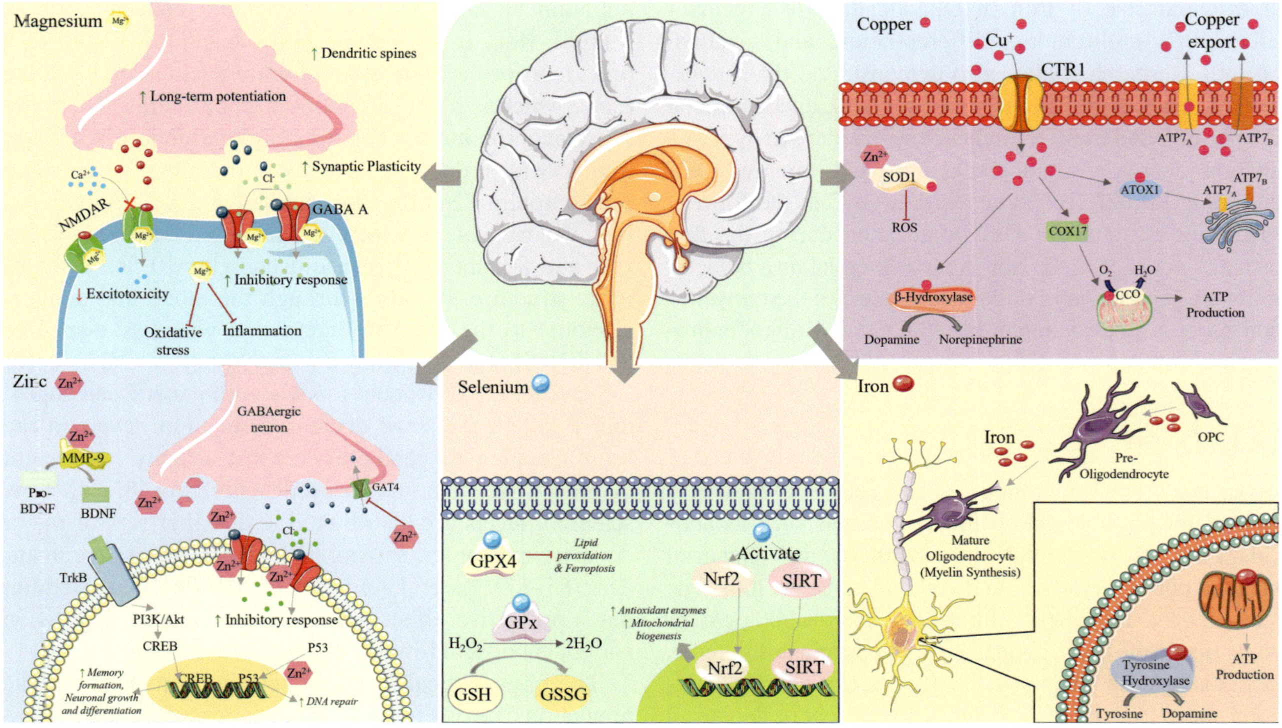

FIGURE 8.1 Mechanism of minerals in brain health.

TABLE 8.1
Deficiency States of Various Minerals and Their Neurological Consequences

Mineral	Deficiency	Neurological abnormalities	Dietary Sources	References
Magnesium (Mg^{2+})	Hypomagnesemia	Migraine, depression, cognitive and memory deficits	Spinach, nuts, seeds, legumes and whole grains	[95–97]
Zinc (Zn^{2+})	Zinc deficiency	Reduced neurogenesis, depression, peripheral neuropathy, Alzheimer's disease	Oysters, meat, seafood, eggs, beans, and nuts	[98]
Iron (Fe^{2+}/Fe^{3+})	Iron-deficiency anemia (serum ferritin <30 μg/L)	Hypomyelination, mood disorder, fatigue, impaired cognition	Meat, seafood, lentils, spinach vitamin C	[99,100]
Copper (Cu^{2+})	Hypocupremia (serum <70 μg/dL)	Myelopathy, ataxia, peripheral neuropathy, neurodegeneration	Shellfish, seeds, nuts, chocolate, whole grain products	[101,102]
Selenium (Se)	Selenium inadequacy (serum <7 μg/dL)	Cognitive and motor deficits Alzheimer's disease	Brazil nuts, seafood, meat, and poultry	[103,104]

they cannot be synthesized in the body due to the absence of the desaturase enzyme. Although the body can synthesize Eicosapentaenoic Acid (EPA), Docosahexaenoic Acid (DHA), and Arachidonic Acid (ARA) using α-linolenic acid and linoleic acid as precursors, the conversion efficiency is relatively low. Therefore, to meet physiological needs, dietary intake of EPA, DHA, and ARA is also necessary.

DHA is the most common PUFA in neural tissue, making up about 12%–14% of the brain's total fatty acids [105]. DHA is an important component of neural cell membranes and plays a crucial role in neurogenesis as well as the maintenance of synaptic plasticity [106]. Studies have reported that DHA supplementation promotes hippocampal neurogenesis, neurite outgrowth, and synapse formation, resulting in enhanced learning and memory [107,108]. DHA has been demonstrated to actively participate in the proliferation and differentiation of neural stem cells [109,110]. DHA administration, either through breast milk or supplement formula, also enhances psychomotor function and improves mental development, memory, attention, and retinal development in infants [111,112].

EPA and DHA are also involved in the regulation of inflammation and exhibit anti-inflammatory effects by reducing the production of pro-inflammatory cytokines and prostaglandins [113,114]. The enzymatic metabolism of omega-3 PUFA by lipoxygenase (LOX12/15), cyclooxygenase (COX2), and cytochrome P450 enzymes produces a range of bioactive lipid mediators like maresins, resolvins, and protectins. These mediators are collectively called specialized pro-resolving mediators (SPMs), which

TABLE 8.2
Evidence from Clinical and Epidemiological Studies on PUFA and Cognition

S. No.	Study Type and Duration	Participants	Intervention	Outcomes	References
1	Double-blind randomized controlled trial	181 infants 1–9 days of age	DHA	Improved cognitive function ↑ Mental development index score	[117]
2	Randomized two arm intervention trial 18 weeks	177 very low birth weight infants	Dietary DHA (50–80 mg/day)	↑ Cognitive and motor scores in BSID-III	[118]
3	Open-label randomized controlled trial 12 weeks	71 adolescent with depression 13–24 of age	Omega-3 PUFA (2,700 mg/day)	↓ Depressive symptoms ↑ cognitive function and memory	[119]
4	Randomized control trial	40 adults with BMI 27–35 kg/m^2	Omega-3 PUFA (1,020 mg/day)	↓ BMI and weight ↑ Cognitive score	[120]
5	Randomized double-blind placebo-controlled trial	217 individuals with mild cognitive impairment	Dietary DHA (2 gm/day)	↑ Cognitive function ↓ Blood Aβ-42 level and expression of Aβ protein mRNA	[121]
6	Randomized double-blind placebo-controlled trial	219 individuals with mild cognitive impaired	Dietary DHA (2 gm/day)	↑ Cognitive function ↓ hippocampal atrophy	[122]
7	Double-blind randomized controlled trial 12 weeks	65 healthy adults 18–29 years of age	DHA-rich fish oil (5:1)	↑ Cerebral blood flow and oxyhemoglobin	[123]

are responsible for maintaining immune homeostasis and exhibiting anti-inflammatory functions in neuron glia cell cultures and brain tissue homogenates [115,116].

Various studies have shown that deficiency in PUFA can lead to cognitive decline, reduced synaptic plasticity, and neuroinflammation. The tissue concentration of omega-6 PUFA relative to omega-3 PUFA is expressed by the omega-6 to omega-3 PUFA ratio, a key factor in overall health. The recommended range for the omega-6/omega-3 PUFA ratio is 1.1–4.1, and keeping it within this range is essential for neurogenesis, synaptic plasticity, reducing inflammation, and preventing neurological disorders. The modern diet often causes the omega-6/omega-3 PUFA ratio to rise to 20:1, mainly due to vegetable oils and excessive processed food consumption. Several studies have shown that this ratio in the cell membranes of inflammatory cells is a key factor in regulating pro- and anti-inflammatory signaling pathways. A high omega-6/omega-3 PUFA ratio can increase inflammation and immune responses, as higher levels of ARA serve as precursors for pro-inflammatory mediators like thromboxane, leukotrienes, and prostaglandins. Conversely, lower omega-3 PUFA levels reduce the production of SPMs, which may further worsen inflammation. Table 8.2 exhibits evidence from clinical and epidemiological studies on PUFA and cognition.

8.5 POLYPHENOLS

Polyphenols are a diverse group of phytochemicals naturally found in fruits, vegetables, tea, cocoa, and various medicinal plants. Polyphenols are known for their powerful neuroprotective properties [124]. These bioactive compounds have antioxidative properties by scavenging free radicals and enhancing the body's own antioxidant defense systems, thereby reducing oxidative stress, which is a key factor in neurodegeneration [125,126]. Polyphenols reduce neuroinflammation by inhibiting pro-inflammatory cytokines and preventing microglial overactivation. They regulate neuronal survival and synaptic plasticity by modulating signaling pathways such as PI3K/Akt, MAPK/ERK, and Nrf2 [127,128]. Finally, polyphenols prevent apoptosis by altering Bcl-2 family proteins and caspase activity, while enhancing mitochondrial integrity [129]. Research indicates that they improve cerebral blood flow and cognition, potentially providing therapeutic benefits in diseases such as Alzheimer's disease (AD), PD, and multiple sclerosis (MS) [130]. Therefore, polyphenols seem to be promising options for functional nutrition and supportive interventions to enhance neuroprotection and preserve brain health [131].

8.5.1 Subclasses of Polyphenols

Polyphenols are a broad group of naturally occurring plant chemicals made up of different phenolic structures, commonly found in plant-based foods and drinks [132]. They exert strong antioxidant, anti-inflammatory, neuroprotective, and cardioprotective effects, partly by influencing oxidative stress, signaling pathways, and gene expression [133]. Structurally, polyphenols can be classified into four main subclasses: flavonoids (such as flavonols, flavones, flavanones, isoflavones, anthocyanidins, and flavanols), phenolic acids (including hydroxybenzoic acids and hydroxycinnamic acids), stilbenes (notably resveratrol), and lignans (which are precursors to enterolignans) [134]. The bioactivity of polyphenols is affected by molecular structure, degree of polymerization, and metabolic transformation by gut microbiota [135]. Polyphenols help prevent and reduce chronic diseases, including neurodegenerative diseases, cancers, diabetes, and cardiovascular diseases, mainly by scavenging ROS, modulating inflammatory mediators, and influencing epigenetic pathways [136].

8.5.2 Flavonoids, Phenolic Acids, Stilbenes, and Lignans

Flavonoids, phenolic acids, stilbenes, and lignans are the most common types of polyphenolic compounds with potential neuroprotective effects [137,138]. Flavonoids are found in citrus fruits, berries, and tea and can affect neuronal signaling, reduce oxidative stress, and influence neuroinflammatory states [139]. Phenolic acids, especially caffeic acid and ferulic acid, have strong free radical scavenging effects, enhance mitochondrial function, and protect neural structure and function [140]. Resveratrol and other stilbenes have anti-inflammatory, antioxidant, and anti-apoptotic effects, often by modulating SIRT1, AMPK, and Nrf2 signaling [141]. Lignans are found in whole grains, seeds, legumes, and some vegetables, and they help preserve brain function by modulating estrogen receptors, boosting antioxidant defenses, and preventing lipid peroxidation [142]. These compounds collaborate to influence neurogenesis, synaptic plasticity, and cerebrovascular health, enabling potential therapeutic effects for preventing or slowing the progression of AD, PD, and MS through multi-targeted molecular approaches [143].

8.5.3 Mechanisms of Neuroprotection

Neuroprotective mechanisms are intricate molecular and cellular processes that try to safeguard neuronal structure and function from harmful insults [144]. Oxidative stress can be reduced by increasing the activity of natural antioxidant enzymes like superoxide dismutase, catalase, and glutathione peroxidase [145]. Neuroprotection involves suppressing neuronal inflammation by inhibiting microglia activation and reducing pro-inflammatory mediators such as TNF-α, IL-1β, and COX-2 [146,147]. Anti-apoptotic mechanisms are mediated through altering the Bcl-2/Bax ratio, inhibiting caspase activation, and stabilizing the mitochondrial membrane potential [148,149]. Furthermore, activating pro-survival signaling pathways like PI3K/Akt and MAPK/ERK supports neuronal repair and synaptic plasticity [150]. Increased cerebral blood flow, blood–brain barrier integrity, and stimulation of neurogenesis also promote functional recovery and protection in neurodegenerative diseases [151].

8.5.4 Antioxidant and Anti-Inflammatory Effects

Embedding antioxidant and anti-inflammatory effects offers crucial mechanisms for safeguarding neuronal homeostasis and preventing neurodegeneration [152]. Antioxidant effects can stem from either directly scavenging ROS or reactive nitrogen species (RNS), or activating endogenous defense pathways (e.g., superoxide dismutase, catalase, glutathione peroxidase, heme oxygenase-1), which help reduce lipid peroxidation, protein peroxidation, and DNA oxidation to preserve cellular integrity [153]. Anti-inflammatory effects are achieved by preventing microglial overactivation, reducing the production of pro-inflammatory cytokines TNF-α, IL-6, and IL-1β, and decreasing the expression of cyclooxygenase-2 (COX-2) and inducible nitric oxide synthase (iNOS) [154]. Inflammation can also be decreased by inhibiting the activity of NF-κB and stimulating the Nrf2 signaling pathways [155]. In sum, antioxidant and anti-inflammatory actions offer important interventions for protecting neurons, enhancing their synaptic capacity, and slowing down the progression of neuronal loss associated with neurodegenerative diseases [156].

8.5.5 Enhancement of Cerebral Blood Flow and Neurogenesis

Boosting blood flow to the brain and promoting neurogenesis are crucial for maintaining cognitive functions and the natural resilience of neurons [157]. Enhanced cerebral perfusion involves an increase in oxygen and nutrients, along with improved removal of metabolic waste, supporting healthy metabolism in neurons. This effect is typically vasodilatory and includes the production of nitric oxide (NO), activation of endothelial or eNOS, and changes in vascular tone [158]. Neurogenesis is the process of neural progenitor cells developing into neurons, proliferating, and integrating into existing neural circuits. Neurogenesis occurs mainly in the hippocampus dentate gyrus [159]. Factors important for neurogenesis include BDNF, vascular endothelial growth factor (VEGF), and glial cell line-derived neurotrophic factor (GDNF) [160]. Increased cerebral blood flow, combined with neurogenesis, supports synaptic plasticity, memory formation, and recovery from neurodegenerative insults. Synergistic interactions with PUFA, vitamins, and minerals enhance neuronal function, support synaptic plasticity, reduce oxidative stress, and contribute to overall cognitive health. [161].

8.5.6 Polyphenols Enhancing Antioxidant Functions of Vitamins

Polyphenols enhance vitamins' antioxidant functions by working together to neutralize ROS and regenerate oxidized vitamin forms. These phytochemicals stabilize free radicals through electron donation, thereby reducing oxidative stress in lipids, proteins, and nucleic acids [162].

Polyphenols can regenerate the reduced forms of vitamins C and E, potentially restoring redox capacity and thereby extending protective properties [163]. They also induce the expression of endogenous antioxidant enzymes, which are indirectly involved in vitamin-mediated neural defense systems. Polyphenols can also enhance vitamin bioavailability and stability by improving intestinal absorption and preventing oxidative destruction [164]. Polyphenols further boost the anti-inflammatory and cytoprotective effects of antioxidant vitamins by modulating signaling pathways, including Nrf2 and NF-κB. This leads to beneficial neuroprotective effects and helps regulate oxidative stress and cellular balance in the central nervous system [165].

8.5.7 Polyphenols Enhance the Bioavailability of Cofactor Minerals

Polyphenols can improve the bioavailability of cofactor minerals by regulating their absorption, transport, and utilization in biological systems. These organic compounds can reduce mineral precipitation and prevent oxidative destruction during transit through the gastrointestinal tract [166]. Specific polyphenols have been shown to influence the intestinal absorption of key cofactor minerals such as magnesium, zinc, iron, and selenium by regulating the expression of metal transporters, mainly divalent metal transporter 1 (DMT1) and other zinc transporters (i.e., ZnT and ZIP). They also protect these minerals from pro-oxidant reactions, enabling them to function normally in enzymatic catalysis [167]. In vivo, polyphenols further improve mineral bioavailability by decreasing inflammation and oxidative stress in the gastrointestinal epithelium [168]. The bioavailability of minerals affects their synergistic properties, boosting the catalytic efficiency of metalloenzymes that support antioxidant defense, neurotransmitter synthesis, and cellular energy metabolism—all essential for proper brain function and protection [169].

8.5.8 Evidence from Animal and Human Studies on Polyphenols and Brain Health

Polyphenols have been shown in animal and human studies to provide significant neuroprotective effects through multiple mechanisms. Preclinical models of neurodegenerative diseases like AD, PD, and MS demonstrate that polyphenol treatment reduces oxidative stress, suppresses neuroinflammation, prevents neuronal death, and enhances synaptic plasticity [143]. These effects result from modulation of signaling pathways including PI3K/Akt, MAPK/ERK, Nrf2, and NF-κB, as well as increased levels of neuroprotective factors such as BDNF and GDNF. In human clinical and epidemiological studies, dietary consumption of polyphenols, fruits, and vegetables has been positively associated with improved cognition, delayed cognitive impairment, and a reduced likelihood of developing a neurodegenerative disease [170]. Clinical studies show that polyphenol-rich therapies enhance cerebral blood flow and improve memory and executive function. In the growing body of research from both animal and human studies, polyphenols are emerging as potential nutraceutical agents to support brain health and reduce the risk of neurodegeneration [171]. Documented synergies between vitamins, polyphenols, PUFAs, and minerals in brain health are given in Table 8.3.

8.6 VITAMINS AND THEIR SYNERGISTIC ROLES

Vitamins play a crucial and complementary role in maintaining neuronal integrity, cognitive function, and overall brain health. Fat-soluble vitamins, including vitamins A, D, E, and K, regulate gene expression, modify neuroinflammation, and protect against oxidative damage by stabilizing membranes and scavenging free radicals [181]. Water-soluble vitamins, including the B-complex and vitamin C, serve as coenzymes in neurotransmitter synthesis, myelination, and energy metabolism. Vitamin D influences cognitive function through neurotrophic factor expression and calcium balance, while vitamin E safeguards neuronal membranes from oxidative damage as a potent lipid-soluble antioxidant and key protector of PUFAs [182]. Vitamins have a cumulative effect and improve bioavailability and regeneration capabilities. For example, vitamin C regenerates oxidized vitamin E, supporting ongoing antioxidant activity in the body [183]. Combined vitamin action boosts polyphenols, cofactor minerals, and polyphenol efficiency, supporting mitochondrial function, synaptic plasticity, and neuronal protection. This multidimensional micronutrient synergy is vital for preventing cognitive decline and fighting neurodegenerative diseases [184].

Brain vitamins are essential micronutrients that support neurons in growth, function, and protection through multiple mechanisms. Vitamin A regulates neurogenesis, synaptic plasticity, and visual signal transduction through retinoid-mediated gene-regulating mechanisms [185]. B-complex vitamins support enzyme functions in neurotransmitter and homocysteine metabolism, as well as in maintaining the myelin sheath, thereby helping to prevent neurotoxicity and cognitive decline [186]. Vitamin C acts as a water-soluble antioxidant to protect neural structures from ROS and promotes iron absorption, which is vital for neurotransmitter metabolism. Vitamin D participates in calcium regulation and modulates neurotrophic factors such as BDNF, while also exhibiting anti-inflammatory effects in the central nervous system. Vitamin E, a lipid-soluble antioxidant, prevents lipid peroxidation of PUFAs in neuronal membranes [187]. Collectively, brain vitamins help preserve neuronal membranes, improve synaptic connections, and offer neuroprotection from neurodegenerative processes, highlighting their importance for brain health and maintaining cognitive function [188].

TABLE 8.3
Documented Synergies Between Vitamins, Polyphenols, PUFAs, and Minerals in Brain Health

Nutrient Combination	Mechanistic Pathways	Observed Neuroprotective/ Cognitive Effects	Key Findings	References
Vitamin E+omega-3 PUFAs (DHA/EPA)	Antioxidant defense via lipid peroxidation inhibition; modulation of membrane fluidity; regulation of Nrf2–ARE pathway	Improved memory, reduced oxidative damage, decreased neuroinflammation	Synergy enhances DHA stability in membranes, protecting synaptic integrity in aging models	[172]
Vitamin C+polyphenols (e.g., EGCG, Quercetin)	Vitamin C regenerates oxidized polyphenols; polyphenols enhance vitamin C uptake; both modulate ROS scavenging and NF-κB inhibition	Reduced neuroinflammation, improved working memory, and increased BDNF	Co-supplementation provides stronger protection against glutamate-induced excitotoxicity	[164]
Vitamin D+magnesium	Magnesium is required for vitamin D activation; both modulate Ca^{2+} homeostasis and neurotrophic signaling	Improved cognitive flexibility, reduced depression risk	Magnesium supplementation enhances serum vitamin D response and neuroprotective effects	[173]
Vitamin B12+folate (B9)	Methylation cycle cofactors; regulate homocysteine metabolism; support myelin synthesis	Slowed cognitive decline, reduced brain atrophy	Synergy lowers plasma homocysteine, preserving white matter integrity	[174]
Polyphenols (resveratrol)+omega-3 PUFAs	Activate SIRT1–PGC1α axis; reduce oxidative stress; modulate synaptic plasticity	Enhanced hippocampal neurogenesis, improved spatial learning	Combined treatment has a greater impact on mitochondrial biogenesis than single compounds	[175]
Zinc+polyphenols	Zinc stabilizes antioxidant enzymes (SOD, CAT); polyphenols prevent zinc-induced oxidative imbalance	Reduced amyloid aggregation, improved synaptic function	Chelating and redox-modulating properties prevent metal-induced neurotoxicity	[176]
Vitamin K+DHA	Modulate sphingolipid metabolism and neuronal membrane composition; anti-inflammatory via reduced microglial activation	Better cognitive scores in the elderly, protection against ischemia-reperfusion injury	Vitamin K supports DHA incorporation into brain phospholipids	[177]
Vitamin A (retinoic acid)+polyphenols	Regulate gene expression via RAR/RXR and Nrf2; modulate synaptic protein synthesis	Improved learning, reduced neuroinflammation in AD models	Retinoids enhance polyphenol-induced antioxidant gene expression	[178]
Selenium+vitamin E	Selenium-dependent GPx works synergistically with vitamin E to neutralize lipid peroxides	Reduced oxidative neuronal injury, improved motor coordination	Combined deficiency exacerbates neurodegeneration; supplementation restores antioxidant balance	[179]
Vitamins B+polyphenols	Support neurotransmitter synthesis; polyphenols modulate MAO activity and synaptic signaling	Improved mood, reduced cognitive fatigue	Synergy enhances dopamine and serotonin levels in hippocampus	[180]

8.6.1 Vitamin B-Complex

Brain function relies heavily on vitamin B because they play a crucial role in energy metabolism and the formation of neurotransmitters through several coenzymatic processes [189,190]. Thiamine (B1) is a cofactor for pyruvate dehydrogenase and α-ketoglutarate dehydrogenase, which facilitate ATP production in neuronal mitochondria; riboflavin (B2) and niacin (B3) support oxidative phosphorylation through flavoproteins and the NAD+/NADH systems; pantothenic acid (B5) is necessary for generating coenzyme A, which ultimately produces acetyl-CoA for energy metabolism and the synthesis of the neurotransmitter acetylcholine; finally, pyridoxine (B6) is a cofactor for aromatic L-amino acid decarboxylase and glutamate decarboxylase, enzymes responsible for regulating the synthesis of serotonin, dopamine, and γ-aminobutyric acid (GABA) [191]. Folate (vitamin B9) and cobalamin (vitamin B12) are important for their roles in one-carbon metabolism, methylation, and myelin production. Together, the activities of the B-complex support optimal neural energy production, maintain proper neurotransmitter balance, and protect against neurodegenerative and cognitive diseases [192].

8.6.2 Vitamin D

Vitamin D has strong immunomodulatory and neuroprotective effects through both genomic and non-genomic pathways. The active form of vitamin D (calcitriol, 1,25-dihydroxyvitamin D_3) triggers gene activation involved in neuronal survival, myelination, and synaptic plasticity in the central nervous system by binding to the vitamin D receptor (VDR) [193]. From an immunological perspective, vitamin D primarily influences both adaptive and innate immune responses by enhancing T regulatory (Treg) function while reducing pro-inflammatory Th1 and Th17 cell activity [194]. Examples include reductions in pro-inflammatory cytokines like IL-6, IL-17, and TNF-α, alongside increases in the anti-inflammatory mediator IL-10. The neuroprotective mechanisms of vitamin D involve decreasing glutamate-mediated neuronal excitotoxicity, regulating intracellular calcium levels, and boosting neurotrophic factors such as nerve growth factor (NGF) and BDNF. Vitamin D also reduces oxidative stress by activating Nrf2 and improves blood–brain barrier integrity, all of which collectively help lower neuroinflammation and defend against neurodegenerative diseases [195].

8.6.3 Vitamin E

Vitamin E is a lipid-soluble antioxidant and is most known for preventing lipid peroxidation of PUFAs in neuronal membranes and structures, which likely plays a significant role in supporting better neural health [196]. In the brain, neuronal membranes and lipoproteins contain α-tocopherol, the most complete and physiologically active form of vitamin E. In addition to reacting with and scavenging lipid peroxyl radicals, vitamin E can prevent oxidative damage caused by these radicals, which is vital for the aging nervous system [187]. By reducing oxidative damage, vitamin E helps preserve membrane fluidity, receptor function, and the efficiency of synaptic transmission [197]. Vitamin E can also prevent oxidative stress-mediated activation of NF-κB and inhibit the production of pro-inflammatory mediators such as COX-2 and iNOS. Furthermore, vitamin E can help maintain mitochondrial function by limiting oxidative damage to complexes in the respiratory chain, allowing mitochondrial function to continue producing ATP in the later stages of neurodegenerative disorders [198]. Studies have shown that vitamin E and vitamin C work together to restore oxidized α-tocopherol to its active form; therefore, greater protective and enhanced benefits can be achieved with lower doses. Vitamin E's role in neuroprotective functions is very important if we want to maintain cognitive abilities and prevent oxidative damage in neurodegenerative disorders [199].

8.6.4 Vitamin C

Vitamin C (ascorbic acid) is a water-soluble vitamin essential for collagen formation, antioxidant defense, and neurotransmitter synthesis in the brain. Vitamin C acts as a cofactor for prolyl and lysyl hydroxylases, which hydroxylate collagen polypeptides to maintain the structural stability of blood vessels and the extracellular matrix, supporting cerebral microvascular integrity [200]. Vitamin C functions in antioxidant defense by scavenging ROS and recycling other antioxidants like α-tocopherol (vitamin E). It also influences redox-sensitive transcription factors, such as Nrf2, increasing endogenous antioxidant enzyme level expression. Vitamin C is a cofactor for dopamine β-hydroxylase, which converts dopamine to norepinephrine [164]. It is equally important in the synthesis of serotonin from tryptophan. When in an active ferrous state, vitamin C enhances iron absorption to support the enzymatic activities necessary for neurotransmission. Overall, vitamin C aids neuronal function, vascular health, and the regulatory mechanisms of redox balance, contributing to neuroprotection and cognitive enhancement [201].

8.6.5 Vitamin K

Vitamin K is a lipid-soluble micronutrient that plays crucial roles in cardiovascular health and neuroprotection by promoting γ-carboxylation of vitamin K-dependent proteins. In the cerebrovascular system, vitamin K activates matrix Gla protein (MGP), which inhibits vascular calcification to help maintain cerebral blood flow and lower the risk of ischemic stroke [202]. Vitamin K is also essential for producing coagulation factors that promote normal hemostasis and decrease neurotoxicity caused by microbleeds. Other vitamin K-dependent proteins in the brain, such as Gas6 and protein S, support neuronal cell survival, inhibit apoptosis, and maintain myelin stability by activating tyrosine kinase receptors [203]. Moreover, vitamin K is crucial for producing sphingolipids, an essential lipid that supports neuronal membrane integrity and synaptic function [204]. Vitamin K also has anti-inflammatory effects and prevents neurodegeneration by reducing inflammation through the inhibition of microglial overactivation and lowering pro-inflammatory cytokines [205]. Current evidence suggests that low vitamin K influences cognitive decline, increases oxidative stress, and raises the risk of neurodegenerative disorders, highlighting the dual role of vitamin K in preserving both vascular integrity and neuronal function [143].

8.6.6 Synergistic Effects of Vitamins with Other Nutrients

Vitamins work together with other nutrients, including PUFAs, polyphenols, and essential minerals, all of which contribute to enhancing brain function and protection [206]. Fat-soluble vitamins A, D, and E are present in neuronal membranes alongside omega-3 PUFAs (DHA), which enhances membrane fluidity (boosting receptor function and synaptic communication) [207]. Vitamin E prevents PUFA from peroxidation, helping to preserve its structural and functional properties, while vitamin C restores

TABLE 8.4
Challenges in Studying and Understanding Nutrient Synergy in Brain Health

Challenge	Description	Scientific Implications	Potential Solutions	References
Complexity of interactions	Nutrients act via overlapping and intersecting biochemical pathways, making it difficult to attribute observed effects to specific combinations.	May lead to over- or underestimation of the role of individual nutrients.	Use factorial RCTs, systems biology modeling, and network-based nutrient interaction analysis.	[210]
Bioavailability variability	Absorption and utilization are influenced by nutrient chemical form, food matrix, co-ingestion, and individual physiology.	Inconsistent efficacy across trials and populations.	Standardize nutrient sources, control dietary backgrounds, and apply bioavailability-enhancing delivery systems.	[211]
Dose–response uncertainty	Optimal synergistic ratios and upper safe limits are poorly defined for many combinations.	Potential for antagonistic interactions or toxicity at high doses.	Conduct dose-ranging synergy trials and meta-analyses to determine optimal ranges.	[212]
Interindividual variability	Genetic polymorphisms (e.g., VDR, MTHFR), gut microbiota composition, age, sex, and metabolic status alter responses to nutrient combinations.	Hinders universal recommendations and public health policy.	Apply nutrigenomics, nutrigenetics, and microbiome profiling for personalization.	[213]
Measurement limitations	Lack of precise biomarkers to quantify nutrient synergy in vivo.	Difficulty in mechanistic confirmation and clinical translation.	Develop multi-omics biomarkers, neuroimaging, and stable isotope tracing.	[214]
Long-term study challenges	Synergistic benefits may require months/years to manifest.	Short studies may miss cumulative effects and neuroprotective outcomes.	Implement longitudinal, multi-year interventions and population-based cohort studies.	[215]
Dietary confounding factors	Whole diets contain numerous compounds that can mask or interact with studied synergies.	Reduces the clarity of causative nutrient effects.	Use controlled feeding designs with precise nutrient dosing.	[216]
Stability and degradation	Some nutrients degrade during storage, processing, or digestion, altering synergy potential.	Lab results may not match real-world intake.	Use microencapsulation, nanocarriers, or protective formulations.	[217]
Ethical and practical constraints	Long-term nutrient restriction or high-dose administration may be unsafe.	Limits the experimental scope in human trials.	Apply safe supplementation protocols, rely on natural dietary enrichment.	[218]
Translational gap	Strong preclinical synergy evidence often fails in clinical validation due to physiological complexity.	Weakens confidence in dietary guidelines.	Align preclinical models with human physiology; bridge with phase 0/II trials.	[219]
Intervention compliance	Adherence to complex multi-nutrient regimens is difficult.	Reduces reproducibility and efficacy.	Use digital adherence tracking, palatable formulations, and simplified dosing.	[220]
Population-specific effects	Cultural diet patterns, baseline status, and environmental exposures influence results.	Limited generalizability of findings.	Conduct multi-center, multi-ethnic trials.	[221]

and regenerates damaged α-tocopherol so it can continue functioning as an antioxidant [208]. Polyphenols, which are polycyclic biosynthetic compounds, enhance vitamin action by restoring or regenerating reduced forms, improving bioavailability, and activating other complementary antioxidant systems (e.g., Nrf2) [209]. Minerals (e.g., magnesium, zinc, iron, selenium) act as cofactors for various vitamin-dependent enzymes, facilitating neurotransmitter synthesis or activation of antioxidant enzymes, and supporting the formation of myelin [164]. For example, vitamin D enhances the utilization of calcium and magnesium, as it influences neuronal excitability and synaptic plasticity. Together, these dietary factors boost cellular resilience, lower oxidative stress, and help reduce neuroinflammatory responses, all of which support cognition and reduce the risk of neurodegeneration [210].

Table 8.4 shows challenges in studying and understanding nutrient synergy in brain health.

8.6.7 Evidence from Clinical Studies on Vitamin Supplementation and Brain Health

Clinical trials provide strong evidence that vitamin supplementation promotes brain health and slows cognitive decline [222]. Randomized controlled trials (RCTs) demonstrate that B-complex vitamin supplementation, with folate, B6, and B12, lowers plasma homocysteine levels, reducing the risk of neurotoxicity and enhancing high cognitive performance, while also delaying brain shrinkage in elderly patients [223]. Additionally, vitamin D supplementation is associated with improved executive function, mood regulation, and reduced neuroinflammation in patients deficient in vitamin D [224]. Vitamin E (α-tocopherol) has been studied for preventing functional decline in AD based on its beneficial effects as a potent lipid-soluble antioxidant [225]. Vitamin C supplementation has also enhanced antioxidant capacity, supporting health and well-being by improving vascular health and neurotransmitter synthesis [225]. Moreover, multiple micronutrient therapies, which include vitamins, PUFAs, and minerals, could have a synergistic effect on memory retention, attention, and synaptic plasticity [226]. Overall results vary depending on dosage, duration, and baseline nutrition, but the clinical findings suggest that focused vitamin supplementation may be a valuable adjunct intervention for improving cognitive function and enhancing neuroprotection [227].

8.7 NUTRITIONAL SYNERGY: MECHANISTIC INSIGHTS

Nutritional synergy refers to the interaction and complementary effects of nutrients that work together to enhance neuroprotection and cognitive function beyond their individual roles in diet or supplements. Synergistic mechanisms of nutrients include supporting antioxidant defense, modulating inflammatory pathways, and maintaining optimal metabolic functions [228]. For example, polyphenols work together with vitamins C and E in redox cycling to reduce oxidized α-tocopherol, which helps maintain lipid membranes. Additionally, omega-3 PUFAs incorporate into neuronal membranes, and the antioxidant effect of vitamin E stabilizes them [229]. In addition to polyphenols, vitamins C and E, and minerals such as magnesium and zinc are also used as cofactors by vitamin-dependent enzymes to produce neurotransmitters, maintain synaptic plasticity, and generate myelin [164]. Synergistic effects can also be achieved through a combination of signaling pathways, such as Nrf2 for antioxidant gene expression and PI3K/Akt for neuronal survival [230]. Overall, the integrated nutritional network improves mitochondrial efficiency, maintains membrane integrity, and reduces oxidative and inflammatory damage to support long-term brain health and guard against dementia [231].

Nutrient synergy in brain health involves the cooperative interaction of vitamins, minerals, PUFAs, polyphenols, and other bioactive compounds that enhance neuronal resilience and neurophysiology [232]. These nutrients interact with each other in nearly all biochemical pathways to enhance antioxidant capacity, manage neuroinflammation, and activate energy metabolism. For an instance of synergy, vitamins C and E work in a regeneration cycle, with vitamin C able to regenerate oxidized α-tocopherol and maintain its ability to protect the membranes [233]. Omega-3 PUFAs, especially DHA, are integrated into neuronal membranes to enhance flexibility and activate receptors, while vitamin E guards against peroxidation [234]. In the realm of minerals, magnesium, zinc, and selenium all function as cofactors for vitamin-dependent enzymes involved in neurotransmitter synthesis, myelin production, and redox regulation. Polyphenols boost the benefits of vitamins E and C by enhancing their absorption and activating protective mechanisms, including Nrf2 and PI3K/Akt [234]. Through effective molecular processes and nutrient synergy, cognitive abilities are enhanced, oxidative and inflammatory disturbances are minimized, and the risk of neurodegenerative diseases is lowered [235].

8.7.1 Mechanisms of Nutrient Interactions

Nutrient interactions in brain health happen through biochemical, molecular, and physiological systems that enable certain nutrients to act together or enhance each other's effects [236]. For instance, antioxidant recycling is a process where vitamin C regenerates oxidized vitamin E to protect lipid membranes, while also boosting the redox potential of the vitamins with the help of polyphenols [237]. Omega-3 PUFAs become incorporated into neuronal membranes, and vitamin E can prevent PUFAs from peroxidizing in membranes, thus maintaining membrane fluidity and neuronal signaling [238]. Minerals such as magnesium, zinc, and selenium serve as cofactors for vitamin-dependent enzymes, which can also boost neurotransmitter production, antioxidant enzyme activity, and the myelination of neurons involved in membrane formation [164]. Nutrient combinations can influence signaling pathways, such as Nrf2-mediated antioxidant gene expression, to boost antioxidant levels, promote antioxidant-antioxidant interactions like NF-κB inhibition for anti-inflammatory effects, and activate PI3K/Akt for neuronal survival [239]. Additionally, some nutrients may improve the bioavailability or stability of other nutrients, increasing their physiological effectiveness. These interconnected systems support neuroprotection, enhance cognitive function, and may reduce the risk of neurodegenerative diseases [240]. The mechanism of nutrient interactions is shown in Figure 8.2.

8.7.2 Cofactor Minerals Enhancing Enzymatic Functions of Vitamins

Cofactor minerals used by vitamins enhance the enzyme activity needed to maximize their activation, stability, and catalytic function in neurophysiological processes [241]. Magnesium functions as a cofactor in enzymes dependent on action of vitamin D to support optimal calcium balance, neuronal excitability, and synaptic plasticity [242]. Zinc is essential for regulating vitamin A transcription through its action-dependent retinoic acid receptors, which promote neurogenesis and visual signaling [243]. Iron supports vitamin C-driven hydroxylation reactions vital for collagen synthesis and dopamine β-hydroxylase activity in neurotransmitter production [244]. Selenium and vitamin

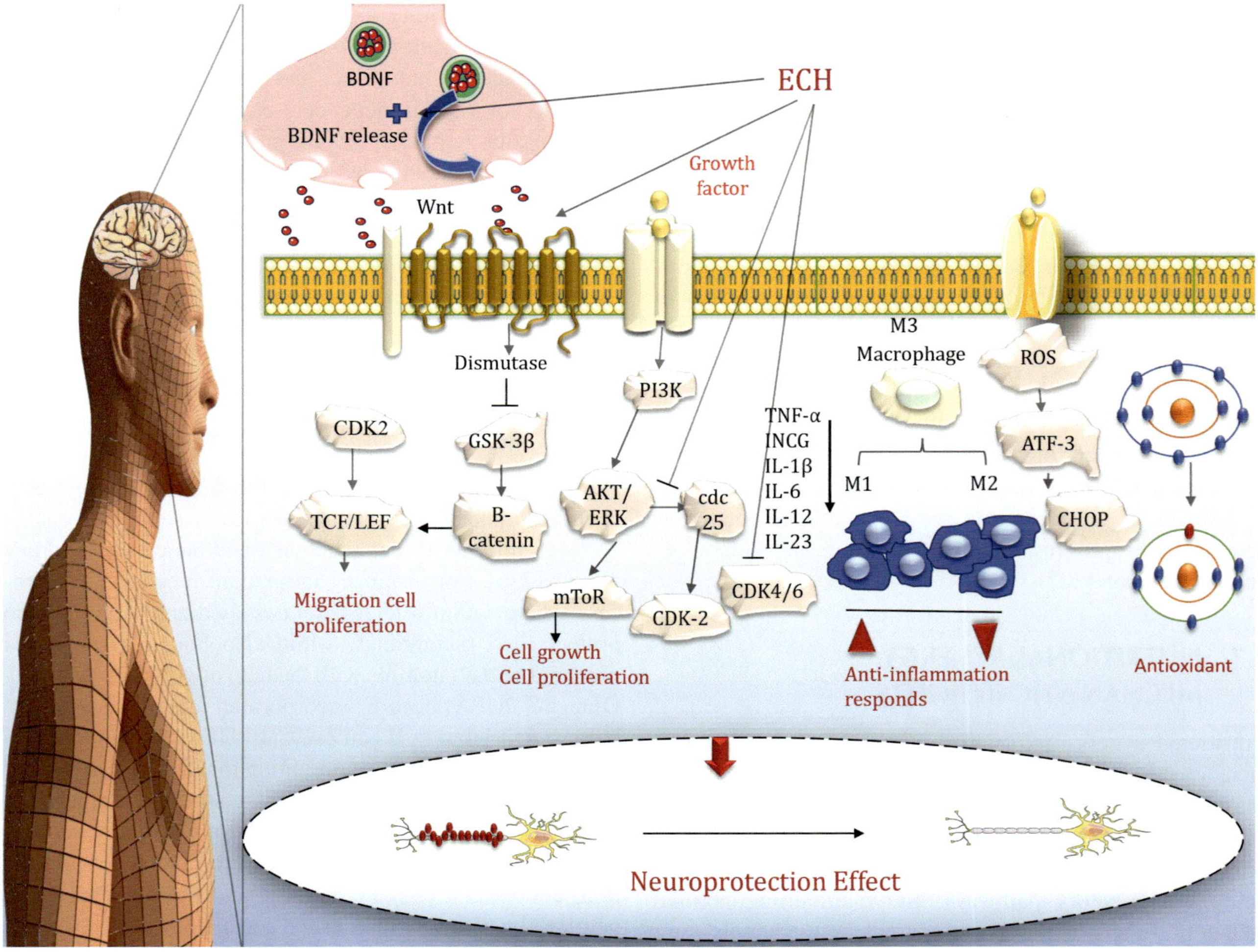

FIGURE 8.2 Mechanism of nutrient interactions.

E work together through glutathione peroxidase activity to prevent neuronal membrane lipid peroxidation [244]. Copper controls monoamine oxidase activity, which is responsible for producing dopamine and norepinephrine, and depends on vitamin C [245]. Each of these minerals functions as a structural or catalytic cofactor to enhance vitamin-dependent enzymatic activities, strengthen antioxidant defenses, improve neurotransmission, and support neuroprotection, cognitive function, and the prevention of neurodegenerative diseases [246].

8.7.3 PUFA and Polyphenols in Anti-Inflammatory and Neuroprotective Pathways

PUFAs and polyphenols act together on anti-inflammatory and neuroprotective pathways through different yet shared molecular mechanisms [247]. Omega-3 PUFAs, especially DHA and EPA, are incorporated into neuronal membranes, enhancing membrane fluidity and altering lipid raft-dependent signaling [248]. They are further converted into SPMs, such as resolvins and protectins, which inhibit microglial activation, NF-κB signaling, and pro-inflammatory cytokines [249]. Polyphenols (flavonoids and stilbenes) inhibit oxidative stress and inflammation by scavenging ROS, activating Nrf2-dependent antioxidant genes, and blocking MAPK and NF-κB pathways. Additionally, polyphenols enhance nitric oxide synthase (eNOS) activity, which increases cerebral blood flow [249]. Together, both PUFAs and polyphenols are associated with positively influencing mitochondrial integrity, maintaining synaptic integrity, and limiting excitotoxicity. This well-studied intersection of nutrition, PUFAs, and polyphenolic compounds contributes to neuronal survival, regulation of neuroimmune mechanisms, and slows the progression of neurodegenerative diseases, particularly AD and PD [250].

8.7.4 Vitamins and Polyphenols in Antioxidant Defense and Mitochondrial Function

Vitamins and polyphenols interact and function within interconnected metabolic pathways to support antioxidant defenses and mitochondrial performance, thereby helping preserve neural integrity [251]. Vitamins C and E mainly function as antioxidants, with vitamin C helping to regenerate destroyed α-tocopherol to protect lipid membrane integrity, and vitamin E preventing the peroxidation of PUFAs within mitochondria and neural membranes [187]. B-complex vitamins, notably riboflavin and niacin, serve as cofactors for flavoproteins and NAD^+/NADH-sensitive enzymes in the electron transport chain to support energy production as ATP [252]. Polyphenols like flavonoids and resveratrol promote mitochondrial biogenesis by activating PGC-1α and SIRT1 pathways while reducing ROS production [253]. Both nutrient groups modulate Nrf2 signaling pathways to promote the production of antioxidant enzymes and suppress redox-sensitive pro-inflammatory pathways such as NF-κB [253]. The combined benefit of vitamins and polyphenols boosts mitochondrial membrane potential, improves the efficiency of oxidative phosphorylation, and protects against apoptosis, offering better support for cognitive function and increasing resilience to neurodegenerative diseases [133].

8.7.5 Combined Effects on Synaptic Plasticity, Cognition, and Neurogenesis

Vitamins and polyphenols work together on synaptic plasticity, cognition, and neurogenesis by influencing molecular and cellular pathways. B-complex vitamins (especially B6, B9, and B12) support neurotransmitter production and methylation processes that are crucial for synaptic functions, while vitamins D and E improve neuronal membrane integrity and receptor signaling [252]. Polyphenols (flavonoids) enhance the expression of BDNF and GDNF in the brain, which promote dendritic development, synaptogenesis, and long-term potentiation—key components of learning and memory [254]. Both food groups are known to also influence PI3K/Akt and MAPK/ERK signaling pathways, which are essential for neuronal survival and plasticity. They promote the proliferation and differentiation of neural progenitor cells in the hippocampus, which is important for neurogenesis [255]. Antioxidant and anti-inflammatory actions will also protect normal synaptic structures from oxidative damage and neuroimmune inflammation. Collectively, all these actions can improve cognitive function, slow the progression of age-related declines, and support recovery in neurodegenerative disorders [256].

8.7.6 Challenges in Studying and Understanding Nutrient Synergy

One of the biggest challenges in studying nutritional synergy is the complexity of biological systems and multiple interactions between dietary components [257]. Mixtures of nutrients can have additive, synergistic, or antagonistic effects, which makes pathway investigation challenging [258]. Bioavailability variations are affected by molecular structure, gut flora, and genetic polymorphisms, thereby further complicating understanding [259]. Another factor undermining the repeatability and transnationality of outcomes is the methodological differences between clinical trials, such as dosage shifts, varied length contrasts, and demographic peculiarities [259]. Another layer of complexity is created by confounding factors such as diet–lifestyle interactions, metabolic status, and environmental influences. Multi-omics, systems biology, and in silico technologies are necessary to analyze the complex networks underlying nutritional synergy [260].

8.8 EVIDENCE FROM EPIDEMIOLOGICAL, PRECLINICAL, AND CLINICAL STUDIES

Extensive epidemiological studies have consistently shown the important role of nutritional interactions in supporting brain health and lowering the risk of neurodegenerative diseases [261]. It has been consistently shown that adherence to Mediterranean-style or Mediterranean-DASH Intervention for Neurodegenerative Delay (MIND) diets is associated with a 20%–35% lower risk of mild cognitive impairment (MCI), dementia, and AD [262,263]. Key dietary components, such as cofactor minerals (e.g., zinc and magnesium), PUFAs, polyphenols, and essential vitamins (B-complex, C, D, and E), work together to support neuronal function, antioxidant defenses, synaptic plasticity, and the regulation of neuroinflammation [264,265]. These nutrients support various biochemical and cellular processes that are essential for cognitive performance and neuroprotection throughout the lifespan [266,267]. Studies have demonstrated that diets rich in these nutrients are closely linked to lower risks of cognitive decline, AD, and PD [268,269]. Zinc is a vital cofactor involved in neurotransmitter synthesis and synaptic signaling, and its deficiency has been linked to impaired cognition and increased risk of neurodegeneration [270,271]. Omega-3 fatty acids, especially EPA and DHA, offer anti-inflammatory and neuroprotective benefits, while polyphenols, strong antioxidants from plants, boost these effects by working together to reduce oxidative stress and neuroinflammation [272,273].

Vitamins B_6, B_{12}, and folate are essential for regulating homocysteine metabolism and maintaining myelin integrity. Elevated homocysteine levels have been linked to cognitive decline and neurodegenerative disorders [274,275]. Vitamin B supplementation decreases homocysteine and slows brain atrophy in MCI, with clinical trials confirming its role in delaying cognitive decline in the elderly [276,277]. In preclinical studies, it has been shown that combined supplementation of these nutrients mitigates neurodegenerative processes, reduces pathological protein aggregation, and promotes synaptic plasticity, highlighting the importance of nutrient synergy [278,279]. The evidence indicates that effective neuroprotection results from complex nutrient interactions rather than from isolated nutrient effects [280]. Besides individual nutrients, epidemiological research highlights the protective role of overall dietary patterns [281]. Large cohort studies and meta-analyses have demonstrated that following the Mediterranean, DASH (Dietary Approaches to Stop Hypertension), and

MIND diets—which are rich in fruits, vegetables, whole grains, nuts, fish, and unsaturated fats—is associated with improved cognitive outcomes and a reduced risk of dementia and age-related neurodegeneration [282,283]. These diets offer a wide variety of bioactive compounds, such as polyphenols, omega-3 fatty acids, B vitamins, and antioxidants, that collectively lower inflammation, fight oxidative damage, and enhance neuroplasticity [284]. Notably, combined intake of omega-3s and B vitamins has been shown to more effectively slow brain atrophy and cognitive decline than either nutrient alone [285,286]. Emerging evidence also highlights the gut–brain axis as a key factor in mediating nutrition-related effects on brain health [287]

Importantly, recent studies highlight that the negative effects of ultra-processed foods on brain health seem to be independent of overall adherence to healthy dietary patterns [287]. For example, high consumption of refined sugars, processed meats, and industrial fats has been linked to faster cognitive decline and higher dementia risk, even in otherwise balanced diets [288]. This highlights the importance of dietary guidelines considering not only nutrient content but also the degree of food processing [289].

Meanwhile, the field of personalized nutrition is gaining momentum, propelled by progress in genomics, metabolomics, and microbiome research [290]. Integrating individual genetic profiles, metabolic markers, and gut microbial composition enables the creation of targeted dietary strategies to enhance brain function and prevent cognitive decline [291,292]. Personalized nutrition provides a promising alternative to the traditional "one-size-fits-all" model, especially for populations with specific metabolic or neurological vulnerabilities [293]. However, the widespread adoption of personalized nutrition is currently hindered by methodological challenges, especially in accurately assessing dietary intake [294]. Most nutrition studies depend on self-report tools like food frequency questionnaires, which are prone to recall bias and misreporting [295]. To address these limitations, researchers are increasingly promoting the use of objective dietary biomarkers, wearable tracking devices, and digital food logs to enhance the accuracy and reliability of dietary data collection [296,297].

Recent multidisciplinary evidence highlights the significance of nutrient synergy, where multiple nutrients collaborate to support brain health and prevent cognitive decline, rather than focusing only on the effects of isolated nutrients [298,299]. This perspective is supported by emerging research on the interaction between dietary quality, gut microbiome composition, and personalized nutritional interventions, which together provide a holistic framework for understanding how nutrition impacts brain function throughout the lifespan [300,301]. These insights emphasize the importance of adopting integrated, life-course dietary strategies that consider the biological complexity of the brain and the changing nutritional and environmental landscape in which it develops [302,303].

8.8.1 Preclinical Studies on Combined Nutrient Supplementation in Animal Models

Recent preclinical research using animal models has offered valuable insights into the potential of multi-nutrient strategies to support brain health [304]. Specifically, combined supplementation with cofactor minerals, PUFAs, polyphenols, and vitamins has been shown to have synergistic effects on neural function and cognitive outcomes [305–307]. These studies provide mechanistic evidence that nutrient interactions improve processes like neurodevelopment, neuroprotection, and cognitive performance, often leading to better results than isolated nutrient interventions [308].

One of the most researched combinations is that of omega-3 PUFAs and B vitamins. In rodent models, the co-administration of DHA and EPA with vitamins B6, B9 (folate), and B12 has been shown to improve memory performance and significantly reduce neuroinflammation [308,309]. For instance, the Fortasyn Connect formulation, which includes DHA, EPA, uridine monophosphate, choline, phospholipids, and antioxidants, enhances synaptic function and cognitive performance in aged and AD mouse models [310,311]. This combination aids phospholipid synthesis and membrane repair, which are essential for preserving synaptic integrity [312]. Similarly, combining PUFAs with vitamins and minerals has shown enhanced neuroprotective effects. In a study by Agrawal et al. 2023, rats exposed to early-life stress demonstrated improved spatial learning [313,314]. They decreased anxiety-like behavior when given a nutrient mix with omega-3 fatty acids, vitamin A, and zinc [315]. These effects were attributed to the anti-inflammatory and antioxidant properties of the nutrients, which worked together to reduce stress-induced neuronal damage [316,317].

Emerging evidence highlights the importance of minerals such as magnesium, zinc, and selenium in supporting neural function and boosting the effects of vitamins and PUFAs [318]. These micronutrients play a vital role in enzymatic activity, synaptic signaling, and antioxidant defense. Preclinical studies show that combined supplementation with folic acid, vitamins B6 and B12, choline, methionine, and zinc can reverse stress-induced cognitive deficits [319,320]. This synergistic action supports methylation, neurotransmitter production, and redox balance, ultimately strengthening cognitive resilience [321]. Polyphenols, known for their potent antioxidant and anti-inflammatory effects, have been shown to work together with PUFAs to support brain health [322]. For example, a study combining DHA with grape-derived polyphenols in mice showed improved memory performance and decreased amyloid-beta accumulation, a hallmark of AD pathology [323]. The combined supplementation modulated neuroinflammatory markers and improved synaptic plasticity more effectively than either nutrient alone [324].

The timing of supplementation also seems to be crucial; nutrient interventions during sensitive developmental periods such as gestation, early postnatal life, and adolescence have demonstrated the most significant benefits [325,326]. In a study by Mahmoudian Dehkordi et al. 2023, perinatal supplementation with choline, folate, and DHA enhanced hippocampal neurogenesis and cognitive outcomes in offspring exposed to maternal stress [327]. These nutrients aided brain development by promoting cell membrane formation and controlling gene expression via epigenetic mechanisms [328]. Furthermore, long-term combined supplementation has been demonstrated to protect against age-related cognitive decline [329]. In aged rats, a multi-nutrient intervention including PUFAs, vitamins B & E, and selenium enhanced memory performance and lowered oxidative stress markers in the brain [330]. The study indicated that the nutrients operated via complementary mechanisms: PUFAs promoted synaptic function, whereas vitamins and minerals boosted antioxidant defenses and mitochondrial activity [331,332].

While the therapeutic potential of nutritional supplementation is widely recognized, it is equally important to consider the risks associated with excessive or unbalanced intake [333]. High doses of specific antioxidants, like vitamin E and beta-carotene, have been associated with higher mortality rates in both animal studies and clinical research [334,335]. These findings emphasize the need for caution in supplementation strategies and highlight the importance of using physiologically relevant doses [336]. Preclinical studies further support the idea that balanced nutrient combinations, rather than isolated high-dose compounds, are crucial to maximize therapeutic efficacy while reducing the risk of toxicity [337,338].

8.8.2 Clinical Trials Exploring Multi-Nutrient Interventions for Neurodegenerative Diseases and Cognitive Decline

The increasing global burden of neurodegenerative disorders, especially AD and MCI, has led to extensive research on preventive and therapeutic strategies aimed at early pathological changes [339,340]. Among these, nutritional interventions, especially multi-nutrient combinations, have gained increasing scientific interest for their potential to influence key neurobiological processes involved in disease progression, including oxidative stress, mitochondrial dysfunction, neuroinflammation, and synaptic degeneration [341–342].

Multi-nutrient formulations are designed to address the complex causes of neurodegeneration by combining essential nutrients, including B-complex vitamins, vitamin D, E, omega-3 fatty acids (DHA and EPA), choline, polyphenols, magnesium, and selenium [343–345]. Unlike single-nutrient approaches, these combinations are designed to create synergistic effects that enhance neuroprotection, regulate neurotransmission, and bolster cellular resilience [346]. Notably, clinical studies evaluating formulations like Fortasyn Connect (used in Souvenaid) have shown modest improvements in memory performance, brain connectivity, and hippocampal atrophy in early AD when used as an adjunctive therapy [347–349].

Recent clinical trials and meta-analyses have examined these interventions across various populations. Combinations that include DHA/EPA with antioxidant vitamins, or B vitamins with magnesium and zinc, have demonstrated potential in maintaining cognitive function, especially when given during early or prodromal stages [350–352]. Preclinical studies further support these findings; for example, synergistic effects were observed with Antarctic krill oil, nobiletin, and L-theanine in aged animal models, leading to reduced amyloid burden and enhanced synaptic integrity [353]. However, outcomes remain inconsistent across trials, with some studies failing to show significant benefits on primary cognitive endpoints [354]. This heterogeneity results from differences in trial design, nutrient composition, dosage, duration, baseline nutritional status, genetic variability (e.g., APOE ε4 carriers), and participant adherence [355,356]. Moreover, differences between individuals in nutrient absorption, microbiota composition, and metabolic profiles further affect efficacy, highlighting the importance of stratified approaches [357,358].

Current research is thus shifting toward precision nutrition, incorporating biomarkers of nutrient status, genomics, and metabolomics to customize interventions [359]. Stratified trial designs and the use of digital cognitive assessments and neuroimaging markers improve the ability to identify subtle yet significant effects of nutritional therapies [360,361]. Despite ongoing challenges, multi-nutrient interventions continue to be a promising complement to traditional therapies for neurodegenerative diseases [362]. Future directions involve standardizing formulations, determining optimal dosing schedules, and conducting long-term, well-powered RCTs [363]. Ultimately, these strategies seek to translate nutritional science into personalized, evidence-based protocols for delaying or reducing neurodegenerative decline [364].

8.8.3 Limitations and Gaps in Current Research

Current research on nutrition and cognition is limited by short trial durations, small homogeneous samples, and methodological inconsistencies, which weaken statistical power and restrict generalizability [365]. Meta-analyses show that only about 40%–45% of omega-3 RCTs exhibit positive cognitive effects, with benefits mainly limited to individuals with MCI rather than cognitively normal adults, despite strong observational links with fish intake and PUFA levels [366,367]. Large, long-term RCTs, including multi-supplement trials lasting up to seven years, show no significant protection against dementia or cognitive decline, often downgraded to low or moderate certainty due to attrition and heterogeneity in design [368,369]. Reliability is further undermined by dependence on self-reported dietary intake, variable cognitive assessments, and the absence of biomarker-based exposure measures, all of which hinder cross-trial synthesis [370,371]. Mechanistic investigation remains limited, with few trials assessing relevant biomarkers such as neurotrophins, pharmacokinetics, or microbiome-derived metabolites [372]. Despite preclinical evidence for nutrient synergy, such as DHA with alpha-lipoic acid, which slows decline only when combined, synergy modeling remains rare in clinical studies [373]. Similarly, the use of omics methods, such as genetic stratification, metabolomics, and microbiome profiling, is rare, which limits personalized insights and conceals variability in response [374]. Polyphenol trials encounter additional challenges because of their very low and inconsistent bioavailability, with limited focus on co-nutrient or matrix effects that could improve absorption [375,376].

8.9 NUTRITIONAL INTERACTIONS IN NEURODEGENERATIVE DISEASES

8.9.1 Role of Nutrient Synergy in AD, PD, and Other Neurodegenerative Conditions

Recent findings highlight the increasing potential of nutrient synergy in fighting neurodegenerative diseases like AD and PD [377]. Combinations of PUFAs, antioxidants, vitamins B, and trace elements may work together to combat oxidative stress, inflammation, and mitochondrial dysfunction, providing increased neuroprotection and helping to slow cognitive decline [344,378,299]. Evidence shows that certain nutrient combinations are more effective together in supporting brain health, reducing neuroinflammation, and preventing cognitive decline [309,379,380]. In AD, malnutrition is strongly linked to cognitive decline and increased disease severity [381]. Diets high in synergistic nutrients, such as the Mediterranean and DASH diets, which focus on fruits, vegetables, whole grains, and healthy fats, reduce oxidative stress and inflammation markers like IL-6, TNF-α, and CRP, and decrease the buildup of Alzheimer-related proteins like beta-amyloid peptides [382–384]. These diets enhance cerebral blood flow and provide neuroprotective effects, especially in the initial stages of AD [385]. Conversely, Western diets rich in saturated fats and cholesterol worsen neurodegeneration by raising oxidative stress and disrupting metabolism and brain blood flow [386,387].

More specifically, combinations of vitamins (B6, B12, and folate) have shown synergistic effects by reducing homocysteine, a neurotoxic amino acid associated with neurodegeneration [388,389]. Clinical trials, such as the VITATOPS study involving 1,400 participants across multiple countries, demonstrated that B vitamin supplementation lowered homocysteine levels by approximately 4 μmol/L and slowed the loss of brain white matter, indicating protective effects against nerve demyelination and reduced neurotoxicity related to homocysteine [390,391]. Other nutrient combinations, such as omega-3 fatty acids with alpha-lipoic acid, have been shown to significantly slow cognitive decline in AD patients compared to omega-3 supplementation alone [392,393].

In PD, nutrition also plays an emerging role. Although PD is mainly characterized by dopaminergic neuron loss and has multifactorial causes, environmental and nutritional factors influence disease risk and progression [394–397]. Certain nutrients with antioxidant and anti-inflammatory properties may provide neuroprotection, while others could worsen neurodegeneration [398]. Research reviews have highlighted promising nutrients and their mechanisms in modulating mitochondrial dysfunction, inflammation, and cellular stress, all key processes in PD pathology [399,400].

More broadly, nutrient synergy supports nervous system health by enhancing cognitive function and managing mood-related symptoms like depression and anxiety through combined effects on neurovascular and neurotransmitter systems [299,401]. For example, blends containing Rhodiola, green tea extract, magnesium, and B vitamins showed greater improvements in stress response and brain electrical activity than individual components [402]. Altogether, the evidence underscores that targeting multiple nutritional pathways and combining key vitamins, antioxidants, and other bioactive compounds can provide a more effective strategy for neurodegenerative diseases than single-nutrient approaches [403,404]. Nutrient synergy helps to address complex multifactorial pathologies by reducing oxidative stress, inflammation, metabolic dysfunction, and vascular insufficiency, which are central to AD, PD, and similar neurological disorders [405].

8.9.2 Mechanistic Insights into How Cofactor Minerals, PUFA, Polyphenols, and Vitamins Mitigate Neurodegeneration

Cofactor minerals, PUFAs, polyphenols, and vitamins exert synergistic neuroprotective effects by targeting key pathological mechanisms involved in neurodegeneration, including oxidative stress, neuroinflammation, mitochondrial dysfunction, and aberrant protein aggregation [406,407]. These bioactive nutrients maintain neuronal health by balancing redox processes, supporting mitochondrial energy production, controlling inflammatory pathways, and stopping the accumulation of neurotoxic proteins [408]. Through these complementary molecular actions, they help maintain cognitive function and slow the progression of age-related neurodegenerative disorders such as AD and PD, offering a promising multi-targeted nutritional strategy for brain health [409,410].

8.9.2.1 Cofactor Minerals

Minerals like iron, zinc, copper, and magnesium are crucial enzymatic cofactors that aid neuronal metabolism, antioxidant defense, and synaptic signaling [411,412]. For example, zinc and magnesium regulate neurotransmitter receptors and ion channels that are essential for synaptic transmission and plasticity [413]. Iron, although essential for mitochondrial function and oxygen transport, must be carefully regulated because excess iron catalyzes the production of harmful ROS through Fenton chemistry, leading to oxidative damage and neuronal death [414,415]. Copper participates in redox reactions, but like iron, in excess, it can promote oxidative stress [416]. Therefore, maintaining optimal levels of these minerals is essential for preserving mitochondrial efficiency, reducing oxidative damage, and maintaining synaptic integrity in the nervous system [417,418].

8.9.2.2 PUFAs

PUFAs, especially omega-3 fatty acids like EPA and DHA, are essential parts of neuronal membrane phospholipids that affect membrane fluidity, receptor activity, and cell signaling [418–420]. These fatty acids demonstrate anti-inflammatory effects by modulating membrane-bound receptors and reducing pro-inflammatory cytokine cascades, thereby lowering chronic neuroinflammation involved in neurodegeneration [421,422]. Furthermore, PUFAs help maintain mitochondrial membrane stability, preventing dysfunction that could impair neuronal energy production and lead to apoptosis [423,424]. They also promote the expression of neurotrophic factors like BDNF and GDNF, which support neuronal survival, differentiation, and synaptic plasticity [425,426].

8.9.2.3 Polyphenols

Polyphenols are plant-derived bioactive compounds, including flavonoids, resveratrol, and curcumin, that protect neurons through multifaceted mechanisms [427]. Their potent antioxidant properties enable them to directly neutralize ROS and RNS, thereby reducing oxidative stress [428,429]. Polyphenols also activate endogenous cellular antioxidant pathways, especially the Keap1/Nrf2/ARE axis, resulting in higher expression of enzymes like glutathione peroxidase, superoxide dismutase, and catalase, which strengthen cellular defenses against oxidative damage [430,431]. In addition to antioxidant properties, polyphenols have anti-inflammatory effects by inhibiting microglial activation and suppressing pro-inflammatory cytokines such as tumor necrosis factor-alpha (TNF-α) and IL-1β [427,432,433]. They further support proteostasis by decreasing the aggregation of misfolded proteins (such as amyloid-beta and hyperphosphorylated tau in AD) and boosting autophagy, which helps remove toxic protein build-up linked to neurodegeneration [434–436].

8.9.2.4 Vitamins

Vitamins serve as essential modulators of neuronal function and protect against neurodegenerative damage. B-complex vitamins (B6, B9/folate, and B12) decrease homocysteine levels, an amino acid associated with vascular problems and neurotoxicity, thus lowering the risk of neurodegeneration [275,437,438]. These vitamins also support neurotransmitter synthesis and maintain myelin sheath integrity, which is crucial for proper nerve conduction. Vitamin E acts as a lipid-soluble antioxidant that prevents lipid peroxidation of neuronal membranes, helping to maintain their fluidity and functionality [439,440]. Vitamin C restores oxidized vitamin E, maintaining antioxidant capacity in the neural environment [441]. Vitamin D modulates neuroinflammation and stimulates neurotrophin expression, and deficiency in vitamin D is increasingly linked to cognitive decline and neurodegenerative conditions [442,443]. Vitamin K also supports neural cell survival and may affect sphingolipid metabolism and anti-inflammatory pathways [444].

8.9.3 Potential Therapeutic Implications for Multi-Nutrient Interventions

Multi-nutrient interventions are a promising therapeutic approach that aligns with the multifactorial nature of chronic diseases, especially those impacting the nervous, metabolic, and immune systems [445,446]. Unlike single-nutrient approaches, multi-nutrient supplementation addresses multiple biochemical pathways at the same time, reflecting the complex nutritional needs of human physiology [299]. This combinatorial strategy increases the likelihood of synergistic effects, where nutrients work together to support shared or complementary mechanisms such as antioxidant defense, anti-inflammatory signaling, mitochondrial function, and neuroprotection [322,377].

In neurodegenerative diseases like AD, emerging evidence indicates that multi-nutrient formulations combining omega-3 fatty acids, B vitamins, choline, uridine, and antioxidants can delay cognitive decline, enhance synaptic function, and decrease brain atrophy [447,344,448]. These effects are especially noticeable when interventions are started in the early or prodromal stages of disease progression [449]. Mechanistically, these nutrients influence essential pathways related to membrane phospholipid synthesis, homocysteine metabolism, oxidative stress, and neuronal repair [450]. Such targeted support of neuronal structure and function provides a useful complement to existing pharmacologic treatments, which often lack disease-modifying effects [451].

In psychiatric disorders, including major depressive disorder, anxiety, and attention-deficit/hyperactivity disorder (ADHD), multi-nutrient interventions have shown measurable benefits in reducing symptoms and improving function [452,453]. Studies have emphasized the importance of broad-spectrum micronutrient support, particularly in individuals with underlying nutritional deficiencies or heightened physiological demands [454]. Improvements in mood regulation, emotional resilience, attention, and behavior are linked to restored neurotransmitter production, decreased neuroinflammation, and improved neurotrophic signaling [455,456].

Beyond neurology and psychiatry, multi-nutrient supplementation has demonstrated therapeutic potential in cardiovascular and metabolic conditions [457,458]. Combining B vitamins, magnesium, coenzyme Q10, and polyphenols has been associated with better endothelial function, reduced homocysteine levels, and improved lipid profiles [459,275]. In maternal health, multi-nutrient supplementation during pregnancy helps lower the risks of low birth weight, preterm delivery, and neural tube defects, especially in resource-limited settings [460]. Additionally, the immune-enhancing and anti-inflammatory effects of combined nutrients have shown promise in supporting recovery from infections, injuries, and ischemic events, especially in older individuals [461].

Despite these promising findings, further research is needed to improve the clinical usefulness of multi-nutrient interventions by identifying ideal compositions, dosing schedules, and timing based on specific disease contexts [462,463]. Personalized nutrition strategies and biomarker-guided applications will probably improve effectiveness and maximize therapeutic benefits in future integrative healthcare models [463]. The role of nutrient synergy in mitigating various diseases is shown in Figure 8.3. Table 8.5 shows synergistic roles of cofactor minerals, PUFA, polyphenols, and vitamins in brain health.

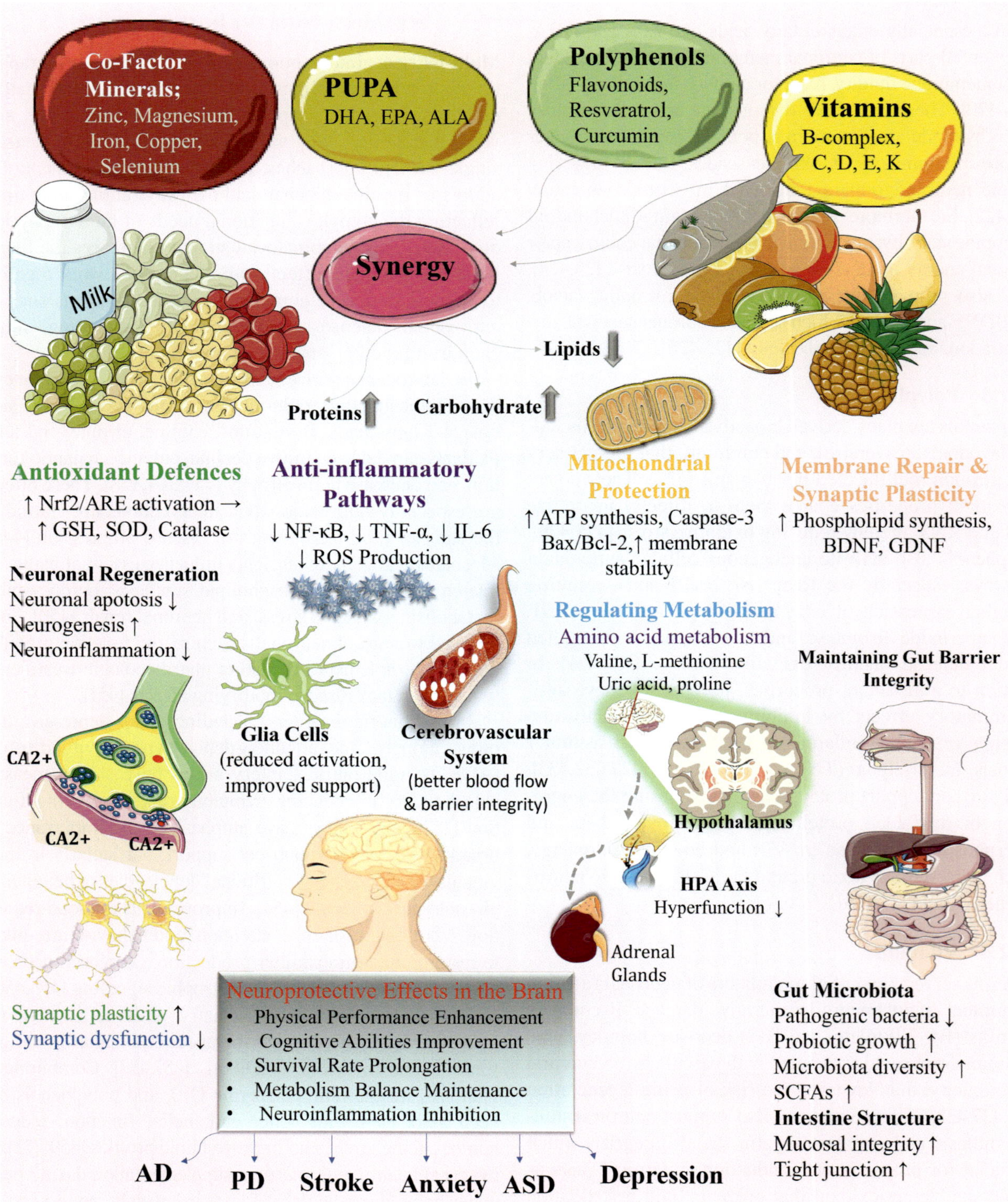

FIGURE 8.3 Role of nutrient synergy in mitigating various diseases.

TABLE 8.5
Synergistic Roles of Cofactor Minerals, PUFA, Polyphenols, and Vitamins in Brain Health

S. No.	Nutrient Group	Key Examples	Neuroprotective Mechanisms	Diseases	Synergistic Interactions	References
1.	Cofactor minerals	Magnesium + vitamin B6	Regulates neurotransmitter synthesis, reduces neuronal excitability	Stress & cognitive function	Vitamin B6 enhances magnesium uptake and utilization	[464]
		Zinc + multi-nutrient supplement	Supports synaptic function, modulates glutamate/GABA balance	Autism spectrum disorder	Synergizes with vitamins, magnesium, and omega-3s for behavioral improvement	[465]
		Zinc	Modulates microglial activation, reduces NLRP3 inflammasome activity	Alzheimer's disease	Balances iron and copper levels to reduce oxidative damage	[466]
		Mg citrate/oxide; with riboflavin & CoQ10	NMDA/CGRP modulation; mitochondrial support	Migraine	Mg + riboflavin + CoQ10 improved migraine burden in RCT (secondary outcomes)	[467]
		High-dose B-vitamin mix	↓Homocysteine; myelin/one-carbon metabolism	MCI/early cognitive decline	B-vitamins × omega-3 (DHA/EPA)—benefit seen only when ω-3 status is high	[275]
		Zinc gluconate/picolinate (adjunct)	Synaptic plasticity; antioxidant via metallothionein	Major depression	Zinc + SSRI improves depressive symptoms versus SSRI + placebo (add-on)	[468]
2.	Polyunsaturated fatty acids (PUFA)	DHA & EPA	Enhances global cognitive function, improves memory accuracy, and supports neural efficiency by reducing prefrontal cortex oxygen demand during tasks Improves reaction time in word recognition, supports synaptic membrane fluidity, and neuronal signaling	Cognition function	Works with DHA for cerebrovascular modulation and combined effects on neural efficiency Synergizes with EPA for brain perfusion benefits; structural role complements EPA's functional effects	[469]
		DHA-enriched diet	Reduces amyloid-β production and plaque burden, modulates APP processing	Tg2576 Alzheimer's mouse model	Dietary DHA significantly decreases Aβ levels and plaque deposition	[470]
		DHA + high EPA	Reduces tau pathology, alters brain PUFA composition	3×Tg-AD mice	Combined DHA and EPA more effectively reduce tau phosphorylation than DHA alone	[471]
		EPA/DHA	Attenuates microglial activation and neuroinflammation	Aβ-infused rodent model	n-3 PUFA mitigates inflammatory markers and glial responses	[472]
		DHA intake	Maintains membrane integrity, slows cognitive decline	Older adults (prospective data)	Regular DHA intake linked to preserved cognitive function over time	[473]
		DHA, EPA	Anti-inflammatory; membrane fluidity; ↑BDNF; ↓amyloid	Alzheimer's disease (clinical & preclinical)	DHA + EPA + antioxidants (vitamin E, curcumin, resveratrol) enhance synaptic protection & Aβ clearance	[474]
		DHA, EPA + B6, B9, B12	One-carbon metabolism + membrane stabilization; ↓homocysteine	Mild Cognitive Impairment (MCI)	B-vitamins × ω-3 slowed brain atrophy in high ω-3 status participants	[475]
		Resolvin D1, protectin D1	Resolution of neuroinflammation: ↑phagocytosis of Aβ	AD and stroke models	Synergy with Nrf2 activators (sulforaphane, curcumin) enhanced antioxidant defense	[116]

(Continued)

TABLE 8.5 (*Continued*)
Synergistic Roles of Cofactor Minerals, PUFA, Polyphenols, and Vitamins in Brain Health

S. No.	Nutrient Group	Key Examples	Neuroprotective Mechanisms	Diseases	Synergistic Interactions	References
3.	Polyphenols	Resveratrol (grape, red wine)	Sirtuin activation, anti-amyloidogenic effects, antioxidant and anti-inflammatory signaling, promotion of neurogenesis	Alzheimer's models: small clinical trials in AD	Shows additive effects with caloric restriction–mimetics and certain PUFAs for mitochondrial support.	[476]
		Curcumin + ferulic acid	Dual inhibition of Aβ aggregation; reduced Aβ-induced neuronal cytotoxicity; improved cell viability	Aβ-challenged neuronal cells (SH-SY5Y)	Curcumin + FA > single agents in suppressing Aβ toxicity	[477]
		Resveratrol + quercetin	Attenuated microglia-mediated neuroinflammation; reduced apoptotic neuronal death	PD microglia-conditioned media → neuronal apoptosis model	Resveratrol + Quercetin co-treatment reduced neuronal apoptosis vs controls	[478]
		Grape seed extract (proanthocyanidins) + resveratrol + Concord grape juice extract (anthocyanins, phenolic acids)	↓ Brain total amyloid; better preservation of cognition; brain-penetrant metabolites protect LTP from oAβ in hippocampal slices	J20 AD transgenic mice	Cocktail produced superior cognitive and neuropathology outcomes vs individual components	[479]
		Baicalein + daidzein	Inhibited Aβ aggregation; ↓ Aβ-induced cytotoxicity; ↑ ER phosphorylation & estrogen-responsive transcription	PC12 neuronal cells (Aβ); MCF-7 ER assays	Cotreatment enhanced both estrogenic signaling and anti-Aβ neuroprotection vs singles	[480]
		EGCG + ferulic acid	↓ Aβ plaques & soluble Aβ; ↑ α-secretase (ADAM10); ↓ BACE1; ↓ astro/microgliosis; ↑ synaptophysin; ↓ oxidative stress; improved cognition	APP/PS1 Alzheimer's mice	The combination outperformed either compound alone in reducing pathology & improving behavioral	[481]
		Resveratrol + curcumin	↓ Neuroinflammation; ↓ GFAP & Iba-1; improved spatial memory; reduced oxidative stress	Aluminum chloride-induced AD-like rats; PC12 cells	Combination is more effective than monotherapy in reducing pathology & behavioral deficits	[482]
		Resveratrol + enriched environment	↓ Oxidative stress; ↑ ERK1/2 signaling modulation; improved neurological recovery	Rat middle cerebral artery occlusion (MCAO)	The combination enhanced functional recovery beyond either alone	[483]
4.	Vitamins	B1 (thiamine), B6, B12, Folate	Lower homocysteine, enhance methylation, support neurotransmitter synthesis	Mild cognitive impairment (MCI), AD (clinical)	B-vitamin complex synergistically slows gray matter atrophy vs. single supplementation	[275]
		Ascorbic acid + α-tocopherol	Antioxidant synergy, lipid peroxidation reduction, ROS scavenging	Alzheimer's disease patients (clinical cohort)	Combined C + E reduces oxidative stress biomarkers and cognitive decline risk more than either alone	[484]

(*Continued*)

TABLE 8.5 (*Continued*)
Synergistic Roles of Cofactor Minerals, PUFA, Polyphenols, and Vitamins in Brain Health

S. No.	Nutrient Group	Key Examples	Neuroprotective Mechanisms	Diseases	Synergistic Interactions	References
		Cholecalciferol + DHA/EPA	Anti-inflammatory, modulates calcium homeostasis, and synaptic plasticity	Cognitive decline, dementia (human observational, intervention)	Vitamin D enhances DHA-mediated neuroprotection, reducing amyloid pathology	[485]
		Menaquinone + cholecalciferol	Regulates sphingolipid metabolism, myelin integrity, and neurovascular health	Aging and dementia models	Vitamin K potentiates vitamin D effects on cognition and synaptic plasticity	[486]
		B vitamin	Support neurotransmitter synthesis, protect neurons, and regulate mood and cognitive function	AD, cognitive decline	Work synergistically with minerals and omega-3s for improved cognitive outcomes, especially in aging	[487]

8.10 TRANSLATIONAL IMPLICATIONS AND DIETARY RECOMMENDATIONS

8.10.1 Importance of Holistic Dietary Approaches for Brain Health

Holistic dietary patterns have gained significant attention in scientific research for their vital role in promoting and maintaining brain health [488]. Unlike interventions focused on isolated nutrients, these comprehensive dietary approaches include a variety of bioactive compounds, such as antioxidants, polyphenols, omega-3 fatty acids, dietary fiber, and anti-inflammatory agents that work together to support cognitive function [489,490]. Among the most extensively studied models are the Mediterranean Diet (MedDiet), the DASH diet, and the MIND diet [491,281]. These dietary patterns have consistently shown neuroprotective benefits across various high-quality studies, including large-scale cohort analyses, RCTs, and systematic reviews, making them important tools in maintaining cognitive health [492,493].

The Mediterranean Diet emphasizes high consumption of fruits, vegetables, whole grains, legumes, nuts, and olive oil, along with a moderate intake of fish, dairy products, and red wine, while limiting red and processed meats [494]. A substantial amount of evidence associates following this dietary pattern with lower risks of cognitive impairment and neurodegenerative diseases, especially AD [261]. These protective effects mainly come from the Med Diet's high levels of anti-inflammatory and antioxidant nutrients, which help reduce neuroinflammation, support synaptic plasticity, and boost neurogenesis [495]. Longitudinal studies show that sustained adherence, especially starting in midlife, can significantly slow memory decline and enhance long-term cognitive outcomes [496].

The DASH diet, originally designed to control hypertension, focuses on consuming fruits, vegetables, whole grains, and low-fat dairy, while restricting saturated fats, sodium, and added sugars [497]. Its cognitive benefits are thought to result from enhancements in cerebrovascular health, especially in controlling blood pressure, a vital element for maintaining cerebral perfusion and brain structure [281,379]. Although these benefits are more noticeable in non-diabetic populations, the DASH diet's wider effect on vascular and metabolic health indirectly supports brain function and resilience [498].

The MIND diet combines aspects of the Mediterranean and DASH diets, placing particular focus on foods that have been independently associated with neuroprotection, such as berries and leafy green vegetables [499,500]. Evidence from prospective studies, including the Rush Memory and Aging Project, suggests that even moderate adherence to the MIND diet significantly slows cognitive decline and reduces the risk of AD more effectively than either of its parent diets alone [501,502]. It provides a clear and practical framework by dividing foods into "brain-healthy" and "unhealthy" groups, thus allowing for more focused dietary interventions [503].

Despite some differences, these three dietary patterns share common features: an emphasis on plant-based foods, healthy fats (especially from olive oil, nuts, and seeds), whole grains, and a reduction in processed foods, red meats, added sugars, and excessive sodium [504,505]. This nutritional framework supports a healthy gut microbiome, decreases systemic inflammation, improves metabolic efficiency, and promotes vascular integrity—all of which are vital for cognitive health and emotional well-being throughout life [506].

In conclusion, strong scientific evidence backs the use of the Mediterranean, DASH, and MIND diets as practical, effective, and safe ways to support cognitive health and reduce the risk of neurodegenerative conditions [281,507]. Further ongoing studies seek to explore how Mediterranean diets improve brain health and how genetic or metabolic differences might inform personalized nutrition strategies [508]. Consistent evidence from various populations supports including these dietary patterns in clinical guidelines and public health policies to combat the increasing global burden of cognitive decline and dementia [509].

8.10.2 Challenges in Achieving Optimal Nutrient Synergy through Diet Alone

Nutrient synergy refers to the phenomenon where the combined effects of multiple nutrients consumed together produce greater health benefits than the sum of their individual actions [299,510]. This concept emphasizes the benefit of consuming a variety of whole foods instead of isolated supplements because nutrients within their natural matrix often interact in complex and advantageous ways [511,512].

A key scientific challenge is understanding the complex influence of the food matrix, which is the physical and structural makeup of foods that impacts how nutrients are released, absorbed, and metabolized [513,514]. The food matrix includes not only the nutrients themselves but also bioactive compounds, fibers, carbohydrates, fats, and proteins, arranged in a complex physical structure [515]. These elements can modify digestion kinetics, enhance or inhibit nutrient bioavailability, and influence physiological outcomes differently than isolated nutrients [516]. For example, consuming whole eggs leads to higher muscle protein synthesis than eating only egg whites, because of the combined effects of nutrients and their food matrix [517].

Within this matrix, nutrient bioavailability is also influenced by the presence of enhancers and inhibitors [518]. Vitamin C can enhance the absorption of non-heme iron, while compounds such as phytates, oxalates, and tannins common in plant-based foods can decrease the absorption of minerals like calcium, iron, and zinc [519,520]. Such antinutritional factors serve as inhibitors within the food matrix, making it more difficult to predict and optimize nutrient absorption from whole foods [521].

Scientifically, studying nutrient synergy is a methodologically challenging endeavor. Whole foods contain thousands of compounds that may interact in numerous ways, which are often poorly characterized [299,522]. Isolating individual nutrient interactions amidst this complexity requires innovative experimental approaches such as cell culture models and organ-on-a-chip technologies, which remain in early stages [523,524]. Traditional reductionist trials that test single or a few nutrients fail to capture the full scope of these interactions [525].

Lastly, socioeconomic and practical barriers restrict achieving optimal nutrient synergy through diet alone [526]. Many populations depend heavily on ultra-processed foods and have low dietary diversity because of economic constraints, limited access to fresh, nutrient-rich foods, cultural preferences, and environmental factors [527,528]. These realities hinder the regular consumption of diverse whole-food combinations needed for nutrient synergy, making consistent intake of beneficial nutrient interactions difficult at the population level [529,299].

In summary, while the whole food matrix and nutrient interactions provide significant health benefits, achieving optimal nutrient synergy through diet alone is limited by complex food structure effects, bioavailability inhibitors, scientific research challenges, and practical dietary and socioeconomic barriers [513,530,531]. Addressing these challenges may call for complementary strategies like food fortification, supplementation, and innovative food technologies, alongside efforts to enhance dietary quality and access to a variety of whole foods [532].

8.10.3 Recommendations for Balanced Intake of Cofactor Minerals, PUFA, Polyphenols, and Vitamins

Balanced intake of cofactor minerals, PUFAs, polyphenols, and vitamins remains essential for supporting enzymatic functions, immune competence, antioxidant defenses, and metabolic balance, as emphasized by authoritative nutritional guidelines and recent research (2023–2025) [533,534]. Cofactor minerals, like magnesium, zinc, copper, manganese, iodine, and molybdenum, are essential enzyme cofactors that influence hormone production, nerve signaling, antioxidant functions, and metabolic processes [535–537]. Recommended daily allowances vary by demographic factors but generally include magnesium (310–420 mg/day), zinc (8–11 mg/day), copper (~900 mcg/day), manganese (1.8–2.3 mg/day), iodine (150 mcg/day), and molybdenum (45 mcg/day) [69,538]. Dietary sources rich in these minerals include nuts, seeds, whole grains, legumes, leafy green vegetables, seafood, and fortified foods [539,540]. Their adequate intake supports cellular metabolism, skin health, and immune system regulation (NIN Dietary Guidelines 2023; WHO vitamin and mineral requirements) [541,542]. For PUFAs, emphasis is on achieving sufficient intake of omega-3 fatty acids, specifically EPA and DHA, which support cardiovascular and cognitive health [543,544]. Current guidelines recommend approximately 250–500 mg per day of combined EPA and DHA, usually obtained by eating two servings of oily fish like salmon, mackerel, or sardines each week [545]. Plant-based omega-3 (ALA) sources such as flaxseeds, walnuts, and canola oil contribute to EPA/DHA but have limited conversion in humans [546]. Importantly, regulating the omega-6 to omega-3 ratio by moderating omega-6-rich seed oil intake helps control chronic inflammation and lower cardiovascular risk [547,548]. Polyphenols are a wide-ranging group of plant chemicals with strong antioxidant and anti-inflammatory effects [549]. Recommended intakes vary but generally range from several hundred milligrams daily, obtained through diverse plant sources such as berries, apples, cruciferous vegetables (broccoli, cauliflower), nuts, tea, coffee, and moderate wine consumption [550,551]. Polyphenols help lower cardiometabolic risks and improve gut microbiome health by supporting beneficial bacteria like Lactobacillus and Bifidobacterium [552]. Their chemical diversity includes subclasses such as flavonoids (flavonols, anthocyanins), phenolic acids, and stilbenes like resveratrol, with ongoing research highlighting their multifunctional roles in human health [553,137]. Regarding vitamins, recommended intakes should include both water-soluble (e.g., vitamin C ~90 mg/day) and fat-soluble vitamins (vitamin D ~20 mcg/day or 800 IU, vitamin A ~900 mcg RAE/day, vitamin E ~15 mg/day) [554,555]. While a balanced diet containing fruits, vegetables, whole grains, dairy, and lean proteins generally meets these requirements, vitamin D supplementation may be necessary in populations with limited sunlight exposure [556,557]. Vitamins support calcium absorption, bone health, immune regulation, and antioxidant activity, which are essential for maintaining overall physiological balance (Nutritional reports 2023; WHO) [558].

8.10.4 Role of Personalized Nutrition and Functional Foods in Brain Health

Brain health is a complex and ever-changing concept influenced by genetic factors, lifestyle choices, environmental factors, and dietary habits [559]. Personalized nutrition, defined as the strategic tailoring of dietary interventions to an individual's genetic makeup, metabolic profile, gut microbiota composition, and lifestyle factors, has become a strong, evidence-based method for enhancing cognitive performance and reducing neurological decline [560–562]. Meanwhile, functional foods enriched with bioactive compounds that go beyond basic nutrition are increasingly seen as essential dietary tools for maintaining and improving neural health [563,564].

The basis of personalized nutrition is the understanding that one-size-fits-all dietary advice cannot reflect the physiological differences among individuals [565,566]. Genetic polymorphisms, such as those in the APOE gene, variations in nutrient metabolism, and differences in gut microbiota significantly influence the neurocognitive effects of dietary components [567]. For example, omega-3 fatty acid supplementation may provide greater cognitive benefits in individuals with certain lipid metabolism genotypes, highlighting the need for precision-based strategies instead of one-size-fits-all dietary guidelines [568,569].

Functional foods complement personalized nutrition by providing targeted neuroprotective compounds [570]. Polyphenol-rich berries, omega-3-fortified dairy products, probiotic-enriched yogurts, and B-vitamins and antioxidant-fortified cereals have shown the ability to reduce oxidative stress, suppress neuroinflammation, and improve synaptic plasticity, which are key mechanisms for cognitive resilience [9,571,572]. These effects are facilitated by antioxidant defenses, anti-inflammatory signaling, enhanced neurotransmitter biosynthesis, and positive modulation of the gut–brain axis [573].

Clinical and observational evidence supports the use of personalized, functional food–based interventions in both preventing and managing neurodegenerative disorders [574]. Diets enriched with curcumin, flavonoids, and vitamin E, tailored to the needs of individuals with mild cognitive impairment, are associated with improved memory performance and delayed progression to AD [575,576]. Similarly, low-glycemic, polyphenol-rich diets may help those at risk of vascular dementia by improving cerebrovascular function [577].

Advances in nutrigenomics have further clarified gene–diet interactions in neurological disease progression, especially through the gut–brain axis, which connects dietary patterns to microbiome composition, neurotransmitter regulation, and neuroinflammatory control [578,579]. Mediterranean- and MIND-style diets, tailored to individual nutrient needs, have been linked to fewer depressive symptoms, better mood regulation, and an improved quality of life in older adults [580,581].

8.11 FUTURE DIRECTIONS AND EMERGING RESEARCH

8.11.1 Emerging Technologies to Study Nutrient Interactions

Recent advances in analytical methodologies, especially metabolomics and nutrigenomics, are transforming the study of nutrient interactions and their importance for brain health. Metabolomics allows detailed, high-resolution profiling of metabolite networks, capturing subtle and dynamic biochemical changes in response to dietary inputs. Nutrigenomics provides mechanistic insight into gene–nutrient relationships, supporting the development of genotype-informed nutritional strategies for neuroprotection. When combined with microbiome profiling and systems biology, these tools offer a multidimensional framework to clarify synergistic or antagonistic effects among nutrients, cofactor minerals, and bioactive compounds. This integrated approach offers significant potential for improving precision nutrition, strengthening evidence-based dietary guidelines, and elucidating molecular mechanisms vital for preventing and managing neurodegenerative diseases.

8.11.2 Areas for Further Research

Future research should focus on integrating metabolomics, nutrigenomics, and microbiome profiling to create comprehensive maps of nutrient–gene–microbe interactions that are relevant to brain health. Long-term cohort studies and carefully planned dietary intervention trials are vital to establishing causal relationships between specific nutritional patterns, metabolic signatures, and neuroprotective results. Exploring the combined effects of cofactor minerals, PUFAs, and bioactive compounds may reveal new strategies for therapy. Additionally, using artificial intelligence (AI) and systems biology modeling could improve the accuracy of predictions for personalized nutrition. Considering population-specific genetic differences will be essential to making precision nutritional approaches applicable for preventing and managing neurodegenerative disorders.

8.11.3 Long-Term Effects of Multi-Nutrient Interventions

Long-term investigations of multi-nutrient interventions are essential for understanding their lasting effects on brain health and their potential to reduce the risk of neurodegenerative diseases. Extended supplementation or dietary modification trials can reveal the cumulative and synergistic effects of combined nutrients, such as cofactor minerals, PUFAs, and bioactive phytochemicals, on cognitive performance, synaptic integrity, and redox balance. These studies should include robust biomarkers to monitor metabolic, molecular, and functional changes, assess interactions with host genetics and microbiome composition, and ensure comprehensive safety evaluations over long periods. The evidence gained will be vital for developing precision nutrition strategies aimed at providing lasting neuroprotection.

8.11.4 Understanding Nutrient–Nutrient Interactions at the Molecular Level

Future studies should define the molecular basis of nutrient–nutrient interactions using an integrated approach combining structural biology, omics technologies, and systems pharmacology. Techniques such as cryo-EM for structural biology, metabolomics to characterize interactions, and systems-based modeling methodologies for network interactions will help identify interactions with competitive, synergistic, or antagonistic effects at transport, receptor, or signaling levels, and predict their bioefficacy outcomes. Ultimately, this will enable the rational design of nutrient compositions that enhance therapeutic potential.

8.11.5 Role of Gut–Brain Axis in Nutrient Synergy

Novel evidence suggests that the gut–brain axis is a key interface for nutrient-driven synergy. Future studies should also explore how different nutrient combinations affect the gut microbiome, metabolite and neurotransmitter pathways, and vagal–humoral signaling networks. We can combine microbiome engineering, neuroimaging, and metabolite profiling to see if nutrient synergy influences cognitive health and neuroprotection, aiming to develop microbiota-targeted nutritional interventions.

8.11.6 Potential for Multi-Nutrient Therapies in Clinical Practice

The translational frontier is to develop clinically validated multi-nutrient therapies that leverage synergistic interactions, maximizing therapeutic effects while minimizing side effects. Personalized nutrition platforms, AI-driven predictive models, and adaptive clinical trials will enable stratification of patients based on their genetic, metabolic, and microbiome profiles. This precision-guided approach could revolutionize strategies for neurodegenerative diseases, metabolic diseases, and psychiatric disorders, shifting from using single-nutrient supplementation to evidence-based multi-nutrient synergistic therapies.

8.12 CONCLUSION

This chapter emphasizes the importance of dietary nutrients in maintaining overall brain health. The evidence presented shows that brain health relies not on a single nutritional factor but on intricate and synergistic interactions between cofactor minerals, PUFA, polyphenols, and vitamins. Adequate levels and synergy among these nutrients are essential for various fundamental processes such as neurogenesis, antioxidant defense, redox activity, neurotransmitter synthesis, and synaptic plasticity. Any deficiency or imbalance in these nutrients negatively impacts brain function and can accelerate oxidative stress, aging, neurodegeneration, and excitotoxicity. Recent studies on targeted combinations of these nutrients highlight the significance of nutritional synergy in neuroprotection, cognition, and memory development. In the future, multi-nutrient and personalized dietary strategies could be developed through integrated technologies such as metabolomics and nutrigenomics. By combining nutritional synergy with precision approaches, such dietary and therapeutic interventions can enhance brain function and reduce cognitive decline and brain-related disorders across diverse populations.

ACKNOWLEDGMENTS

The authors thank Chairman Mr. Parveen Garg, ISF College of Pharmacy, Moga (Punjab), India, for his excellent vision and support.

AUTHOR'S CONTRIBUTIONS

Divya Choudhary (D.C.), MD Nasiruddin Khan (M.D.K.), and Ravi Rana (R.R.) wrote the original draft of the review and contributed to writing and review; Sidharth Mehan (S.M) was responsible for conceptualization and supervision; Zuber Khan (Z.K.), Arun Kumar Sharma (A.K.S.), and Manjeet Kumar (M.K.) conducted the formal analysis and contributed to the editing. All authors agree to be accountable for all aspects of work, ensuring integrity and accuracy.

FUNDING

None.

ETHICS APPROVAL

Not applicable.

CONSENT TO PUBLISH

Authors approve the submission of the publication.

CONSENT TO PARTICIPATE

Not applicable.

DATA AVAILABILITY STATEMENTS

Data sharing is not applicable to this article as no datasets were generated or analyzed during the current study.

REFERENCES

1. P. Mergenthaler, U. Lindauer, G. A. Dienel, and A. Meisel, "Sugar for the brain: the role of glucose in physiological and pathological brain function," *Trends Neurosci.* 2013. doi:10.1016/j.tins.2013.07.001.
2. E. Tourkochristou, C. Triantos, and A. Mouzaki, "The influence of nutritional factors on immunological outcomes," *Front. Immunol.* 2021. doi:10.3389/fimmu.2021.665968.
3. K. C. Kadosh et al., "Nutritional support of neurodevelopment and cognitive function in infants and young children—an update and novel insights," *Nutrients.* 2021. doi:10.3390/nu13010199.
4. T. C. Borge, H. Aase, A. L. Brantsæter, and G. Biele, "The importance of maternal diet quality during pregnancy on cognitive and behavioural outcomes in children: a systematic review and meta-analysis," *BMJ Open.* 2017. doi:10.1136/bmjopen-2017-016777.
5. Y. Mou, E. Blok, M. Barroso, P. W. Jansen, T. White, and T. Voortman, "Dietary patterns, brain morphology and cognitive performance in children: results from a prospective population-based study," *Eur. J. Epidemiol.* 2023. doi:10.1007/s10654-023-01012-5.
6. L. Begdache, N. N. Aslan Çin, and N. M. Jadavji, "Editorial: nutrition and neurobehaviors," *Front. Nutr.* vol. 12, no. 5, 2025. doi:10.3389/fnut.2025.1637728.

7. T. Harayama and T. Shimizu, "Roles of polyunsaturated fatty acids, from mediators to membranes," *J. Lipid Res.* 2020. doi:10.1194/JLR.R120000800.
8. K. Jomova et al., "Essential metals in health and disease," *Chem. Biol. Interact.* 2022. doi:10.1016/j.cbi.2022.110173.
9. R. Naomi et al., "Dietary polyphenols as a protection against cognitive decline: evidence from animal experiments; mechanisms and limitations," *Antioxidants.* 2023. doi:10.3390/antiox12051054.
10. S. Song et al., "Molecular basis for antioxidant enzymes in mediating copper detoxification in the nematode Caenorhabditis elegans," *PLoS One.* 2014. doi:10.1371/journal.pone.0107685.
11. M. H. Kim and M. K. Choi, "Seven dietary minerals (Ca, P, Mg, Fe, Zn, Cu, and Mn) and their relationship with blood pressure and blood lipids in healthy adults with self-selected diet," *Biol. Trace Elem. Res.* 2013. doi:10.1007/s12011-013-9656-1.
12. M. G. Stathopoulou, S. Kanoni, G. Papanikolaou, S. Antonopoulou, T. Nomikos, and G. Dedoussis, "Mineral intake," in Shenolikar, S. (ed.), *Progress in Molecular Biology and Translational Science.* Elsevier, 2012. doi:10.1016/B978-0-12-398397-8.00009-5.
13. A. Dent and R. Selvaratnam, "Measuring magnesium: physiological, clinical and analytical perspectives," *Clin. Biochem.* 2022. doi:10.1016/j.clinbiochem.2022.04.001.
14. Y. H. Ko, S. Hong, and P. L. Pedersen, "Chemical mechanism of ATP synthase: magnesium plays a pivotal role in formation of the transition state where ATP is synthesized from ADP and inorganic phosphate," *J. Biol. Chem.* 1999. doi:10.1074/jbc.274.41.28853.
15. R. Yamanaka, Y. Shindo, and K. Oka, "Magnesium is a key player in neuronal maturation and neuropathology," *Int. J. Mol. Sci.* 2019. doi:10.3390/ijms20143439.
16. D. Dangour, "Book Review: principles of nutritional assessment, 2nd edn, by Rosalind S. Gibson," *Br. J. Nutr.* 2006. doi:10.1079/bjn20061738.
17. S. Van Laecke, "Hypomagnesemia and hypermagnesemia," *Acta Clin. Belg.* 2019. doi:10.1080/17843286.2018.1516173.
18. G. Dilliraj and K. Sumathi, "Significance of serum magnesium levels in the auditory status of type ii diabetes mellitus patients," *Biomed.* 2019. doi:org/10.51248/v39i2.5
19. J. J. DiNicolantonio, J. H. O'Keefe, and W. Wilson, "Subclinical magnesium deficiency: a principal driver of cardiovascular disease and a public health crisis," *Open Heart.* 2018. doi:10.1136/openhrt-2017-000668.
20. I. Slutsky et al., "Enhancement of learning and memory by elevating brain magnesium," *Neuron.* 2010. doi:10.1016/j.neuron.2009.12.026.
21. H. Zhou, G. Q. Bi, and G. Liu, "Intracellular magnesium optimizes transmission efficiency and plasticity of hippocampal synapses by reconfiguring their connectivity," *Nat. Commun.* vol. 15, no. 1, 2024. doi:10.1038/s41467-024-47571-3.
22. F. H. Nielsen, "Relation between magnesium deficiency and sleep disorders and associated pathological changes," in Watson, R. R. (ed.), *Modulation of Sleep by Obesity, Diabetes, Age, and Diet.* Academic Press, 2015. doi:10.1016/B978-0-12-420168-2.00031-4.
23. Z. P. Xu et al., "Magnesium protects cognitive functions and synaptic plasticity in streptozotocin-induced sporadic Alzheimer's model," *PLoS One.* 2014. doi:10.1371/journal.pone.0108645.
24. J. H. F. de Baaij, J. G. J. Hoenderop, and R. J. M. Bindels, "Magnesium in man: implications for health and disease," *Physiol. Rev.* 2015. doi:10.1152/physrev.00012.2014.
25. J. Durlach, P. Bac, M. Bara, and A. Guiet-Bara, "Physiopathology of symptomatic and latent forms of central nervous hyperexcitability due to magnesium deficiency: a current general scheme," *Magnes. Res.* 2000. doi:10.1684/mrh.2000.0016.
26. V. Papadopo and M. Nechifor, "Magnesium in neuroses," in Vink, R. and Nechifor, M. (eds.), *Magnesium in the Central Nervous System.* University of Adelaide Press, 2011. doi:10.1017/UPO9780987073051.021.
27. B. Ryszewska-Pokraśniewicz et al., "Effects of magnesium supplementation on unipolar depression: a placebo-controlled study and review of the importance of dosing and magnesium status in the therapeutic response," *Nutrients.* 2018. doi:10.3390/nu10081014.
28. E. Poleszak, P. Wlaź, A. Wróbel, S. Fidecka, and G. Nowak, "NMDA/glutamate mechanism of magnesium-induced anxiolytic-like behavior in mice," *Pharmacol. Rep.* 2008. doi:10.1017/S146114570800931X.
29. M. Macías-Carballo, S. Rosas-Navarro, M. L. López-Meraz, L. Beltran-Parrazal, and C. Morgado-Valle, "Anxiolytic effect of chronic intake of supplemental magnesium chloride in rat," *Behav. Brain Res.* 2021. doi:10.1016/j.bbr.2021.113460.
30. W. Maret, "Zinc in cellular regulation: the nature and significance of 'zinc signals,'" *Int. J. Mol. Sci.* 2017. doi:10.3390/ijms18112285.
31. B. Chen et al., "Cellular zinc metabolism and zinc signaling: from biological functions to diseases and therapeutic targets," *Signal Transduct. Target. Ther.* 2024. doi:10.1038/s41392-023-01679-y.
32. S. Choi, D. K. Hong, B. Y. Choi, and S. W. Suh, "Zinc in the brain: friend or foe?" *Int. J. Mol. Sci.* 2020. doi:10.3390/ijms21238941.
33. J. Dufner-Beattie, Z. L. Huang, J. Geiser, W. Xu, and G. K. Andrews, "Mouse ZIP1 and ZIP3 genes together are essential for adaptation to dietary zinc deficiency during pregnancy," *Dev Genes Evol.* 2006. doi:10.1002/dvg.20211.
34. S. W. Suh et al., "Decreased brain zinc availability reduces hippocampal neurogenesis in mice and rats," *J. Cereb. Blood Flow Metab.* 2009. doi:10.1038/jcbfm.2009.80.
35. Q. H. Cheng and L. J. Gao, "Maternal serum zinc concentration and neural tube defects in offspring: a meta-analysis," *J. Matern.-Fetal Neonatal Med.* 2022. doi:10.1080/14767058.2020.1860930.
36. A. C. Dey, M. Shahidullah, M. A. Mannan, M. K. Noor, L. Saha, and S. A. Rahman, "Maternal and neonatal serum zinc level and its relationship with neural tube defects," *J. Heal. Popul. Nutr.* 2010. doi:10.3329/jhpn.v28i4.6040.
37. S. D. Gower-Winter, R. S. Corniola, T. J. Morgan, and C. W. Levenson, "Zinc deficiency regulates hippocampal gene expression and impairs neuronal differentiation," *Nutr. Neurosci.* 2013. doi:10.1179/1476830512Y.0000000043.
38. X. Liu and P. Oteiza, "Secondary zinc deficiency via exposure to the environmental toxicant Di-2-ethylhexyl phthalate affects neurogenesis at embryonic day 19," *Curr. Dev. Nutr.* 2020. doi:10.1093/cdn/nzaa057_036.
39. É. D. de Medeiros Rocha, N. J. N. de Brito, M. M. G. Dantas, A. de Araújo Silva, M. das Graças Almeida, and J. Brandão-Neto, "Effect of zinc supplementation on GH, IGF1, IGFBP3, OCN, and ALP in non-zinc-deficient children," *J. Am. Coll. Nutr.* 2015. doi:10.1080/07315724.2014.929511.

40. J. Guo et al., "The influence of zinc supplementation on IGF-1 levels in humans: a systematic review and meta-analysis," *J. King Saud Univ. Sci.* 2020. doi:10.1016/j.jksus.2020.01.018.
41. A. S. Prasad, F. W. J. Beck, S. M. Grabowski, J. Kaplan, and R. H. Mathog, "Zinc deficiency: changes in cytokine production and T-cell subpopulations in patients with head and neck cancer and in noncancer subjects," *Proc. Assoc. Am. Physicians*. vol. 109, pp. 68–77, 1997.
42. M. M. George, K. Subramanian Vignesh, J. A. Landero Figueroa, J. A. Caruso, and G. S. Deepe, "Zinc induces dendritic cell tolerogenic phenotype and skews regulatory T Cell–Th17 balance," *J. Immunol.* 2016. doi:10.4049/jimmunol.1600410.
43. S. Kumar et al., "Zinc-induced polymerization of killer-cell Ig-like receptor into filaments promotes its inhibitory function at cytotoxic immunological synapses," *Mol. Cell.* 2016. doi:10.1016/j.molcel.2016.03.009.
44. S. Hagmeyer, J. C. Haderspeck, and A. M. Grabrucker, "Behavioral impairments in animal models for zinc deficiency," *Front. Behav. Neurosci.* 2015. doi:10.3389/fnbeh.2014.00443.
45. E. C. Crăciun et al., "Evaluation of whole blood zinc and copper levels in children with autism spectrum disorder," *Metab. Brain Dis.* 2016. doi:10.1007/s11011-016-9823-0.
46. Y. Yang et al., "High dose zinc supplementation induces hippocampal zinc deficiency and memory impairment with inhibition of BDNF signaling," *PLoS One*. 2013. doi:10.1371/journal.pone.0055384.
47. S. Zhang et al., "Double-edge sword roles of iron in driving energy production versus instigating ferroptosis," *Cell Death Dis.* 2022. doi:10.1038/s41419-021-04490-1.
48. V. T. Cheli, J. Correale, P. M. Paez, and J. M. Pasquini, "Iron metabolism in oligodendrocytes and astrocytes, implications for myelination and remyelination," *ASN Neuro*. 2020. doi:10.1177/1759091420962681.
49. S. L. Kim, S. Shin, and S. J. Yang, "Iron homeostasis and energy metabolism in obesity," *Clin. Nutr. Res.* 2022. doi:10.7762/cnr.2022.11.4.316.
50. D. J. R. Lane et al., "Cellular iron uptake, trafficking and metabolism: key molecules and mechanisms and their roles in disease," *Biochim. Biophys. Acta Mol. Cell Res.* 2015. doi:10.1016/j.bbamcr.2015.01.021.
51. A. Catapano et al., "Iron metabolism and ferroptosis in health and diseases: the crucial role of mitochondria in metabolically active tissues," *J. Nutr. Biochem.* vol. 140, p. 109888, 2025. doi:10.1016/j.jnutbio.2025.109888.
52. K. Hirota, "An intimate crosstalk between iron homeostasis and oxygen metabolism regulated by the hypoxia-inducible factors (HIFs)," *Free Radic. Biol. Med.* 2019. doi:10.1016/j.freeradbiomed.2018.07.018.
53. R. Evstatiev and C. Gasche, "Iron sensing and signalling," *Gut*. 2012. doi:10.1136/gut.2010.214312.
54. K. Tajima et al., "Mitochondrial lipoylation integrates age-associated decline in brown fat thermogenesis," *Nat. Metab.* 2019. doi:10.1038/s42255-019-0106-z.
55. J. S. Yook et al., "The thermogenic characteristics of adipocytes are dependent on the regulation of iron homeostasis," *J. Biol. Chem.* 2021. doi:10.1016/j.jbc.2021.100452.
56. J. Stiban, M. So, and L. S. Kaguni, "Iron-sulfur clusters in mitochondrial metabolism: multifaceted roles of a simple cofactor," *Biochemistry* (Moscow). 2016. doi:10.1134/S0006297916100059.
57. N. Kitamura et al., "Iron supplementation regulates the progression of high fat diet induced obesity and hepatic steatosis via mitochondrial signaling pathways," *Sci. Rep.* 2021. doi:10.1038/s41598-021-89673-8.
58. R. Crichton, *Iron Metabolism: From Molecular Mechanisms to Clinical Consequences,* 4th edn. Wiley-Blackwell, 2016. doi:10.1002/9781118925645.
59. B. Todorich, J. M. Pasquini, C. I. Garcia, P. M. Paez, and J. R. Connor, "Oligodendrocytes and myelination: the role of iron," *Glia*. 2009. doi:10.1002/glia.20784.
60. E. Stephenson, N. Nathoo, Y. Mahjoub, J. F. Dunn, and V. W. Yong, "Iron in multiple sclerosis: roles in neurodegeneration and repair," *Nat. Rev. Neurol.* 2014. doi:10.1038/nrneurol.2014.118.
61. V. T. Cheli et al., "The divalent metal transporter 1 (DMT1) is required for iron uptake and normal development of oligodendrocyte progenitor cells," *J. Neurosci.* 2018. doi:10.1523/JNEUROSCI.1447-18.2018.
62. V. T. Cheli et al., "Transferrin receptor is necessary for proper oligodendrocyte iron homeostasis and development," *J. Neurosci.* 2023. doi:10.1523/JNEUROSCI.1383-22.2023.
63. Y. Zhang et al., "An RNA-sequencing transcriptome and splicing database of glia, neurons, and vascular cells of the cerebral cortex," *J. Neurosci.* 2014. doi:10.1523/JNEUROSCI.1860-14.2014.
64. Y. X. Li, Q. Guan, Y. H. Chen, H. J. Han, W. C. Liu, and Z. Y. Nie, "Transferrin receptor and ferritin-H are developmentally regulated in oligodendrocyte lineage cells," *Neural Regen. Res.* 2013. doi:10.3969/j.issn.1673-5374.2013.01.001.
65. M. E. Badaracco, E. H. Ortiz, E. F. Soto, J. Connor, and J. M. Pasquini, "Effect of transferrin on hypomyelination induced by iron deficiency," *J. Neurosci. Res.* 2008. doi:10.1002/jnr.21709.
66. B. Khalfaoui-Hassani, P. I. Trasnea, S. Steimle, H. G. Koch, and F. Daldal, "Cysteine mutants of the major facilitator superfamily-type transporter CcoA provide insight into copper import," *MBio*, 2021. doi:10.1128/mBio.01567-21.
67. J. Gale and E. Aizenman, "The physiological and pathophysiological roles of copper in the nervous system," *Eur. J. Neurosci.* vol. 60, no. 1, pp. 3505–3543, 2024. doi:10.1111/ejn.16370.
68. I. F. Scheiber, J. F. B. Mercer, and R. Dringen, "Metabolism and functions of copper in brain," *Prog. Neurobiol.* 2014. doi:10.1016/j.pneurobio.2014.01.002.
69. P. Trumbo, A. A. Yates, S. Schlicker, and M. Poos, "Dietary reference intakes: vitamin A, vitamin K, arsenic, boron, chromium, copper, iodine, iron, manganese, molybdenum, nickel, silicon, vanadium, and zinc," *J. Am. Diet. Assoc.* 2001. doi:10.1016/S0002-8223(01)00078-5.
70. N. M. Boulis and A. Donsante, "A novel antibody to treat SOD1-related amyotrophic lateral sclerosis," *Mol. Ther. Methods Clin. Dev.* 2023. doi:10.1016/j.omtm.2023.03.011.
71. S. D. Boyd, M. S. Ullrich, A. Skopp, and D. D. Winkler, "Copper sources for sod1 activation," *Antioxidants*. 2020. doi:10.3390/antiox9060500.
72. N. Bäck, R. E. Mains, and B. A. Eipper, "PAM: diverse roles in neuroendocrine cells, cardiomyocytes, and green algae," *FEBS J.* 2022. doi:10.1111/febs.16049.
73. J. R. Prohaska and B. Brokate, "Dietary copper deficiency alters protein levels of rat dopamine β-monooxygenase and tyrosine monooxygenase," *Exp. Biol. Med.* 2001. doi:10.1177/153537020122600307.

74. M. R. A. Blomberg, "The structure of the oxidized state of cytochrome c oxidase—experiments and theory compared," *J. Inorg. Biochem.* 2020. doi:10.1016/j.jinorgbio.2020.111020.
75. F. Yang et al., "Copper induces oxidative stress with triggered NF-κB pathway leading to inflammatory responses in immune organs of chicken," *Ecotoxicol. Environ. Saf.* 2020. doi:10.1016/j.ecoenv.2020.110715.
76. C. P. W. Soon et al., "Diacetylbis(N(4)-methylthiosemicarbazonato) copper(II) (Cu II(atsm)) protects against peroxynitrite-induced nitrosative damage and prolongs survival in amyotrophic lateral sclerosis mouse model," *J. Biol. Chem.* 2011. doi:10.1074/jbc.M111.274407.
77. M. T. Huuskonen et al., "The copper bis(thiosemicarbazone) complex CuII(atsm) is protective against cerebral ischemia through modulation of the inflammatory Milieu," *Neurotherapeutics.* 2017. doi:10.1007/s13311-016-0504-9.
78. L. Bica et al., "Neuroprotective copper bis(thiosemicarbazonato) complexes promote neurite elongation," *PLoS One.* 2014. doi:10.1371/journal.pone.0090070.
79. N. D. Solovyev, "Importance of selenium and selenoprotein for brain function: from antioxidant protection to neuronal signalling," *J. Inorg. Biochem.* 2015. doi:10.1016/j.jinorgbio.2015.09.003.
80. N. Seeram, "Polyphenols overview," in Coates, P. M., Betz, J. M., Blackman, M. R., Cragg, G. M., Levine, M., Moss, J., and White, J. D. (eds.), *Encyclopedia of Dietary Supplements*, 2nd edn. CRC Press, 2010. https://doi.org/10.1201/b14669
81. R. F. Burk et al., "Selenoprotein P and apolipoprotein e receptor-2 interact at the blood-brain barrier and also within the brain to maintain an essential selenium pool that protects against neurodegeneration," *FASEB J.* 2014. doi:10.1096/fj.14-252874.
82. R. F. Burk and K. E. Hill, "Regulation of selenium metabolism and transport," *Annu. Rev. Nutr.* 2015. doi:10.1146/annurev-nutr-071714-034250.
83. C. N. Byrns, M. W. Pitts, C. A. Gilman, A. C. Hashimoto, and M. J. Berry, "Mice lacking selenoprotein P and selenocysteine lyase exhibit severe neurological dysfunction, neurodegeneration, and audiogenic seizures," *J. Biol. Chem.* 2014. doi:10.1074/jbc.M113.540682.
84. M. Berry, L. Seale, P. Kremer, A. Ogawa, A. Hashimoto, and M. Pitts, "Targeted deletion of selenium transport and recycling pathways in mice: neurological and metabolic consequences," *FASEB J.* 2015. doi:10.1096/fasebj.29.1_supplement.lb328.
85. J. A. Pasco et al., "Dietary selenium and major depression: a nested case-control study," *Complement. Ther. Med.* 2012. doi:10.1016/j.ctim.2011.12.008.
86. J. Wu et al., "Associations of essential element serum concentrations with autism spectrum disorder," *Environ. Sci. Pollut. Res.* 2022. doi:10.1007/s11356-022-21978-1.
87. X. Zhang, R. P. Liu, W. H. Cheng, and J. H. Zhu, "Prioritized brain selenium retention and selenoprotein expression: nutritional insights into Parkinson's disease," *Mech. Ageing Dev.* 2019. doi:10.1016/j.mad.2019.04.004.
88. L. Xu et al., "Trace elements differences in the depression sensitive and resilient rat models," *Biochem. Biophys. Res. Commun.* 2020. doi:10.1016/j.bbrc.2020.05.228.
89. E. Beytut, S. Yilmaz, M. Aksakal, and S. Polat, "The possible protective effects of vitamin E and selenium administration in oxidative stress caused by high doses of glucocorticoid administration in the brain of rats," *J. Trace Elem. Med. Biol.* 2018. doi:10.1016/j.jtemb.2017.10.005.
90. Z. Huang, A. H. Rose, and P. R. Hoffmann, "The role of selenium in inflammation and immunity: from molecular mechanisms to therapeutic opportunities," *Antioxid. Redox Signal.* 2012. doi:10.1089/ars.2011.4145.
91. X. Liang et al., "Selenium supplementation enhanced the expression of selenoproteins in hippocampus and played a neuroprotective role in LPS-induced neuroinflammation," *Int. J. Biol. Macromol.* 2023. doi:10.1016/j.ijbiomac.2023.123740.
92. S. M. Nelson, X. Lei, and K. S. Prabhu, "Selenium levels affect the IL-4-induced expression of alternative activation markers in murine macrophages," *J. Nutr.* 2011. doi:10.3945/jn.111.141176.
93. Y. Saito, "Selenoprotein P as an in vivo redox regulator: disorders related to its deficiency and excess," *J. Clin. Biochem. Nutr.* 2020. doi:10.3164/jcbn.19-31.
94. A. Mahmoodpoor et al., "The effect of intravenous selenium on oxidative stress in critically ill patients with acute respiratory distress syndrome," *Immunol. Invest.* 2019. doi:10.1080/08820139.2018.1496098.
95. S. Dolati, R. Rikhtegar, A. Mehdizadeh, and M. Yousefi, "The role of magnesium in pathophysiology and migraine treatment," *Biol. Trace Elem. Res.* 2020. doi:10.1007/s12011-019-01931-z.
96. U. Gröber, J. Schmidt, and K. Kisters, "Magnesium in prevention and therapy," *Nutrients.* 2015. doi:10.3390/nu7095388.
97. S. Ray and K. W. Park, "Movement disorders and other neurologic impairment associated with hypomagnesemia: a systematic review," *Neurol. Clin. Pract.* vol. 13, no. 6, 2023. doi:10.1212/CPJ.0000000000200202.
98. M. Achari, C. E. Escamilla-Ocañas, and N. Amjadi, "Neurological symptoms in patients with zinc deficiency: a case series with documented blood levels (5323)," *Neurology*, 2020. doi:10.1212/wnl.94.15_supplement.5323.
99. M. K. Georgieff, "Long-term brain and behavioral consequences of early iron deficiency," *Nutr. Rev.* 2011. doi:10.1111/j.1753-4887.2011.00432.x.
100. K. Zych-Krekora, O. Sylwestrzak, and M. Krekora, "The critical role of iron in pregnancy, puerperium, and fetal development," *J. Clin. Med.* vol. 14, no. 10, pp. 1–13, 2025. doi:10.3390/jcm14103482.
101. H. A. Chami and M. A. K. Hall, "Copper deficiency and polyneuropathy: a case report," *Cureus*, 2022. doi:10.7759/cureus.28261.
102. A. Benkirane et al., "Case report: motor neuron disease phenotype associated with symptomatic copper deficiency: challenging diagnosis and treatment," *Front. Neurol.* 2023. doi:10.3389/fneur.2022.1063803.
103. A. Shahar et al., "Plasma selenium is positively related to performance in neurological tasks assessing coordination and motor speed," *Mov. Disord.* 2010. doi:10.1002/mds.23218.
104. V. S. Reddy, S. Bukke, N. Dutt, P. Rana, and A. K. Pandey, "A systematic review and meta-analysis of the circulatory, erythrocellular and CSF selenium levels in Alzheimer's disease: a metal meta-analysis (AMMA study-I)," *J. Trace Elem. Med. Biol.* 2017. doi:10.1016/j.jtemb.2017.04.005.
105. C. Joffre, A. L. Dinel, M. Chataigner, V. Pallet, and S. Layé, "N-3 polyunsaturated fatty acids and their derivates reduce neuroinflammation during aging," *Nutrients.* 2020. doi:10.3390/nu12030647.
106. D. Cao et al., "Docosahexaenoic acid promotes hippocampal neuronal development and synaptic function," *J. Neurochem.* 2009. doi:10.1111/j.1471-4159.2009.06335.x.

107. C. He, X. Qu, L. Cui, J. Wang, and J. X. Kang, "Improved spatial learning performance of fat-1 mice is associated with enhanced neurogenesis and neuritogenesis by docosahexaenoic acid," *Proc. Natl. Acad. Sci. USA*. 2009. doi:10.1073/pnas.0904835106.
108. H. Y. Kim et al., "N-docosahexaenoylethanolamide promotes development of hippocampal neurons," *Biochem. J.* 2011. doi:10.1042/BJ20102118.
109. M. Katakura et al., "Docosahexaenoic acid promotes neuronal differentiation by regulating basic helix-loop-helix transcription factors and cell cycle in neural stem cells," *Neuroscience*. 2009. doi:10.1016/j.neuroscience.2009.02.057.
110. M. A. Rashid and H. Y. Kim, "N-Docosahexaenoylethanolamine ameliorates ethanol-induced impairment of neural stem cell neurogenic differentiation," *Neuropharmacology*. 2016. doi:10.1016/j.neuropharm.2015.11.011.
111. C. L. Jensen et al., "Effects of early maternal docosahexaenoic acid intake on neuropsychological status and visual acuity at five years of age of breast-fed term infants," *J. Pediatr.* 2010. doi:10.1016/j.jpeds.2010.06.006.
112. C. Campoy et al., "Effects of prenatal fish-oil and 5-methyltetrahydrofolate supplementation on cognitive development of children at 6.5 y of age," *Am. J. Clin. Nutr.* 2011. doi:10.3945/ajcn.110.001107.
113. V. De Smedt-Peyrusse, F. Sargueil, A. Moranis, H. Harizi, S. Mongrand, and S. Layé, "Docosahexaenoic acid prevents lipopolysaccharide-induced cytokine production in microglial cells by inhibiting lipopolysaccharide receptor presentation but not its membrane subdomain localization," *J. Neurochem.* 2008. doi:10.1111/j.1471-4159.2007.05129.x.
114. A. Kawashima, T. Harada, H. Kami, T. Yano, K. Imada, and K. Mizuguchi, "Effects of eicosapentaenoic acid on synaptic plasticity, fatty acid profile and phosphoinositide 3-kinase signaling in rat hippocampus and differentiated PC12 cells," *J. Nutr. Biochem.* 2010. doi:10.1016/j.jnutbio.2008.12.015.
115. S. Hong, K. Gronert, P. R. Devchand, R. L. Moussignac, and C. N. Serhan, "Novel docosatrienes and 17S-resolvins generated from docosahexaenoic acid in murine brain, human blood, and glial cells: autacoids in anti-inflammation," *J. Biol. Chem.* 2003. doi:10.1074/jbc.M300218200.
116. W. J. Lukiw et al., "A role for docosahexaenoic acid-derived neuroprotectin D1 in neural cell survival and Alzheimer disease," *J. Clin. Invest.* 2005. doi:10.1172/JCI25420.
117. J. R. Drover et al., "Cognitive function in 18-month-old term infants of the DIAMOND study: a randomized, controlled clinical trial with multiple dietary levels of docosahexaenoic acid," *Early Hum. Dev.* 2011. doi:10.1016/j.earlhumdev.2010.12.047.
118. S. J. Meldrum, N. D'Vaz, K. Simmer, J. A. Dunstan, K. Hird, and S. L. Prescott, "Effects of high-dose fish oil supplementation during early infancy on neurodevelopment and language: a randomised controlled trial," *Br. J. Nutr.* 2012. doi:10.1017/S0007114511006878.
119. S. Li et al., "Omega-3 supplementation improves depressive symptoms, cognitive function and niacin skin flushing response in adolescent depression: a randomized controlled clinical trial," *J. Affect. Disord.* 2024. doi:10.1016/j.jad.2023.10.151.
120. H. B. Salman, M. A. Salman, and E. Y. Akal, "The effect of omega-3 fatty acid supplementation on weight loss and cognitive function in overweight or obese individuals on weight-loss diet," *Nutr. Hosp.* 2022. doi:10.20960/nh.03992.
121. Y. P. Zhang, Y. Lou, J. Hu, R. Miao, and F. Ma, "DHA supplementation improves cognitive function via enhancing Aβ-mediated autophagy in Chinese elderly with mild cognitive impairment: a randomised placebo-controlled trial," *J. Neurol. Neurosurg. Psychiatry*. 2018. doi:10.1136/jnnp-2017-316176.
122. P. A. Jackson, J. L. Reay, A. B. Scholey, and D. O. Kennedy, "Docosahexaenoic acid-rich fish oil modulates the cerebral hemodynamic response to cognitive tasks in healthy young adults," *Biol. Psychol.* 2012. doi:10.1016/j.biopsycho.2011.10.006.
123. A. Bertelli, M. Biagi, M. Corsini, G. Baini, G. Cappellucci, and E. Miraldi, "Polyphenols: from theory to practice," *Foods*. vol. 10, no. 11, 2021. doi:10.3390/foods10112595.
124. M. Jagadeesan, P. Radhakrishnan, A. Sasikala, M. R. Mallu, and Y. Dinakarkumar, "Bioactive constituents and radical scavenging properties of *Punica granatum* L. root extracts: focus on antioxidant and α-amylase inhibitory activities," *Curr. Res. Nutr. Food Sci. J.* vol. 13, no. Special-Issue-July, pp. 209–218, 2025. doi:10.12944/crnfsj.13.special-issue-july.13.
125. S. Onur, A. Ayhanci, and R. Ertekin, "Antioxidant defense system against oxidative stress," in Selin Temana, C. C. (ed.), *Oxidative Stress and Antioxidant Defense Systems*. Seruven Yayınevi, 2024.
126. B. Chen et al., "Anti-melanogenic activities of Sargassum fusiforme polyphenol-rich extract on α-MSH-stimulated B16F10 cells via PI3K/Akt and MAPK/ERK pathways," *Foods*. vol. 13, no. 22, pp. 1–16, 2024. doi:10.3390/foods13223556.
127. M. Jalouli et al., "Targeting natural antioxidant polyphenols to protect neuroinflammation and neurodegenerative diseases: a comprehensive review," *Front. Pharmacol.* vol. 16, no. January, pp. 1–18, 2025. doi:10.3389/fphar.2025.1492517.
128. P. E. Czabotar and A. J. Garcia-Saez, "Mechanisms of BCL-2 family proteins in mitochondrial apoptosis," *Nat. Rev. Mol. Cell Biol.* 2023. doi:10.1038/s41580-023-00629-4.
129. C. Tang et al., "Inhibition of soluble epoxide hydrolase ameliorates cerebral blood flow autoregulation and cognition in Alzheimer's disease and diabetes-related dementia rat models," *GeroScience*. vol. 47, no. 3, pp. 4429–4449, 2025. doi:10.1007/s11357-025-01550-8.
130. D. A. Kumar, "Brain food: a comprehensive review of nutritional strategies for cognitive enhancement and neuroprotection," *African J. Biomed. Res.* pp. 229–236, 2025. doi:10.53555/ajbr.v28i3s.7602.
131. N. Kothalawala and D. Abeykoon, "Enhancing nutrition & health benefits by utilizing phytochemicals as potent food additives ensure food safety and nutrition safety by utilizing phytochemicals as potent food additives," in Dr. Sumudu Mapa, Dr. R.D.A.A. Rajapaksha, Dr. M.A.P.C. Piyathilaka, Dr. Hasintha Wijesekara, and Prof. Nadarajah Kannan (eds.), *Sustainable Bioeconomy Approaches to Strengthen Economy, Society and the Environment in Sri Lanka*. YSF Thematic Publication, National Science and Technology Commission of Sri Lanka, 2025.
132. G. Santulli, P. Mone, and F. Varzideh, "GLP-1 receptor agonists and SGLT2 inhibitors: new anti-aging tools?" *Future Cardiol.* vol. 21, no. 1, pp. 5–8, 2025. doi:10.1080/14796678.2024.2433381.
133. B. Djordjević, "Phenolics compounds in fruits of different types of berries and their beneficent for human health," *Ann. Univ. Craiova Agric. Mont. Cadastre Ser.* 2023. doi:10.52846/aamc.v53i1.1443.

134. D. Pereira Netto et al., "Major relationships among polyphenols, gut microbiota and sports performance: a systematic review," *MedNEXT J. Med. Heal. Sci.* vol. 6, no. 1, 2025. doi:10.54448/mdnt25108.
135. T. K. Kelly, D. D. De Carvalho, and P. A. Jones, "Epigenetic modifications as therapeutic targets," *Nat. Biotechnol.* vol. 28, no. 10, pp. 1069–1078, 2010.
136. M. A. López-Bascón et al., "Characterization of phenolic compounds in extra virgin olive oil from granada (Spain) and evaluation of its neuroprotective action," *Int. J. Mol. Sci.* vol. 25, no. 9, 2024. doi:10.3390/ijms25094878.
137. O. Oluwole, W. B. Fernando, J. Lumanlan, O. Ademuyiwa, and V. Jayasena, "Role of phenolic acid, tannins, stilbenes, lignans and flavonoids in human health—a review," *Int. J. Food Sci. Technol.* vol. 57, no. 10, pp. 6326–6335, 2022. doi:10.1111/ijfs.15936.
138. Y. Zhang, S. Maejima, K. Matsuzaki, and H. Kishi, "Potential of functional flavonoids in targeting vasospasm through modulation of oxidative stress and SPC-induced signaling pathways," *Front. Pharmacol.* vol. 16, no. June, pp. 1–13, 2025. doi:10.3389/fphar.2025.1594060.
139. S. Chen et al., "Effects of phenolic acids on free radical scavenging and heavy metal bioavailability in kandelia obovata under cadmium and zinc stress," *Chemosphere.* 2020. doi:10.1016/j.chemosphere.2020.126341.
140. Q. Hua and L. Ren, "The SIRT1/Nrf2 signaling pathway mediates the anti-pulmonary fibrosis effect of liquiritigenin," *Chinese Med. (United Kingdom).* 2024. doi:10.1186/s13020-024-00886-1.
141. E. H. M. Hassanein et al., "The promising antioxidant effects of lignans: Nrf2 activation comes into view," *Naunyn. Schmiedebergs. Arch. Pharmacol.* vol. 397, no. 9, pp. 6439–6458, 2024. doi:10.1007/s00210-024-03102-x.
142. A. O. Atim, "Toxic synapses : mechanisms of neurotoxicity and disrupted neuromodulation in CNS disorders," *Euras. Exper. J. Public Health.* vol. 7, no. 3, pp. 23–28, 2025.
143. R. Ma, W. Zhang, L. Chen, and L. Tong, "Exploring the neuroprotective mechanisms of Jiawei Suanzaoren decoction in depression: insights from network pharmacology and molecular docking," *BMC Complement. Med. Ther.* vol. 25, no. 1, 2025. doi:10.1186/s12906-025-04939-2.
144. N. Asatiani, N. Sapojnikova, T. Kartvelishvili, L. Asanishvili, N. Sichinava, and Z. Chikovani, "Blood catalase, superoxide dismutase, and glutathione peroxidase activities in alcohol- and opioid-addicted patients," *Medicina.* vol. 61, no. 2, pp. 1–12, 2025. doi:10.3390/medicina61020204.
145. S. Akhter, H. M. Irfan, Alamgeer, S. Jahan, M. Shahzad, and M. B. Latif, "Nerolidol: a potential approach in rheumatoid arthritis through reduction of TNF-α, IL-1β, IL-6, NF-kB, COX-2 and antioxidant effect in CFA-induced arthritic model," *Inflammopharmacology.* vol. 30, no. 2, pp. 537–548, 2022. doi:10.1007/s10787-022-00930-2.
146. M. Plasay and L. Muslimin, "Melittin, a potential natural toxin with anticancer properties: regulating IL-1β, COX-2 and TNF-α in human colorectal cancer cells WiDr," *Trends Sci.* vol. 21, no. 4, 2024. doi:10.48048/tis.2024.7424.
147. M. N. Khan, D. Choudhary, S. Mehan, Z. Khan, G. Das Gupta, and A. S. Narula, "Molecular mechanisms of GDNF/GFRA1/RET and PI3K/AKT/ERK signaling interplay in neuroprotection: therapeutic strategies for treating neurological disorders," *Neuropeptides*, vol. 111, no. February, p. 102516, 2025. doi:10.1016/j.npep.2025.102516.
148. P. Patel, A. Mendoza, D. J. Robichaux, M. C. Wang, X. H. T. Wehrens, and J. Karch, "Inhibition of the anti-apoptotic Bcl-2 family by BH3 mimetics sensitize the mitochondrial permeability transition pore through bax and bak," *Front. Cell Dev. Biol.* vol. 9, no. December, pp. 1–12, 2021. doi:10.3389/fcell.2021.765973.
149. S. Gupta, A. K. Gupta, S. Mehan, Z. Khan, G. Das Gupta, and A. S. Narula, "Disruptions in cellular communication: molecular interplay between glutamate/NMDA signalling and MAPK pathways in neurological disorders," *Neuroscience.* 2025. doi:10.1016/j.neuroscience.2025.01.021.
150. A. Brunet, S. R. Datta, and M. E. Greenberg, "Transcription-dependent and -independent control of neuronal survival by the PI3K–Akt signaling pathway," *Curr. Opin. Neurobiol.* vol. 11, no. 3, pp. 297–305, 2001. doi:10.1016/S0959-4388(00)00211-7
151. A. Cantero-Téllez et al., "Systemic neuroprotection by chlorogenic acid: antioxidant and anti-inflammatory evaluation in early neurodegeneration induced by 3-nitropropionic acid in mice," *Neurochem. Res.* vol. 50, no. 2, pp. 1–13, 2025. doi:10.1007/s11064-025-04356-4.
152. H. Ma et al., "Melatonin suppresses ferroptosis induced by high glucose via activation of the Nrf2/HO-1 signaling pathway in type 2 diabetic osteoporosis," *Oxid. Med. Cell. Longev.* vol. 2020, 2020. doi:10.1155/2020/9067610.
153. D. Choudhary et al., "Navigating the complexities of neuronal signaling and targets in neurological disorders: from pathology to therapeutics," *Eur. J. Pharmacol.* vol. 995, no. February, p. 177417, 2025. doi:10.1016/j.ejphar.2025.177417.
154. M. Huang et al., "The antidepressant effects of kajiichigoside F1 via activating PPAR-γ/CX3CR1/Nrf2 signaling and suppressing NF-κB/NLRP3 signaling pathways," *Front. Pharmacol.* vol. 16, no. April, pp. 1–14, 2025. doi:10.3389/fphar.2025.1569888.
155. X. Gao, S. Li, C. Cong, Y. Wang, and L. Xu, "A network pharmacology approach to estimate potential targets of the active ingredients of epimedium for alleviating mild cognitive impairment and treating Alzheimer's disease," *Evidence-based Complement. Altern. Med.* 2021. doi:10.1155/2021/2302680.
156. M. Rivagorda et al., "A primary cilia–autophagy axis in hippocampal neurons is essential to maintain cognitive resilience," *Nat. Aging*, vol. 5, no. 3, pp. 450–467, 2025. doi:10.1038/s43587-024-00791-0.
157. A. Sapkota, S. K. Halder, S. Ismael, G. J. Bix, and R. Milner, "Partial eNOS deficiency results in greater levels of vascular inflammation and BBB disruption in response to chronic mild hypoxia," *Int. J. Mol. Sci.* 2025. doi:10.3390/ijms26167902.
158. L. Micheli, M. Caruso, G. D'Andrea, D. Volpe, M. Ceccarelli, and F. Tirone, "Survey of transcriptome analyses of hippocampal neurogenesis with focus on adult dentate gyrus stem cells," *Front. Cell Dev. Biol.* vol. 13, no. May, pp. 1–23, 2025. doi:10.3389/fcell.2025.1605116.
159. K. Faust, P. Vajkoczy, B. Xi, and D. Harnack, "The effects of deep brain stimulation of the subthalamic nucleus on vascular endothelial growth factor, brain-derived neurotrophic factor, and glial cell line-derived neurotrophic factor in a rat model of Parkinson's disease," *Stereotact. Funct. Neurosurg.* vol. 99, no. 3, pp. 256–266, 2021. doi:10.1159/000511121.
160. J. Robinson, N. Chawla, S. Patel, E. Spey, O. McNulty, and G. Kaur, "Neurodevelopmental abnormalities in

down syndrome: assessing structural and functional deficits," *Cureus*, vol. 16, no. 12, pp. 1–11, 2024. doi:10.7759/cureus.76156.

161. N. Jain, R. Shukla, V. Pandey, and A. K. Shukla, "Bioactive compounds for skin and beauty: antioxidants and their anti-aging effects on the skin, collagen boosters and their impact on skin elasticity, natural bioactive compounds for radiant and healthy skin," *Wellness Mol.* vol. 1, no. February, pp. 1–41, 2024.
162. A. O. Badmos, E. O. Oni, A. A. Aladesida, and F. Oluwafemi, "Protective role of *Nigella sativa* oil and vitamins C and E against aflatoxin M1-induced hematology and hepatic toxicity in neonatal rats," *Sci. Rep.* vol. 15, no. 1, pp. 1–16, 2025. doi:10.1038/s41598-025-10778-5.
163. M. Khan, M. Mushtaq, M. Usman, M. A. U. Rahman, and G. Quan, "Oxidative stress-induced cytotoxicity and the role of dietary antioxidants in farm animals: a review," *Adv. Redox Res.* vol. 16, no. August, p. 100138, 2025. doi:10.1016/j.arres.2025.100138.
164. Z. Q. Li et al., "Clostridium dalinum: a novel lactic acid bacterium with excellent antioxidant and anti-inflammatory functions through NF-κB, Keap1-Nrf2-ARE signaling pathways," *Probiotics Antimicrob. Proteins.* no. 0123456789, 2025. doi:10.1007/s12602-025-10655-7.
165. F. P. Mmanda, "Importance of minerals and their bioavailability in boosting aquaculture: a systematic review," *Aquac. Fish.* vol. 5, no. 3, 2025. doi:10.1002/aff2.70067.
166. D. Huang, Z. Zhuo, S. Fang, M. Yue, and J. Feng, "Different zinc sources have diverse impacts on gene expression of zinc absorption related transporters in intestinal porcine epithelial cells," *Biol. Trace Elem. Res.* vol. 173, no. 2, pp. 325–332, 2016. doi:10.1007/s12011-016-0655-x.
167. L.-G. Zhuang, R. Zhang, G.-X. Jin, X.-Y. Pei, Q. Wang, and X.-X. Ge, "Asiaticoside improves diabetic nephropathy by reducing inflammation, oxidative stress, and fibrosis: an in vitro and in vivo study," *World J. Diabetes.* vol. 15, no. 10, pp. 2111–2122, 2024. doi:10.4239/wjd.v15.i10.2111.
168. A. Budzianowska, K. Banaś, J. Budzianowski, and M. Kikowska, "Antioxidants to defend healthy and youthful skin—current trends and future directions in cosmetology," *Appl. Sci.* vol. 15, no. 5, 2025. doi:10.3390/app15052571.
169. R. Balakrishnan, Y. S. Kim, S. Il Kang, and D. K. Choi, "Green oat cognitaven® attenuates mild cognitive impairment by activating the CREB/BDNF/Nrf2/HO-1 pathway and modulating NF-κB/MAPK signaling," *Biomed. Pharmacother.* vol. 189, no. July, p. 118295, 2025. doi:10.1016/j.biopha.2025.118295.
170. M. Fiore, S. Terracina, and G. Ferraguti, "Brain neurotrophins and plant polyphenols: a powerful connection," *Molecules*, vol. 30, no. 12, pp. 1–29, 2025. doi:10.3390/molecules30122657.
171. R. R. Filho et al., "Effects of nonenzymatic and enzymatic antioxidants in cryopreserved canine semen extended with docosahexaenoic acid," *Clin. Theriogenology*, vol. 17, pp. 1–10, 2025. doi:10.58292/ct.v17.12690.
172. I. Bonfitto, G. Moniello, G. Francavilla, A. Bellomo, and S. Dimalta, "Unipolar and bipolar depressed inpatients: correlations with vitamin D and cognitive symptoms," *Eur. Psychiatry.* vol. 67, no. S1, pp. S545–S545, 2024. doi:10.1192/j.eurpsy.2024.1131.
173. T. Shirasawa, "Relationship between decline in cognitive function and homocysteine, folate, and vitamin B12," *ES J. Case Rep.* vol. 4, no. 2, pp. 1–6, 2023. doi:10.59152/esjcr/1042.
174. C. Yang, L. Chen, X. Guo, H. Sun, and D. Miao, "The vitamin D-Sirt1/PGC1α axis regulates bone metabolism and counteracts osteoporosis," *J. Orthop. Transl.* vol. 50, no. July, 2024, pp. 211–222, 2025. doi:10.1016/j.jot.2024.10.011.
175. J. H. Min, H. Sarlus, and R. A. Harris, "Glycyl-l-histidyl-l-lysine prevents copper- and zinc-induced protein aggregation and central nervous system cell death in vitro," *Metallomics.* vol. 16, no. 5, 2024. doi:10.1093/mtomcs/mfae019.
176. J. Zhang et al., "Effects of intermittent theta burst stimulation on the inflammatory response and cerebral blood flow in promoting neurovascular repair after ischemic stroke," *Mol. Brain.* vol. 18, no. 1, pp. 1–13, 2025. doi:10.1186/s13041-025-01222-w.
177. M. A. Meester-Smoor et al., "MN1 affects expression of genes involved in hematopoiesis and can enhance as well as inhibit RAR/RXR-induced gene expression," *Carcinogenesis.* vol. 29, no. 10, pp. 2025–2034, 2008. doi:10.1093/carcin/bgn168.
178. E. Garcia, F. Hernández-Ayvar, R. Rodríguez-Barrera, A. Flores-Romero, C. Borlongan, and A. Ibarra, "Supplementation with vitamin E, zinc, selenium, and copper re-establishes T-cell function and improves motor recovery in a rat model of spinal cord injury," *Cell Transplant.* vol. 31, no. 46, 2022. doi:10.1177/09636897221109884.
179. V. De Peña-arteaga, S. Bertolín, and I. Diez, "Mapping cross-modal functional connectivity of major neurotransmitter systems in the human brain," *Brain Struct. Funct.* vol. 230, pp. 1–16, 2025.
180. A. V. Shvets, K. V. Savichan, and Y. M. Klymenko, "Vitamin therapy in the treatment of mild traumatic brain injury: new horizons of clinical practice," *Ukr. Z. Vijskovoi Med.* vol. 5, no. 4, pp. 187–197, 2024. doi:10.46847/ujmm.2024.4(5)-187.
181. M. H. Rahman, "Effects of Vitamin E supplemented feed on growth performance of fish: a review," *Aquac. Fish.* vol. 7, no. 3, pp. 1–11, 2023. doi:10.24966/aaf-5523/100070.
182. G. N. Varlıhan, Ö. B. Ağralı, H. S. Güngörmek, S. D. Delipınar, L. Kuru, and H. Ö. Özener, "Vitamins K2 and D3 enhance guided bone regeneration in rat calvarial bone defects," *J. Periodontol.* no. July, 2025. doi:10.1002/jper.11372.
183. I. Domuschiev, "Natural molecules to slow down aging," March 2025. doi:10.13140/RG.2.2.31972.28806.
184. S. Petshow et al., "Activity-dependent regulation of Cdc42 by Ephexin5 drives synapse growth and stabilization," *Sci. Adv.* vol. 11, no. 13, pp. 1–13, 2025. doi:10.1126/sciadv.adp5782.
185. S. Gordon et al., "Associations of one-carbon metabolism, related B-vitamins and ApoE genotype with cognitive function in older adults: identification of a novel gene-nutrient interaction," *BMC Med.* vol. 23, no. 1, 2025. doi:10.1186/s12916-025-04276-8.
186. A. Dietrich-Muszalska, and B. Kontek, "Lipid peroxidation in patients with schizophrenia," *Psychiatry Clin. Neurosci.* vol. 64, no. 5, pp. 469–475, 2010.
187. L. Wang et al., "Membrane remodeling by FAM92A1 during brain development regulates neuronal morphology, synaptic function, and cognition," *Nat. Commun.* vol. 15, no. 1, 2024. doi:10.1038/s41467-024-50565-w.
188. A. A. Shuprovich, O. V. Zinich, N. M. Kushnaryova, and K. P. Komisarenko, "Folate cycle activity and neurotransmitter metabolism in the pathogenesis of neuropsychiatric disorders: A literature review," *Clin. Endocrinol. Endocrine Surgery*, vol. 2, no. 90, pp. 53–60, 2025.

189. M. Zhao et al., "Multiomics-based analysis of key genes, metabolites and pathways unveils mechanism associated with social rank in chickens," *Poult. Sci.* vol. 104, no. 7, p. 105192, 2025. doi:10.1016/j.psj.2025.105192.
190. S. Shkurashivska, D. Shkoruta, Y. Seniuk, H. Tokaryk, and L. Kuras, "Thiamine deficiency, clinical manifestations and treatment of Wernicke-Korsakoff syndrome," *Perspect. Innov. Sci.* vol. 5, no. 5(39), 2024. doi:10.52058/2786-4952-2024-5(39)-993-1004.
191. S. S. Godonu, "Role of vitamin B9 (folate) in neuronal plasticity," *Int. J. Sci. R. Tech.* vol. 2, no. January, pp. 77–82, 2025.
192. Z. K. Criss, K. Deans-Fielder, J. C. Fleet, S. Christakos, and N. Shroyer, "Analysis of 1,25-dihydroxyvitamin D genomic action in human enteroids and colonoids reveals multiple regulatory effects of vitamin D in human intestinal physiology," *Front. Endocrinol. (Lausanne).* vol. 16, no. June, pp. 1–15, 2025. doi:10.3389/fendo.2025.1538463.
193. J. Y. Chen, L. J. Shih, M. T. Liao, K. W. Tsai, K. C. Lu, and W. C. Hu, "Understanding the immune system's intricate balance: activation, tolerance, and self-protection," *Int. J. Mol. Sci.* vol. 26, no. 12, pp. 1–19, 2025. doi:10.3390/ijms26125503.
194. B. Zhang, W. Lian, J. Zhao, Z. Wang, A. Liu, and G. Du, "DL0410 alleviates memory impairment in d-galactose-induced aging rats by suppressing neuroinflammation via the TLR4/MyD88/NF-κB pathway," *Oxid. Med. Cell. Longev.* 2021. doi:10.1155/2021/6521146.
195. A. Naudí et al., "Region-specific vulnerability to lipid peroxidation and evidence of neuronal mechanisms for polyunsaturated fatty acid biosynthesis in the healthy adult human central nervous system," *Biochim. Biophys. Acta Mol. Cell Biol. Lipids.* 2017. doi:10.1016/j.bbalip.2017.02.001.
196. G. Gallo, "Advances in life sciences," *Adv. Life Sci.* vol. 2, no. November, 2022. doi:10.22271/int.book.191.
197. P. Mishra, E. Esfahani, P. Fernyhough, and B. C. Albensi, "Effects of 17-beta estradiol on inflammation and mitochondrial function in Alzheimer's disease," *Alzheimer's Dement.* vol. 20, no. S6, pp. 1–2, 2024. doi:10.1002/alz.091551.
198. C. Gokbulut, "The effects of vitamin E (α-tocopherol) on the male reproductive system and infertility," in Duncan, L. T. (ed.), *Advances in Health and Disease*, 2025. Nova Science Publishers.
199. F. Casale, S. Suh, K. Yale, and N. Atanaskova Mesinkovska, "Clinical role of oral vitamin C and E therapy in skin and hair disorders," *Int. J. Cosmet. Dermatol.* 2021. doi:10.55124/ijcd.v1i1.44.
200. S. Shao and L. Yang, "Research progress on traditional Chinese medicine treatment for osteoporosis based on gut microbiota regulation," *J. Contemp. Med. Pract.* vol. 7, no. 7, pp. 67–73, 2025. doi:10.53469/jcmp.2025.07(07).14.
201. B. Galunska, Y. Yotov, M. Nikolova, and A. Angelov, "Extrahepatic vitamin K-dependent gla-proteins–potential cardiometabolic biomarkers," *Int. J. Mol. Sci.* vol. 25, no. 6, 2024. doi:10.3390/ijms25063517.
202. S. Majumder, L. Miele, and R. Majumder, "Proof of concept: protein S as an immune modulatory agent to control pancreatic cancer," *Blood.* 2020. doi:10.1182/blood-2020-136388.
203. T. Zhang et al., "Regulation of synaptic function and lipid metabolism," *Neural Regen. Res.* vol. 21, no. 3, pp. 1037–1057, 2026. doi:10.4103/NRR.NRR-D-24-01412.
204. N. Wu et al., "Lithium chloride exerts anti-inflammatory and neuroprotective effects by inhibiting microglial activation in LPS-induced retinal injury," *Investig. Ophthalmol. Vis. Sci.* vol. 64, no. 3, 2023. doi:10.1167/iovs.64.3.35.
205. R. P. Casaroli-Marano, "Walnuts for the eye and vision," *Acta Ophthalmol.* 2024. doi:10.1111/aos.16465.
206. I. K. Koutsaliaris, D. Pantazi, A. N. Tsouka, O. Argyropoulou, C. C. Tellis, and A. D. Tselepis, "Differential effect of omega-3 fatty acids on platelet inhibition by antiplatelet drugs in vitro," *Int. J. Mol. Sci.* vol. 25, no. 18, 2024. doi:10.3390/ijms251810136.
207. C. E. Okoroama, P. C. Unekwe, L. C. Okoroama, S. U. Okparaoka, and G. C. Akuodor, "Evaluation of the protective role of antioxidants: α-tocopherol, vitamin C, and quercetin, against ibuprofen-induced renal damage in male Wistar rats," *Int. J. Basic Clin. Pharmacol.* 2023. doi:10.18203/2319-2003.ijbcp20232557.
208. Y. Deng et al., "Discovery of a novel chalcone-derived covalent Keap1 binder for mitigating cisplatin-induced mitochondrial dysfunction and nephrotoxicity," *Redox Biol.* vol. 85, no. May, p. 103737, 2025. doi:10.1016/j.redox.2025.103737.
209. D. J. I. Ugli et al., "The influence of polyphenols on calcium dynamics in synaptosomes of model rats with attention deficit hyperactivity disorder of varying ages," *Trends Sci.* vol. 22, no. 9, p. 10434, 2025. doi:10.48048/tis.2025.10434.
210. M. F. Nisar and C. Wan, "Nanotechnology-based delivery systems for enhanced bioavailability of antioxidant compounds in fruits and vegetables," *Curr. Res. Nutr. Food Sci. J.* vol. 13, no. Special-Issue-July, pp. 1–15, 2025. doi:10.12944/crnfsj.13.special-issue-july.01.
211. E. M. Repash, K. M. Pensabene, P. M. Palenchar, and A. L. Eggler, "Solving the problem of assessing synergy and antagonism for non-traditional dosing curve compounds using the DE/ZI method: application to Nrf2 activators," *Front. Pharmacol.* 2021. doi:10.3389/fphar.2021.686201.
212. I. Koutromanos et al., "Integrating gut microbiome and neuroplasticity genomics in alcohol use disorder therapy," *Hum. Genomics.* vol. 19, no. 1, 2025. doi:10.1186/s40246-025-00793-y.
213. J. A. Olin, M. J. Polito, and S. Cruz, "A multi-biomarker approach supports the use of compound-specific stable isotope analysis of amino acids to quantify basal carbon source use in a salt marsh consumer," no. November 2019, 2022. doi:10.1002/rcm.8538.
214. B. Han et al., "Cumulative handgrip strength and longitudinal changes in cognitive function and daily functioning among people aged 50 years and older: evidence from two longitudinal cohort studies," *Arch. Public Heal.* pp. 1–17, 2025. doi:10.1186/s13690-025-01624-1.
215. E. Prabu et al., "Impacts of dietary Saccharomyces cerevisiae fermentation derived postbiotic on growth performance and health status of Pacific white shrimp," *Sci. Rep.* vol. 15, no. 1, pp. 1–14, 2025. doi:10.1038/s41598-025-09556-0.
216. S. P. Vanare, R. K. Singh, J. Chen, and F. Kong, "Double emulsion microencapsulation system for Lactobacillus rhamnosus GG using pea protein and cellulose nanocrystals," *Foods.* vol. 14, no. 5, 2025. doi:10.3390/foods14050831.
217. S. F. Ndunguru et al., "Amino acid supplementation supports growth and reproductive development under dietary restriction," *Front. Anim. Sci.* vol. 6, no. July, pp. 1–13, 2025. doi:10.3389/fanim.2025.1622877.

218. M. Laubach et al., "Clinical translation of 3D-printed patient-specific bone implants: a consensus statement," *Int. J. Surg.* no. July, 2025. doi:10.1097/js9.0000000000002944.
219. J. Alacapa et al., "Engagement with digital adherence technologies as measures of intervention fidelity among adults with drug-susceptible tuberculosis and health care providers: descriptive analysis using data from cluster-randomized trials in five countries," *JMIR Public Heal. Surveill.* vol. 11, no. July, pp. e62881–e62881, 2025. doi:10.2196/62881.
220. N. C. Stemerdink et al., "An empirical dietary inflammatory pattern increases the incidence of the metabolic syndrome in a multi-ethnic Asian population," *Eur. J. Nutr.* vol. 64, no. 6, p. 256, 2025. doi:10.1007/s00394-025-03770-2.
221. A. Cisek-Wo, "Impact of vitamin D status and supplementation on brain-derived neurotrophic factor and mood-cognitive outcomes in adults : a structured impact," *Nutrients.* 2025. doi:10.3390/nu17162655.
222. N. Kataria et al., "Effect of vitamin B6, B9, and B12 supplementation on homocysteine level and cardiovascular outcomes in stroke patients: a meta-analysis of randomized controlled trials," *Cureus.* 2021. doi:10.7759/cureus.14958.
223. S. Hassan et al., "Does vitamin D3 supplementation improve ventilatory function of Vitamin D deficient COPD patients? A randomized controlled trial," *J. Med. Coll. Women Hosp.* vol. 21, no. 1, pp. 13–22, 2025. doi:10.3329/jmcwh.v21i1.79870.
224. Y. L. Feng, M. Xie, J. Tang, W. Huang, Q. Zhang, and S. S. Hou, "Effects of vitamin A on growth performance and tissue retinol of starter White Pekin ducks," *Poult. Sci.* vol. 98, no. 5, pp. 2189–2192, 2019. doi:10.3382/ps/pey571.
225. A. Zajac-Grabiec, K. Sroczynska, and J. Miszczyk, "Synergistic effects of omega-3 polyunsaturated fatty acids and aspirin in the clinical practice: a narrative review," *Curr. Issues Pharm. Med. Sci.* vol. 37, no. 1, pp. 59–67, 2024. doi:10.2478/cipms-2024-0010.
226. P. Ratna Rosalina and S. Maryati, "How does regular cow's milk consumption combined with vitamin D supplementation affect nutritional status and growth in malnourished children ? A systematic review," *Int. J. Med. Sci. Heal. Res.* vol. 12, no. 7, pp. 53–94, 2025. doi:10.70070/k79vr975.
227. D. C. Anih, K. A. Arowora, K. C. Ugwuoke, M. A. Abah, and B. Habibu, "Nutritional modulation of epigenetic changes induced by mycotoxins : a biochemical perspective for at-risk populations in Africa," no. August, 2025. doi:10.17311/sciintl.2025.90.109.
228. P. Aristotelous et al., "The effects of specific omega-3 and omega-6 polyunsaturated fatty acids and antioxidant vitamins on gait and functional capacity parameters in patients with relapsing-remitting multiple sclerosis," *Nutrients.* vol. 13, no. 10, 2021. doi:10.3390/nu13103661.
229. Q. Wu et al., "L-Theanine attenuates oxidative damage induced by heat stress through the PI3K/AKT/Nrf2 signaling pathway in skeletal muscle cells," *Poult. Sci.* vol. 104, no. 6, p. 105140, 2025. doi:10.1016/j.psj.2025.105140.
230. P. Y. Chiu, H. Y. Leung, and K. M. Ko, "Schisandrin B enhances renal mitochondrial antioxidant status, functional and structural integrity, and protects against gentamicin-induced nephrotoxicity in rats," *Biol. Pharm. Bull.* 2008. doi:10.1248/bpb.31.602.
231. B. Șuian and S. Amariei, "Extraction and characterization of bioactive compounds in horseradish (Armoracia rusticana): focus on polyphenols, vitamin C, and fatty acids," *Appl. Sci.* vol. 15, no. 12, 2025. doi:10.3390/app15126534.
232. M. S. Chishti et al., "Fertigation with alpha-tocopherol enhances morphological, physiological, and antioxidant responses in radish (Raphanus sativus L.) under drought stress," *BMC Plant Biol.* vol. 25, no. 1, 2025. doi:10.1186/s12870-025-06052-5.
233. R. Z. Alshahawy, S. M. Safwat, S. S. El-Kholy, S. F. El-basiony, S. El-Desouky, and S. M. Helmi, "Vitamin D and omega-3 fatty acids attenuate MSG-induced neurodegeneration by modulating tau pathology, neuroinflammation, and VDR expression in rats," *Sci. Rep.* vol. 15, no. 1, pp. 1–22, 2025. doi:10.1038/s41598-025-06109-3.
234. K. Singh and A. H. Sange, "Nutrition across the lifespan: a lifelong approach to brain health and cognitive resilience," *Cent. India J. Med. Res.* vol. 4, no. 01, pp. 4–6, 2025. doi:10.58999/cijmr.v4i01.219.
235. J. F. J. Rosit and R. B. Armecin, "Validating nutrient deficiencies in abaca (*Musa Textilis Née* var. *inosa*) via morpho-physiological and biochemical analysis under field conditions," *Int. J. Multidiscip. Appl. Bus. Educ. Res.* vol. 6, no. 6, pp. 2778–2784, 2025. doi:10.11594/ijmaber.06.06.10.
236. M. Eigenschink et al., "Redox properties of human erythrocytes are adapted for Vitamin c recycling," *Front. Physiol.* vol. 12, no. December, pp. 1–12, 2021. doi:10.3389/fphys.2021.767439.
237. E. Davidson, O. Obrosov, L. Coppey, and M. Yorek, "Omega-3 polyunsaturated fatty acids (PUFAs) and diabetic peripheral neuropathy: a pre-clinical study examining the effect of omega-3 PUFAs from fish oil, krill oil, algae or pharmaceutical-derived ethyl esters using type 2 diabetic rats," *Biomedicines.* vol. 13, no. 7, 2025. doi:10.3390/biomedicines13071607.
238. Z. Zhao, Q. Yang, Y. Sun, and X. Ruan, "Unveiling the antioxidant and anti-inflammatory potential of syringic acid: mechanistic insights and pathway interactions," *Front. Pharmacol.* vol. 16, no. July, pp. 1–22, 2025. doi:10.3389/fphar.2025.1615294.
239. V. Ognenoska and G. Pavlovska, "The power of superfoods: unlocking health benefits, nutritional value and sustainable choices," *Knowl. Int. J.* vol. 69, no. 3, pp. 577–580, 2025.
240. M. Asadi, N. Resources, and N. Asadzadeh, "The effect of vitamin and mineral supplement injection during the transition period on inflammatory and antioxidant indices of dairy cows," no. January, 2025. doi:10.22067/ijasr.2025.92312.1239.
241. A. O. Ghoweri et al., "Neuronal Calcium imaging, excitability, and plasticity changes in the Aldh2-/-mouse model of sporadic Alzheimer's disease," *J. Alzheimer's Dis.* vol. 77, no. 4, pp. 1623–1637, 2020. doi:10.3233/JAD-200617.
242. I. Tambones et al., "New structural insights into the control of the retinoic acid receptors RAR/RXR by DNA, ligands and transcriptional coregulators," no. March, 2025. doi:10.1101/2025.03.18.643890.
243. G. Xiao, M. Zhao, Z. Liu, F. Du, and B. Zhou, "Zinc antagonizes iron-regulation of tyrosine hydroxylase activity and dopamine production in Drosophila melanogaster," *BMC Biol.* vol. 19, no. 1, pp. 1–17, 2021. doi:10.1186/s12915-021-01168-0.
244. M. Naoi, Y. Wu, W. Maruyama, and M. Shamoto-Nagai, "Phytochemicals modulate biosynthesis and function of serotonin, dopamine, and norepinephrine for treatment of monoamine neurotransmission-related psychiatric diseases," *Int. J. Mol. Sci.* vol. 26, no. 7, 2025. doi:10.3390/ijms26072916.

245. A. C. Grădinaru and S. Popa, "Vitamin C: from self-sufficiency to dietary dependence in the framework of its biological functions and medical implications," *Life*. vol. 15, no. 2. 2025. doi:10.3390/life15020238.
246. A. V. Skalny, M. Aschner, A. A. Tinkov, and J. Aaseth, Trace elements and neurodegenerative diseases: Focus on Alzheimer's disease, *Nutrients*, vol. 10, no. 4, p. 454, 2018. doi:10.3390/nu10040454
247. S. C. R. Sherratt, R. A. Juliano, C. Copland, D. L. Bhatt, P. Libby, and R. P. Mason, "EPA and DHA containing phospholipids have contrasting effects on membrane structure," *J. Lipid Res*. vol. 62, 2021. doi:10.1016/J.JLR.2021.100106.
248. A. Gracia Aznar et al., "Pro-resolving inflammatory effects of a marine oil enriched in specialized pro-resolving mediators (SPMs) supplement and its implication in patients with post-COVID syndrome (PCS)," *Biomedicines*. vol. 12, no. 10, 2024. doi:10.3390/biomedicines12102221.
249. O. V. Ifejeokwu et al., "Immune checkpoint inhibition perturbs neuro-immune homeostasis and impairs cognitive function," *J. Exp. Clin. Cancer Res*. vol. 44, no. 1, pp. 1–18, 2025. doi:10.1186/s13046-025-03442-3.
250. R. Atallah, J. Gindlhuber, W. Platzer, R. Rajesh, and A. Heinemann, "Succinate regulates endothelial mitochondrial function and barrier integrity," *Antioxidants*. vol. 13, no. 12, pp. 1–21, 2024. doi:10.3390/antiox13121579.
251. A. Han, L. Almeida, N. Anand, J. Paulo, and L. Daher, "Exploring neuropsychiatric manifestations of vitamin B complex deficiencies," no. August, pp. 1–15, 2025. doi:10.3389/fpsyt.2025.1569826.
252. S. K. Wu, L. Wang, F. Wang, and J. Zhang, "Resveratrol improved mitochondrial biogenesis by activating SIRT1/PGC-1α signal pathway in SAP," *Sci. Rep*. vol. 14, no. 1, pp. 1–16, 2024. doi:10.1038/s41598-024-76825-9.
253. Z. Salimi, F. Moradpour, Z. Rashidi, F. Zarei, M. R. Khazaei, and S. M. Ahmadi, "The association of brain-derived neurotrophic factor with long-term potentiation," *Shiraz E Med. J*. vol. 23, no. 2, 2022. doi:10.5812/semj.112278.
254. R. Tan et al., "Leptin promotes the proliferation and neuronal differentiation of neural stem cells through the cooperative action of MAPK/ERK1/2, JAK2/STAT3 and PI3K/AKT signaling pathways," *Int. J. Mol. Sci*. vol. 24, no. 20, 2023. doi:10.3390/ijms242015151.
255. J. Kong, Y. Xie, R. Fan, Q. Wang, Y. Luo, and P. Dong, "Exercise orchestrates systemic metabolic and neuroimmune homeostasis via the brain–muscle–liver axis to slow down aging and neurodegeneration: a narrative review," *Eur. J. Med. Res*. vol. 30, no. 1, 2025. doi:10.1186/s40001-025-02751-9.
256. International American Academy of Biology, "Effect of AI-powered nutrition coaching on glycemic control in adults with type 2 diabetes : a narrative review of controlled trials," *Int. Am. Acad. J. Biol*. 2025. doi:10.59298/IAAJB/2025/1325256.
257. V. Mukamparirwa, S. M. S. Maliondo, and C. P. Mugunga, "Synergistic and antagonistic effects of mixed-leaf litter decomposition on nutrient cycling," *Plants*. vol. 13, no. 22, 2024. doi:10.3390/plants13223204.
258. S. Siddiqui, A. Srivastava, S. Hashmi, and A. Bhadauria, "The interplay between the gut flora and micronutrients: the key to modulating mineral and vitamin bioavailability," *Rev. Food Agric*. vol. 5, no. 1, pp. 54–60, 2024. doi:10.26480/rfna.01.2024.54.60.
259. C Manach, A. Scalbert, C. Morand, C. Rémésy, and L. Jiménez, "Polyphenols: Food sources and bioavailability," *Am. J. Clin. Nutr*., vol. 79, no. 5, pp. 727–747, 2004. doi:10.1093/ajcn/79.5.727
260. L. L. Barnes et al., "Trial of the MIND diet for prevention of cognitive decline in older persons," *N. Engl. J. Med*., 2023. doi:10.1056/nejmoa2302368.
261. M. Fekete et al., "The role of the Mediterranean diet in reducing the risk of cognitive impairement, dementia, and Alzheimer's disease: a meta-analysis," *GeroScience*, vol. 47, no. 3, pp. 3111–3130, 2025. doi:10.1007/s11357-024-01488-3.
262. S. Gillette-Guyonnet, M. Secher, and B. Vellas, "Nutrition and neurodegeneration: epidemiological evidence and challenges for future research," *Br. J. Clin. Pharmacol*. vol. 75, no. 3, pp. 738–755, 2013. doi:10.1111/bcp.12058.
263. M. Sahithi and M. Samskruthi, "Nutritional interventions for the prevention and treatment of neurological disorders such as anxiety, bipolar disorder, depression, epilepsy, multiple sclerosis, and schizophrenia," *J. Neurosci. Neurol. Disord*. 2022. doi:10.29328/journal.jnnd.1001070.
264. P. Zafrilla, J. M. Morillas, J. M. Rubio-Perez, and E. Cantos Villar, "Ingredients for functional drinks in neurodegenerative diseases: a review," *Nat. Prod. Commun*. 2009. doi:10.1177/1934578x0900400508.
265. M. A. Abdulwahab et al., "Nutritional deficiencies and their role in endodontic disease susceptibility and recovery," *Int. J. Community Med. Public Heal*. vol. 12, no. 1, pp. 535–539, 2024. doi:10.18203/2394-6040.ijcmph20244068.
266. J. K. Osei, J. S. Panti, M. N. Kumi, E. O. Brobbey, J. Abalem, and N. O. Appiah, "The role of nutrition in substance abuse recovery: addressing nutritional deficiencies for improved outcomes," *Asian J. Food Res. Nutr*. vol. 4, no. 1, pp. 77–85, 2025. doi:10.9734/ajfrn/2025/v4i1228.
267. M. C. Morris, D. A. Evans, J. A. Schneider, C. C. Tangney, J. L. Bienias, and N. T. Aggarwal, "Dietary folate and vitamins B-12 and B-6 not associated with incident Alzheimer's disease," *J. Alzheimer's Dis*. 2006. doi:10.3233/JAD-2006-9410.
268. L. M. L. De Lau, M. Bornebroek, J. C. M. Witteman, A. Hofman, P. J. Koudstaal, and M. M. B. Breteler, "Dietary fatty acids and the risk of Parkinson disease: the Rotterdam Study," *Neurology*, 2005. doi:10.1212/01.WNL.0000166038.67153.9F.
269. C. J. Frederickson, J. Y. Koh, and A. I. Bush, "The neurobiology of zinc in health and disease," *Nat. Rev. Neurosci*. 2005. doi:10.1038/nrn1671.
270. A. S. Prasad, "Zinc is an antioxidant and anti-inflammatory agent: its role in human health," *Front. Nutr*. 2014. doi:10.3389/fnut.2014.00014.
271. S. C. Dyall, "Long-chain omega-3 fatty acids and the brain: a review of the independent and shared effects of EPA, DPA and DHA," *Front. Aging Neurosci*. 2015. doi:10.3389/fnagi.2015.00052.
272. D. Vauzour, A. Rodriguez-Mateos, G. Corona, M. J. Oruna-Concha, and J. P. E. Spencer, "Polyphenols and human health: prevention of disease and mechanisms of action," *Nutrients*. 2010. doi:10.3390/nu2111106.
273. B. S. Solvik et al., "Dietary intake and biomarkers of folate and cobalamin status in Norwegian preschool children: the FINS-KIDS study," *J. Nutr*. 2020. doi:10.1093/jn/nxaa111.
274. A. D. Smith et al., "Homocysteine and dementia: an international consensus statement," *J. Alzheimers Dis*. 2018. doi:10.3233/JAD-171042.
275. A. D. Smith et al., "Homocysteine-lowering by B vitamins slows the rate of accelerated brain atrophy in mild cognitive impairment: a randomized controlled trial," *PLoS One*. 2010. doi:10.1371/journal.pone.0012244.
276. R. Clarke et al., "Vitamin B12 and folate deficiency in later life," *Age Ageing*. 2004. doi:10.1093/ageing/afg109.

277. D. Vauzour et al., "Nutrition for the ageing brain: towards evidence for an optimal diet," *Ageing Res. Rev.* 2017. doi:10.1016/j.arr.2016.09.010.
278. V. J. Clemente-Suárez, A. Martín-Rodríguez, A. Curiel-Regueros, A. Rubio-Zarapuz, and J. F. Tornero-Aguilera, "Neuro-nutrition and exercise synergy: exploring the bioengineering of cognitive enhancement and mental health optimization," *Bioengineering.* vol. 12, no. 2, pp. 1–59, 2025. doi:10.3390/bioengineering12020208.
279. R. V. Rao et al., "Rationale for a multi-factorial approach for the reversal of cognitive decline in Alzheimer's disease and MCI: a review," *Int. J. Mole. Sci.* 2023. doi:10.3390/ijms24021659.
280. M. A. Ortega et al., "Biological role of nutrients, food and dietary patterns in the prevention and clinical management of major depressive disorder," *Nutrients.* 2022. doi:10.3390/nu14153099.
281. A. C. Van Den Brink, E. M. Brouwer-Brolsma, A. A. M. Berendsen, and O. Van De Rest, "The Mediterranean, dietary approaches to stop hypertension (DASH), and mediterranean-dash intervention for neurodegenerative delay (MIND) diets are associated with less cognitive decline and a lower risk of Alzheimer's disease-a review," *Adv. Nutr.* 2019. doi:10.1093/advances/nmz054.
282. L. Huang et al., "Mediterranean-dietary approaches to stop hypertension intervention for neurodegenerative delay (MIND) diet and cognitive function and its decline: a prospective study and meta-analysis of cohort studies," *Am. J. Clin. Nutr.* 2023. doi:10.1016/j.ajcnut.2023.04.025.
283. G. A. Franco, L. Interdonato, M. Cordaro, S. Cuzzocrea, and R. Di Paola, "Bioactive compounds of the Mediterranean diet as nutritional support to fight neurodegenerative disease," *Int. J. Mole. Sci.* 2023. doi:10.3390/ijms24087318.
284. D. Smith, H. Refsum, A. Oulhaj, C. A. de Jager, and F. Jerneren, "Beneficial interactions between B vitamins and omega-3 fatty acids in the prevention of brain atrophy and of cognitive decline in early stage Alzheimer's disease," *FASEB J.* 2016. doi:10.1096/fasebj.30.1_supplement.407.6.
285. J. M. Bourre, "Effects of nutrients (in food) on the structure and function of the nervous system: update on dietary requirements for brain. Part 1: micronutrients," *J. Nutr. Health Aging.* vol. 10, pp. 377–385, 2006.
286. A. K. R. Abadio et al., "A pharmacophore-based virtual screening approach for the discovery of *Trypanosoma cruzi* GAPDH inhibitors," *J Chem Inf Model.* 2015. doi:10.1002/9783527633326.
287. M. Ohlau, A. Spiller, and A. Risius, "Plant-based diets are not enough? Understanding the consumption of plant-based meat alternatives along ultra-processed foods in different dietary patterns in Germany," *Front. Nutr.* 2022. doi:10.3389/fnut.2022.852936.
288. S. Chandana and N. Kumar Maurya, "Nutritional influences on hormonal homeostasis: exploring mechanisms and implications," *Int. J. Food Sci. Nutr.* vol. 9, no. 2, pp. 1–5, 2024.
289. M. C. Karlsen et al., "Theoretical food and nutrient composition of whole-food plant-based and vegan diets compared to current dietary recommendations," *Nutrients.* 2019. doi:10.3390/nu11030625.
290. S. T. Westreich, I. Korf, D. A. Mills, and D. G. Lemay, "Metatranscriptomics: monitoring gut microbiome activity in response to diet," *FASEB J.* 2016. doi:10.1096/fasebj.30.1_supplement.683.6.
291. K. A. Meyer et al., "Abstract P505: midlife cognitive function is associated with gut microbial species and metabolic pathways: coronary artery risk development in young adults (CARDIA) study," *Circulation.* 2023. doi:10.1161/circ.147.suppl_1.p505.
292. M. J. Dauncey, "Nutrition, the brain and cognitive decline: insights from epigenetics," *Eur. J. Clin. Nutr.* 2014. doi:10.1038/ejcn.2014.173.
293. M. Verma, R. Hontecillas, N. Tubau-Juni, V. Abedi, and J. Bassaganya-Riera, "Challenges in personalized nutrition and health," *Front. Nutr.* 2018. doi:10.3389/fnut.2018.00117.
294. H. Forster, M. C. Walsh, M. J. Gibney, L. Brennan, and E. R. Gibney, "Personalised nutrition: the role of new dietary assessment methods," *Pro. Nutr. Soc.* 2016. doi:10.1017/S0029665115002086.
295. M. N. Ravelli and D. A. Schoeller, "Traditional self-reported dietary instruments are prone to inaccuracies and new approaches are needed," *Front. Nutr.* 2020. doi:10.3389/fnut.2020.00090.
296. A. Skinner, Z. Toumpakari, C. Stone, and L. Johnson, "Future directions for integrative objective assessment of eating using wearable sensing technology," *Front. Nutr.* 2020. doi:10.3389/fnut.2020.00080.
297. S. Romero-Tapiador et al., "AI4FoodDB: a database for personalized e-Health nutrition and lifestyle through wearable devices and artificial intelligence," *Database*, 2023. doi:10.1093/database/baad049.
298. B. J. Mortazavi and R. Gutierrez-Osuna, "A review of digital innovations for diet monitoring and precision nutrition," *J. Diabetes Sci. Technol.* 2023. doi:10.1177/19322968211041356.
299. J. R. Townsend, T. O. Kirby, P. A. Sapp, A. M. Gonzalez, T. M. Marshall, and R. Esposito, "Nutrient synergy: definition, evidence, and future directions," *Front. Nutr.* 2023. doi:10.3389/fnut.2023.1279925.
300. R. Sproten, D. Nohr, and D. Guseva, "Nutritional strategies modulating the gut microbiome as a preventative and therapeutic approach in normal and pathological age-related cognitive decline: a systematic review of preclinical and clinical findings," *Nutr. Neurosci.* 2024. doi:10.1080/1028415X.2023.2296727.
301. G. Ribeiro, A. Ferri, G. Clarke, and J. F. Cryan, "Diet and the microbiota–gut–brain-axis: a primer for clinical nutrition," *Curr. Opin. Clin. Nutr. Metab. Care.* 2022. doi:10.1097/MCO.0000000000000874.
302. D. Raubenheimer et al., "An integrative approach to dietary balance across the life course," *iScience.* 2022. doi:10.1016/j.isci.2022.104315.
303. R. Alruwaili et al., "Diet and memory loss in older adults: a neuroscience perspective," *J. Pioneer. Med. Sci.* vol. 14, no. 04, pp. 7–20, 2025. doi:10.47310/jpms2025140402.
304. Y. I. Asiri, S. S. Moni, M. Ramar, and K. Chidambaram, "Advancing pain understanding and drug discovery: insights from preclinical models and recent research findings," *Pharmaceuticals.* vol. 17, no. 11, 2024. doi:10.3390/ph17111439.
305. O. F. Sultana, R. A. Hia, and P. H. Reddy, "A combinational therapy for preventing and delaying the onset of Alzheimer's disease: a focus on probiotic and vitamin co-supplementation," *Antioxidants.* 2024. doi:10.3390/antiox13020202.
306. C. G. Awuchi and C. O. R. Okpala, "Natural nutraceuticals, especially functional foods, their major bioactive components, formulation, and health benefits for disease prevention: an overview," *J. Food Bioact.* 2022. doi:10.31665/jfb.2022.18317.
307. M. K. Georgieff, S. E. Ramel, and S. E. Cusick, "Nutritional influences on brain development," *Acta Paediatr.* 2018. doi:10.1111/apa.14287.

308. N. Sanadgol, S. S. Zahedani, M. Sharifzadeh, R. Khalseh, G. R. Barbari, and M. Abdollahi, "Recent updates in imperative natural compounds for healthy brain and nerve function: a systematic review of implications for multiple sclerosis," *Curr. Drug Targets*, 2016. doi:10.2174/13894501 18666161108124414.
309. J. Chang et al., "Effects of vitamins and polyunsaturated fatty acids on cognitive function in older adults with mild cognitive impairment: a meta-analysis of randomized controlled trials," *Eur. J. Nutr.* 2024. doi:10.1007/s00394-024-03324-y.
310. N. Van Wijk et al., "Targeting synaptic dysfunction in Alzheimer's disease by administering a specific nutrient combination," *J. Alzheimer's Dis.* 2014. doi:10.3233/JAD-130998.
311. H. Koivisto et al., "Special lipid-based diets alleviate cognitive deficits in the APPswe/PS1dE9 transgenic mouse model of Alzheimer's disease independent of brain amyloid deposition," *J. Nutr. Biochem.* 2014. doi:10.1016/j.jnutbio.2013.09.015.
312. P. N. Pallier et al., "A nutrient combination designed to enhance synapse formation and function improves outcome in experimental spinal cord injury," *Neurobiol. Dis.* 2015. doi:10.1016/j.nbd.2015.09.007.
313. M. Fekete et al., "Improving cognitive function with nutritional supplements in aging: a comprehensive narrative review of clinical studies investigating the effects of vitamins, minerals, antioxidants, and other dietary supplements," *Nutrients.* 2023. doi:10.3390/nu15245116.
314. L. P. Chandravanshi, P. Agrawal, H. W. Darwish, and S. K. Trigun, "Impairments of spatial memory and N-methyl-d-aspartate receptors and their postsynaptic signaling molecules in the hippocampus of developing rats induced by As, Pb, and Mn mixture exposure," *Brain Sci.* 2023. doi:10.3390/brainsci13121715.
315. N. G. Norwitz and U. Naidoo, "Nutrition as metabolic treatment for anxiety," *Front. Psychiatry.* 2021. doi:10.3389/fpsyt.2021.598119.
316. J.-Y. Wang, L.-L. Wen, Y.-N. Huang, Y.-T. Chen, and M.-C. Ku, "Dual effects of antioxidants in neurodegeneration: direct neuroprotection against oxidative stress and indirect protection via suppression of gliamediated inflammation," *Curr. Pharm. Des.* 2006. doi:10.2174/138161206778343109.
317. J. Sharifi-Rad et al., "Multi-target mechanisms of phytochemicals in Alzheimer's disease: effects on oxidative stress, neuroinflammation and protein aggregation," *J. Person. Med.* 2022. doi:10.3390/jpm12091515.
318. R. Businaro, "Food supplements to complement brain functioning: the benefits of a combination of magnesium, folic acid, omega-3 fatty acids and vitamin E," *F1000Research.* 2022. doi:10.12688/f1000research.75856.1.
319. S. E. McKee and T. M. Reyes, "Effect of supplementation with methyl-donor nutrients on neurodevelopment and cognition: considerations for future research," *Nutr. Rev.* 2018. doi:10.1093/nutrit/nuy007.
320. O. J. Onaolapo, A. T. Olofinnade, F. O. Ojo, J. Falade, and A. Y. Onaolapo, "Prepubertal continuous dietary folate fortification enhances the brain function of adult mice by modulating antioxidant status, inflammation, and brain neurotransmitter levels," *Antiinflamm. Antiallergy. Agents Med. Chem.* 2023. doi:10.2174/0118715230249814230925060325.
321. R. A. Bekdash, "Epigenetics, nutrition, and the brain: improving mental health through diet," *Int. J. Mol. Sci.* vol. 25, no. 7, 2024. doi:10.3390/ijms25074036.
322. A. Winiarska-Mieczan, M. Kwiecień, K. Jachimowicz-Rogowska, J. Donaldson, E. Tomaszewska, and E. Baranowska-Wójcik, "Anti-inflammatory, antioxidant, and neuroprotective effects of polyphenols—polyphenols as an element of diet therapy in depressive disorders," *Int. J. Mole. Sci.* 2023. doi:10.3390/ijms24032258.
323. A. Dal-Pan et al., "Cognitive-enhancing effects of a polyphenols-rich extract from fruits without changes in neuropathology in an animal model of Alzheimer's disease," *J. Alzheimer's Dis.* 2016. doi:10.3233/JAD-160281.
324. R. Balakrishnan, S. Azam, D. Y. Cho, I. Su-Kim, and D. K. Choi, "Natural phytochemicals as novel therapeutic strategies to prevent and treat Parkinson's disease: current knowledge and future perspectives," *Oxid. Med. Cell. Longev.* 2021. doi:10.1155/2021/6680935.
325. B. Koletzko et al., "Nutrition during pregnancy, lactation and early childhood and its implications for maternal and long-term child health: the early nutrition project recommendations," *Ann. Nutr. Metab.* 2019. doi:10.1159/000496471.
326. N. E. Marshall et al., "The importance of nutrition in pregnancy and lactation: lifelong consequences," *Am. J. Obstet. Gynecol.* 2022. doi:10.1016/j.ajog.2021.12.035.
327. N. Irvine, G. England-Mason, C. J. Field, D. Dewey, and F. Aghajafari, "Prenatal folate and choline levels and brain and cognitive development in children: a critical narrative review," *Nutrients.* 2022. doi:10.3390/nu14020364.
328. A. del Saz-Lara, M. C. López de las Hazas, F. Visioli, and A. Dávalos, "Nutri-epigenetic effects of phenolic compounds from extra virgin olive oil: a systematic review," *Adv. Nutr.* 2022. doi:10.1093/advances/nmac067.
329. K. R. Gildawie, R. L. Galli, B. Shukitt-Hale, and A. N. Carey, "Protective effects of foods containing flavonoids on age-related cognitive decline," *Curr. Nutr. Rep.* 2018. doi:10.1007/s13668-018-0227-0.
330. Q. Ren, J. Sun, D. Xu, H. Xie, M. Ye, and Y. Zhao, "A dietary supplement containing micronutrients, phosphatidylserine, and docosahexaenoic acid counteracts cognitive impairment in D-galactose-induced aged rats," *Front. Nutr.* 2022. doi:10.3389/fnut.2022.931734.
331. F. Girodon et al., "Effect of a two-year supplementation with low doses of antioxidant vitamins and/or minerals in elderly subjects on levels of nutrients and antioxidant defense parameters," *J. Am. Coll. Nutr.* 1997. doi:10.1080/07315724.1997.10718698.
332. A. N. Khan, R. D. Jawarkar, M. E. A. Zaki, and A. A. Al Mutairi, "Natural compounds for oxidative stress and neuroprotection in schizophrenia: composition, mechanisms, and therapeutic potential," *Nutr. Neurosci.* 2024. doi:10.1080/1028415X.2024.2325233.
333. C. Crawford, C. Boyd, B. Avula, Y. H. Wang, I. A. Khan, and P. A. Deuster, "A public health issue: dietary supplements promoted for brain health and cognitive performance," *J. Altern. Complement. Med.* 2020. doi:10.1089/acm.2019.0447.
334. G. Bjelakovic, D. Nikolova, L. L. Gluud, R. G. Simonetti, and C. Gluud, "Antioxidant supplements for prevention of mortality in healthy participants and patients with various diseases," *Sao Paulo Med. J.* 2015. doi:10.1590/1516-3180.20151332T1.
335. Y. Li, A. Ma, X. Shao, and Z. Du, "Study the effect of antioxidant vitamin E, vitamin C and beta-carotene supplement on erythrocyte functions in elderly person," *Wei Sheng Yan Jiu.* vol. 37, p. 305, 2008.

336. I. C. Li, F. C. Chang, C. C. Kuo, H. T. Chu, T. J. Li, and C. C. Chen, "Pilot study: nutritional and preclinical safety investigation of fermented hispidin-enriched Sanghuangporus sanghuang mycelia: a promising functional food material to improve sleep," *Front. Nutr.* 2022. doi:10.3389/fnut.2021.788965.
337. K. L. Steinmetz and E. G. Spack, "The basics of preclinical drug development for neurodegenerative disease indications," *BMC Neurol.* 2009. doi:10.1186/1471-2377-9-S1-S2.
338. C. E. Zwilling, J. Wu, and A. K. Barbey, "Investigating nutrient biomarkers of healthy brain aging: a multimodal brain imaging study," *NPJ Aging*, vol. 10, no. 1, pp. 1–12, 2024. doi:10.1038/s41514-024-00150-8.
339. H. Zhang, Q. Zhang, and A. A. T. Monfared, "Incidence of Alzheimer's disease (AD) dementia and mild cognitive impairment (MCI) in the United States Medicare population," *Alzheimer's Dement.* 2023. doi:10.1002/alz.077092.
340. B. X. Song et al., "Evaluating the relationship between endostatin and cognition in mild cognitive impairment (MCI), coronary artery disease (CAD), and Alzheimer's disease (AD)," *Alzheimer's Dement.* 2022. doi:10.1002/alz.061846.
341. A. Ścibior, J. Llopis, P. P. Dobrakowski, and T. Męcik-Kronenberg, "Magnesium (Mg) and neurodegeneration: a comprehensive overview of studies on mg levels in biological specimens in humans affected some neurodegenerative disorders with an update on therapy and clinical trials supplemented with selected animal studies," *Int. J. Mol. Sci.* vol. 25, no. 23, pp. 1–28, 2024. doi:10.3390/ijms252312595.
342. C. Thygesen, M. R. Larsen, and B. Finsen, "Proteomic signatures of neuroinflammation in Alzheimer's disease, multiple sclerosis and ischemic stroke," *Expert Rev. Proteomics.* 2019. doi:10.1080/14789450.2019.1633919.
343. I. Ablinger et al., "Interdisciplinary approaches to deal with Alzheimer's disease—from bench to bedside: what feasible options do already exist today?" *Biomedicines.* 2022. doi:10.3390/biomedicines10112922.
344. D. Jansen et al., "Effects of specific multi-nutrient enriched diets on cerebral metabolism, cognition and neuropathology in AβPPswe-PS1dE9 mice," *PLoS One*, 2013. doi:10.1371/journal.pone.0075393.
345. S. N. Rai, P. Singh, H. W. M. Steinbusch, E. Vamanu, G. Ashraf, and M. P. Singh, "The role of vitamins in neurodegenerative disease: an update," *Biomedicines.* 2021. doi:10.3390/biomedicines9101284.
346. A. M. Johnstone, E. Albanese, D. R. Crabtree, B. Dalile, and S. Grabrucker, "Consensus statement on exploring the Nexus between nutrition, brain health and dementia prevention," *Nutr. Metab.* vol. 2, p. 82, 2025.
347. R. Molteni, F. Gómez-Pinilla, and V. Tucci, "Effects of diet on brain plasticity in animal and human studies: mind the gap," *Neural Plast.* 2014. doi:10.1155/2014/563160.
348. R. L. Wieggers et al., "P3-305: an overview of the medical food Souvenaid clinical trial program," *Alzheimer's Dement.* 2013. doi:10.1016/j.jalz.2013.05.1380.
349. M. G. M. Olde Rikkert et al., "Tolerability and safety of Souvenaid in patients with mild Alzheimer's disease: results of multi-center, 24-week, open-label extension study," *J. Alzheimer's Dis.* 2015. doi:10.3233/JAD-141305.
350. C.-C. Tan, C. Wang, and L. Tan, "Efficacy of vitamins B supplementation on mild cognitive impairment and Alzheimer's disease: a systematic review and meta-analysis," *Curr. Alzheimer Res.* 2014. doi:10.2174/156720501166614100114140.
351. P. Y. Lin et al., "Omega-3 fatty acids and blood-based biomarkers in Alzheimer's disease and mild cognitive impairment: a randomized placebo-controlled trial," *Brain. Behav. Immun.* 2022. doi:10.1016/j.bbi.2021.10.014.
352. R. Power et al., "Omega-3 fatty acid, carotenoid and vitamin E supplementation improves working memory in older adults: a randomised clinical trial," *Clin. Nutr.* 2022. doi:10.1016/j.clnu.2021.12.004.
353. C. C. Wang et al., "Antarctic krill oil exhibited synergistic effects with nobiletin and theanine in ameliorating memory and cognitive deficiency in SAMP8 mice: applying the perspective of the sea–land combination to retard brain aging," *Front. Aging Neurosci.* 2022. doi:10.3389/fnagi.2022.964077.
354. M. Fiala, L. Restrepo, and M. Pellegrini, "Immunotherapy of mild cognitive impairment by ω-3 supplementation: why are amyloid-β antibodies and ω-3 not working in clinical trials?," *J. Alzheimer's Dis.* 2018. doi:10.3233/JAD-170579.
355. N. Castellanos-Perilla, M. G. Borda, D. Aarsland, and G. E. Barreto, "An analysis of omega-3 clinical trials and a call for personalized supplementation for dementia prevention," *Expert Rev. Neurother.*. 2024. doi:10.1080/14737175.2024.2313547.
356. C. M. Vyas et al., "Apolipoprotein E and its association with cognitive change and modification of treatment effects of vitamin D3 and omega-3s on cognitive change: results from the in-clinic subset of a randomized clinical trial," *J. Gerontol. Ser. A Biol. Sci. Med. Sci.* 2024. doi:10.1093/gerona/glad260.
357. L. Costantini, R. Molinari, B. Farinon, and N. Merendino, "Impact of omega-3 fatty acids on the gut microbiota," *Int. J. Mol. Sci.* 2017. doi:10.3390/ijms18122645.
358. S. S. F. Putri, I. Irfannuddin, K. Murti, Y. Kesuma, H. Darmawan, and N. Koibuchi, "The role of gut microbiota on cognitive development in rodents: a meta-analysis," *J. Physiol. Sci.* 2023. doi:10.1186/s12576-023-00869-1.
359. J. Keijer et al., "Omics biomarkers and an approach for their practical implementation to delineate health status for personalized nutrition strategies," *Crit. Rev. Food Science Nutr.* 2024. doi:10.1080/10408398.2023.2198605.
360. B. Rodrigues, E. A. Asamane, R. Magalhães, N. Sousa, J. L. Thompson, and N. C. Santos, "The association of dietary patterns with cognition through the lens of neuroimaging—a Systematic review," *Ageing Res Rev.* 2020. doi:10.1016/j.arr.2020.101145.
361. G. Arjmand, M. Abbas-Zadeh, and M. H. Eftekhari, "Effect of MIND diet intervention on cognitive performance and brain structure in healthy obese women: a randomized controlled trial," *Sci. Rep.* 2022. doi:10.1038/s41598-021-04258-9.
362. S. C. Moreira, A. K. Jansen, and F. M. Silva, "Dietary interventions and cognition of Alzheimer's disease patients a systematic review of randomized controlled trial," *Dement. Neuropsychol.* 2020. doi:10.1590/1980-57642020dn14-030008.
363. E. Calderon Martinez et al., "The effects of omega-3, DHA, EPA, Souvenaid® in Alzheimer's disease: a systematic review and meta-analysis," *Neuropsychopharmacol. Rep.* vol. 44, no. 3, pp. 545–556, 2024. doi:10.1002/npr2.12455.
364. S. Ristori, G. Bertoni, E. Bientinesi, and D. Monti, "The role of nutraceuticals and functional foods in mitigating cellular senescence and its related aspects: a key strategy for delaying or preventing aging and neurodegenerative disorders," *Nutr.* vol. 17, no. 11, pp. 1–40, 2025. doi:10.3390/nu17111837.

365. J. F. Trepanowski and J. P. A. Ioannidis, "Perspective: limiting dependence on nonrandomized studies and improving randomized trials in human nutrition research: Why and how," *Adv. Nutr.* 2018. doi:10.1093/ADVANCES/NMY014.
366. B. Andriambelo, M. Stiffel, K. Roke, and M. Plourde, "New perspectives on randomized controlled trials with omega-3 fatty acid supplements and cognition: a scoping review," *Ageing Res. Rev.* 2023. doi:10.1016/j.arr.2022.101835.
367. F. K. Welty, "Omega-3 fatty acids and cognitive function," *Curr. Opin. Lipidol.* vol. 34, no. 1, pp. 12–21, 2023. doi:10.1097/MOL.0000000000000862.
368. J. S. Brainard et al., "Omega-3, omega-6, and polyunsaturated fat for cognition: systematic review and meta-analysis of randomized trials," *J. Am. Med. Direct. Assoc.* 2020. doi:10.1016/j.jamda.2020.02.022.
369. R. J. Kryscio et al., "Association of antioxidant supplement use and dementia in the prevention of Alzheimer's disease by Vitamin E and selenium trial (PREADViSE)," *JAMA Neurol.* 2017. doi:10.1001/jamaneurol.2016.5778.
370. K. Zuniga and E. McAuley, "Considerations in selection of diet assessment methods for examining the effect of nutrition on cognition," *J. Nutr. Heal. Aging*, 2015. doi:10.1007/s12603-014-0566-5.
371. V. E. Hedrick, A. M. Dietrich, P. A. Estabrooks, J. Savla, E. Serrano, and B. M. Davy, "Dietary biomarkers: advances, limitations and future directions," *Nutr. J.* 2012. doi:10.1186/1475-2891-11-109.
372. Y. Y. Ma, X. Li, J. T. Yu, and Y. J. Wang, "Therapeutics for neurodegenerative diseases by targeting the gut microbiome: from bench to bedside," *Transl. Neurodegen.* 2024. doi:10.1186/s40035-024-00404-1.
373. L. Shinto et al., "A randomized placebo-controlled pilot trial of omega-3 fatty acids and alpha lipoic acid in Alzheimer's disease," *J. Alzheimer's Dis.* 2014. doi:10.3233/JAD-130722.
374. G. Molla and M. Bitew, "Revolutionizing personalized medicine: synergy with multi-omics data generation, main hurdles, and future perspectives," *Biomedicines*, vol. 12, no. 12, pp. 1–30, 2024. doi:10.3390/biomedicines12122750.
375. A. Duda-Chodak and T. Tarko, "Possible side effects of polyphenols and their interactions with medicines," *Molecules.* 2023. doi:10.3390/molecules28062536.
376. S. Quesada-Vázquez et al., "Polyphenols and metabolism: from present knowledge to future challenges," *J. Physiol. Biochem.* vol. 80, no. 3, pp. 603–625, 2024. doi:10.1007/s13105-024-01046-7.
377. Y. Hussain, "Neuroprotective effects of phytonutrients at gene level," in Khan, H., Aschner, M. and Mirzaei, H. (eds.), *Phytonutrients and Neurological Disorders: Therapeutic and Toxicological Aspects.* Elsevier, 2023. doi:10.1016/B978-0-12-824467-8.00012-7.
378. T. Albrahim, "The potential role of nutritional components in improving brain function among patients with Alzheimer's disease: a meta-analysis of RCT studies," *Neurosciences*, 2020. doi:10.17712/nsj.2020.1.20190037.
379. M. N. Key and A. N. Szabo-Reed, "Impact of diet and exercise interventions on cognition and brain health in older adults: a narrative review," *Nutrients.* 2023. doi:10.3390/nu15112495.
380. M. Wagner et al., "The association of MIND diet with cognitive resilience to neuropathologies," *Alzheimer's Dement.*, 2023. doi:10.1002/alz.12982.
381. M. M. A. Bouwman, I. Frigerio, N. Reijner, W. D. J. van de Berg, and L. E. Jonkman, "Synaptic density in the hippocampal and parahippocampal subregions and its association with the severity of axonal damage and cognitive decline in Alzheimer's disease," *Alzheimer's Dement.* vol. 20, no. S1, pp. 1–3, 2024. doi:10.1002/alz.092170.
382. A. S. Doorduijn et al., "Associations of AD biomarkers and cognitive performance with nutritional status: the NUDAD project," *Nutrients*, 2019. doi:10.3390/nu11051161.
383. A. Kimura et al., "Malnutrition is associated with behavioral and psychiatric symptoms of dementia in older women with mild cognitive impairment and early-stage alzheimer's disease," *Nutrients*, 2019. doi:10.3390/nu11081951.
384. A. Onu et al., "Integrative Strategies for preventing and managing metabolic syndrome: the impact of exercise and diet on oxidative stress reduction—a review," *Life*, vol. 15, no. 5, 2025. doi:10.3390/life15050757.
385. M. Barbera et al., "Multimodal precision prevention: a new direction in Alzheimer's disease," *J. Prev. Alzheimer's Dis.* 2023. doi:10.14283/jpad.2023.114.
386. A. Więckowska-Gacek, A. Mietelska-Porowska, M. Wydrych, and U. Wojda, "Western diet as a trigger of Alzheimer's disease: from metabolic syndrome and systemic inflammation to neuroinflammation and neurodegeneration," *Ageing Res. Rev.* 2021. doi:10.1016/j.arr.2021.101397.
387. A. Mietelska-Porowska et al., "Induction of brain insulin resistance and alzheimer's molecular changes by western diet," *Int. J. Mol. Sci.* 2022. doi:10.3390/ijms23094744.
388. M. Candito et al., "Influence of meal on plasma folate and vitamin B12, by three methods—and on vitamin B6, homocysteine and red blood cell folate," *Pteridines*, 2005. doi:10.1515/pteridines.2005.16.1.22.
389. C. J. Grobler, W. H. Oldewage-Theron, and J. M. Chalwe, "The effect of vitamins B12, B6 and folate supplementation on homocysteine metabolism in a low-income, urbanised, black elderly community in South Africa," *South African J. Clin. Nutr.* 2022. doi:10.1080/16070658.2021.2003148.
390. M. Cavalieri et al., "B vitamins and magnetic resonance imaging-detected ischemic brain lesions in patients with recent transient ischemic attack or stroke: the Vitamins to prevent stroke (VITATOPS) MRI-substudy," *Stroke*, 2012. doi:10.1161/STROKEAHA.112.665703.
391. G. J. Hankey et al., "Effect of B vitamins and lowering homocysteine on cognitive impairment in patients with previous stroke or transient ischemic attack: a prespecified secondary analysis of a randomized, placebo-controlled trial and meta-analysis," *Stroke*, 2013. doi:10.1161/STROKEAHA.113.001886.
392. L. Shinto et al., "A randomized placebo-controlled pilot trial of omega-3 fatty acids and alpha lipoic acid in Alzheimer's disease," *J. Alzheimer's Dis.*, 2014.
393. A. Oulhaj, F. Jernerén, H. Refsum, A. D. Smith, and C. A. de Jager, "Omega-3 fatty acid status enhances the prevention of cognitive decline by B Vitamins in mild cognitive impairment," *J. Alzheimer's Dis.* 2016. doi:10.3233/JAD-150777.
394. G. Madhaw and N. Kumar, "Dietary and environmental risk factors in Parkinson's disease and Alzheimer's disease: beware of your food plate and your surroundings!" *Ann. Indian Acad. Neurol.* 2023. doi:10.4103/aian.aian_82_23.
395. D. Terenzi, A. K. Muth, and S. Q. Park, "Nutrition and gut–brain pathways impacting the onset of Parkinson's disease," *Nutrients.* 2022. doi:10.3390/nu14142781.

396. C. Boulos, N. Yaghi, R. El Hayeck, G. N. H. A. Heraoui, and N. Fakhoury-Sayegh, "Nutritional risk factors, microbiota and Parkinson's disease: what is the current evidence?," *Nutrients*. 2019. doi:10.3390/nu11081896.
397. S. Kalampokini, A. Becker, K. Fassbender, E. Lyros, and M. M. Unger, "Nonpharmacological modulation of chronic inflammation in Parkinson's disease: role of diet interventions," *Parkinson's Disease*. 2019. doi:10.1155/2019/7535472.
398. J. Wang, Y. Song, Z. Chen, and S. X. Leng, "Connection between systemic inflammation and neuroinflammation underlies neuroprotective mechanism of several phytochemicals in neurodegenerative diseases," *Oxid. Med. Cell. Longev.* 2018. doi:10.1155/2018/1972714.
399. T. He et al., "Mitochondrial dysfunction-targeting therapeutics of natural products in Parkinson's disease," *Front. Pharmacol.* 2023. doi:10.3389/fphar.2023.1117337.
400. H. F. Chiu, K. Venkatakrishnan, and C. K. Wang, "The role of nutraceuticals as a complementary therapy against various neurodegenerative diseases: a mini-review," *J. Trad. Complem. Med.* 2020. doi:10.1016/j.jtcme.2020.03.008.
401. G. Shanbhag, P. Panchal, and K. Menon, "Impact of dietary interventions on anxiety, depression and stress: a narrative review," *Indian J. Public Heal. Res. Dev.* 2019. doi:10.5958/0976-5506.2019.01815.1.
402. N. B. Boyle, J. Billington, C. Lawton, F. Quadt, and L. Dye, "A combination of green tea, rhodiola, magnesium and B vitamins modulates brain activity and protects against the effects of induced social stress in healthy volunteers," *Nutr. Neurosci.* 2022. doi:10.1080/1028415X.2021.1909204.
403. R. Businaro et al., "Therapeutic opportunities for food supplements in neurodegenerative disease and depression," *Front. Nutr.* 2021. doi:10.3389/fnut.2021.669846.
404. H. A. Young et al., "Multi-nutrient interventions and cognitive ageing: are we barking up the right tree?" *Nutr. Res. Rev.* 2023. doi:10.1017/S095442242200018X.
405. A. Agnihotri and O. I. Aruoma, "Alzheimer's disease and Parkinson's disease: a nutritional toxicology perspective of the impact of oxidative stress, mitochondrial dysfunction, nutrigenomics and environmental chemicals," *J. Am. Coll. Nutr.* 2020. doi:10.1080/07315724.2019.1683379.
406. R. A. Arias-Sánchez, L. Torner, and B. Fenton Navarro, "Polyphenols and neurodegenerative diseases: potential effects and mechanisms of neuroprotection," *Molecules*. 2023. doi:10.3390/molecules28145415.
407. A. Maruszak, A. Pilarski, T. Murphy, N. Branch, and S. Thuret, "Hippocampal neurogenesis in Alzheimer's disease: is there a role for dietary modulation?," *J. Alzheimer's Dis.* 2014. doi:10.3233/JAD-131004.
408. A. L. Tardy, E. Pouteau, D. Marquez, C. Yilmaz, and A. Scholey, "Vitamins and minerals for energy, fatigue and cognition: a narrative review of the biochemical and clinical evidence," *Nutrients*. 2020. doi:10.3390/nu12010228.
409. I. Grabska-Kobyłecka et al., "Polyphenols and their impact on the prevention of neurodegenerative diseases and development," *Nutrients*. 2023. doi:10.3390/nu15153454.
410. P. Jeandet, B. Delaunois, A. Aziz, D. Donnez, Y. Vasserot, S. Cordelier, and E. Courot, "Metabolic engineering of yeast and plants for the production of the biologically active hydroxystilbene, resveratrol," *J. Biomed. Biotechnol.* 2012. doi:10.1155/2012/579089.
411. M. Gerlach, D. Ben-Shachar, P. Riederer, and M. B. H. Youdim, "Altered brain metabolism of iron as a cause of neurodegenerative diseases?," *J. Neurochem.* 1994. doi:10.1046/j.1471-4159.1994.63030793.x.
412. A. Górska et al., "Distribution of iron, copper, zinc and cadmium in glia, their influence on glial cells and relationship with neurodegenerative diseases," *Brain Sciences*. 2023. doi:10.3390/brainsci13060911.
413. C. Marchetti, "Interaction of metal ions with neurotransmitter receptors and potential role in neurodiseases," *BioMetals*. 2014. doi:10.1007/s10534-014-9791-y.
414. P. W. Halcrow, N. Kumar, E. Hao, N. Khan, O. Meucci, and J. D. Geiger, "Mu opioid receptor-mediated release of endolysosome iron increases levels of mitochondrial iron, reactive oxygen species, and cell death," *NeuroImmune Pharmacol. Ther.* 2022. doi:10.1515/nipt-2022-0013.
415. Z. Zhao, "Iron and oxidizing species in oxidative stress and Alzheimer's disease," *Aging Medicine*. 2019. doi:10.1002/agm2.12074.
416. K. Maaroufi, "Iron-induced oxidative stress and anxiety-like behavior in Wistar rats following systemic iron administration," *Biol. Trace Elem. Res.* pp. 1–21, 2025. doi:10.1007/s12011-025-04908-3.
417. P. V. B. Reddy, K. V. Rama Rao, and M. D. Norenberg, "The mitochondrial permeability transition, and oxidative and nitrosative stress in the mechanism of copper toxicity in cultured neurons and astrocytes," *Lab. Investig.* 2008. doi:10.1038/labinvest.2008.49.
418. M. Aschner et al., "Mitochondrial pathways of copper neurotoxicity: focus on mitochondrial dynamics and mitophagy," *Front. Mol. Neurosci.* vol. 17, no. December, 2024. doi:10.3389/fnmol.2024.1504802.
419. L. Zhou et al., "Possible antidepressant mechanisms of omega-3 polyunsaturated fatty acids acting on the central nervous system," *Front. Psychiatry*. 2022. doi:10.3389/fpsyt.2022.933704.
420. K. Tanaka, A. A. Farooqui, N. J. Siddiqi, A. S. Alhomida, and W. Y. Ong, "Effects of docosahexaenoic acid on neurotransmission," *Biomole. Therapeut.* 2012. doi:10.4062/biomolther.2012.20.2.152.
421. C. Joffre, C. Rey, and S. Layé, "N-3 polyunsaturated fatty acids and the resolution of neuroinflammation," *Front. Pharmacol.* 2019. doi:10.3389/fphar.2019.01022.
422. E. Zgórzyńska, D. Stulczewski, B. Dziedzic, K. P. Su, and A. Walczewska, "Docosahexaenoic fatty acid reduces the pro-inflammatory response induced by IL-1β in astrocytes through inhibition of NF-κB and AP-1 transcription factor activation," *BMC Neurosci.* 2021. doi:10.1186/s12868-021-00611-w.
423. S. Afshordel et al., "Omega-3 polyunsaturated fatty acids improve mitochondrial dysfunction in brain aging—Impact of Bcl-2 and NPD-1 like metabolites," *Prostaglandins Leukot. Essent. Fat. Acids*, 2015. doi:10.1016/j.plefa.2014.05.008.
424. G. Barrera et al., "Mitochondrial dysfunction in cancer and neurodegenerative diseases: spotlight on fatty acid oxidation and lipoperoxidation products," *Antioxidants*. 2016. doi:10.3390/antiox5010007.
425. J. T. Erickson, T. A. Brosenitsch, and D. M. Katz, "Brain-derived neurotrophic factor and glial cell line-derived neurotrophic factor are required simultaneously for survival of dopaminergic primary sensory neurons in vivo," *J. Neurosci.* 2001. doi:10.1523/jneurosci.21-02-00581.2001.

426. Q. Yan, J. Wang, C. R. Matheson, and J. L. Urich, "Glial cell line-derived neurotrophic factor (GDNF) promotes the survival of axotomized retinal ganglion cells in adult rats: comparison to and combination with brain-derived neurotrophic factor (BDNF)," *J. Neurobiol.* 1999. doi:10.1002/(SICI)1097-4695(19990215)38:3<382::AID-NEU7>3.0.CO;2-5.
427. F. Zhang, J. S. Shi, H. Zhou, B. Wilson, J. S. Hong, and H. M. Gao, "Resveratrol protects dopamine neurons against lipopolysaccharide-induced neurotoxicity through its anti-inflammatory actions," *Mol. Pharmacol.* 2010. doi:10.1124/mol.110.064535.
428. R. Apak, M. Özyürek, K. Güçlü, and E. Çapanoılu, "Antioxidant activity/capacity measurement. 3. Reactive oxygen and nitrogen species (ROS/RNS) scavenging assays, oxidative stress biomarkers, and chromatographic/chemometric assays," *J. Agricult. Food Chem.* 2016. doi:10.1021/acs.jafc.5b04744.
429. M. Salvagno et al., "Oxidative stress and cerebral vascular tone: the role of reactive oxygen and nitrogen species," *Int. J. Mole. Sci.* 2024. doi:10.3390/ijms25053007.
430. Q. Li et al., "Tea polyphenols alleviate hydrogen peroxide-induced oxidative stress damage through the Mst/Nrf2 axis and the Keap1/Nrf2/HO-1 pathway in murine RAW264.7 cells," *Exp. Ther. Med.* 2021. doi:10.3892/etm.2021.10908.
431. Y. Zhou et al., "Recent advances of natural polyphenols activators for keap1-Nrf2 signaling pathway," *Chem. Biodivers.* 2019. doi:10.1002/cbdv.201900400.
432. J. P. E. Spencer, "The interactions of flavonoids within neuronal signalling pathways," *Genes Nutr.* 2007. doi:10.1007/s12263-007-0056-z.
433. V. Novak, B. Rogelj, and V. Župunski, "Therapeutic potential of polyphenols in amyotrophic lateral sclerosis and frontotemporal dementia," *Antioxidants.* 2021. doi:10.3390/antiox10081328.
434. V. Prakash Reddy, P. Aryal, S. Robinson, R. Rafiu, M. Obrenovich, and G. Perry, "Polyphenols in alzheimer's disease and in the gut–brain axis," *Microorganisms.* 2020. doi:10.3390/microorganisms8020199.
435. S. N. A. Bukhari, "Dietary polyphenols as therapeutic intervention for Alzheimer's disease: a mechanistic insight," *Antioxidants.* 2022. doi:10.3390/antiox11030554.
436. C. Moussa et al., "Resveratrol regulates neuro-inflammation and induces adaptive immunity in Alzheimer's disease," *J. Neuroinflammation*, 2017. doi:10.1186/s12974-016-0779-0.
437. S. Seshadri et al., "Plasma homocysteine as a risk factor for dementia and alzheimer's disease," *N. Engl. J. Med.* 2002. doi:10.1056/nejmoa011613.
438. R. Obeid and W. Herrmann, "Mechanisms of homocysteine neurotoxicity in neurodegenerative diseases with special reference to dementia," *FEBS Lett.* 2006. doi:10.1016/j.febslet.2006.04.088.
439. D. O. Kennedy, "B vitamins and the brain: mechanisms, dose and efficacy—A review," *Nutrients.* 2016. doi:10.3390/nu8020068.
440. J. Selhub, "Folate, vitamin B12 and vitamin B6 and one carbon metabolism," *J. Nutr. Health Aging*, 2002. doi:10.1007/s12603-002-0005-4.
441. M. G. Traber and J. F. Stevens, "Vitamins C and E: beneficial effects from a mechanistic perspective," *Free Radic. Biol. Med.* 2011. doi:10.1016/j.freeradbiomed.2011.05.017.
442. S. A. Schoenrock and L. M. Tarantino, "Developmental vitamin D deficiency and schizophrenia: the role of animal models," *Genes, Brain Behav.* 2016. doi:10.1111/gbb.12271.
443. J. P. Kesby, D. W. Eyles, T. H. J. Burne, and J. J. McGrath, "The effects of vitamin D on brain development and adult brain function," *Mol. Cell. Endocrinol.*. 2011. doi:10.1016/j.mce.2011.05.014.
444. G. Ferland, "Vitamin K and the nervous system: an overview of its actions," *Adv. Nutr.* 2012. doi:10.3945/an.111.001784.
445. V. J. Clemente-Suárez, A. Martín-Rodríguez, L. Redondo-Flórez, C. López-Mora, R. Yáñez-Sepúlveda, and J. F. Tornero-Aguilera, "New insights and potential therapeutic interventions in metabolic diseases," *Int. J. Mol. Sci.* 2023. doi:10.3390/ijms241310672.
446. J. G. Borges-Vieira and C. K. S. Cardoso, "Efficacy of B-vitamins and vitamin D therapy in improving depressive and anxiety disorders: a systematic review of randomized controlled trials," *Nutr. Neurosci.* 2023. doi:10.1080/1028415X.2022.2031494.
447. B. S. Baumel, P. M. Doraiswamy, M. Sabbagh, and R. Wurtman, "Potential neuroregenerative and neuroprotective effects of uridine/choline-enriched multinutrient dietary intervention for mild cognitive impairment: a narrative review," *Neurol. Therapy.* 2021. doi:10.1007/s40120-020-00227-y.
448. M. O. W. Grimm, D. M. Michaelson, and T. Hartmann, "Omega-3 fatty acids, lipids, and apoE lipidation in Alzheimer's disease: a rationale for multi-nutrient dementia prevention: thematic review series: ApoE and Lipid Homeostasis in Alzheimer's Disease," *J. Lipid Res.* 2017. doi:10.1016/j.biopsych.2005.05.009.
449. A. H. V. Schapira and E. Tolosa, "Molecular and clinical prodrome of Parkinson disease: implications for treatment," *Nat. Rev. Neurol.* 2010. doi:10.1038/nrneurol.2010.52.
450. T. J. Tracey, F. J. Steyn, E. J. Wolvetang, and S. T. Ngo, "Neuronal lipid metabolism: multiple pathways driving functional outcomes in health and disease," *Front. Mole. Neurosci.* 2018. doi:10.3389/fnmol.2018.00010.
451. S. Salomone, F. Caraci, G. M. Leggio, J. Fedotova, and F. Drago, "New pharmacological strategies for treatment of Alzheimer's disease: focus on disease modifying drugs," *Br. J. Clin. Pharmacol.* 2012. doi:10.1111/j.1365-2125.2011.04134.x.
452. J. J. Rucklidge and B. J. Kaplan, "Broad-spectrum micronutrient treatment for attention-deficit/hyperactivity disorder: rationale and evidence to date," *CNS Drugs*, 2014. doi:10.1007/s40263-014-0190-2.
453. J. J. Rucklidge, M. J. F. Eggleston, J. M. Johnstone, K. Darling, and C. M. Frampton, "Vitamin-mineral treatment improves aggression and emotional regulation in children with ADHD: a fully blinded, randomized, placebo-controlled trial," *J. Child Psychol. Psychiatry Allied Discip.* 2018. doi:10.1111/jcpp.12817.
454. L. M. Bodnar and K. L. Wisner, "Nutrition and depression: implications for improving mental health among childbearing-aged women," *Biol. Psychiatry.* 2005. doi:10.1016/j.biopsych.2005.05.009.
455. H. K. Manji, G. J. Moore, G. Rajkowska, and G. Chen, "Neuroplasticity and cellular resilience in mood disorders," *Mol. Psychiatry.* 2000. doi:10.1038/sj.mp.4000811.
456. R. Machado-Vieira, G. Salvadore, L. Ibrahim, N. Diaz-Granados, and C. Zarate Jr., "Targeting glutamatergic signaling for the development of novel therapeutics for mood disorders," *Curr. Pharm. Des.* 2009. doi:10.2174/138161209788168010.
457. L. Schwingshackl and G. Hoffmann, "Monounsaturated fatty acids and risk of cardiovascular disease: synopsis of the evidence available from systematic reviews and meta-analyses," *Nutrients.* 2012. doi:10.3390/nu4121989.

458. H. Yuan, J. Zhou, H. Li, S. Kang, and S. Park, "A systems biology approach to memory health: integrating network pharmacology, gut microbiota, and multi-omics for health functional foods," *Int. J. Mol. Sci.* vol. 26, no. 14, pp. 1–36, 2025. doi:10.3390/ijms26146698.
459. L. Gao, Q. Mao, J. Cao, Y. Wang, X. Zhou, and L. Fan, "Effects of coenzyme Q10 on vascular endothelial function in humans: a meta-analysis of randomized controlled trials," *Atherosclerosis*. 2012. doi:10.1016/j.atherosclerosis.2011.10.027.
460. Z. A. Bhutta, J. K. Das, A. Rizvi, M. F. Gaffey, N. Walker, S. Horton, P. Webb, A. Lartey, and R. E. Black, "Evidence-based interventions for improvement of maternal and child nutrition: what can be done and at what cost?" *The Lancet*, vol. 382, no. 9890, pp. 452–477, 2013. doi:10.1016/S0140-6736(13)60996-4.
461. S. N. Meydani et al., "Vitamin E and respiratory tract infections in elderly nursing home residents: a randomized controlled trial," *JAMA*. 2004. doi:10.1001/jama.292.7.828.
462. J. M. Johnstone, A. Hughes, J. Z. Goldenberg, A. R. Romijn, and J. J. Rucklidge, "Multinutrients for the treatment of psychiatric symptoms in clinical samples: a systematic review and meta-analysis of randomized controlled trials," *Nutrients*. 2020. doi:10.3390/nu12113394.
463. F. Gomes et al., "Association of nutritional support with clinical outcomes among medical inpatients who are malnourished or at nutritional risk: an updated systematic review and meta-analysis," *JAMA Netw. Open*. 2019. doi:10.1001/jamanetworkopen.2019.15138.
464. L. Noah, L. Dye, B. Bois De Fer, A. Mazur, G. Pickering, and E. Pouteau, "Effect of magnesium and vitamin B6 supplementation on mental health and quality of life in stressed healthy adults: post-hoc analysis of a randomised controlled trial," *Stress Heal*. 2021. doi:10.1002/smi.3051.
465. J. B. Adams et al., "Comprehensive nutritional and dietary intervention for autism spectrum disorder—a randomized, controlled 12-month trial," *Nutrients*. 2018. doi:10.3390/nu10030369.
466. J. Rivers-Auty et al., "Zinc status alters Alzheimer's disease progression through NLRP3-dependent inflammation," *J. Neurosci.* vol. 41, no. 13, pp. 3025–3038, 2021. doi:10.1523/JNEUROSCI.1980-20.2020.
467. C. Gaul, H. C. Diener, and U. Danesch, "Improvement of migraine symptoms with a proprietary supplement containing riboflavin, magnesium and Q10: a randomized, placebo-controlled, double-blind, multicenter trial," *J. Headache Pain*. vol. 16, no. 1, pp. 6–13, 2015. doi:10.1186/s10194-015-0516-6.
468. E. Ranjbar et al., "Effects of Zinc supplementation in patients with major depression: a randomized clinical trial," *Iran. J. Psychiatry*. vol. 8, no. 2, pp. 73–79, 2013.
469. M. J. Patan et al., "Supplementation with oil rich in eicosapentaenoic acid, but not in docosahexaenoic acid, improves global cognitive function in healthy, young adults: results from randomized controlled trials," *Am. J. Clin. Nutr.* vol. 114, no. 3, pp. 914–924, 2021. doi:10.1093/ajcn/nqab174.
470. G. P. Lim et al., "A diet enriched with the omega-3 fatty acid docosahexaenoic acid reduces amyloid burden in an aged Alzheimer mouse model," *J. Neurosci.* vol. 25, no. 12, pp. 3032–3040, 2005. doi:10.1523/JNEUROSCI.4225-04.2005.
471. M. F. O. M. Mi-Mba et al., "Differential impact of eicosapentaenoic acid and docosahexaenoic acid in an animal model of Alzheimer's disease," *J. Lipid Res.* vol. 65, no. 12, pp. 1–18, 2024. doi:10.1016/j.jlr.2024.100682.
472. K. E. Hopperton, M. O. Trépanier, V. Giuliano, and R. P. Bazinet, "Brain omega-3 polyunsaturated fatty acids modulate microglia cell number and morphology in response to intracerebroventricular amyloid-β 1–40 in mice," *J. Neuroinflammation*. vol. 13, no. 1, pp. 1–18, 2016. doi:10.1186/s12974-016-0721-5.
473. A. D. Dangour, V. A. Andreeva, E. Sydenham, and R. Uauy, "Omega 3 fatty acids and cognitive health in older people," *Br. J. Nutr.* vol. 107, no. Suppl. 2, pp. 152–158, 2012. doi:10.1017/S0007114512001547.
474. K. Yurko-Mauro et al., "Beneficial effects of docosahexaenoic acid on cognition in age-related cognitive decline," *Alzheimer's Dement.* 2010. doi:10.1016/j.jalz.2010.01.013.
475. F. Jernerén, A. K. Elshorbagy, A. Oulhaj, S. M. Smith, H. Refsum, and A. D. Smith, "Brain atrophy in cognitively impaired elderly: the importance of long-chain ω-3 fatty acids and B vitamin status in a randomized controlled trial," *Am. J. Clin. Nutr.* 2015. doi:10.3945/ajcn.114.103283.
476. R. S. Turner et al., "A randomized, double-blind, placebo-controlled trial of resveratrol for Alzheimer disease," *Neurology*. vol. 85, no. 16, pp. 1383–1391, 2015. doi:10.1212/WNL.0000000000002035.
477. H. Ohashi et al., "Combined treatment with curcumin and ferulic acid suppressed the aβ-induced neurotoxicity more than curcumin and ferulic acid alone," *Int. J. Mol. Sci.* vol. 23, no. 17, 2022. doi:10.3390/ijms23179685.
478. G. Bureau, F. Longpré, and M. G. Martinoli, "Resveratrol and quercetin, two natural polyphenols, reduce apoptotic neuronal cell death induced by neuroinflammation," *J. Neurosci. Res.* 2008. doi:10.1002/jnr.21503.
479. J. Wang et al., "Targeting multiple pathogenic mechanisms with polyphenols for the treatment of Alzheimer's disease-experimental approach and therapeutic implications," *Front. Aging Neurosci.* vol. 6, no. MAR, pp. 1–10, 2014. doi:10.3389/fnagi.2014.00042.
480. R. C. Y. Choi et al., "Synergistic action of flavonoids, baicalein, and daidzein in estrogenic and neuroprotective effects: a development of potential health products and therapeutic drugs against alzheimer's disease," *Evidence-based Complement. Altern. Med.* vol. 2013, 2013. doi:10.1155/2013/635694.
481. T. Mori, N. Koyama, J. Tan, T. Segawa, M. Maeda, and T. Town, "Combined treatment with the phenolics ()-epigallocatechin-3-gallate and ferulic acid improves cognition and reduces Alzheimer-like pathology in mice," *J. Biol. Chem.* vol. 294, no. 8, pp. 2714–2731, 2019. doi:10.1074/jbc.RA118.004280.
482. A. Zaky, A. Bassiouny, M. Farghaly, and B. M. El-Sabaa, "A combination of resveratrol and curcumin is effective against aluminum chloride-induced neuroinflammation in rats," *J. Alzheimer's Dis.* vol. 60, no. s1, pp. S221–S222, 2017. doi:10.3233/JAD-161115.
483. Q. Su, H. Pu, and C. Hu, "Neuroprotection by combination of resveratrol and enriched environment against ischemic brain injury in rats," *Neurol. Res.* vol. 38, no. 1, pp. 60–68, 2016. doi:10.1080/01616412.2015.1133027.
484. P. P. Zandi et al., "Reduced Risk of Alzheimer disease in users of antioxidant vitamin supplements: the cache county study," *Arch. Neurol.* 2004. doi:10.1001/archneur.61.1.82.
485. E. Sydenham, A. D. Dangour, and W. S. Lim, "Omega 3 fatty acid for the prevention of cognitive decline and dementia," *Sao Paulo Med. J.* vol. 130, no. 6, p. 419, 2012. doi:10.1590/s1516-31802012000600013.

486. N. Presse et al., "Vitamin K status and cognitive function in healthy older adults," *Neurobiol. Aging*. vol. 34, no. 12, pp. 2777–2783, 2013. doi:10.1016/j.neurobiolaging.2013.05.031.
487. T. Kacerova et al., "Role of B vitamins in modulating homocysteine and metabolic pathways linked to brain atrophy: metabolomics insights from the VITACOG trial," *Alzheimer's Dement*. vol. 21, no. 7, pp. 1–17, 2025. doi:10.1002/alz.70521.
488. S. K. Nishi et al., "Mediterranean, DASH, and MIND dietary patterns and cognitive function: the 2-year longitudinal changes in an older Spanish cohort," *Front. Aging Neurosci.* 2021. doi:10.3389/fnagi.2021.782067.
489. L. Barrea et al., "Comprehensive approach to medical nutrition therapy in patients with type 2 diabetes mellitus: from diet to bioactive compounds," *Antioxidants*. 2023. doi:10.3390/antiox12040904.
490. K. B. Schwarz, "Progressive decrease in plasma ω3 and ω6 fatty acids during pregnancy: time course and effects of dietary fats and antioxidant nutrients," *J. Nutr. Environ. Med.* vol. 8, pp. 335–344, 1998.
491. A. M. Ardekani et al., "Evaluating the association between the Mediterranean-DASH Intervention for Neurodegenerative Delay (MIND) diet, mental health, and cardio-metabolic risk factors among individuals with obesity," *BMC Endocr. Disord.* 2023. doi:10.1186/s12902-023-01284-8.
492. S. C. Duplantier and C. D. Gardner, "A critical review of the study of neuroprotective diets to reduce cognitive decline," *Nutrients*. 2021. doi:10.3390/nu13072264.
493. N. Aderinto, G. Olatunji, M. Abdulbasit, T. N. Olajide, and E. Kokori, "Examining the efficacy of the Mediterranean-DASH diet intervention for neurodegenerative delay in mitigating cognitive decline," *Egypt. J. Neurol. Psychiatr. Neurosurg.* 2023. doi:10.1186/s41983-023-00752-1.
494. Á. Hernáez et al., "Increased consumption of virgin olive oil, nuts, legumes, whole grains, and fish promotes HDL functions in humans," *Mol. Nutr. Food Res.* 2019. doi:10.1002/mnfr.201800847.
495. A. Kurowska, W. Ziemichód, M. Herbet, and I. Piątkowska-Chmiel, "The role of diet as a modulator of the inflammatory process in the neurological diseases," *Nutrients*. 2023. doi:10.3390/nu15061436.
496. R. J. Hardman, G. Kennedy, H. Macpherson, A. B. Scholey, and A. Pipingas, "Adherence to a mediterranean-style diet and effects on cognition in adults: a qualitative evaluation and systematic review of longitudinal and prospective trials," *Front. Nutr.*. 2016. doi:10.3389/fnut.2016.00022.
497. B. E. Wickman et al., "Dietary management of heart failure: dash diet and precision nutrition perspectives," *Nutrients*. 2021. doi:10.3390/nu13124424.
498. J. J. Valenzuela-Fuenzalida et al., "Effectiveness of duloxetine versus other therapeutic modalities in patients with diabetic neuropathic pain: a systematic review and meta-analysis," *Pharmaceuticals*, vol. 17, no. 7, 2024. doi:10.3390/ph17070856.
499. S. L. Gardener and S. R. Rainey-Smith, "The dietary approaches to stop hypertension (DASH) and Mediterranean-DASH intervention for neurodegenerative delay (MIND) diets and brain aging," in Riederer, P. and Wirth, M. (eds.), *Factors Affecting Neurological Aging: Genetics, Neurology, Behavior, and Diet*. Academic Press, 2021. doi:10.1016/B978-0-12-817990-1.00048-2.
500. P. Devranis et al., "Mediterranean diet, ketogenic diet or MIND Diet for aging populations with cognitive decline: a systematic review," *Life*. 2023. doi:10.3390/life13010173.
501. M. C. Morris, C. C. Tangney, Y. Wang, F. M. Sacks, D. A. Bennett, and N. T. Aggarwal, "MIND diet associated with reduced incidence of Alzheimer's disease," *Alzheimer's Dement*. 2015. doi:10.1016/j.jalz.2014.11.009.
502. J. Li et al., "Abstract 46: MIND diet-related transcriptomic alterations in the aging human brain is associated with lower dementia risk and slower cognitive decline," *Circulation*, 2023. doi:10.1161/circ.147.suppl_1.46.
503. D. Timlin, J. M. McCormack, M. Kerr, L. Keaver, and E. E. A. Simpson, "The MIND diet, cognitive function, and well-being among healthy adults at midlife: a randomised feasibility trial," *BMC Nutr.* vol. 11, no. 1, 2025. doi:10.1186/s40795-025-01020-6.
504. J. A. Seabrook et al., "Dietary patterns and brain health in middle-aged and older adults: a narrative review," *Nutr.* vol. 17, no. 9, pp. 1–22, 2025. doi:10.3390/nu17091436.
505. S. Mötteli, B. Provaznikova, S. Vetter, M. Jäger, E. Seifritz, and F. Hotzy, "Examining nutrition knowledge, skills, and eating behaviours in people with severe mental illness: a cross-sectional comparison among psychiatric inpatients, outpatients, and healthy adults," *Nutrients*, 2023. doi:10.3390/nu15092136.
506. N. D. Koblinsky, K. A. Power, L. Middleton, G. Ferland, and N. D. Anderson, "The role of the gut microbiome in diet and exercise effects on cognition: a review of the intervention literature," *J. Gerontol. A Biol. Sci. Med. Sci.* 2023. doi:10.1093/gerona/glac166.
507. M. C. Morris et al., "MIND diet slows cognitive decline with aging," *Alzheimer's Dement*. vol. 11, no. 9, pp. 1015–1022, 2015. doi:10.1016/j.jalz.2015.04.011.MIND.
508. Y. Liu, H. Fowler, D. D. Wang, and L. L. Barnes, "Mediterranean-DASH intervention for neurodegenerative delay (MIND) trial: genetic resource for precision nutrition," pp. 1–14, 2025. doi:10.3390/nu17152548.
509. C. Wang, P. Song, and Y. Niu, "The management of dementia worldwide: a review on policy practices, clinical guidelines, end-of-life care, and challenge along with aging population," *Biosci. Trends*, 2022. doi:10.5582/bst.2022.01042.
510. D. R. Jacobs and L. M. Steffen, "Nutrients, foods, and dietary patterns as exposures in research: a framework for food synergy," *Am. J. Clin. Nutr.* 2003. doi:10.1093/ajcn/78.3.508s.
511. M. K. Esquivel, "Nutrition benefits and considerations for whole foods plant-based eating patterns," *Am. J. Lifestyle Med.* 2022. doi:10.1177/15598276221075992.
512. K. J. M. Paulussen et al., "Leucine is more readily oxidized when ingested as an isolated nutrient versus incorporated in its whole-food matrix," *Curr. Dev. Nutr.* 2021. doi:10.1093/cdn/nzab041_031.
513. J. M. Aguilera, "The food matrix: implications in processing, nutrition and health," *Crit. Rev. Food Sci. Nutr.* 2019. doi:10.1080/10408398.2018.1502743.
514. A. I. Mulet-Cabero et al., "The dairy matrix: its importance, definition, and current application in the context of nutrition and health," *Nutr.* vol. 16, no. 17, 2024. doi:10.3390/nu16172908.
515. E. Dungubat, K. Fujikura, M. Kuroda, T. Fukusato, and Y. Takahashi, "Food nutrients and bioactive compounds for managing metabolic dysfunction-associated steatotic liver

disease: a comprehensive review," *Nutr.* vol. 17, no. 13, pp. 1–32, 2025. doi:10.3390/nu17132211.
516. S. L. Turgeon and G. Brisson, "Symposium review: the dairy matrix—Bioaccessibility and bioavailability of nutrients and physiological effects," *J. Dairy Sci.* 2020. doi:10.3168/jds.2019-17308.
517. S. van Vliet et al., "Greater stimulation of postexercise muscle protein synthesis after consumption of whole eggs versus egg whites in healthy young men," *FASEB J.* 2017. doi:10.1096/fasebj.31.1_supplement.164.7.
518. E. Combet and S. R. Gray, "Nutrient-nutrient interactions: competition, bioavailability, mechanism and function in health and diseases," *Proc. Nutr. Soc.* 2019. doi:10.1017/S0029665118002732.
519. N. T. Milman, "A review of nutrients and compounds, which promote or inhibit intestinal iron absorption: making a platform for dietary measures that can reduce iron uptake in patients with genetic haemochromatosis," *J. Nutr. Metab.* 2020. doi:10.1155/2020/7373498.
520. S. R. Lynch and J. D. Cook, "Interaction of vitamin C and iron," *Ann. N. Y. Acad. Sci.* 1980. doi:10.1111/j.1749-6632.1980.tb21325.x.
521. X. Kong, Y. Li, and X. Liu, "A review of thermosensitive antinutritional factors in plant-based foods," *J. Food Biochem.* 2022. doi:10.1111/jfbc.14199.
522. S. Huhn, S. K. Masouleh, A. Villringer, and A. V. Witte, "Components of a Mediterranean diet and their impact on cognitive functions in aging," *Front. Aging Neurosci.* 2015. doi:10.3389/fnagi.2015.00132.
523. M. Cassotta et al., "Organ-on-chip: the future of nutrition research in a one health world," *Food Front.*, vol. 6, no. 3, pp. 1205–1224, 2025. doi:10.1002/fft2.70000.
524. D. Kumar, R. Nadda, and R. Repaka, "Advances and challenges in organ-on-chip technology: toward mimicking human physiology and disease in vitro," *Med. Biol. Eng. Comput.* 2024. doi:10.1007/s11517-024-03062-7.
525. D. R. Jacobs and L. C. Tapsell, "Food synergy: the key to a healthy diet," *Proc. Nutr. Soc.* 2013. doi:10.1017/S0029665112003011.
526. M. K. Nair, L. F. Augustine, and A. Konapur, "Food-based interventions to modify diet quality and diversity to address multiple micronutrient deficiency," *Front. Public Health.* 2016. doi:10.3389/fpubh.2015.00277.
527. F. Leroy et al., "A framework for adequate nourishment: balancing nutrient density and food processing levels within the context of culturally and regionally appropriate diets," *Anim. Front.* vol. 15, no. 1, pp. 10–23, 2025. doi:10.1093/af/vfae032.
528. Z. Houshialsadat, G. Cediel, I. Sattamini, G. Scrinis, and P. Machado, "Ultra-processed foods, dietary diversity and micronutrient intakes in the Australian population," *Eur. J. Nutr.* 2024. doi:10.1007/s00394-023-03245-2.
529. E. Skaffari et al., "Food consumption and nutrient intake of Finnish preschool children according to parental educational level," *Br. J. Nutr.* 2024. doi:10.1017/S0007114523001460.
530. F. Shahidi and Y. Pan, "Influence of food matrix and food processing on the chemical interaction and bioaccessibility of dietary phytochemicals: a review," *Crit. Rev. Food Sci. Nutr.* 2022. doi:10.1080/10408398.2021.1901650.
531. D. Martini, H. da Costa Ribeiro, P. Gately, R. Mattes, R. Re, and D. Bier, "Positive nutrition: shifting the focus from nutrients to diet for a healthy lifestyle," *Eat. Weight Disord.* 2023. doi:10.1007/s40519-023-01580-1.
532. R. Saran, J. Kasar, M. Jha, J. K. Sahu, S. S. Babu, and A. Mutreja, "Building blocks of life: improving nutrition and health outcomes through fortification and breastfeeding in the first 1000 days of a child's life," *Front. Nutr.* vol. 12, no. March, pp. 1–13, 2025. doi:10.3389/fnut.2025.1517247.
533. D. J. Smid, L. Klous, S. B. Ballak, M. Catoire, I. M. De Hoogh, and F. P. M. Hoevenaars, "Exploring the role of nutritional strategies to influence physiological and cognitive mechanisms in cold weather operations in military personnel," *Front. Physiol.* vol. 16, no. February, pp. 1–12, 2025. doi:10.3389/fphys.2025.1539615.
534. J. Li et al., "Dietary Omega-3 PUFAs in metabolic disease research: a decade of omics-enabled insights (2014–2024)," *Nutr.*, vol. 17, no. 11, pp. 1–27, 2025. doi:10.3390/nu17111836.
535. K. O. Soetan, C. O. Olaiya, and O. E. Oyewole, "The importance of mineral elements for humans, domestic animals and plants: a review," *African J. Food Sci.* 2010. doi:10.5897/AJFS.9000140.
536. S. Chanda, B. N. Paul, K. Ghosh, and S. S. Giri, "Dietary essentiality of trace minerals in aquaculture-a review," *Agric. Rev.* vol. 36, no. 2, p. 100, 2015. doi:10.5958/0976-0741.2015.00012.4.
537. A. Frydrych, M. Krośniak, and K. Jurowski, "The role of chosen essential elements (Zn, Cu, Se, Fe, Mn) in food for special medical purposes (FSMPs) dedicated to oncology patients—critical review: state-of-the-art," *Nutrients.* 2023. doi:10.3390/nu15041012.
538. A. Shenkin, "Dietary reference values for vitamin A, vitamin K, arsenic, boron, chromium, copper, iodine, iron, manganese, molybdenum, nickel, silicon, vanadium and zinc," *J. Hum. Nutr. Diet.* 2003. doi:10.1046/j.1365-277x.2003.00439.x.
539. M. Akram et al., "Vitamins and minerals: types, sources and their functions," in *Functional Foods and Nutraceuticals*, 2020. doi:10.1007/978-3-030-42319-3_9.
540. N. S. Bez et al., "Development of a diet quality score and adherence to the Swiss dietary recommendations for vegans," *J. Heal. Popul. Nutr.* 2024. doi:10.1186/s41043-024-00498-3.
541. M. A. H. Lentjes, "The balance between food and dietary supplements in the general population," *Proc. Nutr. Soc.* 2019. doi:10.1017/s0029665118002525.
542. A. Stefanache et al., "Understanding how minerals contribute to optimal immune function," *J. Immunol. Res.* 2023. doi:10.1155/2023/3355733.
543. N. Singh, E. Dove, and D. L. Bhatt, "Omega-3 fatty acids," in Virani, S. S. (ed.), *Clinical Lipidology: A Companion to Braunwald's Heart Disease.* Elsevier, 2023. doi:10.1016/B978-0-323-88286-6.00019-4.
544. K. Napiórkowska-Baran et al., "Immunomodulatory effects of selected non-nutritive bioactive compounds and their role in optimal nutrition," *Cells.* vol. 47, no. 2. 2025. doi:10.3390/cimb47020089.
545. J. K. Bird and M. I. McBurney, "Seafood Intake of Americans and Meeting Omega-3 Consumption Recommendations," *FASEB J.* 2016. doi:10.1096/fasebj.30.1_supplement.1163.6.
546. Scott D. Doughman, Srirama Krupanidhi, and Carani B. Sanjeevi, "Omega-3 fatty acids for nutrition and medicine: considering microalgae oil as a vegetarian source of EPA and DHA," *Curr. Diabetes Rev.* 2007. doi:10.2174/157339907781368968.

547. Y. Zhang et al., "Higher ratio of plasma omega-6/omega-3 fatty acids is associated with greater risk of all-cause, cancer, and cardiovascular mortality: a population-based cohort study in UK Biobank," *Elife*. vol. 12, pp. 1–16, 2024. doi:10.7554/elife.90132.3.
548. P. Sarajlic et al., "Omega-3 to omega-6 fatty acid oxidation ratio as a novel inflammation resolution marker for metabolic complications in obesity," *Nutr. Metab. Cardiovasc. Dis*. 2023. doi:10.1016/j.numecd.2023.03.007.
549. S. Upadhyay and M. Dixit, "Role of polyphenols and other phytochemicals on molecular signaling," *Oxidat. Med. Cell. Longevity*. 2015. doi:10.1155/2015/504253.
550. M. L. Tereschuk, M. D. Baigorí, L. I. C. De Figueroa, and L. R. Abdala, "Flavonoids from Argentine Tagetes (Asteraceae) with antimicrobial activity," *Methods Mol. Biol.*, 2004. doi:10.1385/1-59259-766-1:317.
551. E. M. Yahia, M. E. M. Celis, and M. Svendsen, "The contribution of fruit and vegetable consumption to human health," in Yahia, E. M. (ed.), *Fruit and Vegetable Phytochemicals: Chemistry and Human Health,* 2nd edn. Wiley, 2017. doi:10.1002/9781119158042.ch1.
552. Z. Janis and S. Sharma, "Enhancing gut microbiome focusing on anti-inflammatory foods, dietary fiber, and polyphenols," *BIO Web Conf.* vol. 178, p. 02013, 2025. doi:10.1051/bioconf/202517802013.
553. A. Ullah et al., "Important flavonoids and their role as a therapeutic agent," *Molecules*. 2020. doi:10.3390/molecules25225243.
554. M. Hagan and N. Anyangwe, "Vitamin content in seaweeds: a systematic review on water-soluble and fat-soluble vitamins for adult daily intake," *Funct. Food Sci.* 2023. doi:10.31989/ffs.v3i10.1273.
555. V. P. Prasher and M. P. Janicki, *Physical Health of Adults with Intellectual and Developmental Disabilities*. Springer, 2018. doi:10.1007/978-3-319-90083-4.
556. M. J. Bruins and U. Létinois, "Adequate Vitamin D intake cannot be achieved within carbon emission limits unless food is fortified: a simulation study," *Nutrients*. 2021. doi:10.3390/nu13020592.
557. K. D. Cashman, "Vitamin D deficiency: defining, prevalence, causes, and strategies of addressing," *Calcified Tissue Int*. 2020. doi:10.1007/s00223-019-00559-4.
558. K. Maitland et al., "Children's oxygen administration strategies and nutrition trial (COAST-Nutrition): a protocol for a phase II randomised controlled trial," *Wellcome Open Res*. 2021. doi:10.12688/wellcomeopenres.17123.2.
559. M. J. Dauncey, "Recent advances in nutrition, genes and brain health," in *Proceedings of the Nutrition Society*, 2012. doi:10.1017/S0029665112000237.
560. B. Klimova, S. Dziuba, and A. Cierniak-Emerych, "The effect of healthy diet on cognitive performance among healthy seniors: a mini review," *Front. Human Neurosci*. 2020. doi:10.3389/fnhum.2020.00325.
561. L. Gutierrez et al., "Effects of nutrition on cognitive function in adults with or without cognitive impairment: a systematic review of randomized controlled clinical trials," *Nutrients*. 2021. doi:10.3390/nu13113728.
562. N. V. Matusheski et al., "Diets, nutrients, genes and the microbiome: recent advances in personalised nutrition," *Br. J. Nutr*. 2021. doi:10.1017/S0007114521000374.
563. K. M. Beathard et al., "The impact of nutrition on visual cognitive performance in the nutrition, vision, and cognition in sport study," *Front. Nutr*. 2023. doi:10.3389/fnut.2023.1208890.
564. L. Barros, "Natural antioxidants and human health effects," *Curr. Pharm. Des*. 2020. doi:10.2174/1381612826616200519085826.
565. S. H. Adams et al., "Perspective: guiding principles for the implementation of personalized nutrition approaches that benefit health and function," *Adv. Nutr*. 2020. doi:10.1093/advances/nmz086.
566. B. van Ommen et al., "Systems biology of personalized nutrition," *Nutr. Rev*. 2017. doi:10.1093/nutrit/nux029.
567. T. T. T. Tran et al., "APOE genotype influences the gut microbiome structure and function in humans and mice: relevance for Alzheimer's disease pathophysiology," *FASEB J*. 2019. doi:10.1096/fj.201900071R.
568. B. Z. Wei, L. Li, C. W. Dong, C. C. Tan, and W. Xu, "The relationship of omega-3 fatty acids with dementia and cognitive decline: evidence from prospective cohort studies of supplementation, dietary intake, and blood markers," *Am. J. Clin. Nutr*. 2023. doi:10.1016/j.ajcnut.2023.04.001.
569. P. Barberger-Gateau, C. Samieri, C. Feart, and M. Plourde, "Dietary omega 3 polyunsaturated fatty acids and Alzheimer's disease: interaction with apolipoprotein E genotype," *Curr. Alzheimer Res*. 2011. doi:10.2174/156720511796391926.
570. M. Fekete et al., "Functional foods in modern nutrition science: mechanisms, evidence, and public health implications," *Nutr.* vol. 17, no. 13, pp. 1–35, 2025. doi:10.3390/nu17132153.
571. P. Faccinetto-Beltrán, E. E. Reza-Zaldivar, D. A. Curiel-Pedraza, A. A. Canales-Aguirre, and D. A. Jacobo-Velázquez, "Docosahexaenoic acid (DHA), vitamin D3, and probiotics supplementation improve memory, glial reactivity, and oxidative stress biomarkers in an aluminum-induced cognitive impairment rat model," *ACS Omega*, vol. 9, no. 19, pp. 21221–21233, 2024. doi:10.1021/acsomega.4c01198.
572. S. Lee et al., "Probiotics that ameliorate cognitive impairment through anti-inflammation and anti-oxidation in mice," *Food Sci. Anim. Resour.* 2023. doi:10.5851/kosfa.2023.e22.
573. N. Kurhaluk, P. Kamiński, R. Bilski, R. Kołodziejska, A. Woźniak, and H. Tkaczenko, "Role of antioxidants in modulating the microbiota–gut–brain axis and their impact on neurodegenerative diseases," *Int. J. Mol. Sci.*, vol. 26, no. 8, 2025. doi:10.3390/ijms26083658.
574. E. Porras-García, I. Fernández-Espada Calderón, J. Gavala-González, and J. C. Fernández-García, "Potential neuroprotective effects of fermented foods and beverages in old age: a systematic review," *Front. Nutr.*. 2023. doi:10.3389/fnut.2023.1170841.
575. L. N. Zhu, X. Mei, Z. G. Zhang, Y. Ping Xie, and F. Lang, "Curcumin intervention for cognitive function in different types of people: a systematic review and meta-analysis," *Phytotherapy Res*. 2019. doi:10.1002/ptr.6257.
576. R. Lakhan, M. Sharma, K. Batra, and F. B. Beatty, "The role of vitamin e in slowing down mild cognitive impairment: a narrative review," *Healthcare (Switzerland)*. 2021. doi:10.3390/healthcare9111573.

577. A. Rees, G. F. Dodd, and J. P. E. Spencer, "The effects of flavonoids on cardiovascular health: a review of human intervention trials and implications for cerebrovascular function," *Nutrients*. 2018. doi:10.3390/nu10121852.
578. J. A. Estrada and I. Contreras, "Nutritional modulation of immune and central nervous system homeostasis: the role of diet in development of neuroinflammation and neurological disease," *Nutrients*, vol. 11, no. 5, 2019. doi:10.3390/nu11051076.
579. A. Sureda and X. Capó, "Editorial: unlocking the nexus of bioactive components, nutrition, and nutrigenomics in age-related disorders," *Front. Nutr.*, vol. 12, 2025. doi:10.3389/fnut.2025.1597658.
580. R. Barkhordari, M. Namayandeh, M. Mirzaei, M. H. Sohouli, and M. Hosseinzadeh, "The relation between MIND diet with psychological disorders and psychological stress among Iranian adults," *BMC Psychiatry*, 2022. doi:10.1186/s12888-022-04128-2.
581. K. A. Skarupski, C. C. Tangney, H. Li, D. A. Evans, and M. C. Morris, "Mediterranean diet and depressive symptoms among older adults over time," *J. Nutr. Heal. Aging*, 2013. doi:10.1007/s12603-012-0437-x.

9 Gut–Brain Axis

Microbiota–Vitamin Interactions and Their Role in Neurodegeneration

Monika Kaushik[1,5], *Hemraj Singh Rajput*[2], *Sumit Sharma*[1], *Jamal Basha Dudekula*[3], and *Sagar Pamu*[4]

[1]Amity Institute of Pharmacy, Amity University, Gwalior, India
[2]Department of Pharmacy Sumandeep Vidyapeeth Deemed University, Vadodara, India
[3]Sri Padmavathi School of Pharmacy, Tirupati, India
[4]Institute of Pharmacy, Nirma University, Ahmedabad, India
[5]School of medical and allied sciences, K R Mangalam University, Haryana, India

9.1 INTRODUCTION

9.1.1 Overview of the Gut–Brain Axis

The gut–brain axis (GBA) is a bidirectional network linking the gut and central nervous system through neural, hormonal, immune, and metabolic pathways, with the gut now recognized as a key regulator of cognition, emotion, and neurological health.

A central component of the GBA is the gut microbiota, the densely populated community of trillions of microorganisms residing in the gastrointestinal tract. These microbes not only assist in digestion and immune modulation but also produce neuroactive compounds such as short-chain fatty acids (SCFAs), gamma-aminobutyric acid (GABA), and serotonin precursors, which are capable of influencing CNS function either directly or through the activation of vagal pathways and systemic circulation [1]. In return, the brain can affect gut function via the hypothalamic–pituitary–adrenal (HPA) axis, which modulates gastrointestinal motility, secretion, and barrier integrity under stress or inflammatory conditions.

Recent decades have witnessed a growing appreciation for the interdependency between the gut microbiota and host micronutrient metabolism, particularly vitamins, which serve as cofactors for numerous biochemical processes critical for neural integrity. Vitamins produced or modulated by gut microbes play a role not only in maintaining metabolic homeostasis but also in the pathophysiology of neurodegenerative disorders, including Alzheimer's disease (AD), Parkinson's disease (PD), and multiple sclerosis (MS) [2].

9.1.2 Importance of Microbiota in Brain Health

The gut microbiota—comprising bacteria, viruses, archaea, and fungi—shapes digestion, immunity, and brain health, influencing development, function, and neurodegeneration risk. Early-life colonization aligns with key neurodevelopmental stages, impacting myelination, synaptogenesis, and blood–brain barrier (BBB) formation [3]. Germ-free animal studies—models that lack any microbial flora—have shown profound abnormalities in stress response, neurotransmitter levels, and brain structure, which can be reversed by microbiota reconstitution, thereby underscoring the microbiota's developmental importance [4].

Microbes exert their neuroregulatory effects through multiple mechanisms. One key avenue is the production of microbial metabolites, such as SCFAs (e.g., butyrate, propionate, acetate), which can cross the BBB and modulate neuroinflammation, gene expression, and neurogenesis [5]. Additionally, some microbes can synthesize or modulate the availability of key neurotransmitters and neuromodulators—such as GABA, dopamine, and serotonin—directly within the gut milieu [6]. As approximately 90% of the body's serotonin is produced in the gut, largely under the influence of microbial activity, the implications for mood regulation and mental health are profound.

Moreover, the gut microbiota interacts closely with the immune system, modulating systemic cytokine levels and influencing the CNS through immune signaling pathways. Chronic low-grade inflammation, often associated with gut dysbiosis, is now recognized as a contributing factor in neurodegenerative diseases such as AD and PD [7]. This neuroimmune crosstalk is a key feature of the GBA, linking microbial ecology to neural integrity.

Alterations in microbiota composition—whether due to diet, antibiotics, aging, or disease—can therefore significantly affect brain function. Specific bacterial taxa have been associated with cognitive decline, depression, and MS, suggesting that the microbiota is not merely a passive player but an active determinant of neurological outcomes [8].

DOI: 10.1201/9781003686675-9

9.1.3 Role of Vitamins in Neurophysiology

Vitamins are essential for neurotransmitter synthesis, myelin formation, mitochondrial function, and neuroplasticity, making them critical to CNS health. Deficiencies—especially in B1, B6, B9, and B12—disrupt one-carbon metabolism, DNA methylation, and neurotransmitter production, with folate and B12 deficiencies elevating homocysteine levels and increasing dementia risk [9]. Similarly, vitamin B6 functions in the decarboxylation of glutamate to GABA, directly affecting neuronal excitability and seizure susceptibility [10].

Vitamin D, a secosteroid hormone, has emerged as another critical modulator of brain health. It exerts neuroprotective effects through anti-inflammatory, antioxidant, and neurotrophic mechanisms, influencing the expression of neurotrophins such as nerve growth factor (NGF) and brain-derived neurotrophic factor (BDNF) [11]. Low serum levels of vitamin D have been associated with increased risks of MS, AD, and mood disorders, possibly via impaired calcium homeostasis and immune dysregulation [12].

In addition, fat-soluble antioxidants such as vitamins A, C, and E contribute to redox balance in the CNS, protecting neural tissue from oxidative stress, a key pathological feature in neurodegeneration. Vitamin E, in particular, prevents lipid peroxidation in neuronal membranes, while vitamin C plays a role in catecholamine synthesis and acts synergistically with vitamin E in scavenging reactive oxygen species (ROS) [13].

9.2 THE GUT MICROBIOTA AND ITS FUNCTIONS

9.2.1 Composition and Diversity

The gut microbiota, a vast ecosystem of 10^{13}–10^{14} microbes with a gene pool far exceeding the human genome, supports digestion, immunity, and neurological health. Dominated by Firmicutes and Bacteroidetes—together over 90%—it also includes Actinobacteria, Proteobacteria, and Verrucomicrobia, with composition influencing CNS resilience [14]. The ratio of these dominant phyla, often referred to as the Firmicutes/Bacteroidetes (F/B) ratio, has been associated with various health outcomes, including metabolic and neuropsychiatric disorders. For example, alterations in the F/B ratio have been observed in individuals with autism spectrum disorders, depression, and PD [15].

Microbial diversity—both in terms of richness (number of species) and evenness (distribution)—is considered a marker of gut health. Higher microbial diversity is generally associated with greater ecosystem stability and host resilience, whereas low diversity is linked to dysbiosis, inflammation, and impaired gut barrier function. Importantly, microbial composition is not static; it is influenced by age, diet, antibiotic use, environment, and host genetics. Infants are born with sterile guts, but colonization begins at birth and evolves over time, achieving relative stability by adulthood, though still remaining responsive to external influences [16].

Several microbial taxa have been specifically associated with neurological health. For example, *Lactobacillus* and *Bifidobacterium* species have been shown to modulate stress reactivity, cognitive function, and emotional regulation, potentially through their capacity to produce GABA and SCFAs [17]. *Akkermansia muciniphila*, a mucin-degrading bacterium, has gained attention for its role in maintaining gut barrier integrity and reducing systemic inflammation, with potential implications for neurodegenerative diseases [18].

In contrast, an overabundance of pathobionts—potentially harmful microbes such as certain strains of *Escherichia coli* or *Clostridium difficile*—can exacerbate inflammation and compromise gut–brain communication. This microbial imbalance, or dysbiosis, has been implicated in the pathogenesis of conditions ranging from irritable bowel syndrome (IBS) to AD [19].

9.2.2 Microbiota-Derived Metabolites

The gut microbiota supports host physiology by producing metabolites—such as SCFAs, neurotransmitter-like molecules, bile acids, indoles, phenolics, and vitamins—that signal through the GBA to affect neural, immune, and hormonal pathways. SCFAs like acetate, propionate, and butyrate, generated by fiber-fermenting bacteria (e.g., *Faecalibacterium prausnitzii*, *Roseburia*, *Bacteroides*), not only fuel colonocytes but also strengthen the BBB, modulate microglia, and influence gene expression via epigenetic mechanisms like histone deacetylase (HDAC) inhibition [20]. Butyrate, in particular, has been shown to exhibit anti-inflammatory and neuroprotective properties, making it a candidate for therapeutic investigation in diseases such as AD and depression [5].

In addition to SCFAs, the microbiota produces a range of neuroactive compounds that mirror classical neurotransmitters. For instance, *Lactobacillus* and *Bifidobacterium* species are known to synthesize GABA, *Escherichia* and *Streptococcus* species can generate serotonin and dopamine precursors, while certain *Clostridium* strains produce acetylcholine-like molecules [6]. These microbial metabolites can modulate neural signaling either locally via enteric neurons or systemically by interacting with vagal afferents, immune cells, and endocrine pathways.

Another significant class of microbiota-derived metabolites includes secondary bile acids, which are modified from primary bile acids in the colon. These compounds act as ligands for receptors such as farnesoid X receptor (FXR) and Takeda G protein-coupled receptor 5 (TGR5), thereby influencing neuroinflammatory pathways, energy metabolism, and glucocorticoid homeostasis [21]. Dysregulation of bile acid metabolism has been implicated in cognitive decline and Parkinsonian pathology.

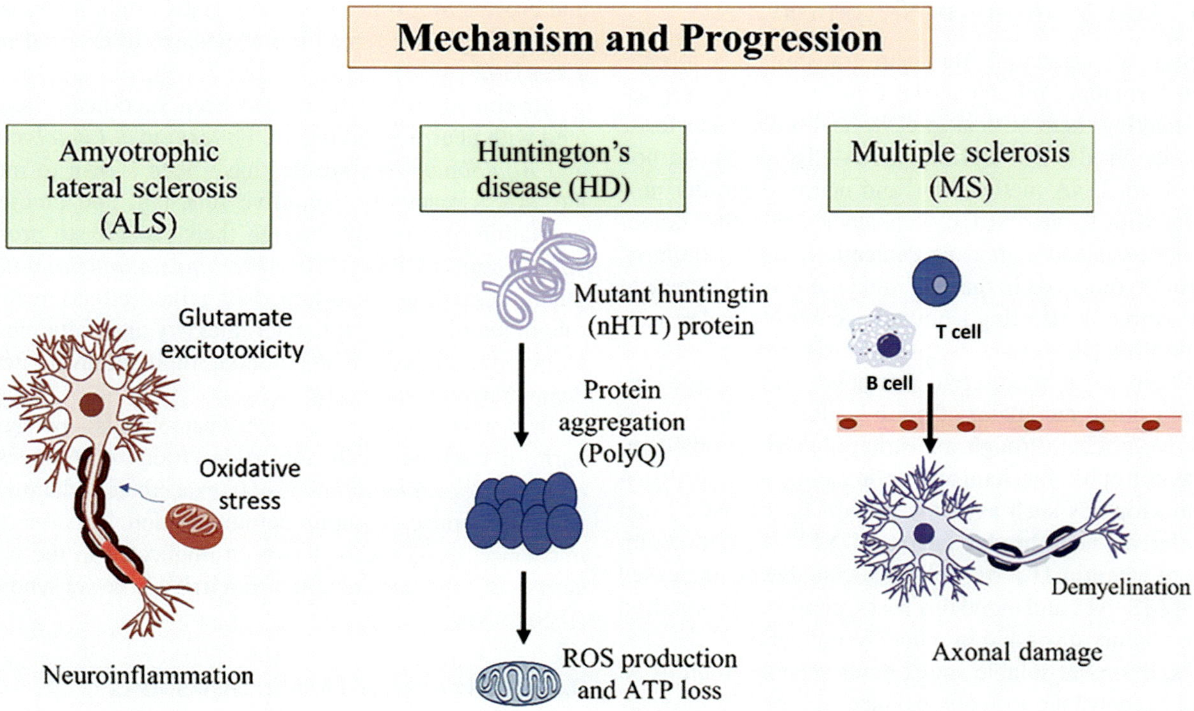

FIGURE 9.1 Gut microbiota and SCFAs modulation of immune functions.

Indole derivatives, produced from tryptophan metabolism by gut bacteria, also play critical roles in gut–brain signaling. Indole-3-propionic acid, for instance, has been shown to possess antioxidant properties and neuroprotective effects, while other tryptophan catabolites influence serotonergic pathways and immune modulation [22]. Altered microbial tryptophan metabolism has been observed in patients with major depressive disorder and MS, further reinforcing the functional link between microbial metabolism and CNS outcomes.

Phenolic compounds, derived from microbial fermentation of polyphenols and amino acids, exert anti-inflammatory and antioxidant effects in both the gut and the brain. These metabolites may inhibit neurotoxic cascades involving ROS and proinflammatory cytokines, which are often elevated in neurodegenerative diseases [23]. Gut Microbiota and SCFAs modulation of immune functions is given in Figure 9.1.

9.2.3 Gut Barrier and Immune Regulation

The intestinal barrier—a multilayered defense of epithelial cells, mucus, tight junctions, immune cells, and microbiota—supports nutrient absorption while blocking pathogens to preserve gut–brain balance. Its epithelial layer relies on tight junction proteins (occludin, claudins, ZO-1), whose integrity is strengthened by microbial metabolites such as the SCFA butyrate [24]. Butyrate not only reinforces tight junction integrity but also modulates mucosal immunity by promoting the differentiation of regulatory T cells (Tregs) and reducing proinflammatory cytokines [25].

The mucus layer, rich in mucin glycoproteins like MUC2, forms a protective barrier above the epithelium. Commensals such as *Akkermansia muciniphila* degrade mucin yet promote its renewal, supporting barrier health. Disruption of this layer can cause "leaky gut," allowing microbial products like LPS into circulation, triggering chronic inflammation and possible neuroinflammation [26].

The gut-associated lymphoid tissue (GALT) houses immune cells that defend against pathogens while promoting tolerance to commensals and dietary antigens. Gut microbes shape T helper cell polarization and Treg balance, influencing systemic inflammation and CNS immune surveillance. Dysbiosis can shift this balance toward proinflammation, contributing to neuroimmune dysregulation in disorders like MS and PD [26].

Furthermore, gut microbes modulate the expression of pattern recognition receptors (PRRs) such as Toll-like receptors (TLRs) and Nucleotide-binding Oligomerization Domain (NOD)-like receptors (NLRs) on immune and epithelial cells [27]. These receptors detect microbial-associated molecular patterns (MAMPs) and initiate immune signaling cascades that can influence neuroinflammatory pathways through the release of cytokines like IL-6, IL-1β, and TNF-α, some of which can cross or affect the BBB [28]. The systemic spillover of gut-derived inflammatory mediators is increasingly recognized as a key driver of neurodegenerative pathologies.

9.3 VITAMINS AND THE GUT MICROBIOME

9.3.1 Vitamin Synthesis by Gut Microbiota

The gut microbiota acts as a biosynthetic organ, producing vitamins—especially B vitamins and vitamin

K—that support neurodevelopment, neurotransmitter synthesis, and neuroprotection. Species within Firmicutes, Actinobacteria, and Bacteroidetes synthesize vitamins like B1, B2, B6, B9, B12, and K2; for instance, *Lactobacillus* and *Bifidobacterium* produce folate and riboflavin, while E. coli and Bacteroides fragilis make vitamin K and biotin. These microbially derived vitamins, absorbed mainly in the colon, provide a unique complement to dietary sources [29].

Importantly, the synthesis of these vitamins is highly strain-specific, not merely species-specific, and is influenced by community structure, substrate availability, and cross-feeding interactions within the gut [30]. For instance, some bacteria possess partial biosynthetic pathways and rely on metabolic cooperation—known as syntrophic metabolism—to complete vitamin production [31]. This microbial interdependence adds a layer of complexity to predicting and manipulating vitamin output within the gut ecosystem.

Microbially synthesized folate (B9) plays a direct role in one-carbon metabolism, influencing DNA methylation patterns and thereby impacting gene expression in the CNS. Vitamin B12, although only produced by microbes, is a crucial cofactor for methionine synthase and methylmalonyl-CoA mutase, enzymes involved in neurotransmitter metabolism and myelin synthesis [32]. A deficiency in cobalamin has been linked with cognitive dysfunction, neuropathy, and psychiatric disturbances, underscoring the relevance of microbial vitamin synthesis to neurological health.

It is also noteworthy that some pathogenic bacteria can exploit vitamin biosynthesis for virulence. For example, *Clostridium difficile* upregulates certain vitamin biosynthetic genes during infection to support its growth in nutrient-depleted environments [33]. Commensal B-vitamin producers can bolster mucosal immunity and aid colonization resistance against pathogens. The bioavailability of these microbially made vitamins depends on colonic absorption, gut permeability, transport mechanisms, and host needs. Antibiotics, diet, and probiotics can all influence the abundance and activity of these vitamin-producing microbes.

9.3.2 Impact of Diet and Supplementation

Diet shapes gut microbiota and vitamin metabolism, influencing neurophysiology via the GBA. Fiber-rich diets favor saccharolytic microbes like *Bifidobacterium*, *Prevotella*, and *Faecalibacterium*, which produce SCFAs and synthesize B vitamins such as B1, B2, B6, B9, and B12 [34]. In contrast, high-fat or high-protein Western-style diets tend to reduce microbial diversity and may suppress populations of beneficial vitamin-synthesizing taxa while promoting inflammatory species such as *Bilophila wadsworthia* and *Enterobacteriaceae* [35].

Importantly, long-term dietary habits have a more substantial and sustained effect on microbiota composition than short-term changes. Yet even acute dietary shifts can rapidly alter microbial gene expression and metabolic output. For example, within 24 hours of a switch to an animal-based diet, studies have noted a decrease in microbial genes associated with folate biosynthesis and an increase in those linked to amino acid fermentation [36]. These microbial shifts may influence host vitamin absorption and bioavailability, particularly for vitamins like B9 and B12 that depend on specific microbial consortia.

Vitamin supplementation introduces another layer of complexity. While oral supplementation can correct systemic deficiencies, it may also disrupt the natural vitamin–microbiota equilibrium, especially at pharmacological doses. For instance, supraphysiological levels of vitamin B12 have been shown to reduce the abundance of *Bacteroides* and increase proinflammatory *Proteobacteria* in some individuals, potentially exacerbating gut inflammation [37]. Similarly, excessive vitamin A intake can alter microbial richness and composition, impacting immune responses and gut integrity [38].

Interestingly, certain probiotic strains have been engineered or selected for their ability to synthesize specific vitamins in vivo. *Lactobacillus plantarum* strains capable of overproducing folate or riboflavin have demonstrated beneficial effects in both animal and human models, not only correcting deficiencies but also enhancing cognitive performance and reducing markers of systemic inflammation [39]. This strategy of microbial biofortification offers an emerging alternative to conventional supplementation, particularly in populations with poor dietary diversity or impaired absorption.

It is also worth noting the bidirectional nature of the relationship: while diet affects microbiota and vitamin synthesis, microbial activity can influence the bioefficacy of dietary vitamins. Microbes may compete with the host for vitamins or convert them into inactive or less bioavailable forms, particularly in dysbiotic conditions. For instance, certain gut bacteria can degrade vitamin B2 into lumichrome and lumiflavin, reducing its availability for host metabolism [40].

9.3.3 Microbiota-Driven Vitamin Modulation

Beyond synthesis, the gut microbiota influences vitamin bioavailability and metabolism, affecting neuroimmune and cognitive health, treatment efficacy, and disease progression. Through biotransformation, microbes can modify vitamins—such as converting dietary K1 into K2, which has greater bioavailability for extra-hepatic, including neural, tissues [41]. Similarly, certain gut microbes can degrade vitamin B2 (riboflavin) into lumichrome, which possesses distinct biological properties [42]. These transformations have implications not only for vitamin absorption but also for downstream metabolic and neurological processes that depend on specific coenzyme forms.

The gut microbiota also modulates vitamin availability through competition and consumption. Some bacterial species actively uptake vitamins from the gut lumen for their own metabolic needs. For example, *Bacteroides fragilis* and *E. coli* possess efficient transport systems for vitamin B12, often outcompeting host epithelial cells, particularly in vitamin-deficient states [43]. Microbial competition can drive systemic deficiencies despite sufficient dietary intake, leading to cognitive decline, neuropathy, or mood disorders, especially in vulnerable groups such as the elderly and individuals with inflammatory bowel disease.

In addition, microbiota composition can influence host gene expression related to vitamin transport and metabolism. Studies have shown that colonization with specific bacteria can upregulate host genes encoding for vitamin transporters such as SLC19A2 (thiamine transporter) and SLC46A1 (folate transporter), thereby enhancing or impairing the systemic uptake of vitamins depending on the microbial context [44]. This suggests a bidirectional feedback loop wherein the microbiota not only supplies and modifies vitamins but also affects the host's capacity to utilize them.

Microbiota-driven vitamin modulation influences epigenetic and immune regulation, with folate and B12 affecting DNA methylation, histone modification, and gene expression in both peripheral and central tissues. Such changes can shape neurodevelopment, synaptic plasticity, and behavior, as seen in animal models where maternal microbiota disruptions alter offspring brain development via reduced methyl donor availability. Drug–nutrient interactions add complexity, as medications like antibiotics, Proton Pump Inhibitors (PPIs), and metformin can disrupt vitamin-producing microbes—metformin notably reducing B12 absorption partly through microbiota-mediated effects—highlighting the need to consider microbial ecology in vitamin-related therapies [45].

9.4 MICROBIOTA–VITAMIN INTERACTIONS IN THE GBA

9.4.1 Neuroactive Compounds from Vitamin Pathways

The gut microbiota and host form a vitamin-linked network that generates neuroactive compounds, with vitamins acting as essential cofactors in neurotransmitter synthesis, methylation, redox balance, and energy metabolism. For example, vitamin B6 (pyridoxal 5′-phosphate) enables aromatic L-amino acid decarboxylase to convert 5-HTP to serotonin and L-DOPA to dopamine—neurotransmitters vital for mood, cognition, and motor control [46].

Gut bacteria such as *Bacteroides* and *Enterococcus* species can synthesize B6 or modulate its levels, thereby influencing neurotransmitter availability indirectly through host enzymatic activity. Altered B6 availability due to dysbiosis or dietary deficiencies has been associated with neuropsychiatric disorders including depression, anxiety, and PD [30].

Folate (vitamin B9), another microbiota-influenced vitamin, plays a pivotal role in one-carbon metabolism—a biochemical pathway essential for DNA synthesis, methylation, and neurotransmitter production. Microbially produced folate contributes to S-adenosylmethionine (SAMe) synthesis, a major methyl donor for epigenetic regulation of gene expression in the brain [47]. Epigenetic changes in genes controlling synaptic plasticity, neural development, and stress responses have been tied to neurodevelopmental disorders, underscoring the far-reaching impact of microbiota–vitamin interactions on brain function.

Vitamin B12, produced solely by microorganisms, is vital for methionine synthase activity, linking folate metabolism with methylation cycles. It is essential for myelin synthesis and maintenance, and its deficiency is associated with cognitive decline, memory loss, and peripheral neuropathy [48]. Interestingly, gut microbes not only produce and consume B12 but also influence its analog diversity, potentially affecting competitive uptake by host tissues and shaping neurochemical outcomes [13].

Another important dimension is the role of vitamin D, whose active form modulates the expression of enzymes involved in neurotransmitter synthesis, such as tyrosine hydroxylase for dopamine and tryptophan hydroxylase for serotonin. Though vitamin D is primarily acquired through sunlight and dietary sources, gut microbiota can affect vitamin D receptor (VDR) signaling, influencing its downstream neuroactive effects [49]. Altered VDR signaling has been associated with mood disorders, schizophrenia, and MS, suggesting a microbiota-dependent modulation of vitamin D action in the CNS.

Antioxidant vitamins—notably vitamin E and vitamin C—also contribute to neuroprotection by mitigating oxidative stress, a hallmark of neurodegenerative diseases. While not directly synthesized by gut microbes, their intestinal absorption and redox cycling may be influenced by microbial metabolites such as hydrogen peroxide, nitric oxide, and SCFAs [50]. These vitamins help preserve the function of neurotransmitter systems and prevent lipid peroxidation in neuronal membranes.

Importantly, the interplay between microbial metabolites and vitamin-dependent pathways creates feedback loops that can amplify or suppress neurochemical signaling. For instance, butyrate—an SCFA produced by fiber-fermenting bacteria—acts as a HDAC inhibitor, affecting the transcription of genes regulated by vitamin B-dependent methylation cycles [51]. This intricate crosstalk between microbial fermentation products and vitamin-regulated enzymes shapes the neurochemical landscape in a context-specific manner.

9.4.2 Communication via the Vagus Nerve and Immune Signaling

The GBA uses the vagus nerve and immune pathways to relay microbial and vitamin-related signals that influence mood, cognition, and neuroinflammation. The vagus nerve, with about 80% afferent fibers, carries sensory input from the gut to the brainstem's nucleus tractus solitarius (NTS) [52]. Microbial metabolites such as SCFAs, secondary bile acids, and vitamin-related neuroactive compounds (e.g., serotonin, GABA) can stimulate enteroendocrine cells (EECs) and enterochromaffin cells (ECCs) in the intestinal lining. These cells, in turn, release signaling molecules like 5-hydroxytryptamine (5-HT) and cholecystokinin, which activate vagal afferents and influence brain circuits involved in emotion regulation and stress responses [53].

Vitamin synthesis and modulation by gut microbes indirectly influence vagal signaling by regulating the availability of neurotransmitter precursors. For example, microbial production of vitamin B6 is essential for the decarboxylation of glutamate into GABA, a major inhibitory neurotransmitter. Elevated GABA levels in the gut lumen can activate GABA receptors on vagal afferents, thereby modulating vagus-mediated pathways implicated in anxiolysis and behavioral control [54]. Experimental models have demonstrated that certain probiotic strains (e.g., *Lactobacillus rhamnosus*) can modulate behavior in a vagus nerve-dependent manner—an effect lost upon vagotomy [55].

Parallel to neural signaling, the immune system forms the second major route through which microbiota–vitamin interactions affect the brain. The GALT constantly samples microbial and dietary antigens, regulating tolerance and inflammatory responses. Vitamins such as A, D, B6, B9, and B12 play pivotal roles in T-cell differentiation, cytokine production, and mucosal immunity. For instance, vitamin A (as retinoic acid) promotes Treg development, while vitamin D dampens proinflammatory Th1/Th17 responses, thereby influencing systemic immune tone [56].

Dysbiosis-induced alterations in microbial vitamin metabolism can tilt this balance toward chronic low-grade inflammation, a state increasingly associated with neurodegeneration. Elevated levels of proinflammatory cytokines—such as IL-6, TNF-α, and IL-1β—can cross the BBB or signal through circumventricular organs to activate microglia and promote neuroinflammation, a key feature in diseases like AD, PD, and MS [54]. Vitamin deficiencies—particularly of B12 and folate—are known to impair immune regulation and elevate homocysteine, further exacerbating neuroinflammatory cascades [57].

Moreover, certain vitamins influence the expression of immune receptors and transporters. Vitamin D, for example, modulates the expression of TLRs on immune and epithelial cells, influencing microbial recognition and shaping downstream inflammatory responses. The microbiota, in turn, can affect the VDR pathway, modulating its transcriptional activity in gut and brain tissues [58].

9.4.3 Influence on Neurotransmitter Synthesis

Microbiota–vitamin interactions help regulate neurotransmitter synthesis—serotonin, dopamine, GABA, acetylcholine, and norepinephrine—by supplying or modulating vitamin-derived cofactors essential for these pathways. Serotonin production, for example, begins with tryptophan, hydroxylated by the B6- and BH4-dependent enzyme tryptophan hydroxylase to form 5-HTP, which is then decarboxylated by the B6-dependent aromatic L-amino acid decarboxylase to yield serotonin [59]. While the majority of serotonin is produced peripherally by ECCs in the gut, its availability is influenced by microbial metabolism. Certain bacteria, such as *Bifidobacterium infantis* and *Escherichia coli*, can modulate tryptophan availability either through synthesis or catabolism into indole derivatives, thereby influencing the serotonin pathway [60].

Vitamin B2 (riboflavin) and vitamin B3 (niacin) are precursors for flavin adenine dinucleotide (FAD) and nicotinamide adenine dinucleotide (NAD^+), respectively—cofactors involved in redox reactions essential to neurotransmitter metabolism and mitochondrial function. Riboflavin-dependent enzymes participate in dopamine and norepinephrine biosynthesis by supporting the activity of monoamine oxidase (MAO) and dopamine beta-hydroxylase [61]. Disruption in these redox processes, whether due to vitamin insufficiency or microbial imbalance, can lead to neurotransmitter deficits associated with depression, anxiety, and cognitive dysfunction.

Dopamine synthesis begins with tyrosine hydroxylation to L-DOPA via BH4-dependent tyrosine hydroxylase, followed by B6-dependent decarboxylation. Microbiota can influence this pathway by producing B6 and modulating BH4 availability through oxidative stress control. Animal studies link microbiota shifts to altered striatal dopamine, and PD—marked by dopaminergic loss—has been associated with dysbiosis and disrupted microbial B-vitamin pathways [62,63].

GABA, the brain's main inhibitory neurotransmitter, is formed by B6-dependent glutamic acid decarboxylase (GAD) converting glutamate. Some *Lactobacillus* and *Bifidobacterium* strains can produce GABA directly, with B6 availability boosting both microbial and host synthesis. Higher GABA levels are linked to anxiolytic and antidepressant effects, making GABA-producing probiotics promising adjuncts for mood disorder treatment [64].

Acetylcholine, essential for learning, memory, and neuromuscular function, is synthesized from choline and acetyl-CoA, a process requiring adequate vitamin B5 (pantothenic acid) to form CoA. Gut bacteria can influence choline metabolism by competing for it or converting it into metabolites like trimethylamine (TMA), which can have neuroinflammatory effects [65,66]. Thus, the interplay between microbiota, vitamin B5 levels, and choline availability may indirectly affect cholinergic neurotransmission and neurodegenerative risk. Figure 9.2 shows the role of different vitamins in neuroprotection.

Furthermore, folate (vitamin B9) and vitamin B12 (cobalamin) are key players in the methionine cycle, which generates SAMe—a universal methyl donor used in the methylation of monoamines, phospholipids, and nucleotides. Impaired methylation, due to deficiencies or microbial imbalance, may alter neurotransmitter turnover, affect receptor expression, and contribute to psychiatric symptoms [66].

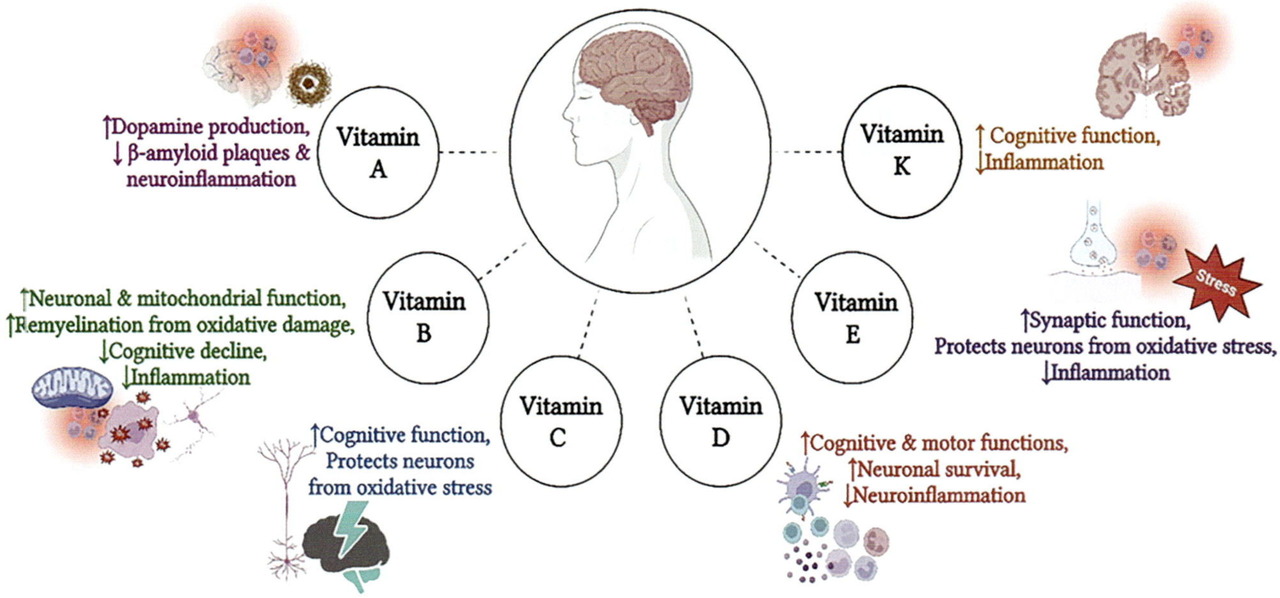

FIGURE 9.2 Role of different vitamins in neuroprotection.

9.5 ROLE OF SPECIFIC VITAMINS IN NEUROPROTECTION

9.5.1 Vitamin B Complex (B1, B6, B9, B12)

The B vitamin complex supports neuronal health, neurotransmission, energy metabolism, and epigenetic regulation, with B1, B6, B9, and B12 offering key neuroprotection. B1 (thiamine) is essential for carbohydrate metabolism and neuronal energy, serving as a coenzyme for mitochondrial enzymes like pyruvate dehydrogenase, alpha-ketoglutarate dehydrogenase, and transketolase [67]. These enzymes are critical in the tricarboxylic acid (TCA) cycle and the pentose phosphate pathway, both of which are vital for ATP production and redox balance in neurons. Thiamine deficiency impairs glucose metabolism, leading to neuronal cell death, oxidative stress, and lactic acidosis, particularly in high-energy-demand regions like the hippocampus and cerebellum. Clinically, thiamine deficiency manifests as Wernicke's encephalopathy and, if untreated, progresses to Korsakoff's syndrome, both marked by confusion, memory deficits, and ataxia [68].

Vitamin B6 is essential for synthesizing neurotransmitters like serotonin, dopamine, GABA, and norepinephrine, acting as a cofactor for enzymes such as GAD and aromatic L-amino acid decarboxylase to maintain brain excitatory–inhibitory balance. It also aids homocysteine metabolism, and its deficiency can cause irritability, depression, cognitive impairment, and seizures, especially in infants and the elderly [9].

Folate (vitamin B9) is a key player in the one-carbon metabolic cycle, enabling the transfer of methyl groups necessary for DNA synthesis, repair, and methylation. This cycle, in conjunction with vitamin B12, is critical for the synthesis of SAMe, the universal methyl donor. Disruptions in folate-dependent methylation pathways can result in aberrant gene expression, impaired neurogenesis, and elevated homocysteine, a neurotoxic amino acid linked to AD, vascular dementia, and stroke risk [30]. Low folate levels have also been linked to mood disorders and neural tube defects in utero, emphasizing its role across the lifespan.

Vitamin B12 is vital for myelin formation, neuronal repair, and methylation, acting with folate in homocysteine-to-methionine conversion for SAMe production. It also supports fatty acid metabolism via methylmalonyl-CoA mutase. Deficiency can cause megaloblastic anemia, cognitive decline, paresthesia, and irreversible neurological damage, with early symptoms sometimes mimicking psychiatric disorders [69].

The gut microbiota modulates the availability of B6, B9, and B12, with B12 produced solely by bacteria but requiring intrinsic factors for ileal absorption—often hindered by dysbiosis, low stomach acid, or metformin use. Microbial pathways also affect folate and B6, with metabolites influencing host vitamin transporters and enzymes, potentially impacting systemic and brain vitamin levels [70].

9.5.2 Vitamin D

Vitamin D, now recognized as a neurosteroid, has receptors in key brain regions linked to memory, mood, and motor control. Through genomic and non-genomic actions, it supports neurotrophic signaling, neurotransmitter synthesis, immune balance, and oxidative stress control, offering neuroprotection across life.

The active form of vitamin D, 1,25-dihydroxyvitamin D3 (calcitriol), regulates the expression of neurotrophic factors including NGF and glial cell line-derived neurotrophic factor (GDNF), which support neuronal survival, axonal growth, and synaptic plasticity [71]. It also modulates the expression of enzymes involved in the synthesis of neurotransmitters such as dopamine, serotonin, and acetylcholine. For instance, vitamin D enhances the transcription of tyrosine hydroxylase, the rate-limiting enzyme in dopamine biosynthesis, implicating it in mood regulation and movement disorders, including PD [72].

One of vitamin D's most potent neuroprotective roles is its anti-inflammatory and immunomodulatory action. It downregulates proinflammatory cytokines such as interleukin (IL)-6, IL-1β, and tumor necrosis factor-alpha (TNF-α), while promoting regulatory T-cell activity, thereby preserving CNS immune homeostasis [73]. Chronic neuroinflammation is a hallmark of neurodegenerative conditions such as AD, MS, and amyotrophic lateral sclerosis (ALS)—disorders in which vitamin D insufficiency is frequently observed [74]. In MS in particular, lower serum vitamin D levels have been associated with greater disease activity, suggesting a potential disease-modifying role.

Vitamin D also acts as an antioxidant, reducing ROS and enhancing the expression of glutathione peroxidase and superoxide dismutase, enzymes critical in protecting neuronal membranes from oxidative damage [75]. In experimental models of AD, vitamin D supplementation has been shown to reduce beta-amyloid accumulation and improve cognitive performance, though human clinical data remain mixed and dosage-dependent.

While vitamin D is primarily synthesized in the skin through UVB exposure, the gut microbiota indirectly influences its bioactivity. Certain microbes modulate VDR expression and signaling, thereby affecting the downstream transcription of vitamin D-responsive genes in both gut and brain tissue [76]. Dysbiosis can impair this interaction, potentially attenuating vitamin D's immunological and neurological effects. Moreover, vitamin D status can reciprocally shape microbial composition—deficiency has been associated with an increased F/B ratio, reduced microbial diversity, and a proinflammatory microbiota profile [77].

9.5.3 Vitamins A, C, and E

Vitamins A, C, and E protect against oxidative stress, neuroinflammation, and barrier damage, bolstering neuronal resilience, immunity, and repair—especially in aging and chronic neuroinflammation. Vitamin A, from retinol, retinyl esters, or carotenoids, is converted to retinoic acid, which binds RARs and RXRs to regulate genes controlling neural patterning, synaptic plasticity, and neurogenesis [78]. RA signaling has been shown to influence the expression of BDNF and axon guidance molecules, making it vital for brain development and repair. In adult neurobiology, RA contributes to memory consolidation and spatial learning, with deficiency linked to cognitive deficits and impaired long-term potentiation [79]. Vitamin A also modulates the immune system, especially at mucosal surfaces such as the GALT. It promotes tolerogenic immune responses by driving the differentiation of Tregs and suppressing proinflammatory Th17 responses, thereby reducing systemic inflammation—a critical factor in diseases like MS and AD [80]. Additionally, retinoic acid affects gut microbiota composition and barrier function, indirectly influencing neuroimmune signaling within the GBA.

Vitamin C (ascorbic acid), a water-soluble antioxidant, is present in high concentrations in the brain, particularly in the cerebrospinal fluid (CSF), hippocampus, and amygdala. It serves as a cofactor for dopamine β-hydroxylase, the enzyme that converts 15 dopamine to norepinephrine, thereby contributing to catecholaminergic signaling [13]. Vitamin C also maintains redox balance by scavenging ROS and regenerating other antioxidants, such as vitamin E. Neurons are particularly vulnerable to oxidative stress due to their high metabolic activity and lipid-rich membranes, making vitamin C an essential shield against lipid peroxidation and mitochondrial damage [81].

Vitamin C supports BBB integrity by regulating tight junctions and collagen synthesis, with deficiency leading to greater permeability and microvascular dysfunction. It may also modulate neuroinflammation by reducing cytokine activity linked to depression and neurodegeneration [82].

Vitamin E, a lipid-soluble antioxidant of tocopherols and tocotrienols, protects neuronal membranes from oxidative damage, stabilizes cell structure, and modulates inflammation via NF-κB and PPAR-γ. In Alzheimer's and Parkinson's models, it may reduce amyloid aggregation, mitochondrial dysfunction, and neuronal death, though clinical outcomes vary by dose and disease stage

From a GBA perspective, these vitamins interact with the intestinal microbiota in multifaceted ways (Table 9.1) [83]. For example, vitamin A and E levels can modulate microbial diversity and barrier function, while microbial fermentation products may affect the absorption and metabolism of these vitamins [84]. Gut dysbiosis may also impair the enterohepatic recycling of vitamin E and oxidative metabolism of carotenoids, potentially diminishing their neuroprotective efficacy.

9.6 MICROBIOTA–VITAMIN DYSBIOSIS AND NEURODEGENERATIVE DISORDERS

The misfolding and aggregation of proteins in neurons have been the primary focus of in-depth research on the pathophysiology of Neurodegenerative Diseases (NDs), such as AD, PD, and MS. Although the precise underlying mechanisms are yet unknown, oxidative stress has also been thought to be directly linked to the formation and development of NDs. Numerous investigations have shown that oxidative stress produced by the microbiome and key vitamins [84] imbalance is closely linked to neurodegeneration. Here, we highlight new connections (Table 9.2) between NDs, oxidative stress, and the gut microbiome, with an emphasis on AD, PD, and MS [88].

TABLE 9.1
Mechanistic Overview of the GBA and Microbiota–Vitamin Interactions

Component	Role in GBA	Vitamin Involvement	Impact on Neurodegeneration	References
Gut microbiota	Modulate nutrient absorption, produce bioactive compounds	Synthesis of vitamins modulation of bioavailability	Dysbiosis reduces vitamin production, increases neuroinflammation	[85]
Vitamins	Act as enzymatic cofactors, antioxidants	B-complex and vitamins D, E, K	Oxidative stress	[86]
Immune signaling	Gut microbiota regulate cytokine production	Vitamins A, D, E modulate inflammatory response	Impairs synaptic function	[7]
Metabolites	Tryptophan metabolites influence brain health	Vitamin cofactors required for metabolic pathways	Altered metabolite profiles associated with Alzheimer's, Parkinson's	[87]

TABLE 9.2
Key Vitamins Influenced by Gut Microbiota and their Role in Neurodegeneration

Vitamin	Microbiota Involvement	Neurological Role	Deficiency Effects in Neurodegeneration	References
Vitamin B1 (thiamine)	Certain gut bacteria synthesize thiamine	Cofactor in carbohydrate metabolism	Cognitive decline	[89]
Vitamin B6 (pyridoxine)	Microbiota regulate availability	Required for neurotransmitter synthesis	Depression	[90]
Vitamin B9 (folate)	Synthesized by gut microbiota	DNA methylation	Increased risk of Alzheimer's due to elevated homocysteine	[91]
Vitamin B12 (cobalamin)	Produced by specific gut bacteria	Myelin synthesis	Demyelination	[92]
Vitamin D	Modulates microbiota composition	Neuroprotection	Higher risk of multiple sclerosis	[93]
Vitamin K	Produced by gut microbiota	Antioxidant role	Increased oxidative stress	[94]
Vitamin E	Microbiota influence absorption	Lipid peroxidation prevention	Accelerated neuronal damage	[95]

9.6.1 AD

AD, the leading cause of dementia, features cognitive decline, amyloid plaques, tangles, and synaptic loss. Beyond genetic risks like APOE ε4, gut microbiota imbalance and vitamin deficiency may contribute via the GBA, promoting inflammation, oxidative stress, and amyloid buildup. AD patients often show reduced microbial diversity, fewer beneficial bacteria (*Bifidobacterium*, *Lactobacillus*), and more proinflammatory taxa (*Escherichia/Shigella*) [96]. These microbial changes lead to elevated levels of lipopolysaccharide (LPS) and trimethylamine-N-oxide (TMAO), both of which can cross the BBB and activate microglia, promoting chronic neuroinflammation and neuronal injury [8]. The resulting inflammatory milieu enhances β-secretase activity, facilitating Aβ deposition and tau hyperphosphorylation.

Compounding this inflammatory burden is vitamin deficiency or dysregulation, particularly of B vitamins (B1, B6, B9, B12), vitamin D, and antioxidants such as vitamins C and E—all of which have neuroprotective functions that are often compromised in AD. Homocysteine, a neurotoxic amino acid whose levels are tightly regulated by folate, B12, and B6, is frequently elevated in AD patients and has been associated with cortical atrophy, cerebrovascular damage, and impaired DNA repair mechanisms [97]. Meta-analyses have shown that homocysteine-lowering therapy using B vitamins can modestly slow cognitive decline in individuals with mild cognitive impairment (MCI), especially when omega-3 fatty acid levels are adequate [9].

Vitamin D deficiency is common in AD, with its active form, calcitriol, promoting Aβ clearance by upregulating degrading enzymes and enhancing microglial phagocytosis. Animal studies show supplementation can lower amyloid burden and improve learning, while VDR polymorphisms and reduced brain VDR expression in AD suggest impaired vitamin D signaling may contribute to disease progression [98].

Oxidative stress—another central feature of AD—is exacerbated by deficiencies in vitamins C and E, both of which act as antioxidants that protect neuronal membranes and mitochondrial DNA from ROS. Vitamin E has been studied in randomized trials, with mixed but suggestive evidence for slowing functional decline in moderate AD when administered at pharmacological doses. However, high doses [99] may carry risks, such as increased bleeding, emphasizing the need for tailored therapy.

9.6.2 PD

PD, marked by dopaminergic neuron loss and both motor and non-motor symptoms, is now considered multisystemic and possibly originating in the gut. Early GI symptoms and α-synuclein aggregates in the enteric nervous system—appearing years before motor signs—support the "gut-first" model, suggesting progression to the brain via the vagus nerve [100]. This model is supported by studies showing that vagotomy reduces the risk of developing PD and that gut microbiota dysbiosis is a consistent feature of the disease, even in early stages [101].

In Parkinson's disease, gut microbiota shifts often include more proinflammatory taxa like *Enterobacteriaceae* and *Desulfovibrio* and fewer SCFA producers such as *Faecalibacterium* and *Prevotella*. This dysbiosis is linked to leaky gut, higher circulating LPS, and increased systemic and neuroinflammation, all of which may hasten α-synuclein misfolding and neurodegeneration [102].

Layered onto this dysbiosis is a growing recognition of vitamin deficiencies, particularly involving B-complex vitamins (notably B1, B6, B9, and B12), vitamin D, and antioxidants. Vitamin B6 (pyridoxine) plays a dual role in PD. It is essential for dopamine synthesis, serving as a cofactor for aromatic L-amino acid decarboxylase, but it also contributes to L-DOPA metabolism. Thus, chronic L-DOPA treatment may reduce B6 levels, creating a feedback loop that can exacerbate motor fluctuations and neurotoxicity [103].

In PD, B12 and folate deficiencies can raise homocysteine levels, contributing to neuronal damage, cognitive decline, motor worsening, and vascular risk. Since L-DOPA metabolism generates homocysteine, inadequate B6, B12, or folate impairs its detoxification. Elevated homocysteine may accelerate PD progression, and some evidence suggests B12 supplementation could slow gait decline, though further studies are needed [104].

Vitamin D may protect against PD by supporting neuroimmune balance, mitochondrial function, and dopaminergic neuron survival. Low serum levels, common in PD, are linked to worse motor symptoms, instability, and higher fall risk. With VDRs in the substantia nigra, vitamin D helps reduce oxidative stress, boost neurotrophic factors, and regulate calcium homeostasis, mitigating neuron vulnerability [105].

Antioxidant vitamins, particularly vitamins C and E, have also been explored for their neuroprotective potential in PD. Oxidative stress is a major driver of dopaminergic neuron death, and both vitamins can scavenge free radicals and protect mitochondrial integrity. However, while observational studies have linked higher antioxidant intake with reduced PD risk, clinical trials have yielded mixed results, suggesting that antioxidants alone may be insufficient once neurodegeneration is established [106].

Importantly, gut microbiota alterations in PD may impair the microbial synthesis of B vitamins, contributing to systemic deficiencies despite adequate dietary intake. Dysbiosis also impacts the enterohepatic circulation and expression of vitamin transporters, further disrupting vitamin absorption. Conversely, certain gut microbes may degrade levodopa in the intestine before it reaches the brain, reducing therapeutic efficacy and increasing the burden on peripheral vitamin-dependent detox pathways [107].

9.6.3 MS and ALS

MS is an autoimmune demyelinating disease in which immune attacks on myelin cause axonal injury, neurodegeneration, and progressive dysfunction. Its development reflects interactions between genetic risk, environmental factors, and immune imbalance, with vitamin D deficiency emerging as one of the most consistent environmental contributors [108].

Vitamin D supports immune tolerance by enhancing Treg differentiation and suppressing proinflammatory Th1 and Th17 activity—both heightened in MS. Higher MS rates occur in low-sunlight regions, and low serum 25(OH) D correlates with increased risk, activity, and relapses. VDR and vitamin D-binding protein gene polymorphisms further link genetic susceptibility with environmental vitamin D status in MS pathogenesis [109].

The gut microbiota also appears to play a critical role in shaping immune responses in MS. Several studies have reported dysbiosis in MS patients, characterized by a decrease in anti-inflammatory commensals like *Bifidobacterium* and *Faecalibacterium*, and an increase in proinflammatory bacteria such as *Akkermansia muciniphila* and *Methanobrevibacter smithii* [110]. This microbial shift may compromise intestinal barrier integrity, enhance systemic immune activation, and ultimately cross-prime autoreactive T cells against CNS antigens.

B vitamins—especially B6, B9, and B12—are vital for myelin integrity, methylation, and homocysteine control in MS. Deficiencies or disrupted microbial metabolism can raise homocysteine, driving oxidative stress and neurovascular damage. B12 deficiency can worsen demyelination and produce MS-like symptoms, complicating diagnosis and treatment [111].

ALS is a progressive motor neuron disease characterized by the degeneration of upper and lower motor neurons, leading to muscle weakness, atrophy, and eventual respiratory failure. Although the cause of ALS is not fully understood, growing evidence points to a multifactorial process involving glutamate excitotoxicity, mitochondrial dysfunction, neuroinflammation, and increasingly, gut–microbiota alterations [112].

In ALS, gut microbiota often show reduced butyrate-producing bacteria and increased pathobionts like Enterobacteriaceae, weakening the gut barrier and enabling proinflammatory molecules (e.g., LPS) to reach the CNS and activate microglia. In mouse models, probiotics, fecal transplants, and *Akkermansia muciniphila* supplementation improved gut integrity, motor function, and survival, partly via nicotinamide-mediated neuroprotection [113].

In ALS, imbalances in vitamins E, D, and B12 have been linked to disease mechanisms. Vitamin E, a lipid-soluble antioxidant, may shield motor neurons from oxidative damage, with higher intake associated with delayed onset in observational studies, though trials remain inconclusive. Vitamin D deficiency is common and tied to faster progression, likely through impaired immune regulation and muscle function [114].

9.7 THERAPEUTIC PERSPECTIVES

9.7.1 Probiotic and Prebiotic Interventions

With the GBA central to neurodegeneration, probiotics and prebiotics offer ways to modulate microbiota–vitamin interactions, restore gut balance, and boost neuroprotection by influencing immunity, neurotransmission, and vitamin production in AD, PD, MS, and ALS. Probiotics—live microbes given in beneficial doses—can replenish healthy gut flora, lower proinflammatory cytokines, and increase neuroactive compounds and vitamin availability [115]. Species from the *Lactobacillus* and *Bifidobacterium* genera are especially well-studied for their ability to produce SCFAs, B vitamins, and GABA—all of which exert neurophysiological effects. These microbes can also influence the host's vitamin transporter expression and modulate barrier integrity, reducing the translocation of endotoxins like LPS into systemic circulation [116].

In clinical studies, multi-strain probiotic supplementation has demonstrated benefits in cognitive performance, emotional regulation, and inflammatory marker reduction in elderly adults and patients with MCI. For example, a randomized controlled trial by Akbari et al. found that a daily probiotic combination of *Lactobacillus acidophilus*, *L. casei*, *Bifidobacterium bifidum*, and *L. fermentum* significantly improved Mini-Mental State Examination (MMSE) scores in patients with AD over 12 weeks [116,117].

Beyond cognitive effects, probiotics also impact vitamin metabolism. *Lactobacillus plantarum* and *Bifidobacterium adolescentis* have been shown to produce folate and riboflavin in the colon, potentially contributing to systemic B-vitamin pools. Some strains have been genetically engineered to enhance in situ vitamin biosynthesis, offering targeted options for correcting subclinical deficiencies in vulnerable populations such as the elderly or those with malabsorption syndromes [40].

Prebiotics are non-digestible food components, such as inulin, FOS, GOS, and resistant starches, that selectively promote beneficial microbes. They boost SCFA production—especially butyrate—which strengthens gut barrier integrity, modulates microglia, and affects brain gene expression via histone acetylation. By fostering vitamin-producing bacteria, prebiotics also support endogenous synthesis of B vitamins and vitamin K2 [118].

Several studies have explored synbiotic approaches, combining probiotics and prebiotics to achieve synergistic effects. In neurodegenerative models, synbiotics have been shown to reduce neuroinflammation, improve motor coordination, and normalize vitamin-dependent metabolic pathways. For example, in a murine model of PD, synbiotic treatment reduced proinflammatory cytokines, preserved dopaminergic neurons, and improved motor behaviors, partly through enhanced butyrate and B-vitamin production [119].

Probiotics and prebiotics can enhance vitamin D signaling by upregulating VDR expression, which supports anti-inflammatory gene activity and strengthens tight junctions—reinforcing gut barrier function and offering potential brain benefits [76,9].

9.7.2 Vitamin-Based Therapies

With neurodegenerative and neuroinflammatory diseases increasingly tied to micronutrient deficiencies—often from gut dysbiosis and aging—vitamin-based therapies are gaining traction as preventive and adjunctive strategies. Beyond correcting deficits, they harness anti-inflammatory, antioxidant, epigenetic, and neurotrophic effects to protect neurons and boost cognition. B-complex vitamins, especially B1, B6, B9, and B12, are central for neurotransmitter synthesis, homocysteine detoxification, and myelin upkeep, with high-dose folate, B6, and B12 shown to markedly lower homocysteine—a risk factor for cognitive decline, stroke, and white matter loss [120]. In patients with MCI, B-vitamin therapy has demonstrated a slowing of brain atrophy, particularly in individuals with high baseline homocysteine and adequate omega-3 fatty acid status [121].

For example, the VITACOG trial reported that a daily regimen of folic acid (0.8 mg), vitamin B12 (0.5 mg), and vitamin B6 (20 mg) slowed the rate of hippocampal shrinkage and cognitive deterioration in older adults with MCI [122]. These findings underscore the importance of early intervention and personalized assessment of nutrient status, as the effects were most pronounced in individuals at the early stages of neurodegeneration.

Vitamin D supplementation has also garnered significant interest due to its roles in neuroimmune regulation, calcium signaling, and neurotrophin expression. Low serum 25(OH) D levels are common in patients with AD, PD, and MS, and are associated with greater disability scores, cognitive impairment, and accelerated progression [123]. Supplementation trials have shown modest improvements in executive function, mood, and motor outcomes, especially when administered early and at doses sufficient to restore serum levels to above 75 nmol/L [124].

Clinical guidelines now increasingly recommend routine vitamin D screening in elderly populations and in individuals with chronic neurological disease, with dosing protocols ranging from 800 to 4000 IU/day, depending on baseline deficiency, comorbidities, and absorption efficiency [100]. Pharmacy professionals play a critical role in monitoring vitamin D therapy, particularly in patients on medications that alter vitamin D metabolism, such as anticonvulsants or corticosteroids.

Vitamin E (particularly alpha-tocopherol) has been studied extensively for its lipid-soluble antioxidant activity, protecting neuronal membranes from oxidative stress—a central driver in neurodegeneration. The TEAM-AD trial showed that 2000 IU/day of vitamin E slowed functional decline in patients with moderate AD, though cognitive outcomes were not significantly affected [125]. Vitamin E may better preserve daily functioning than improve cognition, but high doses carry bleeding risks, particularly for patients on anticoagulants.

In neurodegenerative diseases like ALS and MS, high-dose vitamin B12 (methylcobalamin) therapy has shown potential in slowing disease progression, particularly when initiated early. A recent multicenter trial in Japan demonstrated that intramuscular injections of methylcobalamin (50 mg) significantly delayed motor neuron deterioration in early-stage ALS patients [126]. Similarly, vitamin B1 (as benfotiamine) has been studied for its neuroprotective and anti-glycation effects in AD, with preliminary evidence suggesting improved cognitive outcomes when combined with other nutrients [127].

9.7.3 Fecal Microbiota Transplantation (FMT) and Future Directions

FMT is a potent intervention for severe gut dysbiosis, currently used mainly for refractory *C. difficile* infection. Growing evidence indicates it may also modulate gut–brain signaling, correct vitamin imbalances, and influence neuroinflammation in neurodegenerative diseases. By transferring fecal matter from a healthy donor, FMT restores beneficial microbes, enhances microbial diversity, and repairs functional deficits such as vitamin production and SCFA synthesis. In preclinical models of neurological disease, FMT has shown the ability to:

- Enhance cognitive function,
- Normalize neuroinflammation,
- Increase microbially derived vitamin production, and
- Influence neurotransmitter synthesis pathways [128].

For instance, in AD mouse models, FMT from young or healthy donors improved spatial learning, reduced amyloid plaque burden, and enhanced expression of BDNF, suggesting that gut microbiota modulation may influence synaptic plasticity [129]. Similarly, in PD models, FMT restored motor function, reduced α-synuclein aggregation, and normalized levels of B vitamins and SCFAs, pointing to a potential role in mitigating neurotoxicity [130].

Beyond animal models, early-phase human trials are underway to assess FMT's effects on cognition, motor function, and microbiota composition in neurodegenerative disorders. Although results remain preliminary, some small studies report improvements in sleep, mood, and gastrointestinal symptoms—non-motor features commonly observed in PD and AD [131]. Larger, controlled studies are needed to establish efficacy, safety, and optimal donor–recipient matching criteria.

One of the lesser explored but potentially transformative aspects of FMT lies in its ability to re-establish microbial vitamin synthesis pathways. Dysbiosis can diminish the gut's capacity to produce key micronutrients like biotin, folate, riboflavin, vitamin K2, and nicotinamide, especially when vitamin-depleting taxa dominate the gut environment. FMT can reintroduce taxa such as *Bacteroides*, *Lactobacillus*, and *Bifidobacterium*, which are known contributors to de novo vitamin biosynthesis [132].

9.8 FUTURE DIRECTIONS IN GUT–BRAIN THERAPEUTICS

Future progress in microbiome science, neuropharmacology, and vitamin therapy is likely to focus on:

1. Next-generation probiotics—engineered strains that deliver specific vitamins, neurotransmitters, or anti-inflammatory compounds within the gut.
2. Microbiota-derived postbiotics—bioactive metabolites such as butyrate, GABA, serotonin precursors, and nicotinamide, offering neuroactive benefits without live microbes.
3. Microbiome-informed nutrigenomics—individualized diets and supplementation strategies based on microbiome composition and vitamin-related genetic profiles (e.g., *VDR*, Methylenetetrahydrofolate Reductase *MTHFR*).
4. Refined FMT—deploying defined microbial consortia for targeted roles like vitamin synthesis or immune modulation, reducing safety concerns.
5. Neurorehabilitation integration—combining microbiota-based therapies with physical, cognitive, and pharmacological interventions to enhance neuroregeneration. Therapeutic approaches to GBA modulation are shown in Figure 9.3.

9.9 CURRENT CHALLENGES AND FUTURE RESEARCH

9.9.1 Complexity of Microbiota–Vitamin Interactions

Leveraging the microbiota–vitamin axis for neuroprotection is promising yet complex, as vitamin synthesis, transformation, and absorption vary with microbial composition, host genetics, and environment. Microbes differ in their capacity to produce, convert, or deplete vitamins—e.g., Bifidobacterium adolescents make folate, while others may degrade it—creating unpredictable nutrient availability [133]. Additionally, some microbes rely on host dietary intake for survival and gene expression, meaning their vitamin-producing potential is context-dependent.

The microbiota–vitamin axis is highly dynamic, shaped by fluctuating microbial genes, diet, drugs, age, and disease. Microbes can produce, convert, or deplete vitamins, with effects further modulated by epigenetic changes, receptor signaling, and immune feedback loops. Research is hindered by non-standardized assays, strain-level variability, microbiome diversity, limited longitudinal data, and unclear causality—challenges amplified in neurological disease by

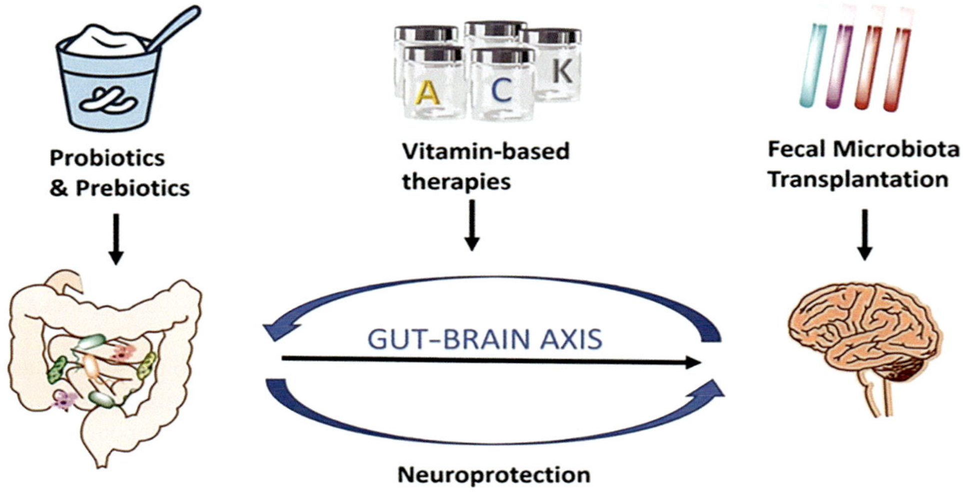

FIGURE 9.3 Therapeutic approaches to GBA modulation.

the BBB and neuroimmune factors. Pharmacotherapies like antibiotics, PPIs, antipsychotics, and L-DOPA can further disrupt microbial vitamin metabolism [134].

Microbiota can influence drug metabolism, alter efficacy and toxicity, while host genetic variants (e.g., MTHFR, VDR, FUT2) interact with microbial function to shape vitamin status—complicating universal treatment strategies. Addressing this requires systems biology, integrating multi-omics, computational modeling, artificial intelligence (AI)-driven prediction, and human-relevant models to map and personalize host–microbe–vitamin interactions.

9.9.2 Personalized Nutrition and Microbiome Medicine

Personalized nutrition and microbiome medicine are shifting neurodegenerative care from generalized management to precision-guided prevention and therapy. By integrating gut microbiota profiles, vitamin metabolism, and host genetics, clinicians can tailor diets, supplements, and microbial interventions to optimize nutrient status and modulate disease risk. Strategies include targeted probiotics, microbiome-based risk prediction, customized nutraceuticals, and dynamic monitoring via metabolomics and AI tools. While challenges remain—such as data interpretation, standardization, cost, equity, and privacy—advances in omics and bioinformatics are paving the way for broader clinical adoption in conditions like AD, PD, and MS [134].

9.9.3 Integrative Omics and Systems Biology Approaches

Understanding the gut–brain–vitamin axis requires integrative omics and systems biology, which capture the dynamic interplay between microbiota, vitamin metabolism, and neurophysiology beyond the limits of reductionist methods. Metagenomics, metabolomics, and epigenomics reveal vitamin-synthesizing pathways, functional metabolites, and host gene regulation, while computational models map causal networks and predict intervention outcomes. Multi-omics signatures hold promise for early diagnosis, risk stratification, and personalized therapies in neurodegenerative disease. Challenges include data complexity, reproducibility, and privacy concerns, but advances in technology and analytics are paving the way for precision, molecularly informed clinical care—where pharmacists and clinicians collaborate to design targeted, microbiota- and vitamin-centered interventions [135].

9.10 CONCLUSION

The gut–microbiota–vitamin axis is a dynamic frontier in neuroscience, showing that brain health depends on interactions among gut microbes, vitamin metabolism, nutrition, immunity, and neural signaling. Microbially derived vitamins, especially B-complex, vitamin D, and antioxidants, support neurotransmitter synthesis, myelin maintenance, and neuroprotection, while dysbiosis and deficiencies contribute to neurodegenerative diseases. Targeted interventions—probiotics, prebiotics, vitamin therapies, and personalized nutrition—offer promise but require precise, individualized application. For pharmacy professionals, integrating microbiome science into patient care can advance precision neurotherapeutics, reframing cognitive health as a process that begins in the gut.

FUNDING

None.

CONFLICT OF INTEREST

None.

CONSENT TO PUBLICATION

All authors have read and agreed to submit this chapter to the publisher.

REFERENCES

1. Cryan JF, O'Riordan KJ, Cowan CSM, Sandhu KV, Bastiaanssen TFS, Boehme M et al. The Microbiota-Gut-Brain Axis. *Physiol Rev.* 2019;99(4):1877–2013.
2. Rowland I, Gibson G, Heinken A, Scott K, Swann J, Thiele I et al. Gut Microbiota Functions: Metabolism of Nutrients and Other Food Components. *Eur J Nutr.* 2018;57(1):1–24.
3. Borre YE, Moloney RD, Clarke G, Dinan TG, Cryan JF. The Impact of Microbiota on Brain and Behavior: Mechanisms & Therapeutic Potential. In: F Cryan & TG Dinan (Eds.), *Advances in Experimental Medicine and Biology* (Vol. 817, pp. 373–403). Springer. 2014.
4. Heijtz RD, Wang S, Anuar F, Qian Y, Björkholm B, Samuelsson A et al. Normal Gut Microbiota Modulates Brain Development and Behavior. *Proc Natl Acad Sci.* 2011;108(7):3047–52.
5. Silva YP, Bernardi A, Frozza RL. The Role of Short-Chain Fatty Acids from Gut Microbiota in Gut-Brain Communication. *Front Endocrinol (Lausanne).* 2020;11:25.
6. Strandwitz P. Neurotransmitter Modulation by the Gut Microbiota. *Brain Res.* 2018;1693:128–33.
7. O'Riordan KJ, Moloney GM, Keane L, Clarke G, Cryan JF. The gut Microbiota-Immune-Brain Axis: Therapeutic Implications. *Cell Rep Med.* 2025;6(3):101982.
8. Vogt NM, Kerby RL, Dill-McFarland KA, Harding SJ, Merluzzi AP, Johnson SC et al. Gut Microbiome Alterations in Alzheimer's Disease. *Sci Rep.* 2017;7(1):13537.
9. Smith AD, Refsum H. Homocysteine, B Vitamins, and Cognitive Impairment. *Annu Rev Nutr.* 2016;36(1):211–39.
10. Field DT, Cracknell RO, Eastwood JR, Scarfe P, Williams CM, Zheng Y et al. High-Dose Vitamin B6 Supplementation Reduces Anxiety and Strengthens Visual Surround Suppression. *Hum Psychopharmacol Clin Exp.* 2022;37(6):e2852.
11. Groves NJ, McGrath JJ, Burne THJ. Vitamin D as a Neurosteroid Affecting the Developing and Adult Brain. *Annu Rev Nutr.* 2014;34(1):117–41.
12. Fernandes de Abreu DA, Eyles D, Féron F. Vitamin D, a Neuro-Immunomodulator: Implications for Neurodegenerative and Autoimmune Diseases. *Psychoneuroendocrinology.* 2009;34:S265–77.
13. Harrison FE, May JM. Vitamin C Function in the Brain: Vital Role of the Ascorbate Transporter SVCT2. *Free Radic Biol Med.* 2009;46(6):719–30.
14. Human Microbiome Project Consortium. Structure, Function and Diversity of the Healthy Human Microbiome. *Nature.* 2012;486(7402):207–14.
15. Sharon G, Sampson TR, Geschwind DH, Mazmanian SK. The Central Nervous System and the Gut Microbiome. *Cell.* 2016;167(4):915–32.
16. Arrieta MC, Stiemsma LT, Amenyogbe N, Brown EM, Finlay B. The Intestinal Microbiome in Early Life: Health and Disease. *Front Immunol.* 2014;5:427.
17. Dinan TG, Cryan JF. Gut Instincts: Microbiota as a Key Regulator of Brain Development, Ageing and Neurodegeneration. *J Physiol.* 2017;595(2):489–503.
18. Everard A, Belzer C, Geurts L, Ouwerkerk JP, Druart C, Bindels LB et al. Cross-Talk between *Akkermansia muciniphila* and Intestinal Epithelium Controls Diet-Induced Obesity. *Proc Natl Acad Sci.* 2013;110(22):9066–71.
19. Vogt NM, Romano KA, Darst BF, Engelman CD, Johnson SC, Carlsson CM et al. The Gut Microbiota-Derived Metabolite Trimethylamine N-Oxide Is Elevated in Alzheimer's Disease. *Alzheimers Res Ther.* 2018;10(1):124.
20. Dalile B, Van Oudenhove L, Vervliet B, Verbeke K. The Role of Short-Chain Fatty Acids in Microbiota–Gut–Brain Communication. *Nat Rev Gastroenterol Hepatol.* 2019;16(8):461–78.
21. Ridlon JM, Kang DJ, Hylemon PB. Bile Salt Biotransformations by Human Intestinal Bacteria. *J Lipid Res.* 2006;47(2):241–59.
22. Agus A, Planchais J, Sokol H. Gut Microbiota Regulation of Tryptophan Metabolism in Health and Disease. *Cell Host Microbe.* 2018;23(6):716–24.
23. Espín JC, González-Sarrías A, Tomás-Barberán FA. The Gut Microbiota: A Key Factor in the Therapeutic Effects of (Poly)Phenols. *Biochem Pharmacol.* 2017;139:82–93.
24. Peng L, Li ZR, Green RS, Holzman IR, Lin J. Butyrate Enhances the Intestinal Barrier by Facilitating Tight Junction Assembly via Activation of AMP-Activated Protein Kinase in Caco-2 Cell Monolayers. *J Nutr.* 2009;139(9):1619–25.
25. Arpaia N, Campbell C, Fan X, Dikiy S, van der Veeken J, deRoos P et al. Metabolites Produced by Commensal Bacteria Promote Peripheral Regulatory T-Cell Generation. *Nature.* 2013;504(7480):451–5.
26. Belkaid Y, Hand TW. Role of the Microbiota in Immunity and Inflammation. *Cell.* 2014;157(1):121–41.
27. Braniste V, Al-Asmakh M, Kowal C, Anuar F, Abbaspour A, Tóth M et al. The Gut Microbiota Influences Blood-Brain Barrier Permeability in Mice. *Sci Transl Med.* 2014;6(263):263ra158..
28. Tarracchini C, Lugli GA, Mancabelli L, van Sinderen D, Turroni F, Ventura M et al. Exploring the Vitamin Biosynthesis Landscape of the Human Gut Microbiota. *mSystems.* 2024;9(10):e00929-24.
29. Rodionov DA, Arzamasov AA, Khoroshkin MS, Iablokov SN, Leyn SA, Peterson SN et al. Micronutrient Requirements and Sharing Capabilities of the Human Gut Microbiome. *Front Microbiol.* 2019;10:1316.
30. O'Leary F, Samman S. Vitamin B12 in Health and Disease. *Nutrients.* 2010;2(3):299–316.
31. Nawrocki KL, Wetzel D, Jones JB, Woods EC, McBride SM. Ethanolamine Is a Valuable Nutrient Source that Impacts *Clostridium difficile* Pathogenesis. *Environ Microbiol.* 2018;20(4):1419–35.
32. Scott KP, Antoine JM, Midtvedt T, van Hemert S. Manipulating the Gut Microbiota to Maintain Health and Treat Disease. *Microb Ecol Health Dis.* 2015;26:25877.
33. David LA, Maurice CF, Carmody RN, Gootenberg DB, Button JE, Wolfe BE et al. Diet Rapidly and Reproducibly Alters the Human Gut Microbiome. *Nature.* 2014;505(7484):559–63.
34. Wu GD, Chen J, Hoffmann C, Bittinger K, Chen YY, Keilbaugh SA et al. Linking Long-Term Dietary Patterns with Gut Microbial Enterotypes. *Science.* 2011;334(6052):105–8.
35. Sokol H, Brot L, Stefanescu C, Auzolle C, Barnich N, Buisson A et al. Prominence of Ileal Mucosa-Associated Microbiota to Predict Postoperative Endoscopic Recurrence in Crohn's Disease. *Gut.* 2020;69(3):462–72.
36. Litvak Y, Byndloss MX, Bäumler AJ. Colonocyte Metabolism Shapes the Gut Microbiota. *Science.* 2018;362(6418):eaat9076.
37. D'Aimmo MR, Mattarelli P, Biavati B, Carlsson NG, Andlid T. The Potential of Bifidobacteria as a Source of Natural Folate. *J Appl Microbiol.* 2012;112(5):975–84.
38. LeBlanc JG, Milani C, de Giori GS, Sesma F, van Sinderen D, Ventura M. Bacteria as Vitamin Suppliers to their Host: A Gut Microbiota Perspective. *Curr Opin Biotechnol.* 2013;24(2):160–8.

39. Conly JM, Stein K. Quantitative and Qualitative Measurements of K Vitamins in Human Intestinal Contents. *Am J Gastroenterol.* 1992;87(3):311–6.
40. LeBlanc JG, Laiño JE, del Valle MJ, Vannini V, van Sinderen D, Taranto MP et al. B-Group Vitamin Production by Lactic Acid Bacteria: Current Knowledge and Potential Applications. *J Appl Microbiol.* 2011;111(6):1297–309.
41. Degnan PH, Barry NA, Mok KC, Taga ME, Goodman AL. Human Gut Microbes Use Multiple Transporters to Distinguish Vitamin B12 Analogs and Compete in the Gut. *Cell Host Microbe.* 2014;15(1):47–57.
42. Stover PJ, Field MS. Trafficking of Intracellular Folates. *Adv Nutr.* 2011;2(4):325–31.
43. Allison J, Kaliszewska A, Uceda S, Reiriz M, Arias N. Targeting DNA Methylation in the Adult Brain through Diet. *Nutrients.* 2021;13(11):3979.
44. Allen GFG, Neergheen V, Oppenheim M, Fitzgerald JC, Footitt E, Hyland K et al. Pyridoxal 5′-Phosphate Deficiency Causes a Loss of Aromatic L-Amino Acid Decarboxylase in Patients and Human Neuroblastoma Cells, Implications for Aromatic L-Amino Acid Decarboxylase and Vitamin B_6 Deficiency States. *J Neurochem.* 2010;114(1):87–96.
45. Kennedy D. B Vitamins and the Brain: Mechanisms, Dose and Efficacy—A Review. *Nutrients.* 2016;8(2):68.
46. Stover PJ. Physiology of Folate and Vitamin B_{12} in Health and Disease. *Nutr Rev.* 2004;62:S3–12.
47. Degnan PH, Taga ME, Goodman AL. Vitamin B 12 as a Modulator of Gut Microbial Ecology. *Cell Metab.* 2014;20(5):769–78.
48. Christakos S, Dhawan P, Verstuyf A, Verlinden L, Carmeliet G. Vitamin D: Metabolism, Molecular Mechanism of Action, and Pleiotropic Effects. *Physiol Rev.* 2016;96(1):365–408.
49. Stilling RM, Dinan TG, Cryan JF. Microbial Genes, Brain & Behaviour: Epigenetic Regulation of the Gut–Brain Axis. *Genes Brain Behav.* 2014;13(1):69–86.
50. Bonaz B, Bazin T, Pellissier S. The Vagus Nerve at the Interface of the Microbiota-Gut-Brain Axis. *Front Neurosci.* 2018;12:49.
51. Bharwani A, Mian MF, Foster JA, Surette MG, Bienenstock J, Forsythe P. Structural & Functional Consequences of Chronic Psychosocial Stress on the Microbiome & Host. *Psychoneuroendocrinology.* 2016;63:217–27.
52. Yano JM, Yu K, Donaldson GP, Shastri GG, Ann P, Ma L et al. Indigenous Bacteria from the Gut Microbiota Regulate Host Serotonin Biosynthesis. *Cell.* 2015;161(2):264–76.
53. Bravo JA, Forsythe P, Chew M V., Escaravage E, Savignac HM, Dinan TG et al. Ingestion of *Lactobacillus* Strain Regulates Emotional Behavior and Central GABA Receptor Expression in a Mouse Via the Vagus Nerve. *Proc Natl Acad Sci.* 2011;108(38):16050–5.
54. Miller JW. Homocysteine and Cognitive Function in the Elderly. *Universa Medicina.* 28(2):106–16. Available from: https://www.univmed.org/ejurnal/index.php/medicina/article/view/236
55. Mora JR, Iwata M, von Andrian UH. Vitamin Effects on the Immune System: Vitamins A and D Take Centre Stage. *Nat Rev Immunol.* 2008;8(9):685–98.
56. Perry VH, Holmes C. Microglial Priming in Neurodegenerative Disease. *Nat Rev Neurol.* 2014;10(4):217–24.
57. Assa A, Vong L, Pinnell LJ, Avitzur N, Johnson-Henry KC, Sherman PM. Vitamin D Deficiency Promotes Epithelial Barrier Dysfunction and Intestinal Inflammation. *J Infect Dis.* 2014;210(8):1296–305.
58. Höglund E, Øverli Ø, Winberg S. Tryptophan Metabolic Pathways and Brain Serotonergic Activity: A Comparative Review. *Front Endocrinol (Lausanne).* 2019;10:158.
59. O'Mahony SM, Clarke G, Borre YE, Dinan TG, Cryan JF. Serotonin, Tryptophan Metabolism and the Brain-Gut-Microbiome Axis. *Behav Brain Res.* 2015;277:32–48.
60. Bottiglieri T. Folate, Vitamin B12, and Neuropsychiatric Disorders. *Nutr Rev.* 2009;54(12):382–90.
61. Desbonnet L, Garrett L, Clarke G, Bienenstock J, Dinan TG. The Probiotic Bifidobacteria Infantis: An Assessment of Potential Antidepressant Properties in the Rat. *J Psychiatr Res.* 2008;43(2):164–74.
62. Barrett E, Ross RP, O'Toole PW, Fitzgerald GF, Stanton C. γ-Aminobutyric Acid Production by Culturable Bacteria from the Human Intestine. *J Appl Microbiol.* 2012;113(2):411–7.
63. Bedarf JR, Hildebrand F, Coelho LP, Sunagawa S, Bahram M, Goeser F et al. Functional Implications of Microbial and Viral Gut Metagenome Changes in Early Stage L-DOPA-Naïve Parkinson's Disease Patients. *Genome Med.* 2017;9(1):39.
64. Wang Z, Klipfell E, Bennett BJ, Koeth R, Levison BS, DuGar B et al. Gut Flora Metabolism of Phosphatidylcholine Promotes Cardiovascular Disease. *Nature.* 2011;472(7341):57–63.
65. Reynolds EH. The Neurology of Folic Acid Deficiency. *Brain.* 2014;137:927–43.
66. Butterworth RF. Thiamin Deficiency and Brain Disorders. *Nutr Res Rev.* 2003;16(2):277–84.
67. Sechi G, Serra A. Wernicke's Encephalopathy: New Clinical Settings and Recent Advances in Diagnosis and Management. *Lancet Neurol.* 2007;6(5):442–55.
68. Stach K, Stach W, Augoff K. Vitamin B6 in Health and Disease. *Nutrients.* 2021;13(9):3229.
69. Allen LH. Causes of Vitamin B_{12} and Folate Deficiency. *Food Nutr Bull.* 2008;29(2_suppl1):S20–34.
70. Eyles DW. Vitamin D: Brain and Behavior. *JBMR Plus.* 2021;5(1):e10419.
71. Garcion E, Wion-Barbot N, Montero-Menei CN, Berger F, Wion D. New Clues about Vitamin D Functions in the Nervous System. *Trends Endocrinol Metab.* 2002;13(3):100–5.
72. Cass WA, Smith MP, Peters LE. Calcitriol Protects against the Dopamine-and Serotonin-Depleting Effects of Neurotoxic Doses of Methamphetamine. *Ann NY Acad Sci.* 2006;1074(1):261–71.
73. Hewison M. Vitamin D and the Immune System: New Perspectives on an Old Theme. *Endocrinol Metab Clin North Am.* 2010;39(2):365–79.
74. Pierrot-Deseilligny C, Souberbielle JC. Vitamin D and Multiple Sclerosis: An Update. *Mult Scler Relat Disord.* 2017;14:35–45.
75. Zhang R, Naughton DP. Vitamin D in Health and Disease: Current Perspectives. *Nutr J.* 2010;9(1):65.
76. Wu S, Zhang Y Guo, Lu R, Xia Y, Zhou D, Petrof EO et al. Intestinal Epithelial Vitamin D Receptor Deletion Leads to Defective Autophagy in Colitis. *Gut.* 2015;64(7):1082–94.
77. Bashir M, Prietl B, Tauschmann M, Mautner SI, Kump PK, Treiber G et al. Effects of High Doses of Vitamin D3 on Mucosa-Associated Gut Microbiome Vary between Regions of the Human Gastrointestinal Tract. *Eur J Nutr.* 2016;55(4):1479–89.

78. Maden M. Retinoic Acid in the Development, Regeneration and Maintenance of the Nervous System. *Nat Rev Neurosci.* 2007;8(10):755–65.
79. Chiang MY, Misner D, Kempermann G, Schikorski T, Giguère V, Sucov HM et al. An Essential Role for Retinoid Receptors RARβ and RXRγ In Long-Term Potentiation and Depression. *Neuron.* 1998;21(6):1353–61.
80. Hall JA, Grainger JR, Spencer SP, Belkaid Y. The Role of Retinoic Acid in Tolerance and Immunity. *Immunity.* 2011;35(1):13–22.
81. Mandl J, Szarka A, Bánhegyi G. Vitamin C: Update on Physiology and Pharmacology. *Br J Pharmacol.* 2009;157(7):1097–110.
82. Heuser G, Vojdani A. Enhancement of Natural Killer Cell Activity and T and B Cell Function by Buffered Vitamin C In Patients Exposed to Toxic Chemicals: The Role of Protein Kinase-C. *Immunopharmacol Immunotoxicol.* 1997;19(3):291–312.
83. Tao Y, Wu M, Su B, Lin H, Li Q, He Y et al. Host–Gut Microbiota Interactions: Exploring the Potential Role of Vitamin B1 and B2 in the Microbiota–Gut–Brain Axis and Anxiety, Stress, and Sleep Quality. *Nutrients.* 2025;17(11):1894.
84. Rizvi S, Raza ST, Ahmed F, Ahmad A, Abbas S, Mahdi F. The Role of Vitamin E in Human Health and Some Diseases. *Sultan Qaboos Univ Med J.* 2014;14(2):e157–65.
85. Benameur T, Porro C, Twfieg ME, Benameur N, Panaro MA, Filannino FM et al. Emerging Paradigms in Inflammatory Disease Management: Exploring Bioactive Compounds and the Gut Microbiota. *Brain Sci.* 2023;13(8):1226. Available from: https://pmc.ncbi.nlm.nih.gov/articles/PMC10452544/
86. Hanna M, Jaqua E, Nguyen V, Clay J. B Vitamins: Functions and Uses in Medicine. *Perm J.* 2022;26(2):89. Available from: https://pmc.ncbi.nlm.nih.gov/articles/PMC9662251/
87. Roth W, Zadeh K, Vekariya R, Ge Y, Mohamadzadeh M. Tryptophan Metabolism and Gut-Brain Homeostasis. *Int J Mol Sci.* 2021;22(6):2973. Available from: https://pmc.ncbi.nlm.nih.gov/articles/PMC8000752/
88. Wang Y, Zhang Z, Li B, He B, Li L, Nice EC et al. New Insights into the Gut Microbiota in Neurodegenerative Diseases from the Perspective of Redox Homeostasis. *Antioxidants.* 2022;11(11):2287.
89. Mrowicka M, Mrowicki J, Dragan G, Majsterek I. The Importance of Thiamine (Vitamin B1) in Humans. *Biosci Rep.* 2023;43(10):BSR20230374. Available from: https://pmc.ncbi.nlm.nih.gov/articles/PMC10568373/
90. Parra M, Stahl S, Hellmann H. Vitamin B6 and Its Role in Cell Metabolism and Physiology. *Cells.* 2018;7(7):84. Available from: https://pmc.ncbi.nlm.nih.gov/articles/PMC6071262/
91. Agostini D, Bartolacci A, Rotondo R, De Pandis MF, Battistelli M, Micucci M et al. Homocysteine, Nutrition, and Gut Microbiota: A Comprehensive Review of Current Evidence and Insights. *Nutrients.* 2025;17(8):1325. Available from: https://pmc.ncbi.nlm.nih.gov/articles/PMC12030302/
92. Calderón-Ospina CA, Nava-Mesa MO. B Vitamins in the Nervous System: Current Knowledge of the Biochemical Modes of Action and Synergies of Thiamine, Pyridoxine, and Cobalamin. *CNS Neurosci Ther.* 2019;26(1):5. Available from: https://pmc.ncbi.nlm.nih.gov/articles/PMC6930825/
93. Sintzel MB, Rametta M, Reder AT. Vitamin D and Multiple Sclerosis: A Comprehensive Review. *Neurol Ther.* 2017;7(1):59. Available from: https://pmc.ncbi.nlm.nih.gov/articles/PMC5990512/
94. Shandilya S, Kumar S, Kumar Jha N, Kumar Kesari K, Ruokolainen J. Interplay of Gut Microbiota and Oxidative Stress: Perspective on Neurodegeneration and Neuroprotection. *J Adv Res.* 2021;38:223. Available from: https://pmc.ncbi.nlm.nih.gov/articles/PMC9091761/
95. Ciarcià G, Bianchi S, Tomasello B, Acquaviva R, Malfa GA, Naletova I et al. Vitamin E and Non-Communicable Diseases: A Review. *Biomedicines.* 2022;10(10):2473. Available from: https://pmc.ncbi.nlm.nih.gov/articles/PMC9599164/
96. Mah E, Bruno RS. Postprandial Hyperglycemia on Vascular Endothelial Function: Mechanisms and Consequences. *Nutr Res.* 2012;32(10):727–40.
97. Zhao Y, Jaber V, Lukiw WJ. Secretory Products of the Human GI Tract Microbiome and their Potential Impact on Alzheimer's Disease (AD): Detection of Lipopolysaccharide (LPS) in AD Hippocampus. *Front Cell Infect Microbiol.* 2017;7:318.
98. Oulhaj A, Refsum H, Beaumont H, Williams J, King E, Jacoby R et al. Homocysteine as a Predictor of Cognitive Decline in Alzheimer's Disease. *Int J Geriatr Psychiatry.* 2010;25(1):82–90.
99. Lương KVQ, Nguyễn LTH. The Role of Vitamin D in Alzheimer's Disease. *Am J Alzheimers Dis Other Demen.* 2013;28(2):126–36.
100. Dysken MW, Sano M, Asthana S, Vertrees JE, Pallaki M, Llorente M et al. Effect of Vitamin E and Memantine on Functional Decline in Alzheimer Disease. *JAMA.* 2014;311(1):33.
101. Braak H, Tredici K Del, Rüb U, de Vos RAI, Jansen Steur ENH, Braak E. Staging of Brain Pathology Related to Sporadic Parkinson's disease. *Neurobiol Aging.* 2003;24(2):197–211.
102. Svensson E, Horváth-Puhó E, Thomsen RW, Djurhuus JC, Pedersen L, Borghammer P et al. Vagotomy and Subsequent Risk of Parkinson's Disease. *Ann Neurol.* 2015;78(4):522–9.
103. Varesi A, Campagnoli LIM, Fahmideh F, Pierella E, Romeo M, Ricevuti G et al. The Interplay between Gut Microbiota and Parkinson's Disease: Implications on Diagnosis and Treatment. *Int J Mol Sci.* 2022;23(20):12289.
104. Barichella M, Cereda E, Pezzoli G. Major Nutritional Issues in the Management of Parkinson's Disease. *Movement Disorders.* 2009;24(13):1881–92.
105. Al-kuraishy HM, Al-Gareeb AI, Elewa YHA, Zahran MH, Alexiou A, Papadakis M et al. Parkinson's Disease Risk and Hyperhomocysteinemia: The Possible Link. *Cell Mol Neurobiol.* 2023;43(6):2743–59.
106. Fullard ME, Duda JE. A Review of the Relationship between Vitamin D and Parkinson Disease Symptoms. *Front Neurol.* 2020;11. doi:10.3389/fneur.2020.00454.
107. Scheider WL, Hershey LA, Vena JE, Holmlund T, Marshall JR, Freudenheim JL. Dietary Antioxidants and other Dietary Factors in the Etiology of Parkinson's Disease. *Movement Disorders.* 1997;12(2):190–6.
108. Maini Rekdal V, Bess EN, Bisanz JE, Turnbaugh PJ, Balskus EP. Discovery and Inhibition of an Interspecies Gut Bacterial Pathway for Levodopa Metabolism. *Science.* 2019;364(6445):eaau6323.
109. Ascherio A, Munger KL, Simon KC. Vitamin D and Multiple Sclerosis. *Lancet Neurol.* 2010;9(6):599–612.
110. Gombash SE, Lee PW, Sawdai E, Lovett-Racke AE. Vitamin D as a Risk Factor for Multiple Sclerosis: Immunoregulatory or Neuroprotective? *Front Neurol.* 2022; 13:796933.

111. Jangi S, Gandhi R, Cox LM, Li N, von Glehn F, Yan R et al. Alterations of the Human Gut Microbiome in Multiple Sclerosis. *Nat Commun.* 2016;7(1):12015.
112. Mohan A, Kumar R, Kumar V, Yadav M. Homocysteine, Vitamin B12 and Folate Level: Possible Risk Factors in the Progression of Chronic Heart and Kidney Disorders. *Curr Cardiol Rev.* 2023;19(4):e090223213539..
113. Hardiman O, Al-Chalabi A, Chio A, Corr EM, Logroscino G, Robberecht W et al. Amyotrophic Lateral Sclerosis. *Nat Rev Dis Primers.* 2017;3(1):17071.
114. Yassin LK, Nakhal MM, Alderei A, Almehairbi A, Mydeen AB, Akour A et al. Exploring the Microbiota-Gut-Brain Axis: Impact on Brain Structure and Function. *Front Neuroanat.* 2025;19:1504065.
115. Wang M, Liu Z, Sun W, Yuan Y, Jiao B, Zhang X et al. Association Between Vitamins and Amyotrophic Lateral Sclerosis: A Center-based Survey in Mainland China. *Front Neurol.* 2020;11:488.
116. Sanders ME, Merenstein DJ, Reid G, Gibson GR, Rastall RA. Probiotics and Prebiotics in Intestinal Health and Disease: From Biology to the Clinic. *Nat Rev Gastroenterol Hepatol.* 2019;16(10):605–16.
117. Akbari E, Asemi Z, Daneshvar Kakhaki R, Bahmani F, Kouchaki E, Tamtaji OR et al. Effect of Probiotic Supplementation on Cognitive Function and Metabolic Status in Alzheimer's Disease: A Randomized, Double-Blind and Controlled Trial. *Front Aging Neurosci.* 2016;8:256.
118. O'Riordan KJ, Collins MK, Moloney GM, Knox EG, Aburto MR, Fülling C et al. Short Chain Fatty Acids: Microbial Metabolites for Gut-Brain Axis Signalling. *Mol Cell Endocrinol.* 2022;546:111572.
119. Perez-Pardo P, de Jong EM, Broersen LM, van Wijk N, Attali A, Garssen J et al. Promising Effects of Neurorestorative Diets on Motor, Cognitive, and Gastrointestinal Dysfunction after Symptom Development in a Mouse Model of Parkinson's Disease. *Front Aging Neurosci.* 2017;9:57.
120. Jernerén F, Elshorbagy AK, Oulhaj A, Smith SM, Refsum H, Smith AD. Brain Atrophy in Cognitively Impaired Elderly: The Importance of Long-Chain ω-3 Fatty Acids and B Vitamin Status in a Randomized Controlled Trial. *Am J Clin Nutr.* 2015;102(1):215–21.
121. Douaud G, Refsum H, de Jager CA, Jacoby R, E. Nichols T, Smith SM et al. Preventing Alzheimer's Disease-Related Gray Matter Atrophy by B-Vitamin Treatment. *Proc Natl Acad Sci.* 2013;110(23):9523–8.
122. Annweiler C, Llewellyn DJ, Beauchet O. Low Serum Vitamin D Concentrations in Alzheimer's Disease: A Systematic Review and Meta-Analysis. *J Alzheimer's Dis.* 2013;33(3):659–74.
123. Rossom RC, Espeland MA, Manson JE, Dysken MW, Johnson KC, Lane DS et al. Calcium and Vitamin D Supplementation and Cognitive Impairment in the Women's Health Initiative. *J Am Geriatr Soc.* 2012;60(12):2197–205.
124. Holick MF. Deficiency of Sunlight and Vitamin D. *BMJ.* 2008;336(7657):1318–9.
125. Kaji R, Imai T, Iwasaki Y, Okamoto K, Nakagawa M, Ohashi Y et al. Ultra-High-Dose Methylcobalamin in Amyotrophic Lateral Sclerosis: A Long-Term Phase II/III Randomised Controlled Study. *J Neurol Neurosurg Psychiatry.* 2019;90(4):451–7.
126. Gibson GE, Hirsch JA, Fonzetti P, Jordan BD, Cirio RT, Elder J. Vitamin B1 (thiamine) and dementia. *Ann NY Acad Sci.* 2016;1367(1):21–30.
127. Sun MF, Zhu YL, Zhou ZL, Jia XB, Xu YD, Yang Q et al. Neuroprotective Effects of Fecal Microbiota Transplantation on MPTP-Induced Parkinson's Disease Mice: Gut Microbiota, Glial Reaction and TLR4/TNF-α Signaling Pathway. *Brain Behav Immun.* 2018;70:48–60.
128. Kim MS, Kim Y, Choi H, Kim W, Park S, Lee D et al. Transfer of a Healthy Microbiota Reduces Amyloid and Tau Pathology in an Alzheimer's Disease Animal Model. *Gut.* 2020;69(2):283–94.
129. Zhao Z, Ning J, Bao X qi, Shang M, Ma J, Li G et al. Fecal Microbiota Transplantation Protects Rotenone-Induced Parkinson's Disease Mice Via Suppressing Inflammation Mediated by the Lipopolysaccharide-TLR4 Signaling Pathway through the Microbiota-Gut-Brain Axis. *Microbiome.* 2021;9(1):226.
130. Xue LJ, Yang XZ, Tong Q, Shen P, Ma SJ, Wu SN et al. Fecal Microbiota Transplantation Therapy for Parkinson's Disease. *Medicine.* 2020;99(35):e22035.
131. Bedani R, Cucick ACC, Albuquerque MAC de, LeBlanc JG, Saad SMI. B-Group Vitamins as Potential Prebiotic Candidates: Their Effects on the Human Gut Microbiome. *J Nutr.* 2024;154(2):341–53.
132. Rossi M, Amaretti A, Raimondi S. Folate Production by Probiotic Bacteria. *Nutrients.* 2011;3(1):118–34.
133. Wilson ID, Nicholson JK. Gut Microbiome Interactions with Drug Metabolism, Efficacy, and Toxicity. *Trans Res.* 2017;179:204–22.
134. Lin CH, Chen CC, Chiang HL, Liou JM, Chang CM, Lu TP et al. Altered Gut Microbiota and Inflammatory Cytokine Responses in Patients with Parkinson's Disease. *J Neuroinflammation.* 2019;16(1):129.
135. Zheng Y, Bonfili L, Wei T, Eleuteri AM. Understanding the Gut–Brain Axis and its Therapeutic Implications for Neurodegenerative Disorders. *Nutrients.* 2023;15(21):4631.

10 Vitamin–Drug and Nutrient–Nutrient Interactions in Neurological Disorders

Falguni Goel[1], *Prateek Gupta*[2], *and Payal Singh*[3]

[1]Department of Pharmaceutical Technology, Meerut Institute of Engineering & Technology (MIET), Meerut, India
[2]Noida Institute of Engineering and Technology (NIET), Greater Noida, India
[3]University Centre for Research & Development (UCRD), Chandigarh University, Punjab, India

ABBREVIATIONS

AD	Alzheimer's Disease
PD	Parkinson's Disease
MS	Multiple Sclerosis
DHA	Docosahexaenoic Acid
MTHFR	Methylene Tetrahydrofolate Reductase

10.1 INTRODUCTION

Neurological disorders are defined by a large number of diseases that affect the central and peripheral nervous system. Alzheimer's disease (AD), Parkinson's disease (PD), multiple sclerosis (MS), epilepsy, strokes, and neuropathies are all examples of neurological disorders [1]. Neurological disorders are normally chronic progressive neurodegeneration with neuronal toxicities and neurotransmission and neuronal survival and neurochemical homeostasis eventually affecting cognition, movements, mood, and quality of life. Neurologic diseases continue to increase in burden and represent some of the highest disability and mortality rates in the world [2]. Although the complexity of their pathogenicity continues to unfold, more specifically the contributions of things like oxidative stress, mitochondrial dysfunction, neuroinflammations, and metabolic disturbances are becoming more crystallized. Although genetics and exposures are expected and recognized contributors, the diet and nutrition are now viewed as oftentimes underappreciated contributors to brain health and risk modifiers of disease [3]. Vitamins and nutrients are critical for optimal function of the brain as they are involved in many biochemical and physiological functions. Vitamins and nutrients may facilitate numerous metabolic functions—in this case, B-vitamins are necessary for one-carbon metabolism, neurotransmitter and myelin synthesis, and vitamin D involves regulation of neuroimmune responses and calcium signaling [4]. Antioxidant vitamins, such as vitamin C and vitamin E, protect neurons against oxidative injury, while minerals, such as magnesium, zinc, and iron, modulate neurotransmitter release, synaptic plasticity, and enzyme activity. Omega-3 fatty acids make up a proportion of the structural components of neuronal membranes and have an anti-inflammatory effect, which is an important contributor to preventing the neurodegenerative process [5]. Deficiencies and imbalances in vitamins and nutrients contribute to the acceleration of disease and pathological processes, while adequate intake or supplementation has been associated with neuroprotection and better clinical outcomes [6].

The drugs used for epilepsy, PD, AD, and MS can affect the absorption, metabolism, and bioavailability of important nutrients and cause second-degree deficiencies [7]. Excess or unmonitored supplementation of vitamins/nutrients can affect the pharmacokinetics and pharmacodynamics of the agents prescribed to our patients, resulting in ineffectiveness of treatment or potential for adverse events [8]. There are antiepileptic drugs that can lower folate and vitamin B12 and lead to cognitive decline in some patients. There has been some vitamin E supplementation in patients with AD, as it can have either favorable or dare I say controversial outcomes, depending on the dose and clinical context [9].

Nutrient–nutrient interactions themselves can show evidence of synergism or antagonism. For instance, showing the bioavailability and consequently the therapeutic effect of many nutrients can demonstrate synergism; for example, omega-3 fatty acids and antioxidant vitamins demonstrated their synergistic effect by retaining cell membranes in the presence of oxidative stress [10]. Other forms of nutrients can demonstrate antagonism; for example, several nutrients could compete for absorption and/or metabolism. Excess zinc can potentially disrupt copper, which is important for neuronal function. With these nutrient interactions, these can certainly add additional layers of complexity for the nutritional therapeutic approach for neurological conditions and confirm the need for a patient-centered, evidence-based approach [11]. It is imperative to understand how vitamins, nutrients, and pharmacotherapy influence one another to design a therapeutic strategy for our neurologically

DOI: 10.1201/9781003686675-10

impaired patients. As the disciplines of neuropharmacology, clinical nutrition, and nutrigenomics move toward convergence, we will have more opportunity to leverage the interactions of diet and pharmacotherapy to optimize patient care [12]. In this chapter, we will analyze clinically relevant vitamin–drug interactions in neurological disorders, nutrient–nutrient interactions, and their therapeutic implications. We want to identify the barriers to and opportunities for incorporating nutrition into comprehensive neurological care.

10.2 MECHANISMS OF VITAMIN–DRUG AND NUTRIENT–NUTRIENT INTERACTIONS

Vitamins and nutrients have interactions with treatment medicines for neurological illnesses that are mediated by a number of mechanistic pathways that determine how those drugs work and how readily those nutrients are available [13]. The nature of those mechanisms may be determined by the pharmacokinetic and pharmacodynamic processes, as well as the possibility of both antagonism and synergism. Individual factors related to genetics, age, and comorbidities also strongly impact the degree and clinical significance of those interactions [14].

Pharmacokinetic interactions are defined as events in which a vitamin or nutrient alters the absorption, distribution, metabolism, or excretion of a drug, or the absorption, distribution, metabolism, excretion (ADME) of a drug alters the intake or usage of a vitamin or nutrient. Many antiepileptic drugs, such as phenytoin, and carbamazepine, induce hepatic cytochrome P450 enzymes, which speed the metabolism of essential nutrients folate, vitamin D, and vitamin K [15]. This reduction in essential nutrient levels can lead to several complications such as megaloblastic anemia, bone demineralization, and coagulopathies, which set patients up for poorer neurological outcomes. Levodopa treatment of patients with PD can lead to disruption of vitamin B6 metabolism and the development of peripheral neuropathies [16]. Nutrients may also affect the absorption of drugs at the gastrointestinal level. Calcium and iron supplements have been shown to chelate to levodopa and certain antibiotics, which may limit both bioavailability and thereby clinical efficacy [17]. Lipid-soluble vitamins, such as A, D, E, and K, require adequate dietary fat in order to be absorbed, so the use of fat-binding agents or drugs that prevent bile secretion can mitigate or even eliminate their absorption. A clear understanding of these pharmacokinetic alterations highlights the fact that nutrient status can enhance side effects of medication or reduce desired therapeutic effects [18].

Pharmacodynamic interactions occur when vitamins and nutrients act directly on the target mechanism of drug action at the level of a receptor, an enzyme, or a signaling pathway. Vitamin E may enhance the neuroprotective effects of the cholinesterase inhibitors used to treat AD by reducing oxidative stress, a prominent source of neuronal damage [19]. An overdose of vitamin B6 could inhibit the action of levodopa in the treatment of PD, since pyridoxine increases peripheral decarboxylation of levodopa before it can enter the brain. Magnesium as an N-methyl-D-aspartate receptor (NMDA) receptor antagonist is capable of increasing the effectiveness of certain antiepileptic medications because it reduces excitotoxicity [20]. High-dose *Ginkgo biloba*, a phytochemical derived from a nutrient, could raise the risk of bleeding when used in conjunction with anticoagulants in patients with vascular dementia or following a stroke. These examples identify modifying effects of nutrient supplementation on drug responses through potentiation of therapeutic effects or unintended adverse effects [21].

The difference between synergistic and antagonistic actions is important for clinical practice. Synergistic effects arise when nutrients and drugs are working together and the combination delivers a better therapeutic effect [22]. Omega-3 fatty acids can enhance mood control in individuals with comorbid severe depression and neurological disorders. Another example includes folate and vitamin B12 supplementation, which can enhance cognitive benefits of cholinesterase inhibitors in dementia by lowering homocysteine levels [23]. Antagonistic effects can result in a reduction in therapeutic effect, or potentially in harm. An excellent clinical case exemplifying this is the antagonism of anticoagulant therapy by vitamin K, as excessive vitamin K can reduce or defeat the therapeutic effect of warfarin increasing risk of thrombotic events in those susceptible to stroke. High levels of zinc supplementation can decrease absorption of copper resulting in secondary neurologic complications including myelopathy [24].

The impact of genetics, age, and associated health conditions makes these interactions even more complex. Polymorphism in drug-metabolizing enzymes, including those that encode cytochrome P450 isoforms, can influence individuals' responses to medications and nutrient supplementation [25]. Age-related changes such as polypharmacy, reduced hepatic and renal clearance, and decreased stomach acid output can raise the likelihood of clinically meaningful interactions. Diabetes, renal insufficiency, and gastrointestinal disorders are examples of comorbidities that might impact nutrient and medication absorption and raise the risk of negative outcomes [26].

Vitamin–drug and nutrient–nutrient interactions in neurological illnesses show how important it is to maintain a fine balance in treatments. If there are potential drug–nutrient interactions, clinicians can either capitalize on them to optimize clinical management or they may overlook them, which could lead to nutrient deficiencies, reduced drug efficacy, or other unfavorable outcomes that further exacerbate the disease's pathophysiology [27]. Mechanistic knowledge is the cornerstone of well-informed clinical decision-making, which will help clinicians make judgments about the usage of nutrients and medications in accordance with the demands of their neurological condition patients [28] (Figure 10.1).

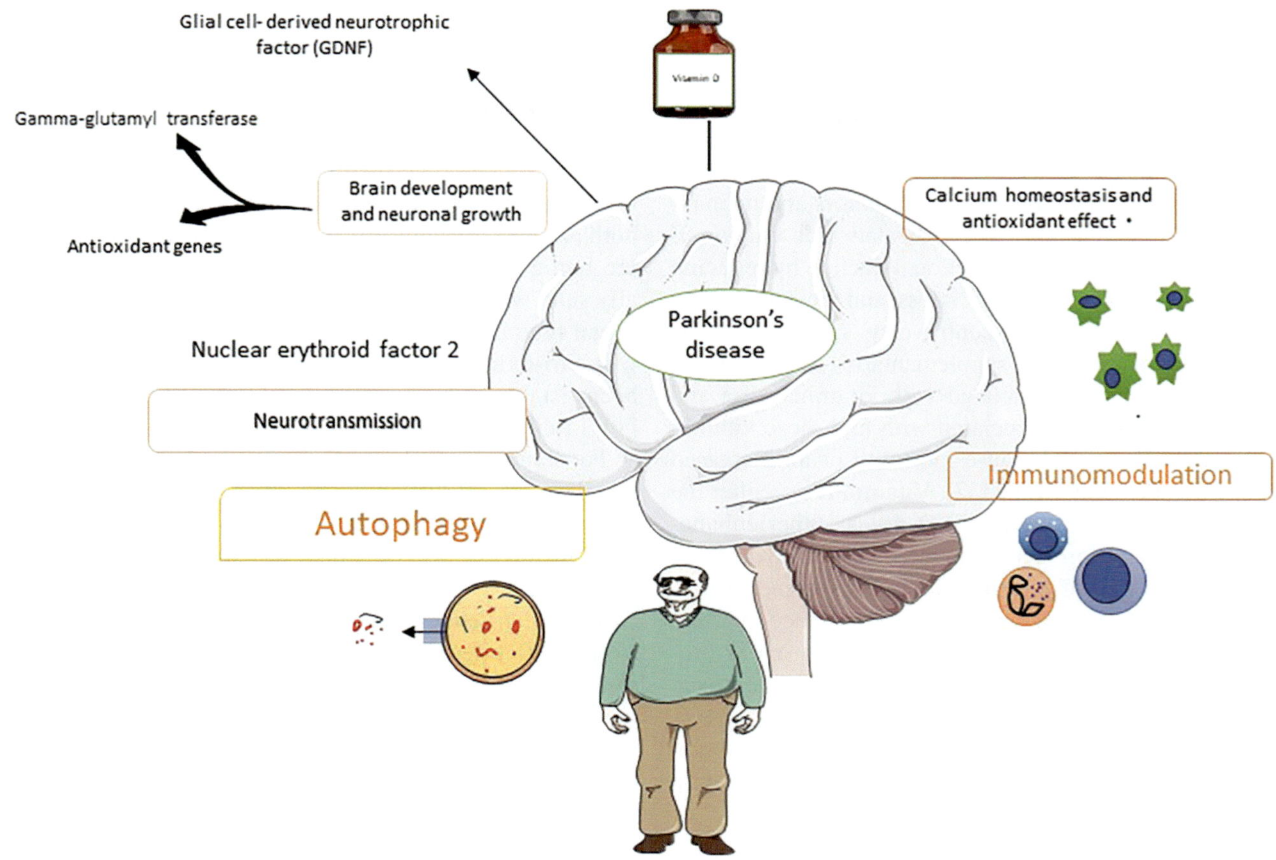

FIGURE 10.1 Mechanism of drug–nutrient interaction in neurological diseases.

10.3 VITAMIN–DRUG INTERACTIONS IN NEUROLOGICAL DISORDERS

The link between vitamins and drugs is particularly significant when it comes to neurologic illnesses because of characteristics like polypharmacy, prolonged therapy, and possible dietary deficiencies. Vitamins serve as enzymatic cofactors, antioxidants, and neuroprotective agents, and their levels can be dramatically affected by drug treatment [29]. Drugs can also modulate the physiological effect of vitamins, in the case of dietary or supplemental intake, which can lead to changes in both the efficacy and safety of the drug being given. The reciprocal interactions are clinically important and can have either a therapeutic or adverse effect in patients with multifactorial neurologic conditions [30].

The relationship between B-vitamins and antiepileptic and PD medications exemplifies this duality. Antiepileptic drugs phenytoin, carbamazepine, and valproic acid have been shown to negatively interfere with folate metabolism and decrease circulating vitamin B12 levels leading to hyperhomocysteinemia, cognitive- and seizure-related dysfunction, and vascular comorbidities [15]. Folate deficiency can adversely impact and worsen the neuropsychiatric aspects of the drugs, including depression in patients with epilepsy. Levodopa interacts with vitamin B6 metabolism. Using pyridoxine, a cofactor required for the decarboxylation of levodopa, will increase peripheral decarboxylation to dopamine and cause the drug no longer to be available for transport across the blood–brain barrier (BBB) [31]. This antagonism reduces desired therapeutic benefits unless it is given with a peripheral decarboxylase inhibitor. Thiamine does have neuroprotective effects in preclinical studies with parkinsonians and diminishes oxidative stress due to medications suggesting that some B-vitamin interactions may offer benefits [32].

The relationship between vitamin D and MS has gained widespread attention in the context of pharmacological management of MS. MS pharmacotherapies, such as interferon-β and corticosteroids, can all alter vitamin D metabolism and bone health, which can lead to secondary osteoporosis [33]. Vitamin D deficits are also associated with higher levels of disease activity and progression in MS, indicating a multifaceted relationship between drug treatment, disease pathology, and nutritional supply. Vitamin D is proposed as a supplementary therapy to positively impact immune modulation, promote the reduction of relapse rates, and improve neurological outcomes, despite limited understanding regarding optimal dosing [34]. The interaction between vitamin D and immunomodulatory therapy suggests the utility of mineral/nutritional monitoring for long-term management of autoimmune neurological diseases [35].

Vitamin E offers another manifestation of a nutrient–drug interaction, especially in cases of AD. Vitamin E is an effective lipid-soluble antioxidant that has been studied as an adjunct to cholinesterase inhibitors and memantine [36]. Clinical trials have indicated that vitamin E supplementation may slow functional decline by reducing neuronal membrane oxidative stress and lipid peroxidation. There is substantial controversy surrounding the effectiveness of vitamin E supplementation as some trials indicate little benefit associated with vitamin E use and also potential dangers associated with high-dose vitamin E, such as increased all-cause mortality or an increased risk of a hemorrhagic stroke [37]. Vitamin E may alter the pharmacodynamics of certain AD drugs, either enhancing a drug's neuroprotective properties or inadvertently increasing the risk of adverse events if used indiscriminately. It helps to emphasize the delicate balance required when combining antioxidant therapy for treating or preventing dementia, along with a pharmacokinetic approach to treatment [38].

Vitamin C, a water-soluble antioxidant, is also thought to interact with neuroprotective drugs. Vitamin C plays a key role in the synthesis of dopamine, aids iron absorption, and neutralizes free radicals associated with neurodegeneration. Vitamin C supplementation may enhance levodopa absorption by promoting gastric acidity and bioavailability of bioactive compounds [39]. Vitamin C has also been investigated as a synergistic agent with neuroprotective agents in reducing oxidative stress in both AD and ischemic stroke. Typically the interactions are benign, and the use of vitamin C has been generally safe, with the caveat that very high doses of vitamin C can interact with some laboratory assays, interactions with pro-oxidative metals, and the safety profile had not been established regarding long-term supplementation [40] (Table 10.1).

Fat-soluble vitamins A, D, E, and K characterize additional facets of nutrient–drug interactions especially in neurodegenerative diseases. Although vitamin A-derived retinoid might affect neuronal development and plasticity, cumulative toxicity may arise from a possible interaction between retinoid and retinoid-based medication [49]. Vitamin D affects PD and AD by influencing calcium signaling and neuronal cell survival. Vitamin K has an increasing involvement in sphingolipid metabolism and neuronal membrane integrity, while vitamin E also interacts with medications that impact cellular oxidative stress

TABLE 10.1
Table Summarizing the Vitamin–Drug Interaction

Vitamin	Neurological Drug or Class	Mechanism of Interaction	Clinical Significance	Practical Recommendation & Monitoring	References
Vitamin B_6 (pyridoxine)	Levodopa (± carbidopa)	Pyridoxine promotes peripheral conversion of levodopa → dopamine; can **reduce levodopa efficacy** unless decarboxylase inhibitor (carbidopa) present.	Worsening of Parkinson's symptoms (if high-dose B_6 given with levodopa alone).	Avoid high-dose pyridoxine supplements in patients on levodopa alone; if on carbidopa/levodopa standard therapy, the effect is often mitigated; discuss any supplements with treating neurologist.	[41]
Vitamin D (and calcium metabolism)	Enzyme-inducing antiepileptic drugs (phenytoin, carbamazepine, phenobarbital, primidone) and some newer ASMs evidence	Induction of hepatic enzymes increases vitamin D catabolism → lower 25(OH)D; reduced calcium absorption → bone demineralization.	Increased risk of **vitamin D deficiency, osteopenia/ osteoporosis, fractures** in people on chronic AED therapy.	Screen baseline and periodically (25-OH-D, calcium, bone density in long-term therapy); supplement vitamin D (and calcium) as indicated per levels. Consider shorter-interval checks after starting/adjusting enzyme-inducing AEDs.	[42]
Vitamin K (dietary/ supplements)	Warfarin (common in stroke/embolism patients)	Vitamin K antagonizes warfarin's anticoagulant effect (and dietary changes in K alter INR)	Variability in anticoagulation control; **reduced warfarin effect** with high vitamin K intake → thrombotic risk; abrupt changes can destabilize INR.	Keep dietary vitamin K consistent; avoid large new vitamin K supplements unless coordinated with anticoagulation clinic and INR monitoring. For new supplements, check INR more frequently until stable.	[43]

(Continued)

TABLE 10.1 *(Continued)*

Table Summarizing the Vitamin–Drug Interaction

Vitamin	Neurological Drug or Class	Mechanism of Interaction	Clinical Significance	Practical Recommendation & Monitoring	References
L-carnitine (and carnitine status)	Valproic acid/valproate (antiepileptic)	Valproate can deplete carnitine (forms valproylcarnitine → urinary loss/ transporter inhibition); carnitine supplementation may correct deficiency and is used in toxicity.	Risk of **hyperammonemia, hepatotoxicity** and related encephalopathy in susceptible patients (esp. young children). L-carnitine is used therapeutically in VPA toxicity/ hyperammonemia.	Monitor clinical/lab signs in high-risk patients (infants, polytherapy, severe GI dysfunction). Consider L-carnitine supplementation if valproate-induced carnitine deficiency or in valproate toxicity as per local protocols.	[44]
Vitamin E (high-dose)	Anticoagulants/ antiplatelets (warfarin, aspirin)	High-dose vitamin E can have antiplatelet/ anticoagulant effects (increases bleeding risk in some reports) and interacts with clotting pathways.	Possible increased bleeding risk (evidence mixed). Caution in patients with hemorrhagic stroke risk or on anticoagulation.	Avoid high-dose vitamin E supplements in anticoagulated patients without specialist advice; monitor for bleeding and check INR if on warfarin after starting high-dose vitamin E.	[45]
St. John's wort (*Hypericum perforatum*) (herbal)	SSRIs, SNRIs, some triptans, anticonvulsants, many CNS drugs	Potent inducer of CYP3A4 and other CYPs + can increase serotonin → **reduces plasma levels of many drugs** or causes **serotonin syndrome** when combined with serotonergic meds.	Reduced efficacy of many CNS drugs (e.g., oral contraceptives, immunosuppressants, some anticonvulsants) and risk of **serotonin syndrome** with antidepressants/triptans.	Advise patients to stop St. John's wort before starting or while on interacting CNS meds; always check herb use. Consult pharmacy for specific drug interactions.	[46]
Vitamin B_{12}/ folate (drug-induced deficiencies)	Long-term PPI use, metformin, some antiseizure drugs (phenobarbital, phenytoin)	Certain drugs reduce absorption or increase clearance of B_{12}/folate → deficiency over time; this can present with neuropathy, cognitive changes.	Neuro symptoms (peripheral neuropathy, cognitive impairment) can be due to vitamin deficiency mimicking or worsening neurological disease.	Monitor B_{12}/folate if on long-term interacting drugs; replace if low. Consider baseline testing in patients with neuropathy or cognitive decline (Note: this is drug → vitamin deficiency rather than classic vitamin → drug).	[47]
Calcium (high dose)	Some oral drugs with narrow absorption windows (levothyroxine is classic; some antibiotics, but less neurology-specific)	Calcium can impair oral absorption of certain drugs if taken simultaneously.	Reduced drug efficacy if taken together (timing problem).	Advise spacing calcium supplements away from medications with absorption concerns (e.g., 2–4 hour separation where applicable).	[48]

pathways. As demonstrated by stroke patients taking warfarin concurrently with anticoagulant medication, vitamin K may pose an intriguing challenge [50]. Overconsumption of vitamin K may counteract the anticoagulant effects of warfarin and raise the risk of thromboembolism, while inadequate vitamin K intake may increase the risk of bleeding. A complex field of medicine, vitamin–drug interactions in neurologic diseases provide both the potential for therapeutic benefit and the possibility of negative side effects. Making clinical judgments requires a grasp of the mechanisms by which specific vitamins affect medication metabolism, drug pharmacodynamics, and disease pathology [51]. In the treatment of patients with neurological disorders, vitamin status monitoring, adequate supplementation, and awareness of the wide range of possible antagonistic or synergistic effects can all help to improve the quality of patient outcomes clinically [39] (Figure 10.2).

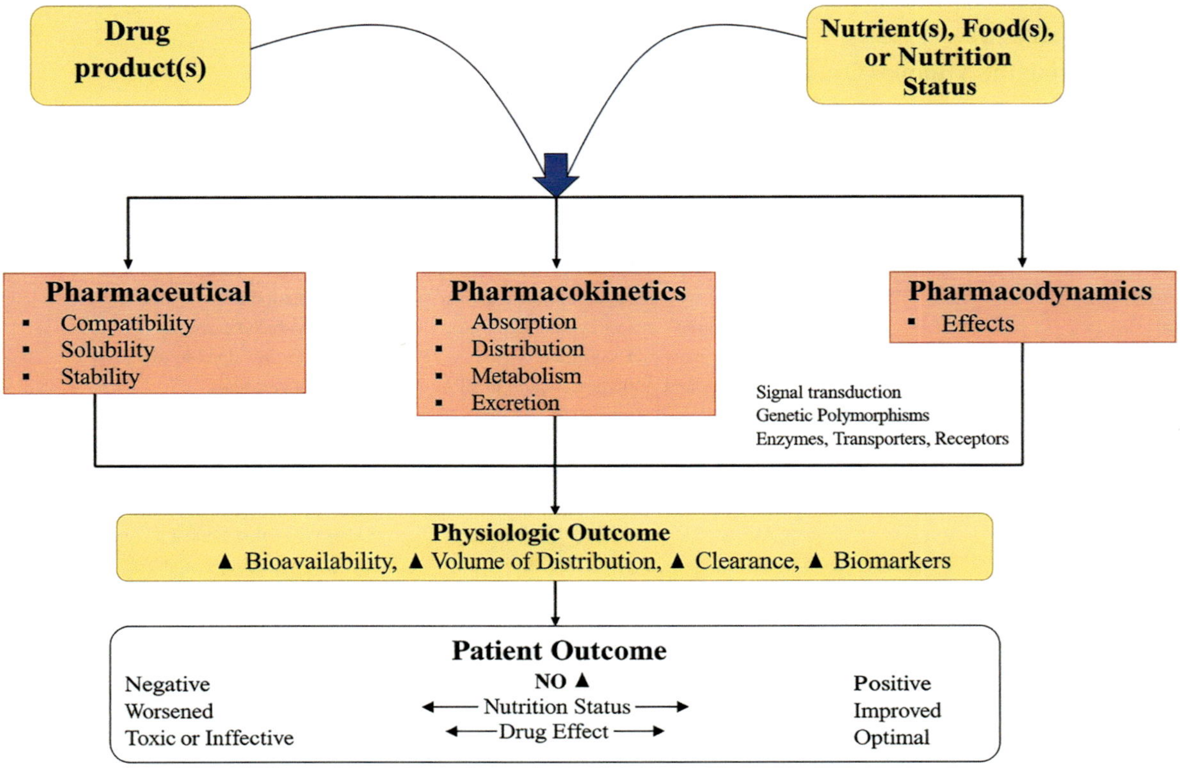

FIGURE 10.2 The flowchart denoting the vitamin–drug interactions on its pharmacokinetics, pharmacodynamics, and physiological outcomes.

10.4 NUTRIENT–NUTRIENT INTERACTIONS IN NEUROLOGICAL DISORDERS

They are part of a complex network of physiological and biochemical interactions with other nutrients that can have a big impact on brain health. Both nutrient–nutrient interactions may result in synergistic benefits or impeded functions that might inform biological processes connected to illness progression and treatment since brain function depends critically on accuracy in metabolic processes and neurotransmitter balance [52]. Given that patients frequently have nutritional status or malabsorption issues related to neurologic disorders, or are not getting enough of particular food groups because of disease pathology or pharmacological therapy, it may be particularly important to recognize nutrient interactions in these conditions [53].

One of the most well-studied nutrient–nutrient interactions in neurology is the regulation of homocysteine metabolism by B-vitamins in coordinated action. Homocysteine is a neurotoxic amino acid associated with vascular dementia, AD, cerebrovascular accident, and cognitive decline [54]. Vitamins B6, B9, and B12 are vital cofactors of one-carbon metabolism pathways responsible for regulating the conversion of homocysteine to methionine or cysteine. Folate and vitamin B12 are involved in the remethylation of homocysteine to methionine, and vitamin B6 is a coenzyme in the transsulfuration pathway for conversion to cystathionine and glutathione [55]. Deficiency of any of these vitamins disrupts the cycle and causes accumulation of homocysteine and, in turn, excitotoxicity, oxidative stress, and/or vascular injury. Studies administering vitamins B6, folate, and B12 found that the combined administration was more effective for lowering homocysteine and maintaining cognitive abilities than individual vitamin administration and that a synergistic relationship is required for neurological protection against elevated homocysteine [23].

Omega-3 fatty acids are believed to be the most abundant fatty acids in neuronal membrane structures; Docosahexaenoic acid (DHA) in particular is believed to be an important contributor to neuronal membrane structure and the built-in ability to influence the synaptogenic plasticity of and neurochemical signaling at synapses, particularly in the face of neuroinflammatory processes [10]. The high degree of unsaturation of the omega-3 fatty acids makes them susceptible to oxidative degradation through lipid peroxidation, which, in turn, can influence their field or prevent their neuroprotective capacity in general. Antioxidant vitamins serve as protective companions. Vitamin E, an antioxidant, is important because it can preserve membrane lipid compositions by scavenging radicals and unintentionally preventing lipid degradation [56]. This protective nutritional association extends beyond preserving the structure and function of neurons; vitamin E, vitamin C, and omega-3 fatty acids can be used as therapeutic agents to treat degenerative diseases such as major depressive disorder, AD, and PD. The connection between antioxidant vitamins and polyunsaturated fatty acids provides compelling evidence for a dual-target dietary intervention aimed at the psychiatric and neurodegenerative domains [57].

Iron is a requirement for dopamine synthesis via the activity of tyrosine hydroxylase, while iron levels must be kept in check because an excess of iron can be toxic because of free radical production in the brain, and tied to neurodegenerative diseases like AD and PD [58]. Zinc can also regulate the signaling associated with glutamate and Gamma-aminobutyric acid (GABA), but too much zinc can cause excitotoxicity, whereas a zinc deficiency is correlated with impaired synaptic plasticity and impaired cognition. Copper is used to make enzymes such as dopamine β-hydroxylase, but copper imbalance can be detrimental through oxidative stress and neurodegeneration related to diseases such as Wilson's disease, but research has reported a link between copper imbalance and other neurodegenerative diseases [59]. Magnesium is reported to behave like a natural NMDA receptor antagonist, protecting against excitotoxicity, and minerals are probably crucial for the balance of neurotransmitters and other brain functions. Copper, zinc, magnesium, and iron are all essential cofactors for enzymes that produce neurotransmitters, regulate oxidative stress, and activate receptors [60].

These minerals not only individually interact with neuronal pathways, but they can also, when combined, affect the absorption and bioavailability of other minerals. Excess dietary zinc can limit the absorption of copper, and both iron and calcium can compete in the intestines to limit uptake of the other and to potentially affect neurotransmitter function in terms of uptake. Therefore, with their complex interactive relationships, it is prudent to maintain a balance of these minerals to support neurotransmitter activity and limit neurotoxicity [61].

Polyphenols are found in plant-derived foods, and they interact with micronutrients in ways that can either promote or hinder their bioavailability. Polyphenols including flavonoids, resveratrol, and catechins, have antioxidant and anti-inflammatory properties that benefit neuroprotection [62]. Polyphenols can also chelate elements, such as iron and zinc, reducing availability and risking deficiency status if intake is excessive or there is dietary imbalance. Polyphenols have the potential to stabilize and promote the bioavailability of certain vitamins and they can provide lipid protection from peroxidation, thereby increasing their bioactivity [63]. Polyphenols can be affected by interactions with vitamins and minerals, suggesting a bidirectional relationship. Synergistic nutrition combinations have drawn more attention due to their possible contribution to neuroprotection and cognitive enhancement. Phosphatidylserine and DHA are two examples of nutrient combinations that have been shown to improve memory and synaptic membrane function [64].

Curcumin is known to have tremendous benefits, but is not well bioavailable; therefore, when curcumin is mixed with piperine, it enhances the bioavailability of curcumin, which in turn allows for increased anti-inflammatory and antioxidant effects in the brain area [65]. B-vitamins, omega-3 fatty acids, and antioxidants have been shown to slow down cognitive decline more effectively than single-nutrient supplements. Certain complex neurological disorders that are caused by multifactorial pathophysiology may be best assessed using multi-nutrient approaches, which account for potential synergistic interactions and offer a more comprehensive cortical neuroprotection effect [66].

The network of interactions between nutrients can either strengthen or diminish neural resilience, depending on context, dosage, and balance. Comprehending these associations not only directs clinical nutrition intervention strategies but also facilitates the possibility of precision nutrition interventions tailored to the metabolic and pathological requirements of neurological patients [67] (Table 10.2).

TABLE 10.2

Table Summarizing the Nutrient–Nutrient Interactions

Nutrient A ↔ Nutrient B	Mechanism of Interaction	Neurological Significance	Practical Recommendation	References
Vitamin B_{12} ↔ folate (B_9)	Folate (when abundant) corrects megaloblastic anemia caused by B_{12} deficiency but does **not** correct B_{12}-related neurological injury.	High folate can "mask" B_{12} deficiency—delaying diagnosis → progression of subacute combined degeneration, peripheral neuropathy, cognitive problems.	Always check B_{12} status before treating macrocytic anemia with high-dose folate. In patients with neuropathy or cognitive decline, measure serum B_{12} (and methylmalonic acid if unclear) before/while supplementing folate.	[68]
Zinc ↔ copper	High zinc intake induces metallothionein in enterocytes which preferentially binds copper → reduced copper absorption and eventual copper deficiency.	Copper deficiency causes **myeloneuropathy**, gait disturbance, sensory ataxia, and hematologic abnormalities clinically similar to B_{12} deficiency.	Avoid prolonged high-dose zinc supplements without monitoring copper. If high zinc is prescribed (e.g., for dermatologic use), monitor serum copper and consider supplementation if levels fall.	[69]

(Continued)

TABLE 10.2 (*Continued*)
Table Summarizing the Nutrient–Nutrient Interactions

Nutrient A ↔ Nutrient B	Mechanism of Interaction	Neurological Significance	Practical Recommendation	References
Iron ↔ calcium	Calcium (and dairy) can inhibit non-heme iron absorption when taken simultaneously by forming insoluble complexes.	Iron deficiency anemia can contribute to fatigue, impaired cognition, and in children, neurodevelopmental delay. Poor iron status can worsen recovery after stroke/brain injury.	Separate iron and calcium supplements/ fortified foods by 1–2 h. Advise patients taking both to stagger doses; give iron with vitamin C to enhance absorption.	[70]
Vitamin C ↔ iron	Vitamin C reduces ferric → ferrous iron and forms soluble complexes → markedly increases non-heme iron absorption.	Improves iron repletion, which supports oxygen delivery to the brain and prevents iron-deficiency-related cognitive/ behavioral issues.	Recommend taking iron supplements with a source of vitamin C (e.g., citrus or 250–500 mg vitamin C) unless contraindicated. Monitor iron indices during therapy.	[71]
Magnesium ↔ vitamin D	Magnesium is a cofactor for enzymes that convert vitamin D to its active forms (25- and 1,25-hydroxyvitamin D). Low Mg impairs vitamin D activation.	Adequate vitamin D is important for neuromuscular function and possibly for mood/cognitive health; Mg deficiency can blunt vitamin D benefits.	In patients treated for vitamin D deficiency who fail to respond, check magnesium. Avoid high-dose vitamin D supplementation without ensuring adequate magnesium intake.	[72]
Omega-3 PUFAs ↔ vitamin E	Polyunsaturated fatty acids (PUFAs) are prone to lipid peroxidation; vitamin E (lipid antioxidant) protects PUFAs and cell membranes. High PUFA intake increases vitamin E requirement.	Omega-3s are neuroprotective/ anti-inflammatory; insufficient antioxidant protection may lead to oxidative stress in neural tissues.	When recommending high-dose fish oil/ omega-3, ensure dietary antioxidant intake is adequate (dietary vitamin E or modest supplementation as advised). Avoid excessive vitamin E dosing in anticoagulated patients.	[73]
Folate ↔ choline/ betaine/B_{12} (methylation network)	Folate, choline/betaine and B_{12} are interlinked in one-carbon metabolism for methylation (DNA, neurotransmitter synthesis). Deficiency in one increases demand for others.	Methylation is vital for myelin maintenance, neurotransmitter synthesis and neurodevelopment; imbalance may affect cognition, mood, and neural repair.	Promote balanced intake of folate, B_{12} and choline (dietary sources). In supplementation, consider whole-network effects especially in pregnancy and neurodevelopmental risk.	[74]
Selenium ↔ iodine	Selenium is required for iodothyronine deiodinases (thyroid hormone activation); selenium deficiency worsens iodine-related thyroid dysfunction.	Thyroid hormones are essential for brain development and cognitive function; combined deficiencies can impair neurodevelopment and cognition.	In regions with borderline iodine status, correct selenium and iodine together as clinically indicated (public-health context). In individuals with thyroid disease, assess both nutrients when clinically relevant.	[75]
Calcium ↔ magnesium	High supplemental calcium can compete with magnesium for absorption; large calcium loads may exacerbate magnesium deficiency in susceptible people.	Magnesium is needed for neuromuscular excitability regulation deficiency can cause cramps, tremor, and worsen migraine or neuromuscular symptoms.	Balance calcium and magnesium in supplementation; prefer dietary sources and stagger large supplemental doses. Check Mg in patients with neuromuscular complaints.	[76]

10.5 CLINICAL EVIDENCE AND CASE STUDIES

Mechanistic understanding, along with trials, observational studies, and case reports, shapes the clinical landscape of vitamin–drug and nutrient–nutrient interactions in neurological diseases. While preclinical and laboratory data suggest that they are biologically plausible, clinical data provide the substantial evidence required for comparisons to patient care and practical alterations that can either confirm or negate the benefits of dietary variables [77]. The importance of nutrition in a team-based approach to neurological care is demonstrated by the growing body of research in the neurological literature that details the beneficial and detrimental effects of vitamin–drug and nutrient–nutrient interactions. Numerous pertinent interactions between vitamins and drugs have been found through clinical investigations [78].

Long-term use in patients with epilepsy of antiepileptic medications phenytoin, carbamazepine, and valproate has consistently shown to be associated with lower serum folate and vitamin B12 levels. Randomized controlled trials have shown folate supplementation can not only correct deficiency, but, in some patients with epilepsy, improve mood and cognitive function, which may reduce neuropsychiatric side effects of antiepileptic therapy [79]. Vitamin B12 and folate supplementation has been shown to reduce levodopa-induced hyperhomocysteinemia and potentially reduce the risk of peripheral neuropathy and vascular complications. Several clinical trials in MS have also investigated the use of vitamin D as an adjunct to immunomodulatory therapy and while improvements in relapse rates and inflammatory markers have been noted the results showed great variance in patients and dosing regimens [80]. Clinical trials investigating vitamin E therapy in AD have shown reduced functional decline in some sub-sets of patients either with or without standard cholinesterase inhibitors, but disagreement about the clinical value of vitamin E therapy due to inconsistencies has been noted. These trials further support the idea that nutrient status should not be perceived as a trivial adjunct, but rather an essential element of therapeutic efficacy or ineffectiveness in neurological disorders [81].

Case reports represent one more level of evidence and often reveal adverse interactions that are not included in controlled studies, such as the impact of vitamin supplementation on drug action as cited in multiple instances in neurological practice [82]. One such example was an upper limit of intake, which was considered too high, identified in a number of cases of pyridoxine, worsening the efficacy of levodopa in patients with PD, returning patients to states of motor symptom until overall medicine intake was reduced. There are also published reports of vitamin K intake that undermined anticoagulant therapy in patients with stroke and dementia [83]. There is also a risk associated with uncontrolled nutrient consumption, especially in patients with polypharmacy status. More common neurotoxic effects observed in clinical case documentation were from high-dose vitamin E where increased risk of hemorrhagic stroke was reported in vulnerable individuals. These reports show that deficiencies will add to neurologic dysfunction, but excess supplementation may be equally harmful if patient levels of vitamins are not adequately authenticated [84].

There is evidence to support the effects of nutrient combinations in the context of neuroprotection. Trials in combined or synergistic approaches have found evidence clearly demonstrating that combinations of B-vitamins can lower homocysteine levels and have a potential impact in slowing cognitive decline compared to monotherapy [23]. In cases of mild cognitive impairment, just combining folate, vitamin B12, and vitamin B6 have demonstrated an association with a slower rate of brain atrophy compared to single-nutrient interventions. Research has shown that omega-3 fatty acids supplemented with antioxidant vitamins, such as C and E, boost memory and executive function [85]. By acting as antioxidants and preventing oxidative damage in the brain, omega-3 supplements protect phosphatidylcholine, a crucial component of cell membranes. Multi-nutrient formulations containing vitamins, minerals, and polyphenols had a minor but statistically significant and clinically relevant impact on mood management, quality of life, and cognitive function in older adults at risk of dementia [86].

The overall clinical evidence presents a conflicting picture: in neurologic intervention, vitamin drug and nutrient–nutrient combinations might present both challenges and opportunities. Deficits caused by long-term medication therapy need to be promptly addressed to prevent problems, and carefully chosen foods may enhance pharmacological intervention and provide neuroprotection [87]. There is a risk of adverse consequences, especially with excessive or unregulated supplementation. Clinical trials, observational studies, and case reports strongly support the necessity of guided supplementation, regular follow-up, and tailored nutritional status evaluation in order to optimize benefits and reduce risks [88].

10.6 THERAPEUTIC IMPLICATIONS AND CHALLENGES

Practice is significantly impacted by the significance of vitamin–drug and nutrient–nutrient interactions in neurological diseases. The effects of nutritional status are now recognized as a significant factor in the development of disease, the efficacy of medications, and the quality of life; it is no longer viewed as a peripheral issue. Applying this knowledge to clinical care may provide difficulties since it calls for prescribing and pharmacologic treatment for personalized therapy and supplements [89].

Precision medicine and personalized nutrition are rapidly taking the lead. The same treatment may have different effects on different people due to genetic polymorphisms, metabolic variations, and special nutritional requirements brought on by disease [90]. B-vitamins' ability to decrease homocysteine and prevent vascular cognitive impairment, for instance, will be impacted by variations in the metabolism of omega-3 fatty acids or folate-metabolizing enzymes like MTHFR. Variability in a patient's reaction to vitamin D treatment may be influenced by polymorphisms in the vitamin D receptor gene, depending on the condition [91]. Precision nutrition, or more personalized nutrition, will take into account such genetic, biochemical, and clinical data to identify nutrient-focused treatments that will produce a net benefit, while assuring that nutritional supplements or foods will not disrupt therapeutic outcomes. Nutrigenomics along with neuropharmacology will lead to a more individualized approach to care with diet and pharmacotherapy developed to maximize the potential to benefit [12].

Despite these potential benefits, over-supplementation and polypharmacy continue to be ongoing threats. Neurological patients have a tendency to take multiple medications for comorbidities and combining these with vitamin or nutrient injections or doses can lead to changes in metabolic pathways [92]. Excessive vitamin E can increase risk of hemorrhage; high doses of pyridoxine can cause sensory neuropathy; and excessive intakes of elements like zinc or iron can decrease the balance of other trace elements and lead to changes that can affect neurological behaviors. The illogical and uncontrolled self-prescription of nutraceuticals and dietary supplements by patients creates additional barriers since the interactions are not acknowledged or considered for treatments [93]. Polypharmacy becomes increasingly problematic in older adults, many of whom have diminished clearance in the kidney and altered absorption, leading to medications with slower half-lives. The enthusiasm for nutraceuticals reduces the knowledge of nutrients as medications. The risks of excess must be considered in addition to deficiencies in therapeutic approaches [94].

Systematic monitoring and understanding of interactions within the parameters of therapeutic practice are also necessary for valuable management. Before patients have symptoms that are clinically evident, baseline evaluations of vitamin and mineral statuses and medication level monitoring may reveal possible imbalances [95]. While monitoring serum 25-hydroxyl vitamin D helps guide supplementation in patients diagnosed with MS or PD, measuring homocysteine and methylmalonic acid levels can detect subclinical B-vitamin deficiencies in patients on long-term antiepileptic medication [31].

Together with food histories and medication reviews, these laboratory tests enable clinicians to track both detectable deficits and the potential for over-supplementation. Considering nutritional status may be necessary for therapeutic medication monitoring because concurrent supplementation can drastically change how a medicine is used and metabolized [53]. According to section 3's definition of "treating," deliberate monitoring necessitates knowledge. Disclosing this with all parties involved is crucial to ensure that therapeutic advantage of nutrients supports patient care without the reduction of pharmacological efficacy [11].

Managing these intricate relationships should not be left to neurologists alone; it will require multidisciplinary teamwork to create, carry out, and monitor interventions in conjunction with dieticians, pharmacists, and doctors [96]. As pharmaceutical care, dieticians will analyze dietary consumption, identify dietary deficits, and create customized dietary plans that manage nutrition alongside medication. Because pharmacists are knowledgeable about how drugs interact with nutrients, they can ensure that dietary supplements don't compromise the safety or effectiveness of medications [97].

Pharmacotherapy, dietary, and health behavior changes to maximize nutritional impact, and the results of dieticians and pharmacists will all be incorporated by neurologists into the overall care of disease [98]. Multidisciplinary care coordination promotes a patient-centered strategy in which pharmacological and nutritional interactions are purposefully planned, risks are reduced, and therapeutic benefits are optimized. Beyond basic supplementation, the therapeutic potential linked to vitamin–drug and nutrient–nutrient interactions can extend to integrated care [99]. Although nutrition has the potential to improve patient outcomes, caution must still be exercised due to the variables of over-supplementation, polypharmacy, and individual variability. Clinical application of these interactions will necessitate an evidence-based, customized, collaborative approach that takes into account the intricacy of neurological illnesses and strikes a balance between medication and nutrition [100].

10.7 FUTURE DIRECTIONS

Vitamin–drug and nutrient–nutrient interactions related to neurological illnesses are rapidly evolving fields with significant opportunities to incorporate nutrition into therapy paradigms. Advances in molecular biology, nutrigenomics, and clinical nutrition are opening the door for the day when pharmacotherapy and dietary approaches will be used in tandem to support brain health [12]. Approaches have focused on addressing deficiencies and avoiding detrimental interactions. The way we treat patients with neurological illnesses may be completely changed by these changes, which also advance the shift toward proactive, individualized, and integrated paradigms [101].

Among the most intriguing new directions in this field of study is nutrigenomics. A strategy to tailor therapies is to investigate the potential effects of genetic variations on nutrition metabolism and response. Different susceptibilities to neurological diseases and responses to treatment have been linked to genetic variations that impact lipid transport proteins, vitamin D receptors, folate metabolism, and other components [102].

People with polymorphisms in methylenetetrahydrofolate reductase may need specific strategies using B-vitamins to lower homocysteine levels and slow cognitive decline. Genetic variation in several genes affecting metabolism of omega-3 fatty acids, etc. may determine how well higher fish oil supplementation will affect cognition or mood regulation [103]. The translational potential of nutrigenomics to enhance personalized dietary interventions in neurology has great theoretical promise and could evolve to personalize dietary interventions as effectively as drug therapy by ensuring nutritional treatment matched with patient genetics, disease state, and the prescribed therapy [12].

New nutraceuticals and bioactive materials are further expanding the toolbox of therapeutic options. Modern vitamins and minerals also now include a variety of phytochemicals, peptides, and lipophilic mediators that originate from plants, which have promising neuroprotective characteristics [104]. Curcumin, resveratrol, ginsenosides, and flavonoids have shown potential in several domains related to neurodegenerative illnesses and have been shown to modulate oxidative stress, inflammation, and amyloid aggregation, significant pathological processes of neurodegenerative diseases. New advancements in the field of nanotechnology and drug delivery systems are addressing the predisposition of poor bioavailability—making it possible to better deliver these different bioactive compounds across BBB [105]. Novel possibilities are not limited to formulations based on lipids. In order to mediate the gut–brain axis and boost neuronal resilience, probiotics and their metabolites are taken into consideration. Since these bioactive substances are still producing data as distinct outcomes in clinical studies, they could be appropriate adjuncts to treatments for conditions including AD, PD, MS, and others [106].

Future directions for integrative strategies for brain health should also include pharmaceutical and nutritional approaches. A complex interaction of variables, such as oxidative stress, neuroinflammation, mitochondrial failure, and vascular pathology, often leads to neurological problems rather than a single mechanism [107]. Rather than addressing each of these issues separately, integrative treatments that combine pharmacological interventions with appropriate nutrient supplementation, dietary adjustments, lifestyle changes, and cognitive tactics are likely to be more successful [108].

Diets rich in omega-3s, polyphenols, and antioxidants would be used in addition to prescription drugs in the hopes of preserving synaptic integrity and delaying the course of illness [109]. We can add another layer to the integrated models by including gut microbiome manipulation with probiotics and prebiotics, as compounds generated from microorganisms efficiently alter brain function. Integrative models indicate a departure from a disease-centered approach, where pharmaceutical control is just as crucial as preserving nutritional balance for health [110].

The treatment of neurological illnesses is anticipated to change as a result of the impending convergence of integrative neurology, nutrigenomics, and nutraceutical development. Future developments will probably include the use of routine metabolic and genetic profiling as a basis for dietary modification prescriptions, sophisticated biogenic compounds that exhibit therapeutic efficacy comparable to pharmaceuticals, and cooperative interdisciplinary care teams putting comprehensive paradigms for brain health into practice [111]. The challenge will be how to incorporate new knowledge into patient-accessible, evidence-based clinical practice while taking patient safety into account when it comes to off-label usage of experimental medicines and excessive supplementation. The next generation of therapeutic interventions will seize the chance to move from treating deficiencies to optimizing brain health through the efficient use of nutrient–drug and nutrient–nutrient connections with rigorous research and intervention [112].

10.8 CONCLUSION

An increasingly acknowledged factor in the occurrence, management, and progression of numerous neurological illnesses is the interplay between vitamins, nutrients, and medications. The brain does require a balanced intake of nutrients and micronutrients for energy metabolism, neurotransmission, and cell death prevention, but treatment inherently upsets this balance through metabolism or depletion. When it comes to reducing secondary toxicity and metabolic inflexibility and fixing underlying biochemical problems that medications are unable to address, nutritional support can increase the effectiveness of medications.

This chapter demonstrates how interactions between vitamins and drugs and between nutrients affect pharmacokinetics and pharmacodynamics in ways that go beyond simple supplementing. When taken with antiepileptic or antiparkinsonian drugs, B-vitamins can cause deficiencies or imbalances if dietary ratios are not maintained. B-vitamins can affect homocysteine metabolism and neuronal integrity. Vitamin D functions as a neuroprotectant and immunomodulator in MS, but it also interacts with disease-modifying treatments in ways that change the rate of the disease. Antioxidants, such as vitamins C and E, work in concert with medications that combat oxidative stress.

Because they improve their capacity to tackle oxidative stressors in disease, fat-soluble vitamins have a combined influence on inflammatory pathways, mitochondrial function, and neuronal survival in AD and other neurodegenerative disorders. Guidelines for B-vitamins, omega-3 fatty acids, minerals, or polyphenols are examples of nutrient–nutrient interactions that may have antagonistic or synergistic effects in dietary strategies. This intricate web of possible interactions only serves to highlight how difficult it can be to plan and implement successful nutritional interventions.

Case studies and clinical trials offer compelling proof that these interactions exist and affect patient outcomes. The positive effects of combining pharmacotherapy with nutrients highlight the possibility of integrative treatment paradigms, while the negative interactions serve as a reminder of the dangers of indiscriminate supplementing and polypharmacy. This double-edged sword emphasizes the necessity of meticulous monitoring, evidence-based and empirically supported techniques, and teamwork and cooperation amongst neurologists, dieticians, and pharmacists in order to provide the best possible care.

The development of precision nutrition may be the key to the future of individualized neurology care, according to the explanation above. The advent of integrative medicine-based therapies, bioactive substances, and nutrigenomics is already changing preconceived notions. We are starting to realize that nutrient deficiencies are no longer only a minor issue because more facets of nutrition will soon be recognized as the main drivers of therapeutic results and long-term brain health. Even though concerns about excessive supplementation, individual variability, and rigorous clinical research are still relevant, we should still be able to advance our search for novel dietary strategies to treat neurological disorders in person-centered care.

Finally, in this debate, we have identified nutrient–nutrient and vitamin–drug connections as both important and unappreciated components of brain health. In order to fully realize this potential, we must abandon the reductionism of nutrients as isolated elements and accept the fact that there are nutrient–drug interactions within the unique biochemical environment of the brain. Neurologists can also offer personalization, holistic care, and specially tailored therapy of illnesses that frequently carry a chronic and historical burden of suffering through the practice of nutrition.

FUNDING

None.

CONFLICT OF INTEREST

None.

CONSENT TO PUBLICATION

All authors have read and agreed to submit the chapter to the publisher.

REFERENCES

1. Mukherjee, S. et al., History, origin and types of neurological disorders, in Jahan S., Siddiqui A.J. (eds.), *Applications of Stem Cells and Derived Exosomes in Neurodegenerative Disorders*. 2023, Springer, p. 1–32.
2. Feigin, V.L. et al., The global burden of neurological disorders: translating evidence into policy. *Lancet Neurology*, 2020. 19(3): p. 255–265.
3. Shehryar, Z., The genetic face of brain diseases, in Koduru T.S., Osmani R.A.M., Singh E., Dutta S. (eds)., *The Neurodegeneration Revolution*. 2025, Elsevier. p. 73–99.
4. Sun, W. et al., From the different pathogenesis of epileptogenesis: vitamins as an adjunctive treatment for epilepsy. *Acta Epileptologica*, 2025. 7(1): p. 38.
5. Thomas, J. et al., Omega-3 fatty acids in early prevention of inflammatory neurodegenerative disease: a focus on Alzheimer's disease. *BioMed Research International*, 2015. 2015(1): p. 172801.
6. Alam, J., Vitamins: a nutritional intervention to modulate the Alzheimer's disease progression. *Nutritional Neuroscience*, 2022. 25(5): p. 945–962.
7. Bhushan, R., F. Goel, and S. Singh, Beyond dopamine: novel therapeutic pathways for Parkinson's disease through receptor signaling. *CNS & Neurological Disorders-Drug Targets*, 2025. 24(6): p. 434–451.
8. Ola, M. et al., Formulation and delivery issues for active ingredients of herbal medicines, nutraceuticals, and cosmetics: comparisons to small-molecule drugs, in Singh A., Kulhari H., Saharan V.A. (eds)., *Formulating Pharma-, Nutra-, and Cosmeceutical Products from Herbal Substances: Dosage Forms and Delivery Systems*. 2025, Wiley, p. 163–191.
9. Browne, D. et al., Vitamin E and Alzheimer's disease: what do we know so far? *Clinical Interventions in Aging*, 2019. 14: p. 1303–1317.
10. Ahmmed, M.K. et al., Marine omega-3 (n-3) phospholipids: a comprehensive review of their properties, sources, bioavailability, and relation to brain health. *Comprehensive Reviews in Food Science and Food Safety*, 2020. 19(1): p. 64–123.
11. Erickson, N. et al., Nutrition care is an integral part of patient-centred medical care: a European consensus. *Medical Oncology*, 2023. 40(4): p. 112.
12. Waheed, A. et al., Nutrigenomics and neurological disorders: exploring diet-brain interactions for cognitive health. *Neurogenetics*, 2024. 26(1): p. 10.
13. Boullata, J.I. and L.M. Hudson, Drug–nutrient interactions: a broad view with implications for practice. *Journal of the Academy of Nutrition and Dietetics*, 2012. 112(4): p. 506–517.
14. Klimek, P., S. Aichberger, and S. Thurner, Disentangling genetic and environmental risk factors for individual diseases from multiplex comorbidity networks. *Scientific Reports*, 2016. 6(1): p. 39658.
15. Safahani, M., H. Aligholi, and A.A. Asadi-Pooya, Management of antiepileptic drug–induced nutrition-related adverse effects. *Neurological Sciences*, 2020. 41(12): p. 3491–3502.
16. Paul, D.A., A.R.M. Qureshi, and A.Q. Rana, Peripheral neuropathy in Parkinson's disease. *Neurological Sciences*, 2020. 41(10): p. 2691–2701.
17. Lenka, A. et al., Practical pearls to improve the efficacy and tolerability of levodopa in Parkinson's disease. *Expert Review of Neurotherapeutics*, 2022. 22(6): p. 489–498.
18. Barrea, L. et al., Coffee consumption, health benefits and side effects: a narrative review and update for dietitians and nutritionists. *Critical Reviews in Food Science and Nutrition*, 2023. 63(9): p. 1238–1261.
19. Dubey, S. and E. Singh, Antioxidants: an approach for restricting oxidative stress induced neurodegeneration in Alzheimer's disease. *Inflammopharmacology*, 2023. 31(2): p. 717–730.
20. Huang, T.-H. et al., The roles of glutamate receptors and their antagonists in status epilepticus, refractory status epilepticus, and super-refractory status epilepticus. *Biomedicines*, 2023. 11(3): p. 686.
21. de Lima Moreira, C.V., R. Saavedra-rodríguez, and J.O. Fajemiroye, Remedies against plausible adverse interactions, in Fajemiroye J.O. (ed.), *Fundamentals of Drug and Non-Drug Interactions: Physiopathological Perspectives and Clinical Approaches*. 2025, Springer, p. 377–400.
22. Li, X. et al., Multifunctional nanoparticle-mediated combining therapy for human diseases. *Signal Transduction and Targeted Therapy*, 2024. 9(1): p. 1.
23. Olaso-Gonzalez, G. et al., Impact of supplementation with vitamins B6, B12, and/or folic acid on the reduction of homocysteine levels in patients with mild cognitive impairment: a systematic review. *IUBMB Life*, 2022. 74(1): p. 74–84.
24. Chen, J.W. et al., Copper deficiency myelopathy mimicking cervical spondylitic myelopathy: a systematic review of the literature with case report. *Spine Journal*, 2024. 24(11): p. 2026–2034.
25. da Silva Reis, A.A. and R. da Silva Santos, Omics approaches to drug and drug-non-drug interactions, in Fajemiroye J.O. (ed.), *Fundamentals of Drug and Non-Drug Interactions: Physiopathological Perspectives and Clinical Approaches*. 2025, Springer, p. 261–290.
26. Abubakar, M. et al., Diabetes, Alzheimer's disease risk factors, and the cafeteria diet: a comprehensive review. *Current Neuropharmacology*, 2025. doi:10.2174/011570159X384737250626094315.
27. Reynolds 3rd, C.F. et al., Mental health care for older adults: recent advances and new directions in clinical practice and research. *World Psychiatry*, 2022. 21(3): p. 336–363.
28. Al-Worafi, Y.M. et al., Evidence-based complementary, alternative and integrated medicine and efficacy and safety: central nervous system infections, in Smith J., Jones R.

(eds.), *Handbook of Complementary, Alternative, and Integrative Medicine*. CRC Press, p. 157–177.
29. Sindhu, R.K. et al., Exploring multifunctional antioxidants as potential agents for management of neurological disorders. *Environmental Science and Pollution Research*, 2022. 29(17): p. 24458–24477.
30. Calderón-Larrañaga, A. et al., Multimorbidity and functional impairment–bidirectional interplay, synergistic effects and common pathways. *Journal of Internal Medicine*, 2019. 285(3): p. 255–271.
31. Phokaewvarangkul, O. et al., Homocysteine, vitamin B metabolites, dopamine-substituting compounds, and symptomatology in Parkinson's disease: clinical and therapeutic considerations. *Journal of Neural Transmission*, 2023. 130(11): p. 1451–1462.
32. Parab, S. et al., Unraveling the mechanisms of hydrophilic vitamins in Alzheimer's and Parkinson's: preclinical and clinical evidence, in Shah A.K., Tappia P.S., Dhalla N.S. (eds.), *Hydrophilic Vitamins in Health and Disease*. 2024, Springer, p. 181–217.
33. Chiodini, I. et al., Treatment options for glucocorticoid-induced osteoporosis. *Expert Opinion on Pharmacotherapy*, 2020. 21(6): p. 721–732.
34. Rebelos, E., N. Tentolouris, and E. Jude, The role of vitamin D in health and disease: a narrative review on the mechanisms linking vitamin D with disease and the effects of supplementation. *Drugs*, 2023. 83(8): p. 665–685.
35. Rak, K. et al., Some immune parameters of term newborns at birth are associated with the concentration of iron, copper and magnesium in maternal serum. *Nutrients*, 2023. 15(8): p. 1908.
36. Narang, J.K. et al., Antioxidants against neurological disorders, in Singh R.B., Patel R.P., Mohapatra S.S. (eds.), *Antioxidants: Nature's Defense Against Disease*. Elsevier, 2025, p. 285–367.
37. Kaye, A.D. et al., Vitamin E (α-Tocopherol): emerging clinical role and adverse risks of supplementation in adults. *Cureus*, 2025. 17(2): p. e78679.
38. Rojas-Solé, C. et al., Integrated approach to reducing polypharmacy in older people: exploring the role of oxidative stress and antioxidant potential therapy. *Redox Report*, 2024. 29(1): p. 2289740.
39. Lopes, M. et al., Food supplement vitamins, minerals, amino-acids, fatty acids, phenolic and alkaloid-based substances: an overview of their interaction with drugs. *Critical Reviews in Food Science and Nutrition*, 2023. 63(19): p. 4106–4140.
40. Alberts, A. et al., Vitamin C: a comprehensive review of its role in health, disease prevention, and therapeutic potential. *Molecules*, 2025. 30(3): p. 748.
41. Beckers, M., B.R. Bloem, and M.M. Verbeek, Mechanisms of peripheral levodopa resistance in Parkinson's disease. *npj Parkinson's Disease*, 2022. 8(1): p. 56.
42. Gaete, P.V., V. Cuellar-Rodríguez, and C.O. Mendivil, Antiseizure medications and bone health. *Neurology and Therapy*, 2025. 14: p. 1–18.
43. Ma, C. et al., Chinese guidelines for the diagnosis and management of atrial fibrillation. *Pacing and Clinical Electrophysiology*, 2024. 47(6): p. 714–770.
44. Gziut, T. and R. Thanacoody, L-carnitine for valproic acid-induced toxicity. *British Journal of Clinical Pharmacology*, 2025. 91(3): p. 636–647.
45. Moorthy, A. and S. Krishna, Management of medical comorbidities in maxillofacial surgery, in Bonanthaya K., Panneerselvam E., Manuel S., Kumar V.V., Rai A. (eds.), *Oral and Maxillofacial Surgery for the Clinician*. 2021, Springer, p. 25–48.
46. Dołoto, A. et al., Interactions of antidepressants with concomitant medications—safety of complex therapies in multimorbidities. *Pharmacological Reports*, 2024. 76(4): p. 714–739.
47. Wakeman, M. and D.T. Archer, Metformin and micronutrient status in type 2 diabetes: does polypharmacy involving acid-suppressing medications affect vitamin B12 levels? *Diabetes, Metabolic Syndrome and Obesity*, 2020: p. 2093–2108.
48. Straub, D.A., Calcium supplementation in clinical practice: a review of forms, doses, and indications. *Nutrition in Clinical Practice*, 2007. 22(3): p. 286–296.
49. Clark, J.N., A. Whiting, and P. McCaffery, Retinoic acid receptor-targeted drugs in neurodegenerative disease. *Expert Opinion on Drug Metabolism & Toxicology*, 2020. 16(11): p. 1097–1108.
50. Zirlik, A. and C. Bode, Vitamin K antagonists: relative strengths and weaknesses vs. direct oral anticoagulants for stroke prevention in patients with atrial fibrillation. *Journal of Thrombosis and Thrombolysis*, 2017. 43(3): p. 365–379.
51. Pedrino, G.R. et al., System overview from physiological and pathological standpoints, in Fajemiroye J.O. (ed.), *Fundamentals of Drug and Non-Drug Interactions: Physiopathological Perspectives and Clinical Approaches*. 2025, Springer, p. 1–34.
52. Wakeman, M., A literature review of the potential impact of medication on vitamin D status. *Risk Management and Healthcare Policy*, 2021. 14: p. 3357–3381.
53. D'Alessandro, C. et al., Interactions between food and drugs, and nutritional status in renal patients: a narrative review. *Nutrients*, 2022. 14(1): p. 212.
54. Cui, L. et al., Relationship among homocysteine, inflammation and cognitive impairment in patients with acute ischemic stroke and transient ischemic attack. *Neuropsychiatric Disease and Treatment*, 2021. 17: p. 3607–3616.
55. Tiwari, D., A. Rani, and H.C. Jha, Homocysteine and folic acid metabolism, in Dubey G.P., Misra K., Kesharwani R.K., Ojha R.P. (eds.), *Homocysteine Metabolism in Health and Disease*. 2022, Springer, p. 3–36.
56. Draper, H.H., Interrelationships of vitamin E with other nutrients, in Packer L. (ed.), *Vitamin E in Health and Disease*. 2023, CRC Press, p. 53–62.
57. Ali, J. et al., The interplay of protein aggregation, genetics, and oxidative stress in Alzheimer's disease: role for natural antioxidants and immunotherapeutics. *Antioxidants*, 2024. 13(7): p. 862.
58. Carocci, A. et al., Oxidative stress and neurodegeneration: the involvement of iron. *Biometals*, 2018. 31(5): p. 715–735.
59. Binesh, A. and K. Venkatachalam, Copper in human health and disease: a comprehensive review. *Journal of Biochemical and Molecular Toxicology*, 2024. 38(11): p. e70052.
60. Singh, R. et al., Recent advances in targeting transition metals (copper, iron, and zinc) in Alzheimer's disease. *Molecular Neurobiology*, 2024. 61(12): p. 10916–10940.
61. Szabo, R., C. Bodolea, and T. Mocan, Iron, copper, and zinc homeostasis: physiology, physiopathology, and nanomediated applications. *Nanomaterials*, 2021. 11(11): p. 2958.
62. Revi, N. and A.K. Rengan, Impact of dietary polyphenols on neuroinflammation-associated disorders. *Neurological Sciences*, 2021. 42(8): p. 3101–3119.
63. Cao, H. et al., Available technologies on improving the stability of polyphenols in food processing. *Food Frontiers*, 2021. 2(2): p. 109–139.

64. Arsecularatne, A. et al., Combination therapy for sustainable fish oil products: improving cognitive function with n-3 PUFA and natural ingredients. *Biomedicines*, 2024. 12(6): p. 1237.
65. Abd El-Hack, M.E. et al., Curcumin, the active substance of turmeric: its effects on health and ways to improve its bioavailability. *Journal of the Science of Food and Agriculture*, 2021. 101(14): p. 5747–5762.
66. Pawłowska, M. et al., Ketogenic metabolism in neurodegenerative diseases: mechanisms of action and therapeutic potential. *Metabolites*, 2025. 15(8): p. 508.
67. Keijer, J. et al., Omics biomarkers and an approach for their practical implementation to delineate health status for personalized nutrition strategies. *Critical Reviews in Food Science and Nutrition*, 2024. 64(23): p. 8279–8307.
68. Hoffbrand, A.V., Megaloblastic anaemia, in Mead A.J., Laffan M.A., Collins G.P., Hay D., Hoffbrand A.V. (eds.), *Hoffbrand's Postgraduate Haematology*, 2025, Wiley, p. 201–221.
69. Castellanos-Sinco, H. et al., Megaloblastic anaemia: folic acid and vitamin B12 metabolism. *Revista Médica Del Hospital General De México*, 2015. 78(3): p. 135–143.
70. Kolarš, B. et al., Iron deficiency and iron deficiency anemia: a comprehensive overview of established and emerging concepts. *Pharmaceuticals*, 2025. 18(8): p. 1104.
71. Moreno, J. et al., Strawberries: antioxidant properties, health benefits and innovative technologies. *Strawberries*, 2014: p. 189–213.
72. Erem, S., A. Atfi, and M.S. Razzaque, Anabolic effects of vitamin D and magnesium in aging bone. *Journal of Steroid Biochemistry and Molecular Biology*, 2019. 193: p. 105400.
73. Aware, C. et al., Algae as food and nutraceuticals, in Bajpai P., Bulfone S.A., Giri T.K., Anal A.K. (eds.), *Algal Biorefineries and the Circular Bioeconomy*. 2022, CRC Press, p. 295–337.
74. Plumptre, L.A., Maternal one-carbon nutrient status and effects on DNA methylation and hydroxymethylation in newborn infants. PhD Diss, 2016, University of Toronto, Canada.
75. Espino-Vázquez, A.N., F.C. Rojas-Castro, and L.M. Fajardo-Yamamoto, Implications and practical applications of the chemical speciation of iodine in the biological context. *Future Pharmacology*, 2022. 2(4): p. 377–414.
76. Stanojević, M. et al., The impact of chronic magnesium deficiency on excitable tissues—translational aspects. *Biological Trace Element Research*, 2025. 203(2): p. 707–728.
77. Minich, D.M. and J.S. Bland, Personalized lifestyle medicine: relevance for nutrition and lifestyle recommendations. *The Scientific World Journal*, 2013. 2013(1): p. 129841.
78. Gröber, U., J. Schmidt, and K. Kisters, Important drug-micronutrient interactions: a selection for clinical practice. *Critical Reviews in Food Science and Nutrition*, 2020. 60(2): p. 257–275.
79. Gan, Y. et al., Precision diagnosis and treatment of vitamin metabolism-related epilepsy. *Acta Epileptologica*, 2024. 6(1): p. 27.
80. Serag, I. et al., Role of vitamin D as adjuvant therapy on multiple sclerosis: an updated systematic review and meta-analysis of randomized controlled trials. *European Journal of Medical Research*, 2025. 30(1): p. 736.
81. Badaeva, A.V. et al., Perspectives on neuronutrition in prevention and treatment of neurological disorders. *Nutrients*, 2023. 15(11): p. 2505.
82. Fekete, M. et al., Improving cognitive function with nutritional supplements in aging: a comprehensive narrative review of clinical studies investigating the effects of vitamins, minerals, antioxidants, and other dietary supplements. *Nutrients*, 2023. 15(24): p. 5116.
83. Branco, D.R. et al., Direct oral anticoagulants vs vitamin K antagonist on dementia risk in atrial fibrillation: systematic review with meta-analysis. *Journal of Thrombosis and Thrombolysis*, 2023. 56(3): p. 474–484.
84. Arshad, M.T. et al., Functional, nutraceutical, and pharmacological properties of black seed. *Food Science & Nutrition*, 2025. 13(8): p. e70725.
85. Power, R. et al., Omega-3 fatty acid, carotenoid and vitamin E supplementation improves working memory in older adults: a randomised clinical trial. *Clinical Nutrition*, 2022. 41(2): p. 405–414.
86. Almahal, Z.H. et al., Molecular perspective of dietary influences on the gut microbiome alongside neurological health: exploring the gut-brain axis. *ACS Chemical Neuroscience*, 2025. 16(11): p. 1996–2012.
87. Fekete, M. et al., The role of the Mediterranean diet in reducing the risk of cognitive impairement, dementia, and Alzheimer's disease: a meta-analysis. *Geroscience*, 2025. 47: p. 1–20.
88. Mattioli, A.V. et al., A personalized approach to vitamin D supplementation in cardiovascular health beyond the bone: an expert consensus by the Italian National Institute for Cardiovascular Research. *Nutrients*, 2024. 17(1): p. 115.
89. Squassina, A. et al., Realities and expectations of pharmacogenomics and personalized medicine: impact of translating genetic knowledge into clinical practice. *Pharmacogenomics*, 2010. 11(8): p. 1149–1167.
90. Marcum, J.A., Nutrigenetics/nutrigenomics, personalized nutrition, and precision healthcare. *Current Nutrition Reports*, 2020. 9(4): p. 338–345.
91. Ammar, M. et al., Variability in response to vitamin D supplementation according to vitamin D metabolism related gene polymorphisms in healthy adults. *European Journal of Clinical Nutrition*, 2023. 77(2): p. 189–194.
92. Babikir, H.E. and P. Singh, Neurology of nutritional disorders, in Salih M.A.M. (ed.), *Clinical Child Neurology*. 2020, Springer, p. 483–527.
93. Mohammad Sadeghi, H. et al., Drug repurposing for the management of depression: where do we stand currently? *Life*, 2021. 11(8): p. 774.
94. Visen, A. et al., Nutraceuticals as a natural alternative for preventive and proactive health care, in Singh R.B., Watanabe S., Isaza A.A. (eds.), *Functional Foods and Nutraceuticals in Metabolic and Non-Communicable Diseases*. 2022, Elsevier, p. 603–618.
95. Berger, M.M., D. Talwar, and A. Shenkin, Pitfalls in the interpretation of blood tests used to assess and monitor micronutrient nutrition status. *Nutrition in Clinical Practice*, 2023. 38(1): p. 56–69.
96. Wills, O. et al., The role of multidisciplinary MS care teams in supporting lifestyle behaviour changes to optimise brain health among people living with MS: a qualitative exploration of clinician perspectives. *Health Expectations*, 2024. 27(2): p. e14042.
97. Harnett, J.E. et al., Advancing the pharmacist's role in promoting the appropriate and safe use of dietary supplements. *Complementary Therapies in Medicine*, 2019. 44: p. 174–181.

98. McCabe-Sellers, B.J., Position of the American dietetic association: integration of medical nutrition therapy and pharmacotherapy. *Journal of the American Dietetic Association*, 2010. 110(6): p. 950–956.
99. van den Berg, H., Assessment of the B-vitamin status: some biochemical and clinical aspects, in den Boer N.C., Heiden C., Leijnse B., Souverijn J.H.M. (eds.), *Clinical Chemistry: An Overview*. 1989, Springer, p. 545–555.
100. Thacharodi, A. et al., Revolutionizing healthcare and medicine: the impact of modern technologies for a healthier future: a comprehensive review. *Health Care Science*, 2024. 3(5): p. 329–349.
101. Bloem, B.R. et al., Integrated and patient-centred management of Parkinson's disease: a network model for reshaping chronic neurological care. *The Lancet Neurology*, 2020. 19(7): p. 623–634.
102. Moretti, R., M.E. Morelli, and P. Caruso, Vitamin D in neurological diseases: a rationale for a pathogenic impact. *International Journal of Molecular Sciences*, 2018. 19(8): p. 2245.
103. Pilecky, M. et al., Omega-3 PUFA profoundly affect neural, physiological, and behavioural competences–Implications for systemic changes in trophic interactions. *Biological Reviews*, 2021. 96(5): p. 2127–2145.
104. Pekdemir, B. et al., Mechanisms and potential benefits of neuroprotective agents in neurological health. *Nutrients*, 2024. 16(24): p. 4368.
105. Rawal, S.U., B.M. Patel, and M.M. Patel, New drug delivery systems developed for brain targeting. *Drugs*, 2022. 82(7): p. 749–792.
106. Wijeweera, G. et al., Therapeutic implications of some natural products for neuroimmune diseases: a narrative of clinical studies review. *Evidence-Based Complementary and Alternative Medicine*, 2023. 2023(1): p. 5583996.
107. Facecchia, K. et al., Oxidative toxicity in neurodegenerative diseases: role of mitochondrial dysfunction and therapeutic strategies. *Journal of Toxicology*, 2011. 2011(1): p. 683728.
108. Mao, J.J. et al., Integrative oncology: addressing the global challenges of cancer prevention and treatment. *CA: A Cancer Journal for Clinicians*, 2022. 72(2): p. 144–164.
109. Baklola, M., Nutritional psychiatry: the present state of the evidence: nutrition and psychiatric disorders, in Skene C., Phelps J., Anderson R. (eds.), *An Evidence-Based Approach to Understanding the Diet-Brain Connection*, 2024: p. 485–505.
110. Maizes, V., D. Rakel, and C. Niemiec, Integrative medicine and patient-centered care. *Explore*, 2009. 5(5): p. 277–289.
111. Ghallab, Y.K. and O.S. Elassal, Biochemical and neuropharmacology of psychiatric disorders, in Mohamed W., Kobeissy F. (eds.), *Nutrition and Psychiatric Disorders: An Evidence-Based Approach to Understanding the Diet-Brain Connection*. 2024, Springer, p. 25–47.
112. Tomada, I. and J.P. Andrade, Science-based anti-ageing nutritional recommendations, in Neves D. (ed.), *Anti-Ageing Nutrients: Evidence-based Prevention of Age-Associated Diseases*. 2015, Wiley, p. 333–390.

11 Vitamins in Alzheimer's and Parkinson's Disease

Mechanistic and Therapeutic Perspectives

Falguni Goel[1], Kuldeep Singh[1], and Sachchida Nand Rai[2]

[1]Department of Pharmaceutical Technology, Meerut Institute of Engineering & Technology (MIET), Meerut, India
[2] University Centre for Research & Development (UCRD), Chandigarh University, Mohali, Punjab, India Abstract

ABBREVIATIONS

AD	Alzheimer's disease
PD	Parkinson's disease
RAR	retinoic acid receptors
RXR	retinoid X receptors
APP	amyloid precursor protein
NMDA	N-methyl-D-aspartate
BBB	blood–brain barrier
MTHFR	methylenetetrahydrofolate reductase
ROS	reactive oxygen species

11.1 INTRODUCTION

Neurodegenerative diseases have become some of the most challenging medical and societal issues in the 21st century, with Alzheimer's disease (AD) and Parkinson's disease (PD) as the most common and devastating examples [1]. Neurodegenerative diseases are characterized by the progressive dysfunction and loss of neurons, usually resulting in serious decline in cognition, motor control, and lessening quality of life [2]. As populations across the globe enjoy the benefits of increased life expectancy, the rate of neurodegenerative diseases has increased tremendously, creating not only a medical challenge, but also a socioeconomic challenge [3]. Current estimates indicate that tens of millions of individuals across the globe are affected by neurodegenerative diseases, and this number continues to increase, with the predicted continued increase identifiably straining healthcare systems and caregiver networks [3]. While advances in molecular neuroscience have been steady, treatment options fall significantly short. There are currently targeted pharmacological options, but they can provide only limited symptomatic relief and they do not significantly limit the underlying mechanisms and factors that ultimately lead to neuronal degeneration [4]. Therefore, there is a compelling need to develop new strategies targeting the mechanism of disease, delay the onset of disease, or slow the progression of the disease.

Nutritional neuroscience is increasingly being recognized as a relevant and complementary approach to traditional therapy. The brain is often considered the most metabolically active part of the body that is prone to oxidative and inflammatory damage that is reactive to a limited array of nutrients to support its architecture and function [5]. In addition, it has been shown that nutritional status can and does impact brain health along the lifespan, while disruptions of nutrient balance can relate to processes involved with inflammation, accelerated aging, and neurodegeneration. Of all the dietary factors that have been investigated, vitamins are highlighted in particular because of their essential biochemical and physiological nature [6]. Vitamins can be seen as cofactors in enzyme-mediated reactions, but in addition to this, they regulate neurotransmitter production and metabolism and homocysteine metabolism, stabilize cell membranes, and act as antioxidants [7].

Lower levels of B vitamins, especially folate and cobalamin, lead to increases in homocysteine, which has been identified as a risk factor in neurotoxicity and vascular function in the brain [8]. Low vitamin D level has also been associated with disruption of synaptic plasticity, modulation dysfunction in immune response, and higher prevalence of AD and PD [9]. Low levels of vitamin C and vitamin E also contribute to reduced ability of the brain to deal with oxidative stress, which is a well-known factor in neuron injury in these diseases [10]. Conversely, adequate or optimized vitamin levels have been correlated with improved resistance to neurodegeneration, better cognitive outcomes, and better preservation of gross motor function—making a compelling argument for the study of using vitamins as therapies [11].

The mechanism by which vitamins may have neuroprotective effects has been clarified by an increasing amount of experimental and clinical data during the past 20 years [12]. Numerous bioenergetic processes, including mitochondria, neurotransmission, redox balance, and gene regulatory activity, can be impacted by vitamins [13]. The main elements of AD and PD pathogenesis, tau phosphorylation,

DOI: 10.1201/9781003686675-11

alpha-synuclein toxicity, and Aβ aggregation, can all be impacted by a variety of vitamins [14]. Vitamins can also affect vascular support and inflammatory cascades, which contributes to the systemic advantages of neuronal survival and maintenance [15]. In addition to their synergistic effects with one another, vitamins can also have synergistic effects with minerals, polyphenols, and other drugs, which opens up the possibility of using combinations as therapeutic interventions [16]. Vitamins by themselves cannot produce neurological benefits.

Vitamins' molecular foundation, accessibility as dietary and supplement components, and relative safety in comparison with other medications all contribute to their potential as therapies [16]. Vitamins are a desirable alternative for prevention because they are dietary and supplementary substances that can be customized [17]. It is crucial to remember that although numerous studies have produced encouraging findings, the outcomes are not all the same. While some clinical trials have shown that vitamin supplementation improves motor results and cognitive decline, other investigations have shown modest or equivocal benefits [18]. These discrepancies can be explained by changes in supplementation dosage, bioavailability, intervention length, hereditary variables, and disease stage.

A unique multi-domain frontier that connects nutrition, neurology, and innovative clinical practice is the investigation of vitamins in the setting of AD and PD [19]. As science advances, vitamins are increasingly being seen as participants in a disease-specific pathway rather than as regulators or supportive nutrients [20]. The molecular and clinical consequences of vitamins in neurodegenerative illnesses will be covered in this chapter. Together with food as risk research, vitamins' preclinical and clinical relationships to neuroprotection will be discussed in this chapter, along with possible future directions for their use in this area. It is reasonable to remind readers to think systemically and across multiple domains in terms of a modern brain health perspective and to consider possibilities that incorporate pharmacological engagement along with nutrition and lifestyle engagement in our pursuit of neurodegeneration prevention, given the nature of the issues. Even with the best of intentions, vitamin research tends to slide into purely pharmacological constructs.

11.2 PATHOPHYSIOLOGY OF AD AND PD

The intricate pathophysiology of AD and PD includes interactions between environmental exposures, genetic predispositions, and a progressive degenerative cellular process that eventually results in neuronal death [21]. The main clinical signs of both conditions are memory loss and cognitive decline in AD and movement impairments in PD [22]. The molecular processes that underlie the illnesses are very similar: neuroinflammation, protein misfolding, mitochondrial dysfunction, and oxidative stress [23]. A growing and non-exclusive understanding of how both illnesses share similar cellular deficiencies, which may result in similar treatment approaches, is reflected in the use of vitamins as potential cellular modulators of resilience [24].

One of the main mechanisms of neuronal injury in each of these etiologies is oxidative stress [25]. The brain is particularly vulnerable to reactive oxygen species (ROS), which are by-products of essential cellular metabolic activities, in part because of its high oxygen consumption in relation to its bulk [26]. Endogenous antioxidant systems can maintain redox homeostasis under physiological conditions. However, if oxidative stress is allowed to increase, Lipid peroxidase (LPO), protein oxidation, and DNA alterations will take place, which may encourage neuronal damage during a diseased state and degenerate the neurodegenerative environment [27]. Mitochondria are at the heart of this process, as they are the powerhouses of the cell. Impaired mitochondrial metabolism has been repeatedly observed in AD and PD, mostly with regard to mitochondrial function in respiration chain complexes, ATP synthesis, and entirely unregulated calcium handling, which forces neurons into damaging environmental settings [28].

Another pathological feature is neuroinflammation. Microglia and astrocytes, which typically protect the neural tissue, can also signal neuronal death by continuously releasing pro-inflammatory cytokines, nitric oxide, and reactive oxygen intermediates once activated [29]. Acute inflammation may be protective; however, chronic neuroinflammation is harmful, especially when mediated by the secondarily oxidative stress pathway mechanism, which destabilizes synaptic connections and enhances the development of protein aggregation [30]. The inflammatory environment is dynamic and interacts with misfolded proteins to worsen toxicity and neurodegeneration in a mutually aggravating cycle [31].

The amyloid cascade theory has influenced how people explain how AD works. Amyloid-beta peptides build up as soluble oligomers due to abnormal metabolism of amyloid precursor proteins (APPs), which ultimately results in insoluble plaque formations [32]. These aggregates can cause a persistent inflammatory response by blocking the function of a single synapse that would otherwise provide sufficient neural signals. The tau protein also plays a role in these disorders; when the tau protein is hyperphosphorylated, neurofibrillary tangles form [33]. Effects on microtubule stability are intrinsically connected to the amyloid-beta and tau diseases; they cause axonal transport and synaptic integrity failure, which results in widespread neuronal death [34]. Beyond structural alterations and protein aggregation, the pathophysiology of AD is caused by essentially defective synapses, which are evident in certain neurotransmitter system dysfunction and cholinergic deficiencies that underlie the cognitive difficulties associated with AD [35].

PD is largely characterized by the primary degeneration of dopaminergic neurons within the substantia nigra pars compacta. The degeneration of dopaminergic inputs into the striatum results in a constant state of striatal dopamine depletion and its associated motor manifestations [36]. Critical to the disease process in PD is the aggregation and misfolding of alpha-synuclein, a presynaptic protein that aggregates within neurons as Lewy bodies [37]. The aggregation of alpha-synuclein, in addition to abnormal neuronal signal transduction, corrupts mitochondrial function, synaptic vesicle trafficking, and proteasomal degradation, which eventually leads to the death of the neuron [38]. Mitochondrial dysfunction is evident in PD, evidenced by specific deficits in the electron transport chain, likely complex I, and increased production of ROS [39]. Together, mitochondrial dysfunction, alpha-synuclein aggregation, and oxidative stress form a continuous "cyclic" pathogenesis contributing to the progressive degeneration of dopaminergic signaling [38]. The additional toll of neuroinflammation, propagated by activated glial cells and persistent inflammatory mediators, further contributes to dopaminergic vulnerability on top of the impact of alpha-synuclein dysfunction [40].

Although AD and PD have different proteinopathies, the mechanistic pathways are shared resulting in neurodegeneration. Both stressors, oxidative stress, mitochondrial dysfunction, and neuroinflammation drive the pathology for both diseases [41]. The pathogenic features of both disorders include protein aggregation and misfolding. Despite having distinct protein substrates, these disorders ultimately share gliosis, proteostasis failure, and synaptic dysfunction [42]. Clinically and anatomically, the two illnesses remain distinct: While PD mostly impacts the nigrostriatal system, which is in charge of motor activities, AD is biased in the hippocampus and cortical area, leading to cognitive and memory impairment [43]. This indicates that while there are treatment strategies that can still target specific disease mechanisms, addressing some of the overlapping pathogenic pathways may also have advantages across the neurodegenerative continuum [44].

The development of treatments is significantly impacted by the identification of the shared pathways in AD and PD. In all disease processes, treatment success may be achieved through modulating inflammation, reducing oxidative stress, and restoring mitochondrial function [45]. Because vitamins can alter mitochondrial metabolism, inflammatory signaling pathways, and antioxidant defense, they are in a unique position to influence these overlapping pathophysiological events [46]. This viewpoint on AD and PD highlights the similarities and differences between the two conditions' underlying mechanisms while giving us a framework for starting to understand the biology and mechanism of the diseases [47]. It also acknowledges that vitamins may be useful mechanistic agents and therapeutic agents in the neurodegeneration of both conditions (Figure 11.1).

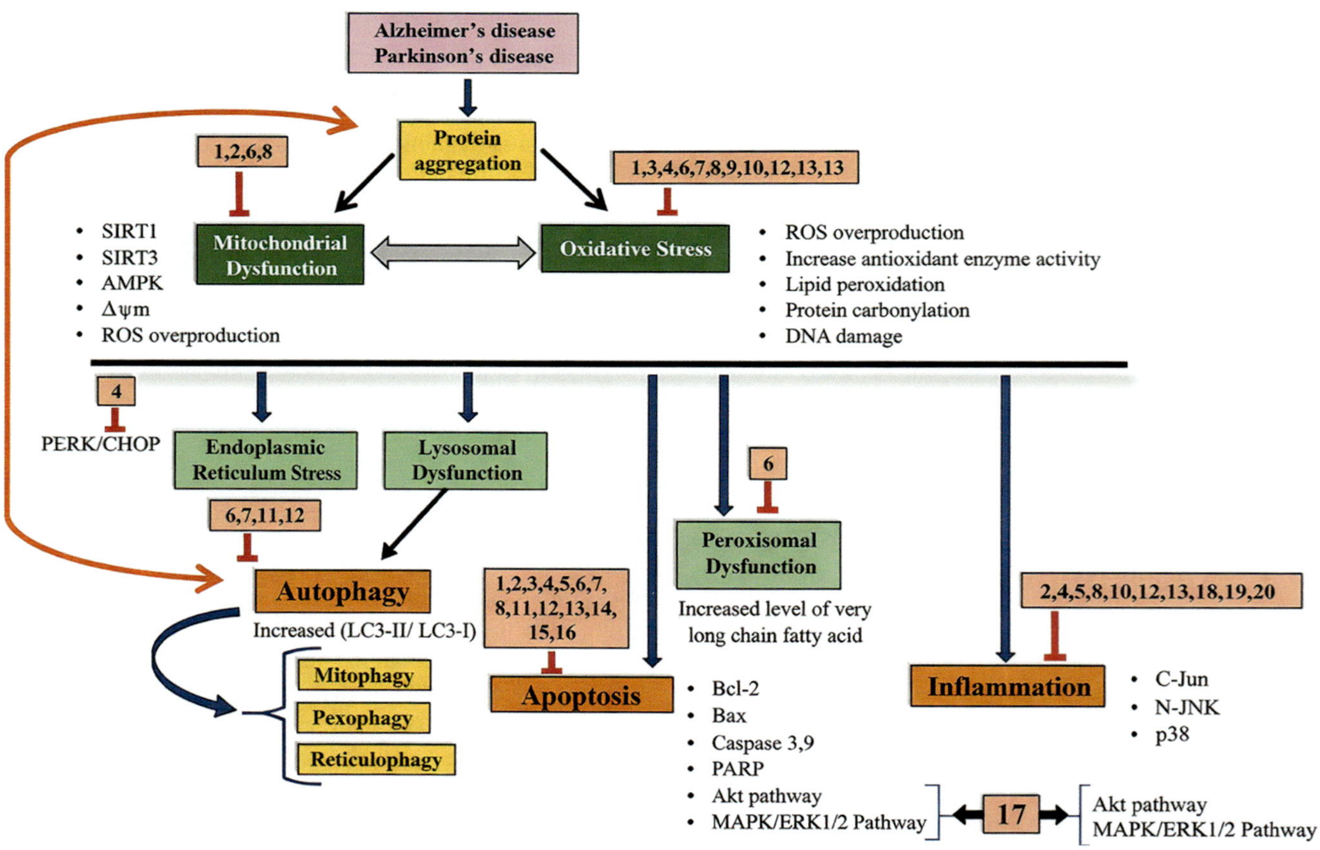

FIGURE 11.1 Schematic representation highlighting the primary molecular pathways causing AD and PD.

11.3 VITAMINS AND THE NERVOUS SYSTEM: AN OVERVIEW

Vitamins are minuscule substances that the body needs in order to function properly. They are necessary for some aspects of maintaining the integrity, shape, and function of nerve tissue [48]. Vitamins have functions as antioxidants, cellular signaling pathways, and gene expression modulators in addition to being necessary cofactors in biochemical reactions and processes [49]. One of the body's most metabolically active tissues is the nervous system, and it takes time for this metabolism to maintain neurotransmission, synaptic plasticity, and other cell functions until cell death [50]. Additionally, certain vitamin imbalances, deficiencies, or toxicities can have far-reaching effects. On the visible spectrum, excessive vitamin deficiency can lead to neurodegeneration, while subtle decreases can result in covert cognitive impairments [51]. Together with their designation, biochemical interactions or functions, and risk-protectiveness, vitamins in their entirety offer information about their unique micronutritional role in neuroprotection and pathophysiological process delay in AD and PD.

Vitamins have been divided into two categories based on their solubility: fat-soluble and water-soluble [52]. Fat-soluble vitamins include A, D, E, and K, which can be taken up in the diet with the lipids, stored in the liver and adipose tissue, and maintain long-lasting stores [53]. Vitamin solubility, or lipophilicity, allows fat-soluble vitamins to enter into cell membranes, where they have structural and functional roles, most notably in maintaining membrane stability and preventing lipid peroxidation [54]. The water-soluble vitamins include the B-complex and vitamin C. Water-soluble vitamins are not stored to any large extent in the body and must be supplied regularly through the diet to maintain sufficient levels in the body [52]. Water-soluble vitamins mainly appear to act as cofactors or coenzymes for metabolic pathways related to energy production, amino acid metabolism, and neurotransmitter production [55]. The distinction in solubility of the vitamins helps to illustrate the different, yet complementary, functions of the vitamins in supporting neuronal physiology and promoting resistance to disease [56].

Together with their designation, biochemical interactions or functions, and risk-protectiveness, vitamins in their entirety offer information about their unique micronutritional role in neuroprotection and pathophysiological process delay in AD and PD. The metabolic pathways necessary to maintain neuronal function depend heavily on vitamins [57]. The tricarboxylic acid cycle, oxidative phosphorylation, and one-carbon metabolism all depend on the B vitamins, which links nutrition to ATP synthesis and maintaining neuronal energy homeostasis [57]. The essential enzymes for glucose metabolism and neurotransmitter synthesis, pyruvate dehydrogenase and α-ketoglutarate dehydrogenase, depend on thiamine (vitamin B1) as a cofactor [58]. The neurotransmitters dopamine, serotonin, and gamma-aminobutyric acid, all of which are important modulators of emotion and cognition, are produced in part by pyridoxine (vitamin B6) [59]. Significant progress has been made in understanding how disruptions in the homocysteine pathway result in neuronal toxicity, DNA damage, and abnormalities in brain blood flow [60]. Folate (vitamin B9) and cobalamin (vitamin B12) also regulate homocysteine metabolism. As of right now, niacin (vitamin B3) has been linked to neurodegeneration because of its role in the metabolism of NAD+ and NADP+, redox, and DNA repair [61]. The example of these biological processes shows how vitamins facilitate neuronal metabolism and may be able to alter disease pathways in AD and PD (Table 11.1).

Vitamins influence neurotransmission and synaptic functioning. For example, vitamin C is an antioxidant that prevents the catecholamines from breaking down oxidatively and is a cofactor for the enzyme dopamine β-hydroxylase, which catalyzes the conversion of dopamine to norepinephrine [68]. The "neurosteroid vitamin," vitamin D, controls the release of neurotransmitters and calcium signaling [69]. It also affects the expression of neurotrophic factors including brain-derived neurotrophic factor and nerve growth factor. By attaching to retinoic acid receptors (RAR), vitamin A and its derivatives, retinoids, alter gene expression [62]. Memory functions, synaptic plasticity, and neurodevelopment are all influenced by vitamin A and retinoid activity [70]. These impacts on synaptic functions show that vitamins are positive modulators of the molecular machinery that drives neural plasticity and transmission, rather than merely secondary or accessory.

Vitamins' anti-inflammatory and antioxidant properties have also proved significant, particularly in relation to neurological diseases. AD and PD are caused by oxidative stress, which occurs when there are too many ROS. [71]. The antioxidant vitamin C, for instance, inhibits the oxidative breakdown of catecholamines and is a cofactor for the enzyme dopamine β-hydroxylase, which catalyzes the conversion of dopamine to norepinephrine [72]. Vitamin D, sometimes known as the "neurosteroid vitamin," regulates calcium transmission and neurotransmitter release. It also affects the expression of neurotrophic factors including brain-derived neurotrophic factor and nerve growth factor [73]. By attaching to RAR, vitamin A and its derivatives, retinoids, alter gene expression. Memory functions, synaptic plasticity, and neurodevelopment are all influenced by vitamin A and retinoid activity [62]. These impacts on synaptic functions show that vitamins are positive modulators of the molecular machinery that drives neural plasticity and transmission, rather than merely secondary or accessory [74].

These antioxidant and anti-inflammatory roles are important as they tackle some of the most fundamental

TABLE 11.1
Vitamins and their Roles in the Nervous System

Vitamin	Classification	Key Functions in the Nervous System	Deficiency Consequences	Relevance to Neurodegeneration	References
Vitamin A (retinoids)	Fat-soluble	Regulates neurodevelopment, neuroplasticity, and gene expression; modulates amyloid and tau pathways	Impaired vision, cognitive dysfunction	Dysregulation linked to AD pathology	[62]
Vitamin B1 (thiamine)	Water-soluble	Coenzyme in energy metabolism (glucose utilization); neuronal function and neurotransmission	Wernicke–Korsakoff syndrome, memory loss	Energy failure contributes to AD and PD	[61]
Vitamin B6 (pyridoxine)	Water-soluble	Cofactor in neurotransmitter synthesis (dopamine, serotonin, GABA); homocysteine regulation	Irritability, seizures, cognitive decline	Hyperhomocysteinemia associated with dementia	[63]
Vitamin B9 (folate)	Water-soluble	One-carbon metabolism, DNA synthesis and repair, epigenetic regulation	Megaloblastic anemia, depression, cognitive decline	Low folate linked with AD progression	[61]
Vitamin B12 (cobalamin)	Water-soluble	Myelination, methylation reactions, neurotransmitter regulation	Peripheral neuropathy, memory loss, dementia	Deficiency associated with cognitive decline in AD and PD	[61]
Vitamin C (ascorbic acid)	Water-soluble	Potent antioxidant; cofactor in dopamine synthesis; maintains synaptic plasticity	Scurvy, fatigue, impaired cognition	Neuroprotection via antioxidant and anti-inflammatory effects	[64]
Vitamin D	Fat-soluble	Regulates calcium homeostasis, neurotransmission, and immune modulation	Osteomalacia, muscle weakness, mood disorders	Deficiency linked with higher AD and PD risk	[24]
Vitamin E (tocopherols/ tocotrienols)	Fat-soluble	Antioxidant defense against lipid peroxidation; membrane stabilization	Neuropathy, impaired balance, oxidative damage	Protects against oxidative stress in AD and PD	[65]
Vitamin K	Fat-soluble	Involved in sphingolipid metabolism, neuronal membrane integrity, and cognitive function	Coagulation defects, possible cognitive decline	Emerging role in synaptic function and cognition	[66]
Niacin (vitamin B3)	Water-soluble	Precursor of NAD^+, supports mitochondrial energy metabolism, DNA repair	Pellagra, dementia, depression	NAD^+ metabolism critical in neuroprotection	[61]
Biotin (vitamin B7)	Water-soluble	Cofactor for carboxylases; maintains myelin and neuronal metabolism	Dermatitis, neurological symptoms	Supports neuronal energy and myelination	[67]

drivers of the disease: by lowering oxidative stress, engaging mitochondria, and inhibiting chronic inflammation, vitamins provide a more hospitable environment for neurons to survive and function [75]. Vitamins do not just play prevention roles; there are multiple studies showing that supplementation can limit or even reverse damage to the nervous system by enhancing the utility of pharmacological treatments and reversing deficiency effects in susceptible individuals [76].

The classification, metabolic processes, neurotransmitter-influencing effects, and protective properties of vitamins emphasize how important they are for maintaining the health of the nervous system [77]. They function as stress protectors against oxidative and inflammatory damage, biochemical catalysts, and structural stabilizers, and their actions place them at the intersection of neurodegeneration and nutrition [78]. Vitamins can act as molecular modulators and treatments for disease conditions like AD and PD, where overlapping chronic inflammation, oxidative damage, and mitochondrial dysfunction work together to destabilize brain health [15]. Thus, vitamins not only provide further information on the biology of the brain, but their many functions also teach us how to make pharmacological and nutritional interventions to improve clinical benefit by delaying or changing the progression of neurodegeneration [79] (Figure 11.2).

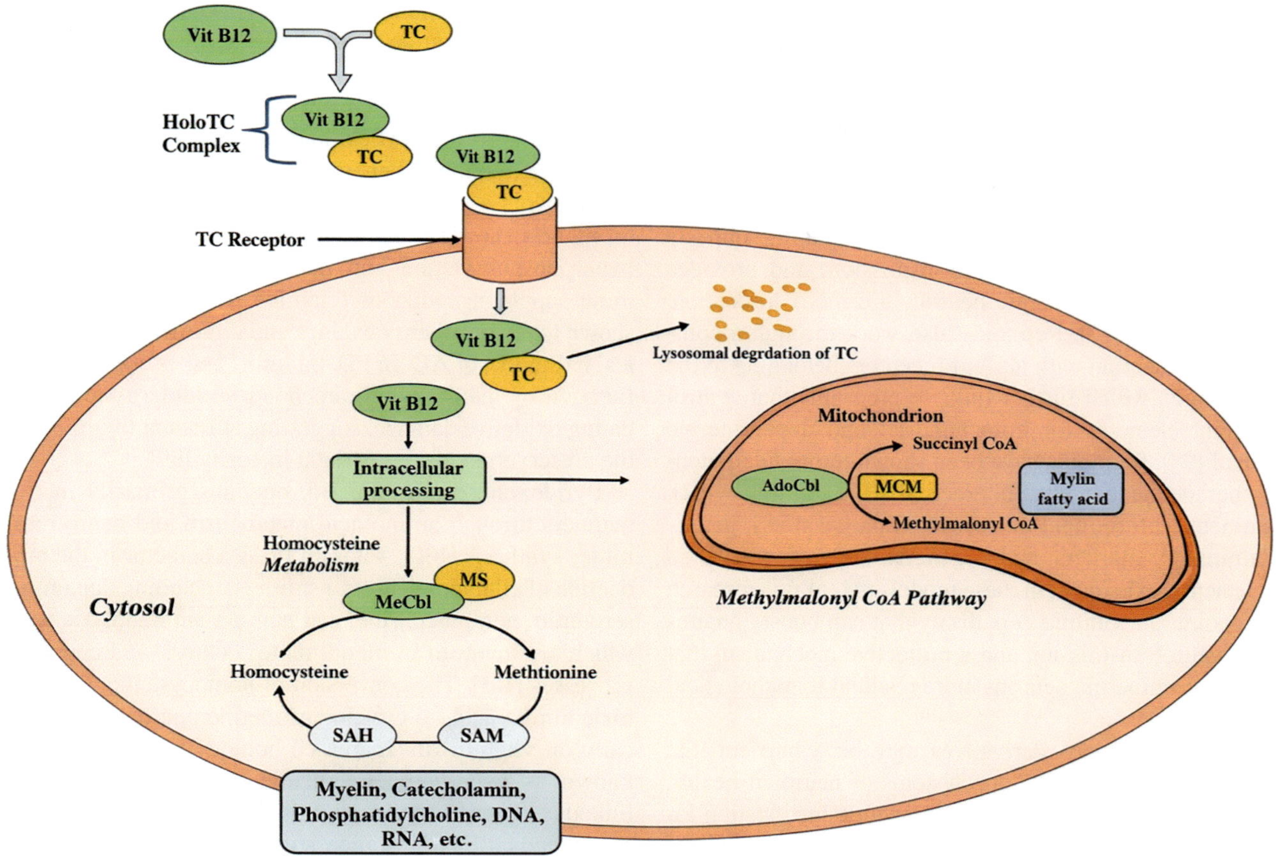

FIGURE 11.2 Depiction of the key functions of vitamins in maintaining healthy nervous processes.

11.4 VITAMIN A AND RETINOIDS

Vitamin A and its active derivatives, collectively called retinoids, serve an important role in the central nervous system well beyond their classical contributions to vision and epithelial health [80]. In the nervous system, vitamin A operates mainly through its metabolites, including retinoic acid, which act as potent regulators of gene transcription through binding with nuclear receptors such as RAR and retinoid X receptors (RXR) [81]. These processes enable vitamin A to impact a myriad of processes necessary for neuronal survival, differentiation, and plasticity. The roles of retinoids in neurodevelopment are particularly well established, as they regulate neuronal patterning, axonal growth, and synaptic creation [82]. In the adult brain, retinoic acid continues to support synaptic plasticity, memory consolidation, and the modulation of long-term potentiation, an important cellular correlate of learning and cognition [83].

Vitamin A has gained recent interest for its role in modulating two of the most weighted pathological features in AD: amyloid-beta aggregation and tau pathology [84]. Experimental data have confirmed that retinoic acid signaling has the capacity to affect APP expression and processing, favoring non-amyloidogenic processing, which suppresses the aggregation of neurotoxic amyloid-beta peptides [85]. Retinoids can potentially halt amyloid buildup not only by delivery but by regulation of the key enzymes implicated in amyloidogenic cleavage that affects the physiology of protective processing compared to pathogenic processing [86]. Vitamin A influences tau protein. Retinoic acid inhibits tau aggregation into neurofibrillary tangles through modulation of kinases and phosphatases implicated in tau phosphorylation [87]. Collectively, these mechanistic factors argue retinoid signaling plays a critical role in the maintenance of protein homeostasis within the brain and that disruption of retinoid pathway signaling may accelerate progression of AD pathology [88].

Additional evidence for this connection comes from results obtained from animal experimental models. Studies on vitamin A deprivation in animals show severe cognitive impairment, impaired synaptic plasticity, and elevated amyloid load, all of which are characteristics of AD disease [89]. It has been demonstrated that retinoic acid therapy or vitamin A supplements can alleviate cognitive impairment, lower amyloid burden, and improve learning and memory function [79]. Retinoids have been shown in experiments to decrease inflammation and oxidative stress at the cellular level, protecting neurons from damage caused by amyloid-beta oligomers [51]. The body of research clearly supports vitamin A's neuroprotective effects in AD, both directly through its impact on the tau and amyloid pathways and indirectly through its anti-inflammatory and antioxidant properties [90].

More people are increasingly appreciating vitamin A's role in PD, particularly in relation to its role in neuroprotection and dopaminergic transmission [91]. In order to maintain dopaminergic homeostasis within the nigrostriatal system, retinoic acid regulates the expression of dopamine receptors, transporters, and the full suite of enzymes necessary for dopamine synthesis [92]. Additionally, it has been shown that the imbalance in retinoid signaling impacts synaptic plasticity, dopamine transmission, and provides evidence that dopaminergic neurons are more susceptible to degeneration [93]. Retinoids also work through a complicated connection with nuclear receptors belonging to the RXR family, which further links to programs that control important elements for mitochondrial and dopaminergic survival [94]. Retinoids have been shown to provide potential therapeutic benefits in preclinical PD animals [62]. Treatment with retinoid decreased oxidative stress, neuroinflammatory markers, and degeneration brought on by a single neurotoxic insult in these models. All of this research could point to a vitamin A pathway that can boost dopaminergic neurotransmission and a protective mechanism that makes dopaminergic neurons more resilient to pathological stresses [62].

Vitamin A and its derivatives may be a nutrient that is ingrained in the regulatory systems of neuronal health, according to the molecular and therapeutic justification for its use in neurodegeneration [15]. One of the most potent neuroprotective substances, vitamin A supports neurodevelopment, maintains synaptic plasticity, alters the metabolism of tau and amyloid, and controls dopamine receptor signaling [64]. The more general translational significance may be significant with adequate dietary intake of vitamin A and with retinoid-based therapies, which offer therapeutic and preventative interventions for AD and PD, even though much of the most convincing evidence comes from experimental and preclinical findings [95].

11.5 VITAMIN B COMPLEX

With their roles in energy metabolism, neurotransmitter synthesis, DNA methylation, and myelination, the B-vitamin complex is a collection of varied, water-soluble micronutrients that sit at the intersection of brain health [61]. Each B vitamin supports distinct but connected metabolic pathways that support cognitive function and neuronal survival. Given the brain's high metabolic load, deficits in B vitamins have been repeatedly linked to mood swings, accelerated neurodegenerative processes, and cognitive loss [15]. They can also negatively impact neuronal function. By connecting nutritional status to molecular pathophysiology and clinical symptoms, the B-vitamin complex acts as a gauge of prospective susceptibility or resilience [96].

Thiamine (vitamin B1) is a crucial component in neuronal energy metabolism. As a coenzyme for enzymes such as pyruvate dehydrogenase, α-ketoglutarate dehydrogenase, and transketolase, thiamine plays an important role in the glucose utilization process, ATP production, and the pentose phosphate pathway [61]. Glucose is the preferred energy substrate for the brain so thiamine deficiency affects neuronal survival in a very short time [97]. In AD, decreased thiamine-dependent enzyme activity has been documented in the cortical and hippocampal regions that contribute to impaired energy metabolism, oxidative stress, and amyloid and tau accumulation [98]. With experimental models showing that thiamine deficiency promotes tau hyperphosphorylation and neuroinflammation, and thiamine supplementation or thiamine derivatives have been shown to restore energy balance and diminish the pathological hallmarks of AD. In PD, thiamine assists dopaminergic function by maintaining mitochondrial integrity and mitigating oxidative damage, suggesting sufficient thiamine for the preservation of nigrostriatal integrity [99].

Pyridoxine, or vitamin B6, operates primarily in two manners through amino acid metabolism and neurotransmitter synthesis [100]. Vitamin B6 is a cofactor in the production of all major neurotransmitters, including dopamine, serotonin, norepinephrine, and gamma-aminobutyric acid, which are involved in mood, motor control, and cognitive processes [101]. The conversion of homocysteine, a neurotoxic amino acid, to cysteine or methionine depends critically on vitamin B6. Increased homocysteine levels, also known as hyperhomocysteinemia syndrome, have been linked to oxidative stress, vascular impairment, and the risk of dementia [8]. For instance, studies have linked increased homocysteine and improved tau and amyloid pathology in AD to increased motor symptom severity and cognitive impairment in PD [102]. Vitamin B6 is therefore essential for neuroprotection since it not only maintains neurotransmitter balance but also controls the risk of neurotoxicity associated with hyperhomocysteinemia [76].

Folate, often known as vitamin B9, is essential for carbon metabolism. Additionally, the nutrient plays a role in methylation processes, prolonged DNA synthesis, gene control, and epigenetic programming [103]. In order to produce S-adenosylmethionine, the universal methyl donor for DNA, RNA, and protein methylation, homocysteine must be remethylated back to methionine, which requires folate [104]. If an individual is insufficient in folate, we can certainly expect mammalian methylation capacity to be disrupted, leading to aberrant gene expression, poor neuronal DNA repair mechanisms, and increased oxidative stress [105]. Individuals with AD and folate deficiency demonstrate hippocampal atrophy along with faulty synaptic plasticity and an increased amyloid burden [106]. Further, epigenetic studies in AD suggest folate deficiency, or lack of methylation capacity, may indirectly accelerate amyloid pathology by changing regulation of genes involved in the degradation or processing of amyloid, and by most certainly permissively altering the methylation landscape of facilitating tau (i.e., phosphorylation) [107]. There is limited research examining a more general understanding of the changes that would occur in protein methylation in the event of folate deficiency. From our point of view, we would consider that folate is at least mechanistically important to the maintenance of dopaminergic survival, through regulating methylation balance as well as toxic homocysteine accumulation in PD. Hence, folate is not only important as a metabolic regulator, but also an epigenetic regulator of neurodegeneration [108].

Cobalamin, or vitamin B12, is essential for myelin integrity, neuronal regeneration and repair, and one-carbon metabolism. It is involved in converting homocysteine to methionine, and in the production of methylmalonyl-CoA, which are essential for DNA synthesis and fatty acid metabolism [109]. Vitamin B12 deficiency causes demyelination, axonal degeneration, and impairment in neurotransmission, all of which share features with neurodegenerative diseases [15]. For example, reduced B12 levels in AD correlate with impairments in memory, cortical atrophy, and increased homocysteine. Improving cognition in people with reduced B12 levels has been shown with supplementation although results in advanced stage diseases appear mixed [110]. In PD, cobalamin deficiency may worsen motor symptoms and peripheral neuropathy, especially among patients on long-term levodopa therapy that may lead to increased homocysteine levels and worsen B12 depletion [111].

Deficiency of the B vitamins generates a biochemical environment of disturbed energy metabolism, homocysteine accumulation, oxidative stress, disturbed methylation, and disturbed neurotransmitter concentrations, and each of these alterations favors neurodegeneration that encourages amyloid deposition, tau hyper-phosphorylation, dopaminergic neuronal loss, and synaptic loss [112]. Clinical trials have provided additional evidence for this relationship affirming across populations that those who have low levels of B vitamins; the evidence is strongest for thiamine, folate, and cobalamin have greater cognitive impairment and more rapid progress of their dementia and poorer clinical outcomes in PD. The intake of single or supplementary B vitamins, using particular supplementation protocols of the B-vitamin complex, have had different effects, and the more regular benefits had been found to be associated with deficiency in B vitamins, or in targeted populations where increased levels of homocysteine were observed [113]. The specific windows of opportunity for any B vitamin supplementation or other therapeutic methods to AD are yet to be established, but in spite of this, there is enough evidence to include B vitamins as either active players in the preservation of neuronal structure and function and modifiable processes that can reduce or slow the course and outcomes of both the AD and PD pathologies.

11.6 VITAMIN C (ASCORBIC ACID)

The brain, unlike other parts of the body, maintains exceptionally high concentrations of vitamin C, particularly in neurons, where it acts not only as a free radical scavenger but also as a cofactor in several essential enzymatic processes. The unique electronic properties of vitamin C, such as a high reducing capacity and a capacity to form complexes with metals, provide an essential aspect of the brain's defense against oxidative and excitotoxic injury [75]. Notably, in terms of AD and PD several studies have examined the mechanistic and therapeutic value of vitamin C against oxidative stress, mitochondrial dysfunction, and protein aggregation contributing to neuronal degeneration [114].

Vitamin C's main distinguishing property is its ability to neutralize ROS and reactive nitrogen species which protects oxidatively modified lipids, proteins, and nucleic acids [115]. Neuronal membranes, being composed of lipids, are particularly vulnerable to lipid peroxidation, where vitamin C works with vitamin E to protect membrane integrity [116]. Ascorbic acid can rejuvenate oxidized vitamin E, thus acting to provide an antioxidant defense via a synergistic antioxidant cycle, to protect the neuron from oxidative stress [117]. Vitamin C acts as a metal chelator, binding redox-active metals that catalyze the formation of highly reactive hydroxyl radicals in Fenton chemistry [118]. Metal-chelating properties are especially relevant in neurodegenerative disorders, where dysfunctional metal homeostasis has been implicated in amyloid β aggregation in AD and alpha-synuclein pathology PD [119]. Vitamin C's role in reducing metal-catalyzed oxidative injury is also an important aspect in the role of iron and copper in neuronal homeostasis and reducing the burden of protein aggregation [120].

Vitamin C has a direct impact on neurotransmitter biosynthesis and also regulates synaptic function. It acts as a necessary cofactor for dopamine β-hydroxylase, the enzyme that catalyzes the conversion of dopamine to norepinephrine as part of the catecholamine biosynthesis process [121]. Additionally, there is evidence that vitamin C regulates the release of signals and the reuptake of neurotransmitters at synaptic terminals, regulating glutamatergic and dopaminergic transmission [122]. In animals, some pharmacological evidence indicates vitamin C regulates the activity of N-methyl-D-aspartate (NMDA) receptors, conferring neuroprotection against excitotoxicity that is believed to underlie aspects of both AD and PD pathology [123]. Vitamin C contributes directly to protection of synaptic communication and preservation of redox balance within the synaptic cleft that promotes ideal neurotransmitter signaling responsible for memory, cognition, and motor control [64].

Studies in both animals and humans have shown that vitamin C may protect nerve cells from action by free radicals and may protect people with neurodegenerative diseases [15]. In the case of AD, vitamin C reduced amyloid-beta aggregation, reduced oxidative stress, and improved cognition in experimental models of the disease. In these models, it has been established that vitamin C supplementation is able to return plasma and cerebrospinal fluid levels of vitamin C to normal levels [124]. Low plasma and cerebrospinal fluid levels have been found in people with AD, and these patients showed greater cognitive decline and had higher levels of oxidative stress markers along with lower levels of antioxidant markers [125]. Vitamin C not only increases antioxidant defenses, but also improves their cholinergic and glutamatergic signaling repair, which are disrupted in a neurodegenerative disease state. Similarly, vitamin C provides neuroprotection in experimental models of PD where it reduced oxidative injury, altered the metabolism of neurotoxin dopamine, and ameliorated neuroinflammation in cell cultures [126]. Experimental evidence also suggests that ascorbic acid can reduce neuronal vulnerability due to neurotoxic environments [127]. MPTP and the naturally occurring rotenone are frequently used to model neuronal physiology observed in PD. Researchers have reported that lower levels of circulating vitamin C were related to increased motor impairments and faster progression of the disease in patients, despite no associations between vitamin C status and specific drugs used to manage motor impairment [128]. These findings suggest vitamin C could modify dopaminergic vulnerability.

Vitamin C may also affect certain systemic contributors to neurodegeneration. Its anti-inflammatory properties inhibit microglial activation and resultant cytokine release and chronic neuroinflammation to protect neurons from chronic injury [129]. Its effects on systemic vascular health and endothelial function further improve cerebral blood flow and the integrity of the BBB, which are frequently affected in both AD and PD [130]. These systemic benefits serve to support the use of vitamin C as a comprehensive neuroprotective agent.

In spite of evidence supporting the clinical value of vitamin C supplementation, various obstacles preclude the translation of vitamin C supplementation to a regular clinical benefit. Because of the BBB saturation of transport pathways, large levels of an oral supplemental drug do not always correspond to proportionately high brain concentrations [131]. Bioavailability and absorption are strictly restricted, and this is similar to other medications of low therapeutic ratio [132]. Additionally, individuals are likely to differ in important ways, such as their baseline nutritional state, the stage of their disease, and their genetic predispositions. These factors can all affect treatment allocations or outcomes, and they can literally muddy the results of any clinical trial. Nonetheless, there are enough data to conclude that getting enough vitamin C is crucial for maintaining brain function, and people who have vitamin C deficiency or increased oxidative stress may benefit most from supplements [133].

Vitamin C is a unique neuroprotective chemical that has anti-inflammatory, metal homeostasis, antioxidant defense, and neurotransmitter production properties [123]. It has a role in both prevention and treatment because of its therapeutic actions, which directly conflict with some of the main pathogenic pathways underlying the development of AD and PD. Vitamin C is a viable option for nutritional therapy of neurodegeneration as we continue to investigate its effects on clinical populations' metabolic and ideal dosage.

11.7 VITAMIN D

Widely known as a hormone that promotes bone health and calcium metabolism, vitamin D is also an essential neurosteroid that has a major impact on the central nervous system [134]. The main underlying theme of all the traditional roles of vitamin D in bone physiology is the control of cellular functions such as calcium signaling, synaptic plasticity, neuronal development, and neuro-immuno-regulation, which is somewhat relevant in neurodegenerative diseases like AD and PD [135].

Since calcium regulates neurotransmission, synaptic plasticity, and long-term potentiation all processes essential for learning and memory managing calcium homeostasis is a crucial part of vitamin D's role in the nervous system [135]. A known mechanism of neuronal damage in various neurodegenerative disorders, including delirium, AD, and PD, is a change in calcium signaling [136]. This is because intracellular accumulation of hazardous calcium levels can cause excitotoxicity, mitochondrial malfunction, and cell death [137]. To accurately control intracellular calcium flux in neurons, vitamin D action modifies the expression of calcium-binding proteins, voltage-gated calcium channels, and neurotransmitter release mechanisms [138]. In order to prevent excitotoxicity from damaging synapses and to preserve the integrity of synaptic connections that depend on neurotransmission impacts, vitamin D regulates calcium homeostasis [139]. Vitamin D plays a crucial role in the degradation of neurons, including dopaminergic neurons in PD and cortical pyramidal cells in AD [9].

The immunomodulatory properties of vitamin D are equally critical, especially since it tackles chronic neuroinflammation, one hallmark of neurodegeneration [139]. Activated microglia and astrocytes damage nearby neurons by releasing inflammatory cytokines, such as interleukin-1, interleukin-6, and tumor necrosis factor-alpha, further exacerbating AD and PD [140]. Vitamin D represses pro-inflammatory cytokines, such as interleukin-6, tumor necrosis factor-alpha, and interferon-gamma, while promoting anti-inflammatory cytokines, such as interleukin-10. Furthermore, vitamin D can decrease overactive microglia and subsequent neuroinflammatory responses [141]. Vitamin D treatment has been shown to reduce activation of microglia induced by amyloid-beta—which would reduce neurotoxic inflammation effects—and protect dopaminergic neurons from inflammation [142]. Vitamin D can confer neuroimmune regulatory properties that can reduce the vicious cycle of oxidative stress and inflammation and the neurodegenerative processes of neurodegeneration [143].

Epidemiology supports a definitive link between vitamin D deficiency and neurodegeneration risk. Low serum levels of 25-hydroxyvitamin D are consistently related to poorer cognitive performance, increased rate of cognitive decline, and higher risk of AD [144]. Vitamin D deficiency is also associated with higher risk of PD. For example, there is evidence that individuals with chronically low vitamin D are significantly more likely to develop parkinsonism symptoms [145]. In addition, the severity of parkinsonism, including more severe motor impairments, more postural instability, and a more rapid progression of non-motor symptoms such as depression and sleep disturbances, is inversely correlated with vitamin D concentrations [146]. These relationships highlight that vitamin D has significance not only as a biomarker of disease progression but also as a modifiable risk factor [147].

Clinical investigations concerning the therapeutic use of vitamin D supplementation have yielded some positive, albeit sometimes inconsistent, outcomes. In AD, supplementation has been shown to improve cognitive performance, reduce amyloid burden, and improve synaptic function, predominantly in those who were low in vitamin D status on baseline testing [148]. In PD, findings from clinical trials indicate that vitamin D supplementation may slow the progression of motor symptoms, and reduce fall risk, likely due to combined effects on dopamine neuron survival, and muscle strength and balance control [146]. However, there is variability in clinical trials outcomes that may be attributable to differences in dose, vitamin D status at baseline, duration of intervention, and stage of disease [149]. Importantly, some studies would suggest that vitamin D may work best as an adjunct treatment, rather than an isolated intervention, potentially complementing pharmacological therapies by providing resilience to neurons and reducing treatment complications associated with the disease [150].

Although there are many opportunities to further optimize vitamin D as a therapeutic option in neurodegenerative diseases, there are also limitations. There are a variety of factors that could affect how responsive a person will be to vitamin D supplementation, including genetic polymorphisms in the VDR gene, interindividual differences in vitamin D metabolism, and variability in BBB transport. In addition, excessive vitamin D may lead to hypercalcemia and vascular calcification [151]. Therefore, doses of vitamin D need to be carefully calibrated in long-term trials. Currently available evidence seems to suggest that maintaining a minimal physiological level compared to high pharmacological doses is likely the most beneficial and safest approach [152].

Vitamin D has multi-faceted neuroprotective effects that directly oppose the main pathogenic mechanisms of AD and PD, including the dysregulation of calcium, inflammation of the nervous system, and synaptic compromise [153]. Epidemiologically, deficiency of vitamin D is highlighted as a consistent risk factor, while clinical evidence has demonstrated the efficacy of vitamin D as an adjunct therapy in producing positive cognitive and motor outcomes [154]. While it is evident that additional large-scale and longer trials are necessary to refine dosage strategies and assess efficacy across the various stages of the disease, the literature increasingly suggests vitamin D is an important nutritional factor with both mechanistic and therapeutic implications in neurodegeneration. Making sure your vitamin D intake is an easy and practical way to support overall brain health and also possibly lessen the burden of AD and PD [155].

11.8 VITAMIN E (TOCOPHEROLS AND TOCOTRIENOLS)

Vitamin E, a collective name for a series of fat-soluble compounds, containing tocopherols and tocotrienols, is one of the most powerful natural antioxidants in the body [156]. Of these, α-tocopherol has received the most attention; however, recent studies suggest tocotrienols also have distinct neuroprotective properties [157]. The importance of vitamin E in the nervous system is due primarily to its effects on oxidative stress, membrane stabilization, and modulation of signaling pathways involved in neuronal survival [49]. The human brain is rich in polyunsaturated fatty acids and uses a large proportion of oxygen, meaning it is particularly vulnerable to oxidative damage [158]. Lipid peroxidation of neuronal membranes is well-described in both AD and PD; thus vitamin E is a key nutritional factor with both mechanistic and therapeutic relevance [62].

The antioxidant properties of vitamin E are almost entirely due to the direct oxidation of vitamin E by lipid peroxyl radicals and by terminating lipid peroxidation chain reactions [159]. In addition to stabilizing free radicals, vitamin E enhances the integrity of myelin sheaths, mitochondrial structures, and neuronal cell membranes [160]. This helps preserve neural communication and cellular function. In addition to regulating the development of oxidative stress-fighting enzymes like SOD and GSH, which support endogenous redox processes and defense systems, vitamin E specifically inhibits the loss of cellular function brought on by lipid peroxidation [161]. Through its control of enzymes, vitamin E may both scavenge oxidative damages and promote resilience against them. Since oxidative stress is an initiator and amplifying driver of amyloid-beta and tau hyperphosphorylation aggregation and/or dopaminergic neuronal death, enzymatic capability is crucial for neuropsychiatric disorders and in a neurodegenerative environment [162].

Studies show that vitamin E has neuroprotective benefits in relation to AD's tau and amyloid pathologies. According to experimental research, vitamin E reduces amyloid-beta toxicity through a likely two-pronged mechanism that inhibits oxidative changes that heighten aggregation and facilitates clearance through enzymatic routes (e.g., neprilysin activity). Vitamin E decreases tau hyperphosphorylation as likely a function of regulation on kinase and phosphatase activities, which maintains cytoskeletal stability and axonal transport [163]. Vitamin E also protects synaptic membranes from lipid peroxidation in addition to stabilizing anti-aggregation effects; synaptic function is key to memory and cognition [164]. In transgenic mice models of AD, vitamin E supplementation has been shown to decrease amyloid pathology motor deficits, resulting in favorable degrees of spatial memory as assessed by Morris water maze and tendon activity to some extent [165].

Clinical research has provided further corroboration for vitamin E in the context of AD, although the results can be somewhat heterogeneous. The initial observational studies show a relationship between a higher dietary intake or plasma concentrations of vitamin E with a decreased risk of cognitive decline and the incidence of AD [166]. The most notable clinical trials, such as the AD Cooperative Study, demonstrated that high doses of supplemental α-tocopherol could delay functional decline for patients with moderate cases of AD, especially for basic activities of daily living. Unfortunately, other clinical trials have not demonstrated significant cognitive benefits [167]. This variability needs to be viewed in the context of vitamin E's therapeutic action and how complex it can be based on several factors, such as the patient's stage of the disease, the dosage of vitamin E administered, and the bioavailability of vitamin E based on which forms of vitamin E were used in the intervention. Interestingly, although less literature has documented tocotrienols as a source of vitamin E, there is emerging evidence that tocotrienols may have superior neuroprotective effects as indicated by the fact that tocotrienols are better distributed in the brain and may produce additional beneficial anti-inflammatory responses [168].

In PD, oxidative stress contributes to dopaminergic neuronal loss; in particular, the metabolization of dopamine in the substantia nigra produces high levels of ROS [36]. Vitamin E can protect against oxidative injury to dopaminergic neurons and also preserve some function of mitochondria when effective antioxidant mechanisms are overwhelmed [169]. There is animal evidence that vitamin E is protective due to its antioxidant capacity in preclinical parkinsonian models influenced by neurotoxins including MPTP and 6-hydroxydopamine [126]. These models may decrease lipid peroxidation and also improve motor function. There is also a recent interest in the potential effects of tocotrienols that appears to promote dopaminergic in some studies or decrease neuroinflammation, as tocotrienols have better antioxidant capacity than vitamin E [170]. All of these antioxidant approaches may enhance dopaminergic cells through different mechanisms.

Clinical studies exploring vitamin E supplementation in PD have produced inconsistent results. The large DATATOP study reported no evidence of delaying the introduction of levodopa therapy in early Parkinson's patients using α-tocopherol supplementation, which is part of the basis for skepticism about its role in the management of PD [171]. On the other hand, secondary analyses and smaller studies have shown some benefit in balance and motor indices and quality of life with the addition of vitamin E with small amounts of other antioxidants or neuroprotective agents. It is also now gaining traction that vitamin E may exhibit synergism with coenzyme Q10, vitamin C, and/or polyphenols, which may enhance its therapeutic effect in the multifactorial pathophysiology of PD [172].

Vitamin E is an important antioxidant protector in the brain: it is involved in the prevention of oxidative stress, modulation of amyloid and tau pathology, and protection of dopaminergic neurons [123]. While preclinical data provide compelling evidence of the neuroprotective effect of vitamin E, clinical data are more difficult to interpret and the results inconclusive. The variation in clinical outcomes likely arises from variations in dose, type of vitamin E, and individual differences across patients. More importantly, the incorporation of tocopherols and tocotrienols may provide greater neuroprotection than α-tocopherol alone, which could be further verified clinically [173]. A simple way to maintain adequate vitamin E is to consume snacks and meals that contain nuts, legumes, and whole grains, as well as dietary oils and soybeans. Nutritional supplementation with vitamin E will be a promising aspect of a multidimensional therapeutic approach in the management of AD and PD [174].

11.9 EMERGING INSIGHTS: OTHER VITAMINS AND COFACTORS

While vitamins A, B-complex, C, D, and E have been studied for their potential impact on neurodegenerative disease, other micronutrients and cofactors like vitamin K, niacin (vitamin B3), and biotin have received limited attention. These often neglected compounds are important for brain metabolism, neurotransmission, mitochondrial homeostasis and neuroprotection [154]. Their emerging importance in AD and PD calls for a broadened view of nutritional neuroscience beyond the vitamins historically analyzed, as deficiencies or dysregulation of these nutrients may be instrumental in the onset or progression of the disease.

Most recognized for its role in blood clotting, vitamin K has recently received attention for its functions in the central nervous system [175]. Vitamin K is involved in co-translational protein modification through γ-carboxylation, which makes calcium binding possible and maintains homeostasis [176]. Importantly, vitamin K regulates and is involved in the biosynthesis of sphingolipids, vital components of neuronal membranes critical for synaptic integrity, myelin structure, and cell signaling [177]. There is now abundant evidence associating cognitive decline with vitamin K deficiency and impaired memory performance, with observational studies linking lower dietary vitamin K intake to increased risk of AD [178]. The mechanisms of how vitamin K influences brain function include reducing oxidative stress and neuroinflammation in part through the activation of proteins, including Gas6 and protein S. Gas6 and protein S signal a series of events that influence common cell survival pathways and inhibit inflammatory cascades in the brain [179]. Additional preclinical studies suggest that vitamin K is also directly related to contributing to mitochondrial function and can also act in mitigating excitotoxicity that is heavily implicated in both Alzheimer's and Parkinson's pathology [180]. While clinical evidence supporting the role of vitamin K in cognitive health and neuroprotection remains limited, these mechanistic perspectives suggest that vitamin K may play an important role in cognitive health and warrants further investigation.

Niacin or vitamin B3 is another nutrient that is gaining attention for its possible role in neurodegenerative diseases [15]. Niacin is a precursor for nicotinamide adenine dinucleotide (NAD+), which is involved in fundamental cellular processes like energy metabolism, mitochondrial respiration, and DNA repair [181]. In the brain aging process, NAD+ levels decrease, which deteriorates mitochondrial function, downregulates sirtuin activity, and increases susceptibility toward oxidative stress, all of which are characteristic of AD and PD [182]. Increased availability of NAD+ through supplementation of niacin or its derivatives can elicit sirtuin pathways restoration in support of neuronal survival, increased mitochondrial biogenesis, and attenuation of neuroinflammation [183]. Mouse models of AD have shown that increasing NAD+, improves synaptic plasticity, reduces accumulation of amyloid-beta, preserves cognition [123]. In mouse models of PD, niacin supplementation has been correlated with increased dopaminergic neuronal survival, changes in alpha-synuclein aggregation, and improved motor behaviors [184]. Clinical studies are still in their infancy but several pilot studies had existential benefits of biological active NAD+ precursors, nicotinamide riboside and nicotinamide mononucleotide, which are NAD+ precursors on mitochondrial health and progression toward neurodegeneration. Since NAD+ metabolism is a central mechanism that leads to cellular resilience, niacin remains a mechanism-driven substantial candidate for dietary nutritional intervention in neurodegenerative disorders [185].

As a water-soluble vitamin of the B-complex, biotin, sometimes referred to as vitamin B7, serves as a cofactor for carboxylase enzymes that are involved in the creation of fatty acids, the breakdown of amino acids, and the production of gluconeogenesis [186]. The fact that biotin is mostly associated with metabolic and skin/nail health suggests that its neurophysiologic potential is gaining traction. The integrity of myelin and its role in glia and neuron energy production are only two of the many vital roles that biotin plays in nervous system function [187]. The field of biotin supplementation has recently been pushed to find more practical treatments for a number of progressing neurological conditions, most notably neurological illnesses. There is a small but encouraging body of research on AD and PD. Maintaining appropriate biotin levels may be crucial for brain health, as biotin deficiency has also been linked to cognitive impairment, unexplained fatigue, and neuropathy symptoms [188]. Biotin supplementation may offer some potential for improved neural resilience to oxidative stress in animal research, as well as for mitochondrial stability and the improvement of homeostasis at one or more neurotransmitter sites [189]. There are a number of variations in the functional outcomes of these investigations, which seem to be primarily observations with minimal clinical intervention with AD and PD patients. Future study to offer biotin behavior as a supplementary nutrition therapy may be justified by the substantial molecular similarities for dysregulation with these disorders.

The growing field of vitamin-driven neuroprotection will undoubtedly continue to be illuminated by these new discoveries. Although modal mechanics that impact mitochondrial health, synaptic properties, neurotransmitter occupancy/expression, and cellular resilience play a significant mechanistic role on the reversal features of the pathophysiology of neurodegenerative diseases, vitamins K, B3, and biotin are not typically in the news when discussing micronutrients [190]. Notably, vitamins K, B3, and biotin seem to influence cellular mechanisms that support neuronal survival and work in concert with other vitamins that influence mechanistic pathways (such as vitamins C, D, and E), suggesting both possible therapeutic benefits and potential side effects [48].

Clearly, testing of the efficacy, safety, and dosing regimen of vitamin K, niacin, and biotin supplementation in AD and PD need to be tested systematically [191]. Also, given the multifactorial nature of such diseases, the complex interactions between vitamins and diet, as well as gene-determined variants, and their effect on pharmacotherapy too should be examined [192]. These approaches will likely help in achieving maximum therapeutic response. Finally, considering the neuroprotective effects of many of these vitamins that are not frequently recognized may help open new fields and opportunities within nutritional neuroscience that permit and propel every patient to advance activity as they delay, lessen, and/or possibly modify the manner of progression of neurodegenerative disorders (Table 11.2).

TABLE 11.2

Emerging Roles of Other Vitamins and Cofactors in Neurodegeneration

Vitamin	Physiological Role in the Nervous System	Neuroprotective Mechanisms	Evidence in AD and PD	References
Vitamin K	Regulates synthesis of sphingolipids; supports neuronal membrane integrity and synaptic function	Anti-apoptotic effects, regulation of oxidative stress, contribution to cognitive function	Low vitamin K levels correlated with cognitive decline and AD progression; emerging links with PD remain under investigation	[193]
Niacin (vitamin B3)	Precursor of NAD^+ and $NADP^+$; essential for mitochondrial energy metabolism and DNA repair	Enhances mitochondrial function, improves cellular bioenergetics, modulates neuroinflammation	NAD^+ boosting strategies (e.g., nicotinamide riboside) show potential in AD and PD models by improving neuronal survival and function	[194]
Biotin (vitamin B7)	Cofactor for carboxylases involved in fatty acid synthesis and energy metabolism; maintains myelin integrity	Promotes myelination, supports axonal health, regulates gene expression via biotinylation	High-dose biotin explored in neurodegenerative and demyelinating disorders; limited but promising evidence in cognitive decline and PD	[188]

(Continued)

TABLE 11.2 (*Continued*)

Emerging Roles of Other Vitamins and Cofactors in Neurodegeneration

Vitamin	Physiological Role in the Nervous System	Neuroprotective Mechanisms	Evidence in AD and PD	References
Coenzyme Q10 (ubiquinone)	Component of the mitochondrial electron transport chain; antioxidant defense	Reduces oxidative stress, improves ATP production, stabilizes mitochondrial function	Clinical trials in PD show modest benefits on motor function; role in AD remains under study	[195]
Alpha-lipoic acid	Cofactor in mitochondrial enzyme complexes; regenerates other antioxidants (Vitamin C, E, glutathione)	Potent antioxidant, improves glucose metabolism, reduces neuroinflammation	Shown to slow cognitive decline in AD patients; experimental evidence supports neuroprotection in PD	[196]
L-carnitine	Facilitates transport of fatty acids into mitochondria for β-oxidation	Enhances mitochondrial energy metabolism, reduces oxidative stress	Acetyl-L-carnitine supplementation associated with cognitive improvement in AD and mild cognitive impairment	[197]

11.10 SYNERGISTIC AND COMBINED EFFECTS OF VITAMINS

As individuals, vitamins each have a specific neuroprotective mechanism of action attributable to antioxidant, anti-inflammatory, or metabolic effects, yet, a growing body of research indicates that combined and/or synergistic effects may have more influence on neurodegeneration than single agents [64]. This is not surprising given the multifactorial pathophysiology of both AD and PD is concurrent interference in oxidative balance, mitochondrial function, protein aggregation, neuroinflammation, and neurotransmitter balance; since no single compound can potentially interfere with all concurrent processes, taking multivitamins, in conjunction with phytochemicals and minerals, and in combination therapies with pharmacological interventions, may create the most holistic neuroprotective approach [198].

The idea of multivitamin supplementation in neurodegeneration is built on the premise that different vitamins will hit complementary molecular pathways [199]. For instance, vitamin E, which is a lipid-soluble antioxidant, acts to protect the neuronal membrane from lipid peroxidation, while vitamin C, which is a water-soluble antioxidant, will regenerate oxidized vitamin E and thus keep its antioxidant potency [117]. Vitamins such as folate, vitamin B6, and vitamin B12 will control homocysteine metabolism, reducing the risk of vascular damage and excitotoxicity [200]. Likewise, niacin will make more NAD+ available to support DNA repair molecules and mitochondrial activity. Vitamin K would promote sphingolipid metabolism and synaptic integrity, while vitamin D, which has immunomodulatory qualities, will improve calcium homeostasis and reduce neuroinflammation [175]. These vitamins offer complementary and overlapping protection against the many different pathogenic insults that are common in neurodegenerative diseases. Although some of the evidence is conflicting, possibly due to formulation, dosage, and research design, clinical trials have indicated that taking multivitamins may be linked to decreased cognitive decline, slower disease progression, and improved quality-of-life outcomes [201].

Another important area of research that has gained traction is the interaction of vitamins with minerals and phytochemicals. Minerals including iron, magnesium, zinc, and selenium are essential for enzymes, antioxidant defense, and neural communication [202]. For instance, zinc and B vitamins play roles in both DNA methylation and neurotransmitter synthesis, while selenium-dependent glutathione peroxidases cannot respond fully to oxidative damage without the assistance of vitamins E and C [203]. Unless the metabolism of these various micronutrients is impaired as in most neurodegenerative diseases, the utilization of combined supplements can restore equilibrium and increase neuronal resilience. Similarly, phytochemicals, especially polyphenols, flavonoids, and carotenoids, often enhance the antioxidant and anti-inflammatory effects of vitamins. For example, curcumin, resveratrol, and quercetin have been reported to enhance the effect of vitamin C and E in limiting the activity of ROS, countering amyloid burden, and modulating inflammatory signal transduction pathways [204]. Lutein and β-carotene carotenoids act to stabilize neuronal membrane structure in conjunction with fat-soluble vitamins A, D, and E against free-radical and light-activated oxidative injury [205]. These combined uses of vitamins with minerals and phytochemicals likely create a holistic antioxidant and neuroprotective sieve, which may be more potent than supplementation as single entities to help delay or ameliorate the course of neurodegeneration [206].

An additional significant dimension in the synergy of vitamins involves enhancing conventional pharmacological therapies for AD and PD. Current pharmacotherapy for AD, which includes acetylcholinesterase inhibitors or the NMDA receptor antagonist, or the use of levodopa and dopamine agonists for PD, only pharmacodynamically relieve or treat neurological symptoms of dementia or parkinsonism but do not modify either chronic condition [207]. It is theoretically possible that nutritional interventions with vitamins would assist pharmacotherapy with benefits or lessen side effects. There are suggestions. Vitamin D may promote the health of dopamine neurons or increase sensitivity to levodopa [208]. Furthermore, antioxidants (such as

TABLE 11.3
Synergistic and Combined Effects of Vitamins in Neuroprotection

Combination	Mechanistic Synergy	Implications in AD	Implications in PD	Clinical Insights	References
Vitamin C+vitamin E	Vitamin C regenerates oxidized Vitamin E; dual protection in aqueous and lipid compartments; enhanced antioxidant defense.	Reduces amyloid-induced oxidative damage and lipid peroxidation in neuronal membranes.	Protects dopaminergic neurons from oxidative stress and improves mitochondrial stability.	Combined supplementation shows better protection than individual vitamins.	[64]
B6+B9 (folate)+B12	Act together in one-carbon metabolism; regulate DNA methylation, neurotransmitter synthesis, and homocysteine clearance.	Lowers homocysteine, improves cognitive function, and slows brain atrophy.	Reduces L-DOPA-induced hyperhomocysteinemia, protecting neurons.	Clinical trials suggest combined supplementation more effective than single vitamins.	[213]
Vitamin D+vitamin K	Joint regulation of calcium homeostasis and neuronal signaling; reduction of vascular calcification.	Supports synaptic health and reduces risk of cognitive impairment.	Prevents neuronal calcium overload and enhances neuroprotection.	Suggested as a dual-supplementation approach for brain and vascular health.	[214]
Vitamin D+polyphenols (e.g., resveratrol, curcumin)	Synergistic anti-inflammatory and immunomodulatory effects; improved antioxidant response.	Reduces neuroinflammation and amyloid burden.	Attenuates microglial activation and protects motor neurons.	Represents an integrative diet–nutrient approach.	[9]
Vitamin E+cholinesterase inhibitors	Vitamin E reduces oxidative stress, complementing pharmacological cholinergic enhancement.	Enhances cognitive outcomes in AD patients receiving AChE inhibitors.	Limited evidence, but may improve neuronal survival alongside dopaminergic therapy.	Clinical studies show additive benefits in AD.	[215]
B-vitamins+L-DOPA	B6, B9, and B12 prevent L-DOPA-induced homocysteine accumulation, reducing toxicity.	Indirect benefit by protecting vascular health in dementia with Parkinsonism.	Improves safety and tolerability of long-term L-DOPA therapy.	Important in clinical management of PD patients on chronic medication.	[213]

vitamins E and C) may reduce oxidative stress, especially when they are up against serious oxidative damage caused by dopamine metabolism [169]. In a similar vein, niacin and other NAD+ analogs may help patients receiving medication therapy by inducing mitochondrial trophism, improving cellular energy metabolism, and maybe enhancing naïve cellular resilience [209]. Vitamins and medication therapies may also result in lower dosages of medications, which may make treating drug-induced dyskinesias in patients easier [210]. This would lessen the negative side effects that one may expect from long-term Lewy body dementia treatment.

Although combined vitamin methods have potential, their application in clinical settings presents difficulties. Certain vitamins' bioavailability depends on dietary fat consumption and intersubject metabolic variability, especially for the fat-soluble vitamins A, D, and E [211]. Furthermore, since the pro-oxidant activity of many vitamins, particularly vitamins E and A, increases with concentration, taking too much of them is concerning [68]. This is particularly concerning for establishing safe and effective concentrations. Genetic factors that can influence an individual's response to supplementing include variations in vitamin D receptors, folate metabolism genes, or NAD+ biosynthesis pathways [212]. This supports the idea that neurodegenerative illnesses can be treated with precise nutrition (Table 11.3).

11.11 CLINICAL EVIDENCE AND TRANSLATIONAL PERSPECTIVES

One of the primary challenges in treating AD and PD is turning new scientific discoveries about vitamin neuroprotection into meaningful therapeutic advantages [216]. Experimental research has consistently demonstrated the antioxidative, anti-inflammatory, and neuromodulatory qualities of vitamins; nevertheless, clinical research has been uneven and has often produced contradictory results. Within the framework of the complexity of neurodegenerative diseases, with their many interacting pathways, heterogeneous patient types, and lengthy disease duration that can affect effects, the divergent opinions show a relationship and disconnect between preclinical promise and clinical reality [217]. We may help minimize the hype and optimism of incorporating vitamin-based treatments into best practices by confining ourselves to the clinical data obtained from observational studies, randomized controlled trials, and interventional research [218].

High food intake or blood levels of vitamins, including B-complex, C, D, and E, are frequently associated with a lower risk of cognitive impairment and dementia, according to epidemiological research [219]. For instance, deficits in folate, vitamin B12, and vitamin D are consistently linked to greater homocysteine levels, a higher prevalence of cognitive decline, and a quicker progression to Alzheimer's dementia, according to large population-based cohorts [220]. Higher consumption of antioxidants, such as vitamin E and C, for instance, is linked to a lower risk of AD, particularly in communities that follow what is known as a rich plant food or Mediterranean diet. Results from interventional trials have been less reliable. For example, trials supplementing B vitamins, mainly folate, B6, and B12, have demonstrated efficacy in lowering homocysteine; nonetheless, it has not always improved cognitive performance, indicating that changing the course of the disease, once the disease pathology is established, is less likely to be changed solely by decreasing homocysteine [221]. Although it still doesn't always improve memory outcomes, vitamin E administration has been shown to have prospective benefits in slowing functional decline in those with mild-to-moderate AD. The large dosages that may have long-term effects are a cause for caution. In part, additive evidence that links vitamin D deficiency to amyloid deposition and neuroinflammation raised awareness and interest in the supplement. However, because clinical trials are typically underpowered and yield inconclusive results, this second level of questioning evidence is not conclusive [222].

The outcomes of clinical trials on vitamin supplementation for PD have also been inconsistent. Increased consumption of the antioxidant vitamins C and E in the diet is associated with a lower incidence of PD, according to observational studies [223]. According to longitudinal studies, people with higher plasma vitamin D had a lower chance of acquiring PD and a slower rate of disease progression. This suggests that vitamin D has immunomodulatory and neuroprotective properties [224]. A lower risk of PD was also linked to vitamin B6, possibly as a result of its role in homocysteine detoxification and dopamine metabolism. But controlled experiments showed a more nuanced picture. Vitamin E supplementation, a plausible suspect for neuroprotection in PD, did not significantly alter the course of the disease in two large clinical trials [225]. Contrary to previous preclinical model results, vitamin C supplementation has not demonstrated a clear protective impact. More recently, vitamin D supplementation has been studied as an adjuvant; some studies have found no benefit, while others have shown improvements in motor function and balance. Because of its role in NAD+ metabolism, mitochondrial function, and anti-inflammatory signaling, niacin supplementation is also noteworthy. Early-stage clinical trials suggest that niacin may help with motor symptoms and exhaustion in PD [226]. Collectively, these results encourage us to think about including vitamin bioavailability, illness stage concerns, and interactions with pharmaceutical interventions into clinical trial design [227].

There are several important limitations and inconsistencies that demonstrate the barriers to translating vitamin supplementation to neurodegenerative diseases [56]. First, methodological heterogeneity is a significant obstacle to translating nutrition to brain health due to variation in dose, formulation, duration of supplementation, and the populations studied. Second, many clinical trials are underpowered and fail to account for the prolonged latency of the pathologies associated with neurodegenerative diseases where progressive and irreversible neuronal injury occurs in an asymptomatic phase prior to clinical symptoms [228]. Studies of vitamin supplementation that are initiated after substantial neuropathological development may fail because of the high degree of neuronal injury that occurred prior to any interventions; this may relate to much of the observational studies demonstrating stronger protective associations with brain health than interventional trials. Third, vitamin bioavailability, absorption, and pharmacokinetics can vary significantly between individuals and are affected by diet [229], gastrointestinal route of absorption, liver and central nervous system metabolism, possible interactions with other nutrients, and genetic polymorphism of each vitamin or mineral supplement. For example, the methylenetetrahydrofolate reductase (MTHFR) gene influences the metabolism of folate and may modify the efficacy of B-vitamin supplementation in lowering homocysteine levels. Similarly, variants in vitamin D receptor genes may also influence individual response to vitamin D therapy under neurodegenerative states. Excessive intake of selective vitamins can be damaging. As previously stated, high doses of vitamin E supplements have been correlated with increased cardiovascular risk. These data provide reasonable support for dosing specificity and patient stratification in future applications [230].

Novel approaches in personalized nutrition and precision medicine may provide a complementary way to optimize vitamin-based interventions in the context of neurodegeneration [231]. Instead of embarking on a "cookie-cutter" supplementation model, precision approaches incorporate the impact of genetics, dietary habits, microbiome, and metabolic profiles when creating interventions. For example, people who have high levels of homocysteine and MTHFR polymorphisms may derive a greater cognitive benefit from targeted supplementation of folate and B-vitamins [232]. Similarly, patients who have VDR polymorphisms and/or chronic vitamin D deficiency may require Dr. Schmidt's anticipated doses of vitamin D to see effects. Multi-omics, including metabolomics, nutrigenomics, and epigenomics, will assist in establishing biomarkers of vitamin deficiency and help project patient subgroups that will benefit the most from interventions. Multi-omics may also assist

TABLE 11.4
Clinical Evidence and Translational Perspectives of Vitamins in AD and PD

Vitamin(s)	Study Type	Findings in AD	Findings in PD	Translational Barriers	References
Vitamin E	Randomized controlled trials (RCTs), observational studies	Slows functional decline, improves antioxidant defense; mixed evidence on cognition.	Reduces oxidative stress, modest benefits in motor symptoms.	High doses associated with bleeding risk; inconsistent clinical outcomes.	[201]
Vitamin D	Epidemiological studies, supplementation trials	Deficiency linked to increased AD risk; supplementation shows improved cognition in some trials.	Low levels associated with motor severity; supplementation may improve balance and mood.	Heterogeneity in dosage, baseline vitamin D status, and trial design.	[201]
B vitamins (B6, B9, B12)	Interventional studies, meta-analyses	Lower homocysteine and brain atrophy; improved cognition in MCI and early AD.	Mitigate L-DOPA-induced hyperhomocysteinemia; may protect dopaminergic neurons.	Variable efficacy depending on baseline deficiency; lack of large-scale long-term trials.	[237]
Vitamin C	Observational studies, supplementation	Higher plasma levels associated with reduced AD risk; antioxidant role in amyloid and tau pathology.	Protective against oxidative stress in dopaminergic neurons; supports dopamine synthesis.	Limited RCTs; bioavailability and dosing remain challenges.	[238]
Vitamin A/ retinoids	Preclinical, small human studies	Modulates amyloid and tau pathology; supports synaptic plasticity.	Enhances dopaminergic signaling; neuroprotective effects in animal models.	Limited human clinical evidence; potential toxicity at high doses.	[239]
Vitamin K	Emerging observational studies	Associated with better cognitive function and reduced dementia risk.	Potential role in neuronal calcium regulation and survival.	Sparse clinical data; requires translational validation.	[240]
Combinational therapy	Clinical trials, integrative interventions	Some studies suggest improved cognition and reduced progression when combined with antioxidants or standard therapies.	May improve quality of life and motor function in PD patients.	Confounding lifestyle factors; lack of standardized formulations.	[241]

in creating synergistic effects by implementing vitamin supplementation with pharmacotherapy, lifestyle changes, and phytochemical-rich diets [233]. Personalized nutrition approaches may therefore present opportunities to prevent and/or delay the onset of disease and/or improve drug efficacy in patients diagnosed with AD or PD [234].

There is reassurance, but also caution, when considering the clinical evidence on the potential roles of vitamins in AD and PD [235]. Observational studies demonstrate a consistent protective relationship between being sufficiently vitamin sufficient and reduced risk of neurodegeneration; however, interventional studies have not frequently shown similar outcomes in symptomatic individuals, thus highlighting the significance of trial timing, dosage, population variety, and methodology in future studies. Importantly, we intend that personalized and precision approaches measuring and observing vitamins levels in guided stratified patients through clinical trials may help realize the true potential of vitamins in neurodegenerative disease in the future [236]. We hope that vitamins may move into specific approaches in preventing and managing neurodegeneration and integrate vitamin therapy into established practice and proven neuroprotective approaches to provide a better opportunity for future patients (Table 11.4) [235].

11.12 FUTURE DIRECTIONS AND CHALLENGES

Although there has been significant progress in defining the mechanistic role of vitamins in AD and PD, converting this evidence into an actionable, practical therapy remains a daunting obstacle [242]. The multifactorial nature of neurodegeneration, variability in nutritional status between individuals, and the intricacy of clinical trial design contribute to variability in outcomes. In the next wave of research, there are several areas of focus, including a biomarker-driven approach, alternative delivery systems, holistic approaches to therapy, and translational barriers [243].

Precision medicine provides an opportunity to customize vitamin supplementation based on specific biomarkers of nutritional status, genetic susceptibility, and disease risk [236]. For example, an elevated homocysteine level in the plasma could lead to B vitamin supplementation, and similarly a low serum 25-hydroxyvitamin D concentration could inform vitamin D interventions. Further, metabolomics, lipidomics, and proteomics have contributed considerably to the development of novel biomarkers that could link vitamin availability to mitochondrial health, oxidative stress and inflammatory responses. Biomarker-driven strategies would ensure that supplementation is personalized and targeted, and more likely to confer clinical benefit by limiting individuals from over- to under-supplementing vitamins [244].

One limitation of traditional vitamin supplementation is the poor bioavailability, rapid metabolism, and difficulty crossing the BBB [245]. Nanotechnology provides a promising solution by enhancing the targeted delivery of vitamins, including vitamin E, vitamin D, and derivatives of the B-complex, directly to the brain. Certain nanoformulations, including liposomes, solid lipid nanoparticles, polymeric nanocarriers, and dendrimers, have shown improved vitamin stability, controlled release, better BBB permeability [246]. These nanotherapeutics may allow the administration of lower doses while exerting higher efficacy, limiting systemic side effects, and maximizing neuroprotective potential. Seeking out future studies that complement the therapeutic benefits of vitamins with nanomedicine may provide a revolutionary paradigm for treating AD and PD [247].

Given the complex pathology of AD and PD, it is unlikely that vitamins alone can yield sufficient disease modification. Evidence suggests that therapeutic directions that constitute a combination of dietary habits, lifestyle changes, and approved medicine provide greater modification of diseases. Some dietary habits of high levels of protein, polyphenols, and unsaturated fatty acids have been shown to yield less risk of neurodegeneration [248].

There are several gaps that continue to delay the clinical translation of vitamin-based research in the context of neurodegeneration [231]. First, studies involving long-term, large-scale, randomized controlled trials with standardized dosages and forms of vitamins have yet to be completed, making definitive conclusions difficult. Second, heterogeneity in patient populations will greatly affect the conclusions of study arms related to patients' genetic polymorphisms, chronic comorbidities, and environmental exposure variables. Third, potential safety issues arising from chronic high-dose vitamin supplementation must be duly addressed [249]. With limited knowledge about the beneficial bidirectional interaction of vitamins with gut microbiota, immune regulation, and epigenetic modification, there is an open-ended question to be addressed in these fields [250].

11.13 CONCLUSION

Vitamins are fundamentally involved in brain health and resistance to neurodegeneration, which is why they are mentioned in AD and PD. The last few decades have seen significant advancements in molecular neuroscience, which have demonstrated that pathogenic cascades are caused by vitamin imbalances or deficiencies. Beyond the traditional roles that vitamins play in metabolism, mechanistic studies have demonstrated that they are also signaling molecules, neuroprotectants, and epigenetic regulators.

Vitamins control all facets of neuronal survival, according to mechanistic studies. Vitamin A and its derivatives influence tau and amyloid disease and are essential for neurodevelopment. Energy metabolism, methylation, homocysteine balance, and myelin integrity all depend on B vitamins.

Vitamin C plays a significant role as an effective antioxidant and as a cofactor for dopamine biosynthesis. Vitamin D impacts calcium homeostasis and has immunomodulatory roles. Vitamin E is a lipid-soluble antioxidant that prevents the peroxidation of lipids in neuronal membranes to prevent amyloid and tau toxicity. There are also emerging functions for vitamin K, niacin, and biotin that enhance the case for vitamins being broader in their cofactor roles in neuroprotection. These avenues of research suggest that vitamins may be multi-targeted modulators of the pathophysiology of AD and PD.

Vitamins have the potential to treat AD and PD because they can alter multiple disease processes simultaneously. Although vitamin supplementation has been shown to reduce oxidative stress, improve cognition, and enhance dopaminergic signaling in clinical trials, the results have been inconsistent, primarily because of variations in study designs, dosing, bioavailability, participants, and disease stage at the start of the intervention. Additionally, the considerable patient variability and possibly poor brain penetration of conventional supplements are drawbacks. This suggests that new delivery methods, such as nanocarriers, are required. To address safety issues related to long-term high-dose supplementation, the therapeutic window that maximizes benefit must be quantified.

It will be necessary to combine an integrative and precision-based approach in future research. Interventions could use nanotechnology to deliver vitamins to the brain and target tissues, or they could use biomarkers to develop personalized supplementing regimens. To treat the complex conditions of AD and PD simultaneously, a combination of dietary and vitamin changes, lifestyle modifications, and medication therapy may be required to achieve additive or synergistic results. In order to connect and integrate mechanistic knowledge into practical healthcare prospects, integrative solutions that bridge fields such as systems biology, clinical pharmacology, neuroscience, and nutrition are being called for.

At the intersection of diet and neurodegeneration, vitamins play a critical role. They represent nutritional neuroscience's best chance to update AD and PD preventative and treatment strategies. However, in order to achieve this aim, current gaps must be filled by innovative approaches, teamwork, and precision in clinical and research procedures. Using the intersection of lifestyle, medication, and vitamins to create holistic or whole brain approaches to neurodegenerative illness and brain health represents a path forward.

FUNDING

None.

CONSENT TO PUBLICATION

All authors have read the manuscript and agree to submit the chapter to the CRC press.

CONFLICT OF INTEREST

None.

REFERENCES

1. Marcos-Rabal, P. et al., Neurodegenerative diseases: A multidisciplinary approach. *Current Pharmaceutical Design*, 2021. **27**(30): p. 3305–3336.
2. Kagawad, P., P. Bhandurge, and S. Gharge, *Diagnosis and Management of Neurodegenerative Diseases*, in T. S. Koduru, R. A. M. Osmani, E. Singh, and S. Dutta (eds.), *The Neurodegeneration Revolution*. Elsevier, 2025. p. 101–114.
3. Memudu, A.E., B.A. Olukade, and G.S. Alex, Neurodegenerative diseases: Alzheimer's disease. Integrating neuroimaging, in I. Chatterjee and N. Moradikor (eds.), *Computational Neuroscience, and Artificial Intelligence*. CRC Press, 2024. p. 128–147.
4. Lamptey, R.N. et al., A review of the common neurodegenerative disorders: Current therapeutic approaches and the potential role of nanotherapeutics. *International Journal of Molecular Sciences*, 2022. **23**(3): p. 1851.
5. Rae, C.D. et al., Brain energy metabolism: A roadmap for future research. *Journal of Neurochemistry*, 2024. **168**(5): p. 910–954.
6. Tourkochristou, E., C. Triantos, and A. Mouzaki, The influence of nutritional factors on immunological outcomes. *Frontiers in Immunology*, 2021. **12**: p. 665968.
7. Aragão, M.Â. et al., Revitalising riboflavin: unveiling its timeless significance in human physiology and health. *Foods*, 2024. **13**(14): p. 2255.
8. Mohan, A. et al., Homocysteine, vitamin B12 and folate level: Possible risk factors in the progression of chronic heart and kidney disorders. *Current Cardiology Reviews*, 2023. **19**(4): p. 66–83.
9. Savran, Z. et al., Vitamin D and neurodegenerative diseases such as multiple sclerosis (MS), Parkinson's disease (PD), Alzheimer's disease (AD), and amyotrophic lateral sclerosis (ALS): A review of current literature. *Current Nutrition Reports*, 2025. **14**(1): p. 77.
10. Yang, N. et al., Antioxidants targeting mitochondrial oxidative stress: Promising neuroprotectants for epilepsy. *Oxidative Medicine and Cellular Longevity*, 2020. **2020**(1): p. 6687185.
11. Foula, W.H. and W.M. Foad, Human nutritional neuroscience: Fundamental issues, in A. Kumar and S. Ramesh (eds.), *Nutrition and Psychiatric Disorders: An Evidence-Based Approach to Understanding the Diet-Brain Connection*. Springer, 2024. p. 49–67.
12. Mallah, K. et al., Anti-inflammatory and neuroprotective agents in clinical trials for CNS disease and injury: Where do we go from here? *Frontiers in Immunology*, 2020. **11**: p. 2021.
13. Casanova, A. et al., Mitochondria: It is all about energy. *Frontiers in Physiology*, 2023. **14**: p. 1114231.
14. Han, J. et al., Parkinson's disease dementia: Synergistic effects of alpha-synuclein, tau, beta-amyloid, and iron. *Frontiers in Aging Neuroscience*, 2021. **13**: p. 743754.
15. Kumar, R.R. et al., Role of vitamins in neurodegenerative diseases: A review. *CNS & Neurological Disorders – Drug Targets*, 2022. **21**(9): p. 766–773.
16. Lopes, M. et al., Food supplement vitamins, minerals, amino-acids, fatty acids, phenolic and alkaloid-based substances: An overview of their interaction with drugs. *Critical Reviews in Food Science and Nutrition*, 2023. **63**(19): p. 4106–4140.
17. Alamgir, A., Vitamins, nutraceuticals, food additives, enzymes, anesthetic aids, and cosmetics, in S. K. Sharma and R. K. Verma (eds.), *Therapeutic Use of Medicinal Plants and their Extracts: Phytochemistry and Bioactive Compounds*. Springer, 2018. Vol. 2, p. 407–534.
18. Guest, N.S. et al., International society of sports nutrition position stand: Caffeine and exercise performance. *Journal of the International Society of Sports Nutrition*, 2021. **18**(1): p. 1.
19. Jafari, R.S. and V. Behrouz, Nordic diet and its benefits in neurological function: A systematic review of observational and intervention studies. *Frontiers in Nutrition*, 2023. **10**: p. 1215358.
20. Guasch-Ferré, M., H.S. Dashti, and J. Merino, Nutritional genomics and direct-to-consumer genetic testing: an overview. *Advances in Nutrition*, 2018. **9**(2): p. 128–135.
21. Migliore, L. and F. Coppedè, Gene–environment interactions in Alzheimer disease: The emerging role of epigenetics. *Nature Reviews Neurology*, 2022. **18**(11): p. 643–660.
22. Aarsland, D. et al., Parkinson disease-associated cognitive impairment. *Nature Reviews Disease Primers*, 2021. **7**(1): p. 47.
23. Cenini, G., A. Lloret, and R. Cascella, Oxidative stress in neurodegenerative diseases: From a mitochondrial point of view. *Oxidative Medicine and Cellular Longevity*, 2019. **2019**(1): p. 2105607.
24. Brown, R.A., Psychoneuroimmunological implications of growing nutrient deficiencies and imbalances: Omegas, iodine, vitamin D, minerals, in N. Rezaei and N. Yazdanpanah (Eds.), *PsychoNeuroImmunology: Integration of Psychology, Neurology, and Immunology*. Springer, 2025. Vol. 1, p. 247–329.
25. Bai, R. et al., Oxidative stress: The core pathogenesis and mechanism of Alzheimer's disease. *Ageing Research Reviews*, 2022. **77**: p. 101619.
26. Metcalfe, N.B. and M. Olsson, How telomere dynamics are influenced by the balance between mitochondrial efficiency, reactive oxygen species production and DNA damage. *Molecular Ecology*, 2022. **31**(23): p. 6040–6052.
27. Hassan, S.S.u. et al., The neuroprotective effects of fisetin, a natural flavonoid in neurodegenerative diseases: Focus on the role of oxidative stress. *Frontiers in Pharmacology*, 2022. **13**: p. 1015835.
28. Borbolis, F., C. Ploumi, and K. Palikaras, Calcium-mediated regulation of mitophagy: Implications in neurodegenerative diseases. *npj Metabolic Health and Disease*, 2025. **3**(1): p. 4.
29. Yang, Q.Q. and J.W. Zhou, Neuroinflammation in the central nervous system: Symphony of glial cells. *Glia*, 2019. **67**(6): p. 1017–1035.
30. Heloterä, H. and K. Kaarniranta, A linkage between angiogenesis and inflammation in neovascular age-related macular degeneration. *Cells*, 2022. **11**(21): p. 3453.
31. Kumar, P., Intersections of heavy metal toxicity, protein misfolding, and neurodegenerative disorders in humans, in N. Gogia, S.K. Singh, and V.D. Jaganath (eds.), *Protein Misfolding in Neurodegenerative Diseases*. Elsevier, 2025. p. 413–461.
32. Madhu, P. and S. Mukhopadhyay, Distinct types of amyloid-β oligomers displaying diverse neurotoxicity mechanisms in Alzheimer's disease. *Journal of Cellular Biochemistry*, 2021. **122**(11): p. 1594–1608.
33. Nisa, F.Y. et al., Role of neurotoxicants in the pathogenesis of Alzheimer's disease: A mechanistic insight. *Annals of Medicine*, 2021. **53**(1): p. 1479–1504.

34. Cario, A. and C.L. Berger, Tau, microtubule dynamics, and axonal transport: New paradigms for neurodegenerative disease. *Bioessays*, 2023. **45**(8): p. 2200138.
35. Chen, Z.-R. et al., Role of cholinergic signaling in Alzheimer's disease. *Molecules*, 2022. **27**(6): p. 1816.
36. Trist, B.G., D.J. Hare, and K.L. Double, Oxidative stress in the aging substantia nigra and the etiology of Parkinson's disease. *Aging Cell*, 2019. **18**(6): p. e13031.
37. Gómez-Benito, M. et al., Modeling Parkinson's disease with the alpha-synuclein protein. *Frontiers in Pharmacology*, 2020. **11**: p. 356.
38. Calabresi, P. et al., Alpha-synuclein in Parkinson's disease and other synucleinopathies: From overt neurodegeneration back to early synaptic dysfunction. *Cell Death & Disease*, 2023. **14**(3): p. 176.
39. Li, J.-L. et al., Mitochondrial function and Parkinson's disease: From the perspective of the electron transport chain. *Frontiers in Molecular Neuroscience*, 2021. **14**: p. 797833.
40. Tansey, M.G. et al., Inflammation and immune dysfunction in Parkinson disease. *Nature Reviews Immunology*, 2022. **22**(11): p. 657–673.
41. Herman, F. et al., Suppression of presymptomatic oxidative stress and inflammation in neurodegeneration by grape-derived polyphenols. *Frontiers in Pharmacology*, 2018. **9**: p. 867.
42. Kumari, M., A comprehensive review of the ubiquitin-proteasome system in synapse remodeling and neurodegenerative diseases, in N. Gogia, S. K. Singh and V. D. Jaganath (eds.), *Proteostasis: Investigating Molecular Dynamics in Neurodegenerative Disorders*. Springer, 2025. p. 297–327.
43. Boonstra, J.T. et al., Neuroimaging detectable differences between Parkinson's disease motor subtypes: A systematic review. *Movement Disorders Clinical Practice*, 2021. **8**(2): p. 175–192.
44. Merjane, J. et al., Molecular mechanisms of amyotrophic lateral sclerosis as broad therapeutic targets for gene therapy applications utilizing adeno-associated viral vectors. *Medicinal Research Reviews*, 2023. **43**(4): p. 829–854.
45. Aborode, A.T. et al., Targeting oxidative stress mechanisms to treat Alzheimer's and Parkinson's disease: A critical review. *Oxidative Medicine and Cellular Longevity*, 2022. **2022**(1): p. 7934442.
46. Lauridsen, C., From oxidative stress to inflammation: Redox balance and immune system. *Poultry Science*, 2019. **98**(10): p. 4240–4246.
47. Bloomingdale, P. et al., Hallmarks of neurodegenerative disease: A systems pharmacology perspective. *CPT: Pharmacometrics & Systems Pharmacology*, 2022. **11**(11): p. 1399–1429.
48. Kumar, M. et al., Micronutrients throughout the life cycle: Needs and functions in health and disease. *Current Nutrition & Food Science*, 2024. **20**(1): p. 62–84.
49. Zingg, J.M., Vitamin E: regulatory role on signal transduction. *IUBMB Life*, 2019. **71**(4): p. 456–478.
50. Bordone, M.P. et al., The energetic brain: A review from students to students. *Journal of Neurochemistry*, 2019. **151**(2): p. 139–165.
51. Sharma, V. and K.R. Aran, Unraveling the molecular mechanisms of vitamin deficiency in Alzheimer's disease pathophysiology. *Aging and Health Research*, 2025: p. 100226.
52. Khaneghah, A.M. et al., Water-soluble vitamins, in F. J. Barba, J. M. A. Saraiva, G. Cravotto, and J. M. Lorenzo (eds.), *Innovative Thermal and Non-Thermal Processing, Bioaccessibility and Bioavailability of Nutrients and Bioactive Compounds*. Elsevier, 2019. p. 241–266.
53. Sugandhi, V.V. et al., Pharmacokinetics of vitamin dosage forms: A complete overview. *Food Science & Nutrition*, 2024. **12**(1): p. 48–83.
54. Gaikwad, J. et al., Nutraceuticals potential of fat-soluble vitamins. Vitamins as Nutraceuticals: *Recent Advances and Applications*, 2023: p. 107–128.
55. Shabbir, A. et al., Vitamin deficiencies and drug-metabolizing enzymes: Challenges and strategies, in M. S. H. Akash and K. Rehman (eds.), *Biochemistry of Drug Metabolizing Enzymes*. Elsevier, 2022. p. 477–505.
56. Changkakoti, L. et al., Exploration of the role of vitamins in preventing neurodegenerative diseases: Comprehensive review on preclinical and clinical findings. *Current Neuropharmacology*, 2024. **23**: p. 547–563.
57. Calderón-Ospina, C.A. and M.O. Nava-Mesa, B Vitamins in the nervous system: Current knowledge of the biochemical modes of action and synergies of thiamine, pyridoxine, and cobalamin. *CNS Neuroscience & Therapeutics*, 2020. **26**(1): p. 5–13.
58. Parkhomenko, Y., A. Vovk, and Z. Protasova, Vitamin B1 and the pyruvate dehydrogenase complex, in R. Szymańska and M. Kowalski (eds.), *Molecular Nutrition*. Elsevier, 2020. p. 185–206.
59. Śliwiński, W. and O. Gawlik-Kotelnicka, Circulating B vitamins metabolites in depressive disorders-connections with the microbiota-gut-brain axis. *Behavioural Brain Research*, 2024. **472**: p. 115145.
60. Nevmerzhytska, N. et al., Mechanisms of toxic effects of homocysteine on the nervous system. *Neurophysiology*, 2019. **51**(5): p. 379–387.
61. Han, A. et al., Exploring neuropsychiatric manifestations of vitamin B complex deficiencies. *Frontiers in Psychiatry*, 2025. **16**: p. 1569826.
62. Behl, T. et al., Therapeutic insights elaborating the potential of retinoids in Alzheimer's disease. *Frontiers in Pharmacology*, 2022. **13**: p. 976799.
63. Onaolapo, O.J. and A.Y. Onaolapo, B vitamins: Pyridoxal phosphate and parkinsonism, in C. R. Martin, V. B. Patel, and V. R. Preedy (eds.), *Vitamins and Minerals in Neurological Disorders*. Elsevier, 2023. p. 527–542.
64. Adhikary, K. et al., Neuroprotective abilities of vitamins, micronutrients, antioxidants, and nutraceuticals, in R. N. Chaurasia, S. Ohia, and D. Bagchi (eds.), *A Review on Diverse Neurological Disorders*. Elsevier, 2024. p. 633–649.
65. Fatima, G. et al., Role of vitamin-E as a vital nutrient in health and diseases. *Indian Journal of Clinical Biochemistry*, 2025. 40: p. 1–14.
66. Alisi, L. et al., The relationships between vitamin K and cognition: A review of current evidence. *Frontiers in Neurology*, 2019. **10**: p. 239.
67. Kannan, S., J. Balakrishnan, and P. Nagarajan, Vitamin B7 (Biotin) and its role in hair, skin and nail health, in A. K. Shah, P. S. Tappia, and N. S. Dhalla (eds.), *Hydrophilic Vitamins in Health and Disease*. Springer, 2024. p. 233–252.
68. Kaźmierczak-Barańska, J. et al., Two faces of vitamin C—Antioxidative and pro-oxidative agent. *Nutrients*, 2020. **12**(5): p. 1501.
69. Liu, Y. et al., Neuroprotective roles of vitamin D: bridging the gap between mechanisms and clinical applications in cognitive decline. *International Journal of Molecular Sciences*, 2025. **26**(15): p. 7146.
70. Dumetz, F. et al., Neuronal morphology and synaptic plasticity in the hippocampus of vitamin A deficient rats. *Nutritional Neuroscience*, 2022. **25**(4): p. 779–790.

71. Manosso, L.M. et al., Vitamin E for the management of major depressive disorder: Possible role of the anti-inflammatory and antioxidant systems. *Nutritional Neuroscience*, 2022. **25**(6): p. 1310–1324.
72. Vissers, M.C. and A.B. Das, Ascorbate as an enzyme cofactor, in J. Lykkesfeldt and H. Frei (eds.), *Vitamin C*. CRC Press, 2020. p. 71–98.
73. Mansouri, F. et al., Protective effects of vitamin D on learning and memory deficit induced by scopolamine in male rats: the roles of brain-derived neurotrophic factor and oxidative stress. *Naunyn-Schmiedeberg's Archives of Pharmacology*, 2021. **394**(7): p. 1451–1466.
74. Bryson, A., C. Reid, and S. Petrou, Fundamental neurochemistry review: GABAA receptor neurotransmission and epilepsy: Principles, disease mechanisms and pharmacotherapy. *Journal of Neurochemistry*, 2023. **165**(1): p. 6–28.
75. Shohag, S. et al., Perspectives on the molecular mediators of oxidative stress and antioxidant strategies in the context of neuroprotection and neurolongevity: An extensive review. *Oxidative Medicine and Cellular Longevity*, 2022. **2022**(1): p. 7743705.
76. Nogueira-de-Almeida, C.A. et al., Neuronutrients and central nervous system: A systematic review. *Central Nervous System Agents in Medicinal ChemistryChemistry-Central Nervous System Agents*, 2023. **23**(1): p. 1–12.
77. Meher, A.K., B. Acharya, and P.K. Sahu, Probiotics: Bridging the interplay of a healthy gut and psychoneurological well-being. *Food Bioengineering*, 2024. **3**(1): p. 126–147.
78. Deepika et al., Crosstalk between copper, Alzheimer's disease, and melatonin. *BioMetals*, 2025: p. 1–40.
79. Alam, J., Vitamins: A nutritional intervention to modulate the Alzheimer's disease progression. *Nutritional Neuroscience*, 2022. **25**(5): p. 945–962.
80. Thirunavukarasu, A.J., A.C. Ross, and R.M. Gilbert, Vitamin A, systemic T-cells, and the eye: Focus on degenerative retinal disease. *Frontiers in Nutrition*, 2022. **9**: p. 914457.
81. Jin, P. et al., Nuclear receptors in health and disease: Signaling pathways, biological functions and pharmaceutical interventions. *Signal Transduction and Targeted Therapy*, 2025. **10**(1): p. 228.
82. Xie, G. et al., Vitamin A and its influence on tumour extracellular matrix. *Discover Oncology*, 2025. **16**(1): p. 16.
83. Zhao, F. et al., Brain–immune interaction mechanisms: Implications for cognitive dysfunction in psychiatric disorders. *Cell Proliferation*, 2022. **55**(10): p. e13295.
84. Khan, M.S. et al., Pathology, risk factors, and oxidative damage related to type 2 diabetes-mediated Alzheimer's disease and the rescuing effects of the potent antioxidant anthocyanin. *Oxidative Medicine and Cellular Longevity*, 2021. **2021**(1): p. 4051207.
85. Wang, Y. et al., Long-term exposure to advanced lipid peroxidation end products impairs cognitive function through microbiota-gut-brain axis. *Food Chemistry*, 2024. **461**: p. 140864.
86. Sharma, A., A. Mannan, and T.G. Singh, Unraveling novel mechanisms of ATP-binding cassette (ABC) transporter in insulin resistance-induced amyloidogenesis. *Metabolic Brain Disease*, 2025. **40**(5): p. 218.
87. Pandey, S.N. et al., Therapeutic approaches of nutraceuticals in the prevention of Alzheimer's disease. *Journal of Food Biochemistry*, 2022. **46**(12): p. e14426.
88. Cheng, J. et al., The emerging roles of protein homeostasis-governing pathways in Alzheimer's disease. *Aging Cell*, 2018. **17**(5): p. e12801.
89. Li, P. et al., Impact of vitamin A on aged people's cognition and Alzheimer's disease progression in an animal model. *npj Science of Food*, 2025. **9**(1): p. 67.
90. Papagiouvannis, G. et al., Multi-target directed compounds with antioxidant and/or anti-inflammatory properties as potent agents for Alzheimer's disease. *Medicinal Chemistry*, 2021. **17**(10): p. 1086–1103.
91. Khan, A.U. et al., Awareness and current knowledge of Parkinson's disease: A neurodegenerative disorder. *International Journal of Neuroscience*, 2019. **129**(1): p. 55–93.
92. McMahon, D.G. and J.E. Dowling, Neuromodulation: Actions of dopamine, retinoic acid, nitric oxide, and other substances on retinal horizontal cells. *Eye and Brain*, 2023: p. 125–137.
93. Zhao, Y. et al., Acupuncture for Parkinson's disease: Efficacy evaluation and mechanisms in the dopaminergic neural circuit. *Neural Plasticity*, 2021. **2021**(1): p. 9926445.
94. Xu, R. et al., Retinoid X receptor heterodimers in hepatic function: Structural insights and therapeutic potential. *Frontiers in Pharmacology*, 2024. **15**: p. 1464655.
95. O'Connor, C., P. Varshosaz, and A.R. Moise, Mechanisms of feedback regulation of vitamin A metabolism. *Nutrients*, 2022. **14**(6): p. 1312.
96. Johnstone, A.M. et al., Consensus statement on exploring the Nexus between nutrition, brain health and dementia prevention. *Nutrition & Metabolism*, 2025. **22**(1): p. 82.
97. Boyko, A. et al., Severe spinal cord injury in rats induces chronic changes in the spinal cord and cerebral cortex metabolism, adjusted by thiamine that improves locomotor performance. *Frontiers in Molecular Neuroscience*, 2021. **14**: p. 620593.
98. Zainuddin, M.-S. et al., Epigenetic modulations and nutrient interactions in Alzheimer's disease: unveiling potential therapeutic pathways. *Nutritional Neuroscience*, 2025: p. 1–23.
99. Gibson, G.E. et al., Pharmacological thiamine levels as a therapeutic approach in Alzheimer's disease. *Frontiers in Medicine*, 2022. **9**: p. 1033272.
100. Zhu, J.P. et al., Age-dependent alteration in metabolism of vitamin B6, neurotransmitters, and amino acids after 4′-O-methylpyridoxine administration in rats. *Journal of Food Science*, 2022. **87**(1): p. 466–480.
101. Kumar, S. et al., Nutrition, neurotransmitters, and behavior, in R. Smith and L. Johnson (eds.), *Nutrition and Psychiatric Disorders*. Springer, 2022. p. 89–108.
102. Smith, A.D. and H. Refsum, Homocysteine–from disease biomarker to disease prevention. *Journal of Internal Medicine*, 2021. **290**(4): p. 826–854.
103. Fareed, M.M. et al., The role of vitamins in DNA methylation as dietary supplements or neutraceuticals: A systematic review. *Current Molecular Medicine*, 2023. **23**(10): p. 1012–1027.
104. Liang, J. et al., Polyphenols, epigenetics, and methionine metabolism: Unlocking therapeutic potential. *Critical Reviews in Food Science and Nutrition*, 2025: p. 1–16.
105. Pappalardo, X.G. and V. Barra, Losing DNA methylation at repetitive elements and breaking bad. *Epigenetics & Chromatin*, 2021. **14**(1): p. 25.
106. Shaikh, A. and H. Roy, Folate deprivation induced neuroinflammation impairs cognition. *Neuroscience Letters*, 2023. **807**: p. 137264.
107. Bou Ghanem, A. et al., Exploring the complexities of 1C metabolism: Implications in aging and neurodegenerative diseases. *Frontiers in Aging Neuroscience*, 2024. **15**: p. 1322419.

108. Mohd Murshid, N., F. Aminullah Lubis, and S. Makpol, Epigenetic changes and its intervention in age-related neurodegenerative diseases. *Cellular and Molecular Neurobiology*, 2022. **42**(3): p. 577–595.
109. Guéant, J.-L. et al., Causes and consequences of impaired methionine synthase activity in acquired and inherited disorders of vitamin B12 metabolism. *Critical Reviews in Biochemistry and Molecular Biology*, 2022. **57**(2): p. 133–155.
110. Prajjwal, P. et al., Association of Alzheimer's dementia with oral bacteria, vitamin B12, folate, homocysteine levels, and insulin resistance along with its pathophysiology, genetics, imaging, and biomarkers. *Disease-a-Month*, 2023. **69**(5): p. 101546.
111. Rekik, A. et al., Parkinson's disease and vitamins: A focus on vitamin B12. *Journal of Neural Transmission*, 2024. **131**(12): p. 1495–1509.
112. Shahid, M., M.R. Siddiqui, and I. Sajid, Ageing and oxidative stress: How they are related to developing neurodegenerative disorders, in A. Kumar and S. Verma (eds.), *Medicinal Plants for the Management of Neurodegenerative Diseases*. CRC Press, 2024. p. 20–48.
113. Olaso-Gonzalez, G. et al., Impact of supplementation with vitamins B6, B12, and/or folic acid on the reduction of homocysteine levels in patients with mild cognitive impairment: A systematic review. *IUBMB Life*, 2022. **74**(1): p. 74–84.
114. Hamid, M. et al., A quantitative meta-analysis of vitamin C in the pathophysiology of Alzheimer's disease. *Frontiers in Aging Neuroscience*, 2022. **14**: p. 970263.
115. Tumilaar, S.G. et al., A comprehensive review of free radicals, oxidative stress, and antioxidants: Overview, clinical applications, global perspectives, future directions, and mechanisms of antioxidant activity of flavonoid compounds. *Journal of Chemistry*, 2024. **2024**(1): p. 5594386.
116. Pitocco, D. et al., Evaluating red blood cells' membrane fluidity in diabetes: Insights, mechanisms, and future aspects. *Diabetes/Metabolism Research and Reviews*, 2025. **41**(1): p. e70011.
117. Sindhu, R.K. et al., Exploring multifunctional antioxidants as potential agents for management of neurological disorders. *Environmental Science and Pollution Research*, 2022. **29**(17): p. 24458–24477.
118. Gaglani, P. et al., A pro-oxidant property of vitamin C to overcome the burden of latent Mycobacterium tuberculosis infection: A cross-talk review with Fenton reaction. *Frontiers in Cellular and Infection Microbiology*, 2023. **13**: p. 1152269.
119. Akanji, M.A. et al., Redox homeostasis and prospects for therapeutic targeting in neurodegenerative disorders. *Oxidative Medicine and Cellular Longevity*, 2021. **2021**(1): p. 9971885.
120. Singh, R. et al., Recent advances in targeting transition metals (Copper, iron, and Zinc) in Alzheimer's disease. *Molecular Neurobiology*, 2024. **61**(12): p. 10916–10940.
121. Gale, J. and E. Aizenman, The physiological and pathophysiological roles of copper in the nervous system. *European Journal of Neuroscience*, 2024. **60**(1): p. 3505–3543.
122. Tienda, A.A., F.E. Harrison, and J.M. Wilcox, Vitamin C transport deficiency alters striatal dopamine gene expression and metabolism in YAC128 Huntington disease mice. *Genes, Brain and Behavior*, 2025. **24**(3): p. e70023.
123. Samanta, S., S. Chakraborty, and D. Bagchi, Pathogenesis of neurodegenerative diseases and the protective role of natural bioactive components. *Journal of the American Nutrition Association*, 2024. **43**(1): p. 20–32.
124. Franzoni, F. et al., Oxidative stress and cognitive decline: The neuroprotective role of natural antioxidants. *Frontiers in Neuroscience*, 2021. **15**: p. 729757.
125. Nie, Y. et al., Lipid metabolism and oxidative stress in patients with Alzheimer's disease and amnestic mild cognitive impairment. *Brain Pathology*, 2024. **34**(1): p. e13202.
126. Rahimmi, A. et al., The neuroprotective roles of dietary micronutrients on Parkinson's disease: A review. *Molecular Biology Reports*, 2022. **49**(8): p. 8051–8060.
127. Moretti, M. and A.L.S. Rodrigues, Functional role of ascorbic acid in the central nervous system: A focus on neurogenic and synaptogenic processes. *Nutritional Neuroscience*, 2022. **25**(11): p. 2431–2441.
128. Dixit, S. et al., Vitamin C in neurological function and neurodegenerative disease, in M.C. Vissers and A.B. Das (eds.), *Vitamin C*. CRC Press, 2020. p. 183–211.
129. Zhang, X.-Y. et al., Vitamin C alleviates LPS-induced cognitive impairment in mice by suppressing neuroinflammation and oxidative stress. *International Immunopharmacology*, 2018. **65**: p. 438–447.
130. Song, K. et al., Oxidative stress-mediated blood-brain barrier (BBB) disruption in neurological diseases. *Oxidative Medicine and Cellular Longevity*, 2020. **2020**(1): p. 4356386.
131. Baringer, S.L., I.A. Simpson, and J.R. Connor, Brain iron acquisition: An overview of homeostatic regulation and disease dysregulation. *Journal of Neurochemistry*, 2023. **165**(5): p. 625–642.
132. Hao, T. et al., Advanced oral breviscapine sustained-release tablets for improved ischemic stroke treatment. *Biomaterials*, 2025. **316**: p. 123030.
133. Plevin, D. and C. Galletly, The neuropsychiatric effects of vitamin C deficiency: A systematic review. *BMC Psychiatry*, 2020. **20**(1): p. 315.
134. Norlin, M., Effects of vitamin D in the nervous system: Special focus on interaction with steroid hormone signalling and a possible role in the treatment of brain cancer. *Journal of Neuroendocrinology*, 2020. **32**(1): p. e12799.
135. Bivona, G. et al., Vitamin D and the nervous system. *Neurological Research*, 2019. **41**(9): p. 827–835.
136. Winek, K., H. Soreq, and A. Meisel, Regulators of cholinergic signaling in disorders of the central nervous system. *Journal of Neurochemistry*, 2021. **158**(6): p. 1425–1438.
137. Verma, M., Z. Wills, and C.T. Chu, Excitatory dendritic mitochondrial calcium toxicity: Implications for Parkinson's and other neurodegenerative diseases. *Frontiers in Neuroscience*, 2018. **12**: p. 523.
138. Hanna, R.M. et al., Calcium transport in the kidney and disease processes. *Frontiers in Endocrinology*, 2022. **12**: p. 762130.
139. Menéndez, S.G. and W. Manucha, Vitamin D as a modulator of neuroinflammation: Implications for brain health. *Current Pharmaceutical Design*, 2024. **30**(5): p. 323–332.
140. Xiao, S.-Y. et al., Possible neuropathology of sleep disturbance linking to Alzheimer's disease: Astrocytic and microglial roles. *Frontiers in Cellular Neuroscience*, 2022. **16**: p. 875138.
141. Qiu, Y.-M. et al., Immune cells in the BBB disruption after acute ischemic stroke: Targets for immune therapy? *Frontiers in Immunology*, 2021. **12**: p. 678744.

142. Allegra, A. et al., Post-chemotherapy cognitive impairment in hematological patients: Current understanding of chemobrain in hematology. *Expert Review of Hematology*, 2020. **13**(4): p. 393–404.
143. Solleiro-Villavicencio, H. and S. Rivas-Arancibia, Effect of chronic oxidative stress on neuroinflammatory response mediated by CD4+ T cells in neurodegenerative diseases. *Frontiers in Cellular Neuroscience*, 2018. **12**: p. 114.
144. Wu, H. et al., Correlation between serum 25 (OH) D and cognitive impairment in Parkinson's disease. *Journal of Clinical Neuroscience*, 2022. **100**: p. 192–195.
145. Fullard, M.E. and J.E. Duda, A review of the relationship between vitamin D and Parkinson disease symptoms. *Frontiers in Neurology*, 2020. **11**: p. 454.
146. Hafiz, A.A., The neuroprotective effect of vitamin D in Parkinson's disease: Association or causation. *Nutritional Neuroscience*, 2024. **27**(8): p. 870–886.
147. Infante, M. et al., Low vitamin D status at admission as a risk factor for poor survival in hospitalized patients with COVID-19: An Italian retrospective study. *Journal of the American Nutrition Association*, 2022. **41**(3): p. 250–265.
148. Shea, M.K., A.Y. Xuan, and S.L. Booth, Vitamin D, Alzheimer's disease and related dementia. *Advances in Food and Nutrition Research*, 2024. **109**: p. 185–219.
149. Jayedi, A., A. Rashidy-Pour, and S. Shab-Bidar, Vitamin D status and risk of dementia and Alzheimer's disease: A meta-analysis of dose-response. *Nutritional Neuroscience*, 2019. **22**(11): p. 750–759.
150. Digvijaya et al., Vitamin D fortification: A promising approach to overcome drug resistance and tolerance in therapeutic interventions. *Scientifica*, 2024. **2024**(1): p. 9978076.
151. Medeiros, J.F.P. et al., The impact of vitamin D supplementation on VDR gene expression and body composition in monozygotic twins: Randomized controlled trial. *Scientific Reports*, 2020. **10**(1): p. 11943.
152. Avila, A.M. et al., An FDA/CDER perspective on nonclinical testing strategies: Classical toxicology approaches and new approach methodologies (NAMs). *Regulatory Toxicology and Pharmacology*, 2020. **114**: p. 104662.
153. Zia, B. et al., The multifaceted role of neuroprotectin D1: Physiological, pathophysiological, and pharmacological insights in neurodegenerative diseases. *Current Neuropharmacology*, 2025.
154. Raggi, A. and R. Ferri, The role of vitamins in neurodegeneration: A brief review of mechanisms, clinical evidence, and therapeutic perspectives. *Psychogeriatrics*, 2025. **25**(4): p. e70071.
155. Maggi, S. et al., The role of nutrition and the Mediterranean diet on the trajectories of cognitive decline. *Experimental Gerontology*, 2023. **173**: p. 112110.
156. Jain, P. et al., tocopherols and tocotrienols: The essential vitamin E, in C. B. Betim Cazarin, J. L. Bicas, G. M. Pastore, and M. R. Marostica (eds.), *Bioactive Food Components Activity in Mechanistic Approach*. Elsevier, 2022. p. 139–154.
157. Biswas, N. and S. Khanna, Tocotrienol vitamin E and neurodegenerative disorders, in R. Pandey and P. Singh (eds.), *Antioxidants and Functional Foods for Neurodegenerative Disorders*. CRC Press, 2021. p. 229–239.
158. Islam, F. et al., Functional roles and novel tools for improving-oxidative stability of polyunsaturated fatty acids: A comprehensive review. *Food Science & Nutrition*, 2023. **11**(6): p. 2471–2482.
159. Szymańska, R. et al., Vitamin E: Structure and Forms, in J. Smith and L. Johnson (eds.), *Molecular Nutrition*. Elsevier, 2020. p. 67–90.
160. Shabani, Z. et al., Evaluation of the neuroprotective effects of Vitamin E on the rat substantia nigra neural cells exposed to electromagnetic field: An ultrastructural study. *Electromagnetic Biology and Medicine*, 2021. **40**(3): p. 428–437.
161. Ali, S.S. et al., Understanding oxidants and antioxidants: Classical team with new players. *Journal of Food Biochemistry*, 2020. **44**(3): p. e13145.
162. Zhang, W. et al., Phospholipids and peroxisomes in ferroptosis: The therapeutic target of acupuncture regulating vascular cognitive impairment and dementia. *Frontiers in Aging Neuroscience*, 2025. **17**: p. 1512980.
163. Cowan, C.M., M.A. Sealey, and A. Mudher, Suppression of tau-induced phenotypes by vitamin E demonstrates the dissociation of oxidative stress and phosphorylation in mechanisms of tau toxicity. *Journal of Neurochemistry*, 2021. **157**(3): p. 684–694.
164. Mendes, O., Role of vitamin D and antioxidant functional foods in the prevention and treatment of Alzheimer's disease pathology, in J. Smith and L. Johnson (eds.), *Antioxidants and Functional Foods for Neurodegenerative Disorders*. CRC Press, 2021. p. 203–218.
165. Fan, Y.-G. et al., Vitamin D deficiency exacerbates Alzheimer-like pathologies by reducing antioxidant capacity. *Free Radical Biology and Medicine*, 2020. **161**: p. 139–149.
166. Power, R. et al., Omega-3 fatty acid, carotenoid and vitamin E supplementation improves working memory in older adults: A randomised clinical trial. *Clinical Nutrition*, 2022. **41**(2): p. 405–414.
167. Pelczarski, M. et al., The role of α-tocopherol in the prevention and treatment of Alzheimer's disease. *Molecular and Cellular Biochemistry*, 2025. **480**(6): p. 3385–3398.
168. Musa, A.F., C.J. Min, and C.G. Yap, Tocotrienols: Mother Nature's gift to the brain, in E. Brown and S. Lee (eds.), *Role of Micronutrients in Brain Health*. Springer, 2022. p. 69–88.
169. Rasool, A. et al., Oxidative stress and dopaminergic metabolism: A major PD pathogenic mechanism and basis of potential antioxidant therapies. *CNS & Neurological Disorders-Drug Targets-CNS & Neurological Disorders*, 2024. **23**(7): p. 852–864.
170. Naomi, R. et al., An interactive review on the role of Tocotrienols in the neurodegenerative disorders. *Frontiers in Nutrition*, 2021. **8**: p. 754086.
171. O'Bryant, S.E. et al., Analysis of a precision medicine approach to treating Parkinson's disease: Analysis of the DATATOP study. *Parkinsonism & Related Disorders*, 2022. **94**: p. 15–21.
172. Atale, N. and V. Rani, Nutraceutical synergism as a combinational therapy in health and diseases, in J. Smith and L. Chen (eds.), *Advancing Biotechnology: From Science to Therapeutics and Informatics: Technological Advancements in Biosciences and Biotechnology*. Springer, 2025. p. 301–311.
173. Parameswari, R. et al., Vitamin E (tocopherol): Implications in cardiovascular health and neuroprotection, in M. Kumar and P. Rao (Eds.), *Lipophilic Vitamins in Health and Disease*. Springer, 2024. p. 233–257.

174. Zaaboul, F. and Y. Liu, Vitamin E in foodstuff: Nutritional, analytical, and food technology aspects. *Comprehensive Reviews in Food Science and Food Safety*, 2022. **21**(2): p. 964–998.
175. Xie, Y. et al., Vitamin K: Infection, inflammation, and auto-immunity. *Journal of Inflammation Research*, 2024: p. 1147–1160.
176. Shen, G. et al., The molecular basis of FIX deficiency in hemophilia B. *International Journal of Molecular Sciences*, 2022. **23**(5): p. 2762.
177. Tracey, T.J. et al., Neuronal lipid metabolism: Multiple pathways driving functional outcomes in health and disease. *Frontiers in Molecular Neuroscience*, 2018. **11**: p. 10.
178. Wang, A. et al., Association of dietary vitamin K intake with cognition in the elderly. *Frontiers in Nutrition*, 2022. **9**: p. 900887.
179. Diachenko, A.I. et al., The role of vitamin K in the development of neurodegenerative diseases. *Biochemistry (Moscow)*, 2024. **89**(Suppl 1): p. S57–S70.
180. Mankovska, I. et al., Interplay of oxidative stress and mitochondrial dysfunction in Alzheimer's and Parkinson's diseases: Mechanisms and treatment strategies. *Neurophysiology*, 2025: p. 1–11.
181. Jung, M. et al., Nicotinamide (niacin) supplement increases lipid metabolism and ROS-induced energy disruption in triple-negative breast cancer: Potential for drug repositioning as an anti-tumor agent. *Molecular Oncology*, 2022. **16**(9): p. 1795–1815.
182. Chen, X., W. Lu, and D. Wu, Sirtuin 2 (SIRT2): Confusing roles in the pathophysiology of neurological disorders. *Frontiers in Neuroscience*, 2021. **15**: p. 614107.
183. Sharma, A. et al., Potential synergistic supplementation of NAD+ promoting compounds as a strategy for increasing healthspan. *Nutrients*, 2023. **15**(2): p. 445.
184. Seamon, M. et al., Niacin for Parkinson's disease. *Clinical and Experimental Neuroimmunology*, 2020. **11**(1): p. 47–56.
185 Hikosaka, K. et al., Implications of NAD metabolism in pathophysiology and therapeutics for neurodegenerative diseases. *Nutritional Neuroscience*, 2021. **24**(5): p. 371–383.
186. Munteanu, C., S.M. Mârza, and I. Papuc, The immunomodulatory effects of vitamins in cancer. *Frontiers in Immunology*, 2024. **15**: p. 1464329.
187. Sharma, T. et al., Targeting oligodendrocyte dynamics and remyelination: Emerging therapies and personalized approaches in multiple sclerosis management. *Current Neurovascular Research*, 2024. **21**(4): p. 359–417.
188. Nemazannikova, N. et al., Is there a link between vitamin B and multiple sclerosis? *Medicinal Chemistry*, 2018. **14**(2): p. 170–180.
189. Neufeld, L.M. et al., Advancing nutrition science to meet evolving global health needs. *European Journal of Nutrition*, 2023. **62**(Suppl 1): p. 1–16.
190. Kokande, A.M. et al., Overview and recent advances of vitamins, in S. Meena and P. Kumar (eds.), *Preventive and Therapeutic Role of Vitamins as Nutraceuticals*. Springer, 2024. p. 253–279.
191. Hajhashemy, Z. et al., Practical supplements for prevention and management of migraine attacks: A narrative review. *Frontiers in Nutrition*, 2024. **11**: p. 1433390.
192. Jat, G.S. and T.K. Behera, Harshawardhan Choudhary, and K. Padmanabha, Challenges of traditional breeding in watermelon, in S. K. Dutta, P. Nimmakayala, and U. K. Reddy (eds.), *The Watermelon Genome*. Springer, 2023, p. 85.
193. Emekli-Alturfan, E. and A.A. Alturfan, The emerging relationship between vitamin K and neurodegenerative diseases: A review of current evidence. *Molecular Biology Reports*, 2023. **50**(1): p. 815–828.
194. Maiese, K., Nicotinamide as a foundation for treating neurodegenerative disease and metabolic disorders. *Current Neurovascular Research*, 2021. **18**(1): p. 134–149.
195. Alcaín, F.J. et al., Coenzyme Q and age-related neurodegenerative disorders: Parkinson and Alzheimer diseases, in M. Kumar and P. Rao (eds.), *Coenzyme Q in aging*. Springer, 2020. p. 241–268.
196. Dos Santos, S.M. et al., Mitochondrial dysfunction and alpha-lipoic acid: Beneficial or harmful in Alzheimer's disease? *Oxidative Medicine and Cellular Longevity*, 2019. **2019**(1): p. 8409329.
197. Verma, N. et al., Ameliorative effect of acetyl L-carnitine in Alzheimer's disease via downregulating of homocysteine levels in hyperhomocysteinemia induced cognitive deficit in mouse model. *Drug Metabolism Letters*, 2021. **14**(3): p. 219–231.
198. Raja, S. et al., Future trends and directives for research on phytochemicals in neurological diseases, in A. Smith and B. Johnson (eds.), *NeuroPhytomedicine*. CRC Press, 2024. p. 221–240.
199. Esposito, S. et al., The role of diet in multiple sclerosis: A review. *Nutritional Neuroscience*, 2018. **21**(6): p. 377–390.
200. Zhou, L. et al., Effects of vitamin B12 deficiency on risk and outcome of ischemic stroke. *Clinical Biochemistry*, 2023. **118**: p. 110591.
201. Fekete, M. et al., Improving cognitive function with nutritional supplements in aging: A comprehensive narrative review of clinical studies investigating the effects of vitamins, minerals, antioxidants, and other dietary supplements. *Nutrients*, 2023. **15**(24): p. 5116.
202. Kumar, A. et al., Magnesium (Mg2+): Essential mineral for neuronal health: From cellular biochemistry to cognitive health and behavior regulation. *Current Pharmaceutical Design*, 2024. **30**(39): p. 3074–3107.
203. Zoidis, E. et al., Selenium-dependent antioxidant enzymes: Actions and properties of selenoproteins. *Antioxidants*, 2018. **7**(5): p. 66.
204. Walia, V. et al., Delineation of neuroprotective effects and possible benefits of antioxidants therapy for the treatment of Alzheimer's diseases by targeting mitochondrial-derived reactive oxygen species: Bench to bedside. *Molecular Neurobiology*, 2022. **59**(1): p. 657–680.
205. Meena, N. et al., Neuroprotective potential of carotenoids, in A. Kumar and S. Patel (eds.), *Carotenoids: Trends and Advances*. Springer, 2024. p. 259–278.
206. Oyovwi, M.O. et al., Potential roles of natural antioxidants in modulating neurodegenerative disease pathways. *Molecular Neurobiology*, 2025: p. 1–31.
207. Chopade, P. et al., Alzheimer's and Parkinson's disease therapies in the clinic. *Bioengineering & Translational Medicine*, 2023. **8**(1): p. e10367.
208. Turcu-Stiolica, A. et al., The impact of the dietary intake of vitamin B12, folic acid, and vitamin D3 on homocysteine levels and the health-related quality of life of levodopa-treated patients with Parkinson's disease—A pilot study in Romania. *Diagnostics*, 2024. **14**(15): p. 1609.
209. Tragni, V. et al., Personalized medicine in mitochondrial health and disease: molecular basis of therapeutic approaches based on nutritional supplements and their analogs. *Molecules*, 2022. **27**(11): p. 3494.

210. Ward, K.M. and L. Citrome, Antipsychotic-related movement disorders: Drug-induced parkinsonism vs. tardive dyskinesia—Key differences in pathophysiology and clinical management. *Neurology and Therapy*, 2018. **7**(2): p. 233–248.
211. Stacchiotti, V. et al., Metabolic and functional interplay between gut microbiota and fat-soluble vitamins. *Critical Reviews in Food Science and Nutrition*, 2021. **61**(19): p. 3211–3232.
212. Minich, D.M. et al., Is melatonin the "next vitamin D"? A review of emerging science, clinical uses, safety, and dietary supplements. *Nutrients*, 2022. **14**(19): p. 3934.
213. Phokaewvarangkul, O. et al., Homocysteine, vitamin B metabolites, dopamine-substituting compounds, and symptomatology in Parkinson's disease: Clinical and therapeutic considerations. *Journal of Neural Transmission*, 2023. **130**(11): p. 1451–1462.
214. Mohanad, M. et al., Neuroprotective effects of vitamin D in an Alzheimer's disease rat model: Improvement of mitochondrial dysfunction via calcium/calmodulin-dependent protein kinase kinase 2 activation of Sirtuin1 phosphorylation. *Biofactors*, 2024. **50**(2): p. 371–391.
215. Foudah, A.I. et al., Anticholinergic effect of resveratrol with vitamin E on scopolamine-induced Alzheimer's disease in rats: Mechanistic approach to prevent inflammation. *Frontiers in Pharmacology*, 2023. **14**: p. 1115721.
216. Pritam, P. et al., Antioxidants in Alzheimer's disease: Current therapeutic significance and future prospects. *Biology*, 2022. **11**(2): p. 212.
217. Tanaka, M., Parkinson's disease: Bridging gaps, building biomarkers, and reimagining clinical translation. *Cells*, 2025. **14**(15): p. 1161.
218. Hahn, S.L. et al., Introducing dietary self-monitoring to undergraduate women via a calorie counting app has no effect on mental health or health behaviors: Results from a randomized controlled trial. *Journal of the Academy of Nutrition and Dietetics*, 2021. **121**(12): p. 2377–2388.
219. Ali, M.A. et al., Dietary vitamin B complex: Orchestration in human nutrition throughout life with sex differences. *Nutrients*, 2022. **14**(19): p. 3940.
220. Venkatraghavan, V. et al., Progression along data-driven disease timelines is predictive of Alzheimer's disease in a population-based cohort. *Neuroimage*, 2021. **238**: p. 118233.
221. Koklesova, L. et al., Homocysteine metabolism as the target for predictive medical approach, disease prevention, prognosis, and treatments tailored to the person. *EPMA Journal*, 2021. **12**(4): p. 477–505.
222. Houtvast, D.C. et al., Underpowered studies in muscle metabolism research: Determinants and considerations. *Clinical Nutrition Espen*, 2024. **64**: p. 334–343.
223. Talebi, S. et al., Dietary antioxidants and risk of Parkinson's disease: A systematic review and dose–response meta-analysis of observational studies. *Advances in Nutrition*, 2022. **13**(5): p. 1493–1504.
224. Lasoń, W. et al., The vitamin D receptor as a potential target for the treatment of age-related neurodegenerative diseases such as Alzheimer's and Parkinson's diseases: A narrative review. *Cells*, 2023. **12**(4): p. 660.
225. Pantzaris, M. et al., Neuroaspis PLP10™, a nutritional formula rich in omega-3 and omega-6 fatty acids with antioxidant vitamins including gamma-tocopherol in early Parkinson's disease: A randomized, double-blind, placebo-controlled trial. *Clinical Neurology and Neurosurgery*, 2021. **210**: p. 106954.
226. Wakade, C. et al., Low-dose niacin supplementation improves motor function in US Veterans with Parkinson's disease: A single-center, randomized, placebo-controlled trial. *Biomedicines*, 2021. **9**(12): p. 1881.
227. Wimalawansa, S.J., Enhancing the design of nutrient clinical trials for disease prevention—A focus on vitamin D: A systematic review. *Nutrition Reviews*, 2025. **83**(7): p. e1740–e1781.
228. Benatar, M. et al., Preventing amyotrophic lateral sclerosis: Insights from pre-symptomatic neurodegenerative diseases. *Brain*, 2022. **145**(1): p. 27–44.
229. Paul-Chima, U.O., C.N. Ugwu, and E.U. Alum, Integrated approaches in nutraceutical delivery systems: Optimizing ADME dynamics for enhanced therapeutic potency and clinical impact. *RPS Pharmacy and Pharmacology Reports*, 2024. **3**(4): p. rqae024.
230. Rychter, A.M. et al., Antioxidant effects of vitamin E and risk of cardiovascular disease in women with obesity–A narrative review. *Clinical Nutrition*, 2022. **41**(7): p. 1557–1565.
231. Avram, O.-E. et al., Modifiable nutritional biomarkers in autism spectrum disorder: A systematic review and meta-analysis of vitamin D, B12, and homocysteine exposure spanning prenatal development through late adolescence. *International Journal of Molecular Sciences*, 2025. **26**(9): p. 4410.
232. Rizzo, G. and A. Marino, Cognitive impairment and micronutrients: Vitamin B12, folate, and homocysteine and implications for dementia, in K. Onaolapo and O.J. Onaolapo (eds.), *Vitamins and Minerals in Neurological Disorders*. Elsevier, 2023. p. 29–46.
233. Rahm, C., Medicine and health: Integrative approaches to post-covid syndrome and parasitic involvement. *Clinical Medicine Engineering and Life Sciences*, 2024. **2**(3): p. 1–10.
234. Svob Strac, D. et al., Personalizing the care and treatment of Alzheimer's disease: An overview. *Pharmacogenomics and Personalized Medicine*, 2021: p. 631–653.
235. Rai, S.N. et al., The role of vitamins in neurodegenerative disease: An update. *Biomedicines*, 2021. **9**(10): p. 1284.
236. Pokushalov, E. et al., Biomarker-Guided dietary supplementation: A narrative review of precision in personalized nutrition. *Nutrients*, 2024. **16**(23): p. 4033.
237. Bakeberg, M., Assessment of the determinants of cognitive decline and related symptoms in Parkinson's disease. Doctoral thesis, The University of Western Australia, 2021.
238. Gogna, T., B.E. Housden, and A. Houldsworth, Exploring the role of reactive oxygen species in the pathogenesis and pathophysiology of Alzheimer's and Parkinson's disease and the efficacy of antioxidant treatment. *Antioxidants*, 2024. **13**(9): p. 1138.
239. Lobine, D. et al., Potential of medicinal plants as neuroprotective and therapeutic properties against amyloid-β-related toxicity, and glutamate-induced excitotoxicity in human neural cells. *Current Neuropharmacology*, 2021. **19**(9): p. 1416–1441.
240. Ma, L.-Z. et al., Serum calcium predicts cognitive decline and clinical progression of Alzheimer's disease. *Neurotoxicity Research*, 2021. **39**(3): p. 609–617.
241. Duarte-Jurado, A.P. et al., Antioxidant therapeutics in Parkinson's disease: Current challenges and opportunities. *Antioxidants*, 2021. **10**(3): p. 453.
242. Parab, S. et al., Unraveling the mechanisms of hydrophilic vitamins in Alzheimer's and Parkinson's: Preclinical and clinical evidence, in A. K. Shah, P. S. Tappia, and N. S.

Dhalla (eds.), *Hydrophilic Vitamins in Health and Disease*. Springer, 2024. p. 181–217.
243. Zafar, S., Translating biomarker discovery into clinical practice: Challenges and opportunities. *Journal of Translational Research*, 2024. **1**(01): p. 19–30.
244. Candebat, C.L., Optimizing dietary specifications for yellowtail kingfish, Seriola lalandi: Requirements and interactions of sulfur amino acids and taurine. Doctoral Thesis, James Cook University, 2021.
245. Zhang, Y., C. Lv, and G. Zhao, Ways to enhance the bioavailability of polyphenols in the brain: A journey through the blood-brain barrier. *Food Reviews International*, 2022. **38**(supl): p. 812–828.
246. Crintea, A. et al., Nanocarriers for drug delivery: An overview with emphasis on vitamin D and K transportation. *Nanomaterials*, 2022. **12**(8): p. 1376.
247. Krsek, A. and L. Baticic, Nanotechnology-driven therapeutic innovations in neurodegenerative disorders: A focus on Alzheimer's and Parkinson's disease. *Future Pharmacology*, 2024. **4**(2): p. 352–379.
248. Franco, G.A. et al., Bioactive compounds of the Mediterranean diet as nutritional support to fight neurodegenerative disease. *International Journal of Molecular Sciences*, 2023. **24**(8): p. 7318.
249. Moses, G., The safety of commonly used vitamins and minerals. *Australian Prescriber*, 2021. **44**(4): p. 119.
250. Intiso, D. et al., Bowel disorders in patients with severe acquired brain injury receiving neurorehabilitation: Results from a national survey in Italy. *Frontiers in Neurology*, 2025. **16**: p. 1474360.

12 Vitamins in Other Neurodegenerative Disorders (ALS, Huntington's Disease, and MS)

Challenges and Future Opportunities

Sagar Pamu[1], *Hemraj Singh Rajput*[2], *Sazal Patyar*[3], *Palakurthi Yanadaiah*[3], *Sumit Sharma*[4], *and Monika Kaushik*[4,5]

[1]Institute of Pharmacy, Nirma University, Ahmedabad, India
[2]Department of Pharmacy, Sumandeep Vidyapeeth Deemed University, Vadodara, India
[3]School of Pharmaceutical Sciences, Lovely Professional University, Phagwara, India
[4]Amity Institute of Pharmacy, Amity University Madhya Pradesh, Gwalior, India
[5]School of medical and allied sciences, K R Mangalam University, Haryana, India

12.1 INTRODUCTION

Neurodegenerative diseases such as amyotrophic lateral sclerosis (ALS), multiple sclerosis (MS), and Huntington's disease (HD) are characterized by a continuous decrease in neuronal function and structure, leading to severe neurological impairment and disability. Despite advances in understanding their complex pathophysiology, effective treatments remain limited, often addressing symptoms rather than underlying causes. Nutrition, particularly vitamin status, has gained increasing recognition as a modifiable factor influencing disease onset, progression, and patient quality of life. Vitamins are essential for numerous neural processes, including energy metabolism, myelin maintenance, antioxidant defense, and immune regulation. This chapter explores the roles of vitamin deficiencies and supplementation in ALS, HD, and MS, integrating current knowledge on pathophysiology, clinical evidence, translational challenges, and emerging opportunities for personalized nutritional interventions in neurodegenerative disease management.

12.2 OVERVIEW OF NEURODEGENERATIVE DISORDERS BEYOND AD AND PD

In this overview, let us understand first what is meant by neurodegenerative disorders. It can be said that there is a heterogeneous group of diseases characterized by the progressive dysfunction and damage of neurons within the central nervous system (CNS) and hence termed as a neurodegenerative disorder. The common occurrence of Parkinson's disease (PD) and Alzheimer's disease (AD) makes them extensively studied neurodegenerative disorders. There are other major neurodegenerative conditions such as ALS, MS, and HD. These disease conditions pose substantial clinical and societal challenges, each with distinct pathological and clinical features [1,2]. Let us explore neurodegenerative disorders beyond AD and PD.

ALS is a relentlessly progressive neurodegenerative disease that predominantly affects upper and lower motor neurons, which eventually leads to muscle weakness, atrophy, and paralysis [3,4]. Most cases reported of ALS are not associated with any kind of genetic factor, while 5%–10% cases can be linked with distinct gene mutations including superoxide dismutase (SOD)1, G9orf72, and TARDBP. Pathological hallmarks include cytoplasmic aggregates of misfolded proteins such as TDP-43, mitochondrial dysfunction, and glutamate-mediated excitotoxicity. Cognitive impairment, especially frontotemporal dysfunction, is increasingly recognized as part of the ALS spectrum [5–7].

HD is an autosomal dominant disease (i.e., there is a 50% chance that the offspring will receive this gene, and it will also show its characteristic). HD is caused by the repetition of the CAG trinucleotide in the huntingtin (HTT) gene. The disease shows a distinct triad of clinical characteristics, i.e., notable chorea, cognitive impairment, and psychiatric disturbances. Neuropathologically, HD exhibits progressive loss in the cerebral cortex and striatum, resulting in significant neuronal dysfunction and aggregation of protein. Onset of disease typically occurs in the middle of adulthood, but the presence of disease can be recognized in the juvenile phase [8–10].

MS stands apart from the classic neurodegenerative disease conditions because of its association with autoimmune etiology. MS involves immune-mediated demyelination and subsequent neurodegeneration within the CNS. Hallmark pathological findings include multifocal inflammatory

DOI: 10.1201/9781003686675-12

lesions, blood–brain barrier (BBB) breakdown, and formation of demyelinating plaques. The variability in disease progression is very high, and disease relapse and remission episodes are frequent and are progressive in nature. Genetic susceptibility and factors such as vitamin D deficiency and viral exposures have shown a significant association [11–14].

When ALS, HD, and MS are looked at individually, they seem to be less common than AD and PD, but their collective implication poses a considerable challenge. Mortality rate is high in ALS, reported within a few years of disease onset. HD leads to progressive disability that may last for decades. The leading cause of neurological impairment in adults is MS. The unique challenges faced from these disease conditions are in their diagnosis, management, and long-term care, resulting in profound functional decline and significant psychosocial implications for patients and caregivers [5,10,12,13].

12.3 IMPORTANCE OF NUTRITIONAL NEUROSCIENCE

The fundamental relation of health and nutrition has been known since the beginning of health science, but what makes it important now is the emerging interdisciplinary field that examines the complex relationship between dietary factors, micronutrients, and health or in the context of this chapter brain health. Emphasis has been given to the roles of nutrition in the progression and prevention of neurological and neurodegenerative diseases. Scientific evidence has revealed the influence of diet and nutrients on the neurological development, neurotransmission, synaptic plasticity, and overall resilience of the CNS. Deficiency of key vitamins, minerals, and essential fatty acids has been associated with accelerated cognitive decline, neuronal injury, and susceptibility to neurodegenerative disorders such as MS, ALS, and HD [15,16].

Studies have indicated healthy dietary patterns, such as the Mediterranean diet and calorie control, exert protective effects against cognitive impairment and the onset of neurodegenerative diseases. Malnutrition, insulin resistance, and low body mass index are associated with increased dementia risk and mortality in older adults. Nutritional interventions can influence disease progression via multiple mechanisms, including modulation of the gut–microbiota–brain axis, reduction in neuroinflammation, and enhancement of neurotrophic signaling [15,17,18].

Deficiencies of micronutrients such as vitamins B, C, D, and E were known to be associated with both pathogenesis and severity of neurodegeneration. If we take the example of vitamins B12 and D, the deficiency of both has been observed in patients suffering from AD, MS, and ALS. The supplementation of these micronutrients has shown mixed results of improvement in both cognitive function and reduction in neuroinflammatory markers. There is a need for more research to establish the claimed antioxidative, anti-inflammatory, neuroprotective, and metabolic pathway-related mechanisms [19,20].

12.4 PATHOPHYSIOLOGY OF ALS, HUNTINGTON'S DISEASE, AND MS

The neurodegenerative diseases such as MS, ALS, and HD represent distinct but overlapping disorders characterized by progressive neuronal dysfunction and loss. Although each disease features unique clinical manifestations and pathological hallmarks, common mechanisms such as oxidative stress, protein misfolding, mitochondrial dysfunction, neuroinflammation, and impaired cellular homeostasis underlie their progression. Understanding the specific molecular and cellular pathways that drive these disorders is critical to developing targeted therapies. This section provides an overview of the pathophysiological processes driving ALS, HD, and MS, elucidating their differences in genetic basis, immunological involvement, and neuropathological features while highlighting key shared pathways that contribute to neurodegeneration.

12.5 AMYOTROPHIC LATERAL SCLEROSIS (ALS): MECHANISMS AND PROGRESSION

ALS is a devastating neurodegenerative disorder characterized by progressive degeneration of upper and lower motor neurons, resulting in muscle weakness, atrophy, spasticity, and eventual paralysis. The clinical course is typically relentless, with median survival from symptom onset being 2–5 years, primarily due to respiratory failure [3,21].

Approximately 5%–10% of ALS cases are familial and associated with mutations in genes such as C9ORF72, FUS, SOD1, and TARDBP, while the majority are sporadic. These mutations disrupt a variety of cellular functions, including protein homeostasis, RNA metabolism, and cytoskeletal integrity. The most common pathological feature in both familial and sporadic ALS is cytoplasmic aggregation of misfolded proteins, particularly TDP-43, which is present in over 95% of cases [3,21].

Excitotoxicity due to dysfunctional glutamate homeostasis is a central theme in ALS pathogenesis. Impaired uptake of glutamate by astrocytic EAAT2 transporters leads to excessive glutamate in the synaptic cleft, causing intracellular calcium overload and downstream neuronal injury [3,21].

Mitochondrial abnormalities, including impaired bioenergetics and reactive oxygen species (ROS) overproduction, are recognized contributors to neuronal injury. Mutations in SOD1 and other genes impair antioxidant defenses, further exacerbating oxidative stress within motor neurons [21].

The hallmark of ALS is neuroinflammation and is perpetuated by the activation of astrocytes and microglia. These glial cells release proinflammatory cytokines, chemokines, and other cytotoxic mediators, contributing both to motor neuron degeneration and disease progression [22].

Defects in axonal transport due to cytoskeletal abnormalities contribute to the failure of neuronal maintenance and connectivity. Mutant proteins, including FUS and TDP-43, disrupt both nuclear and cytoplasmic transport mechanisms, leading to defective RNA metabolism and protein trafficking [21].

12.6 HUNTINGTON'S DISEASE: GENETIC AND CELLULAR INSIGHTS

HD is an autosomal dominant, neurodegenerative disorder, fundamentally characterized by progressive motor dysfunction, psychiatric disturbances, and cognitive decline. The condition is caused by a mutation in chromosome 4, i.e., repetition of the CAG trinucleotide in exon 1 of the HTT gene, leading to an elongated polyglutamine (polyQ) tract in the HTT protein. CAG repeat lengths of 36 or more are associated with pathogenesis, and longer repeat expansions correlate inversely with age of onset and disease progression [8,10,23–25].

The mutant HTT protein undergoes aberrant conformational changes, rendering it prone to misfold, aggregate, and accumulate in neuronal nuclei and cytoplasm, particularly within the striatum and cortex regions most affected in HD. These aggregates can sequester transcription factors, i.e., CREB binding protein, p53, specificity protein 1, and impair the ubiquitin-proteasome system, disrupting cellular proteostasis. PolyQ aggregation also creates a toxic environment that impairs mitochondrial biogenesis, promotes autophagic and lysosomal dysfunction, and drives oxidative stress [8,23,26–29].

HD pathogenesis also involves dysregulated gene expression mediated by mutant HTT's imbalance of histone acetylation and deacetylation, a process modulated by sirtuins (SIRT), affecting factors like brain-derived neurotrophic factor (BDNF) and mitochondrial resilience. This leads to perturbations in neuronal survival, synaptic transmission, and cellular metabolism. Inflammation and microglial activation further exacerbate neuronal injury [8,23].

Clinically, HD's average onset is in mid-adulthood, but juvenile HD occurs in 10%–20% of cases, often presenting with rigidity and seizures rather than chorea. Early features can include subtle cognitive and psychiatric changes, followed by the classic triad of chorea, cognitive decline, and mood disturbances. Over time, patients develop severe dementia, profound motor impairment, and require comprehensive care until death, usually 15–20 years after onset [8,10,25].

12.7 MULTIPLE SCLEROSIS: IMMUNE-MEDIATED DEMYELINATION

MS is a chronic disorder of the CNS, i.e., immune-mediated, characterized by inflammation, neuronal demyelination, and degeneration of neuron and axon. Its pathogenesis involves genetic susceptibility, environmental influences, and dysregulation of both innate and adaptive immune responses [11,30].

The hallmark pathological feature of MS is the demyelinating plaque formation within the CNS, primarily in the white matter but also involving grey matter and the spinal cord. These lesions show perivascular immune cell infiltration, BBB damage, and loss of oligodendrocytes, the cells responsible for producing and maintaining myelin. Autoreactive myelin-specific CD4+ T lymphocytes (especially T helper 1 and T helper 17 populations) are activated in the periphery and cross the BBB, triggering cascades of inflammation, cytokine release (e.g., IFN-γ, IL-17), and direct cytotoxicity to myelin and neurons [11,30].

B cells also play an important role in MS pathogenesis through antibody production, antigen presentation, and release of proinflammatory factors. Intrathecal synthesis of immunoglobulins, evident as oligoclonal bands in cerebrospinal fluid, is a widely used diagnostic marker. Microglia and macrophages, the resident and infiltrating immune cells of the CNS, further amplify inflammation and contribute to tissue injury and chronic lesion development [11,22,30].

MS exhibits several clinical patterns, including relapsing-remitting (the most common, featuring acute episodes of neurological dysfunction interspersed with periods of remission), secondary progressive, and primary progressive forms, each with distinct clinical trajectories and rates of progression. Over time, demyelinated lesions fail to remyelinate, leading to irreversible axonal and neuronal loss and the accumulation of permanent neurological disability [11,30].

Susceptibility to MS is strongly associated with certain HLA class II alleles, especially HLA-DRB1*15:01, but over 200 genetic variants have been implicated. Factors such as Epstein–Barr virus infection, childhood obesity, low vitamin D levels, and cigarette smoking also modulate disease risk or activity [11,30].

Chronic inflammation, mitochondrial dysfunction, glutamate-mediated excitotoxicity, and oxidative stress are increasingly recognized as drivers of the progressive neurodegenerative phase of MS. Grey matter demyelination and neuroaxonal loss correlate with cognitive dysfunction and progressive disability [11,22]. Mechanism and progression of ALS, HD, and MS are shown in Figure 12.1.

12.8 VITAMIN DEFICIENCY AND NEUROLOGICAL DYSFUNCTION

Vitamin deficiencies have increasingly been recognized as important contributors to neurological dysfunction and the progression of neurodegenerative diseases. The CNS relies heavily on adequate vitamin levels to maintain cellular metabolism, myelin integrity, neurotransmitter synthesis, and antioxidant defenses. Deficits in essential vitamins such as B-complex vitamins, vitamin D, and antioxidant vitamins C and E can exacerbate neuronal

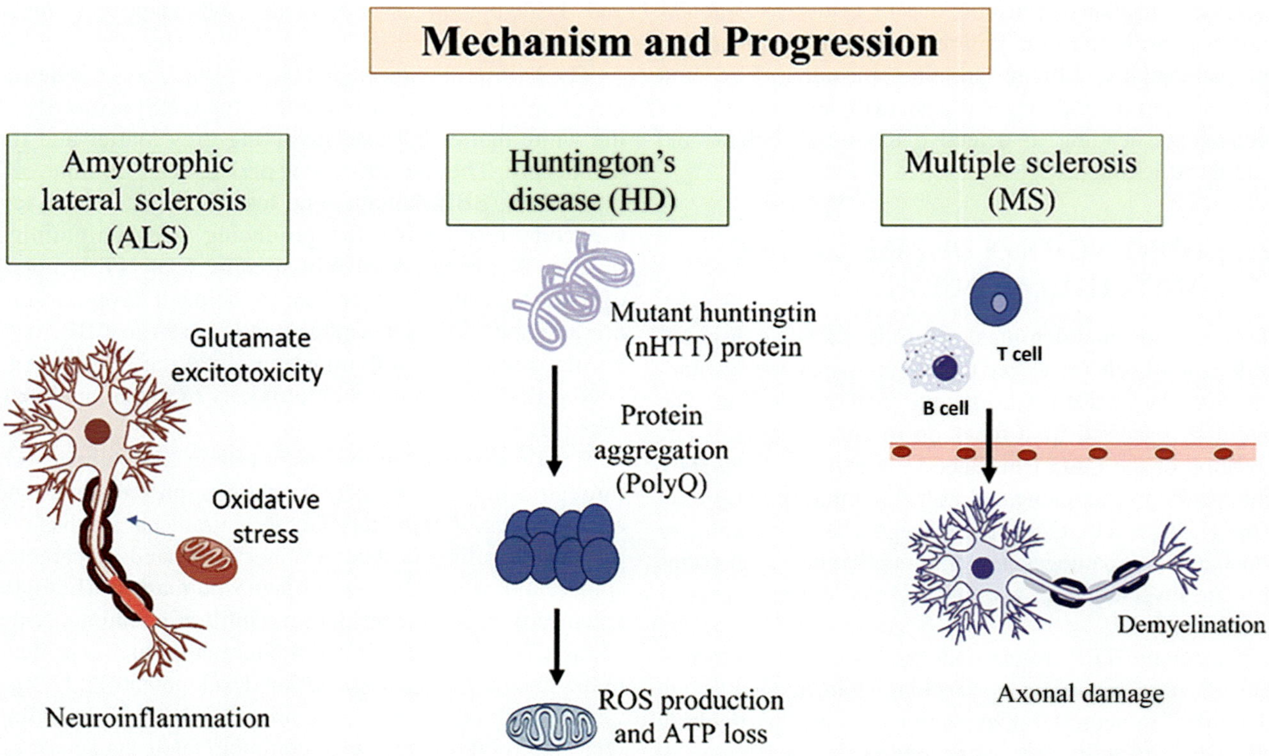

FIGURE 12.1 Mechanism and progression of amyotrophic lateral sclerosis, Huntington disease, and multiple sclerosis.

vulnerability through mechanisms involving oxidative stress, mitochondrial impairment, inflammation, and impaired repair processes. This section sheds light on the complex relationship between vitamin deficiency and neurological dysfunction, focusing on how inadequate nutrient availability disrupts neuronal health and contributes to the pathogenesis and progression of disorders like ALS, HD, and MS.

12.9 MECHANISMS OF VITAMIN DEFICIENCY IN NEURODEGENERATION

Vitamin deficiencies are increasingly implicated as both contributors to the onset and accelerators of progression in neurodegenerative disorders, including MS, ALS, and HD. Published research underscores several fundamental mechanisms by which vitamin insufficiency disrupts neural homeostasis and promotes neurodegeneration.

Deficiencies in B vitamins, especially B12 and folate, are linked to elevated homocysteine levels, which are neurotoxic and promote oxidative stress, excitotoxicity, and vascular injury. This is particularly relevant in disorders such as ALS and MS, where increased homocysteine resulting from poor B vitamin status impairs cerebral metabolism and promotes neuronal injury [20,31]. Vitamin-mediated modulation of α-synuclein accumulation and mitochondrial dysfunction in PD is given in Figure 12.2.

Vitamin D deficiency impairs the synthesis of glial-cell-line-derived neurotrophic factor (GDNF) and nerve growth factor (NGF), both essential for neuron differentiation, maturation, and survival. Reduced vitamin D also increases neuroinflammation through dysregulated cytokine production by microglia, leading to further neurodegeneration, as demonstrated in animal and clinical studies of AD and MS [20,32]. The protective role of vitamins in microtubule stability and cognitive decline is given in Figure 12.3.

Vitamins C and E are key antioxidants. Their deficiency leads to increased oxidative stress and mitochondrial dysfunction in neurons, processes consistently associated with the pathology of AD, ALS, and HD. Deficits in these vitamins hinder the quenching of ROS, resulting in increased neuronal vulnerability and facilitating the accumulation of pathological protein aggregates [20,31,33].

Vitamin B12 deficiency compromises the integrity of the myelin sheath, making axons prone to demyelination and degeneration, which is especially critical in MS. Vitamin C deficiency impairs the BBB and increases neuronal susceptibility to amyloid plaque deposition and tau phosphorylation mechanisms implicated in both dementia and other neurodegenerative diseases [31].

Some vitamins, particularly vitamin A and B group members, regulate the formation of amyloid and tau protein aggregates, modulate synaptic plasticity, and influence neurogenesis. Deficiency in these vitamins exacerbates aggregation of neurotoxic proteins and contributes to synaptic loss and cognitive deficits [20].

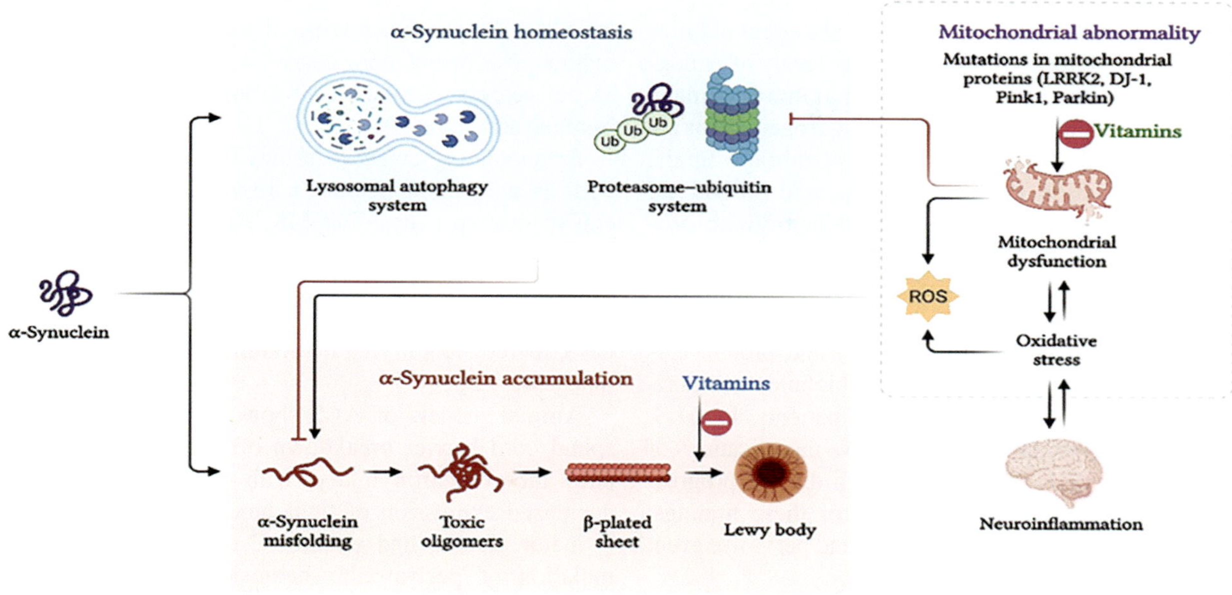

FIGURE 12.2 Vitamin-mediated modulation of α-synuclein accumulation and mitochondrial dysfunction in Parkinson's disease.

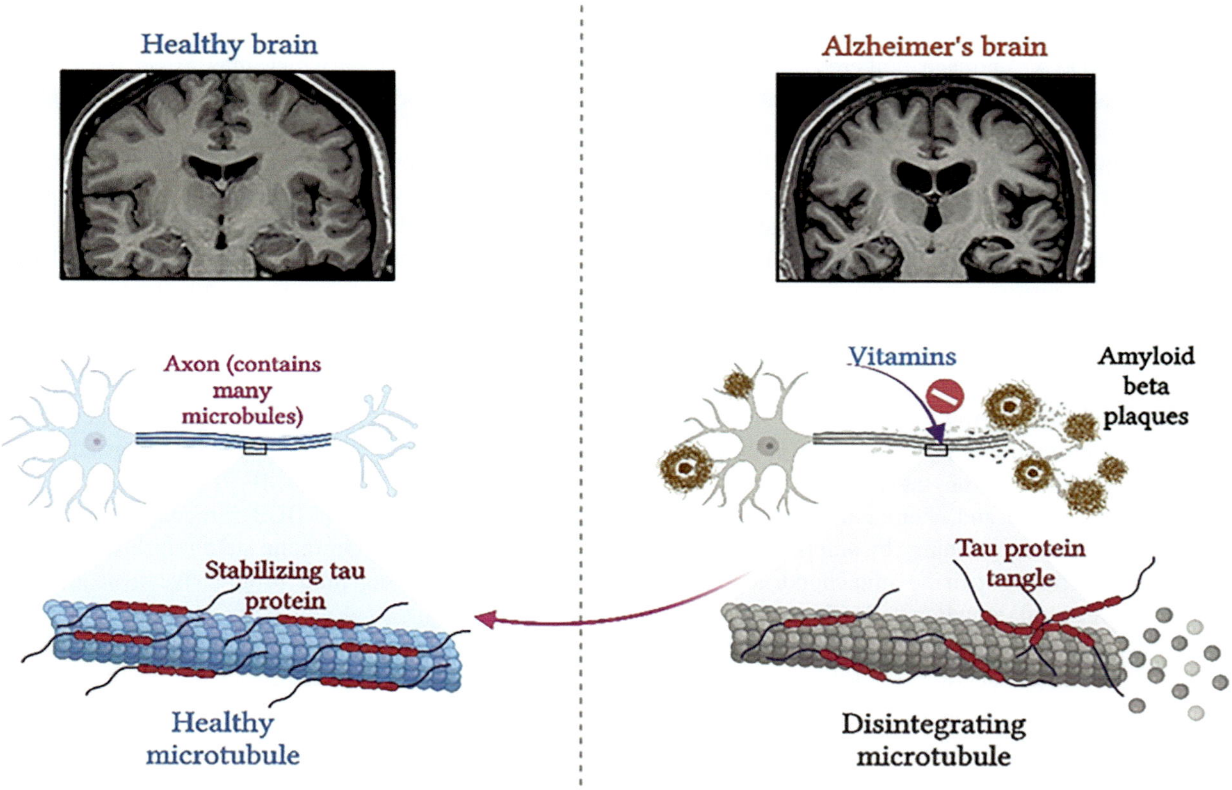

FIGURE 12.3 Protective role of vitamins in microtubule stability and cognitive decline.

12.10 OXIDATIVE STRESS, MITOCHONDRIAL DYSFUNCTION, AND INFLAMMATION

The inflammation, mitochondrial dysfunction, and oxidative stress are intimately intertwined with the pathogenesis of neurodegenerative diseases, including ALS, MS, and HD. Published research provides compelling evidence that these factors act both independently and synergistically to promote neuronal injury, synaptic loss, and progressive neurodegeneration.

An imbalance between the formation of ROS and cellular antioxidant defenses leads to oxidative stress. The

CNS is the most susceptible system in the body, because of its high rate of oxygen consumption, abundant polyunsaturated fatty acids, and relatively low levels of endogenous antioxidants, hence receiving maximum damage in a short duration. Superoxide and hydrogen peroxide are ROS that induce peroxidation of membrane lipids, protein carbonylation, deoxyribonucleic acid (DNA) and RNA damage, and the aggregation of neurotoxic proteins all of which contribute to neuronal dysfunction and death [34–36].

Research demonstrates that biomarkers of oxidative stress, such as malondialdehyde (MDA) and protein carbonyls, are consistently elevated in the biofluids and postmortem tissues of ALS, HD, and MS patients. In ALS, increased lipid peroxidation and oxidative modifications of proteins and nucleic acids correlate with disease progression and severity. Animal models mirror these findings, highlighting oxidative stress as an early and pervasive event in disease evolution [35,36].

Mitochondria, the powerhouses of the cell, are essential for ATP production, calcium homeostasis, and apoptosis regulation. Mitochondrial dysfunction has emerged as a critical factor in neurodegeneration. Defects may include reduced activity of respiratory chain complexes, increased mitochondrial DNA damage, altered mitochondrial dynamics (fusion and fission), and impaired mitophagy, the removal of defective mitochondria [34,37–39].

In ALS, familial mutations such as SOD1 disrupt mitochondrial function by increasing ROS generation, impairing biogenesis, and diminishing energy output. Similarly, HD is marked by compromised mitochondrial bioenergetics and increased sensitivity to mitochondrial toxins, while mitochondrial injury and energetic deficits also play a vital role in chronic inflammation and neurodegeneration in MS [34,37–39].

Mitochondrial dysfunction and oxidative stress are tightly linked with neuroinflammation, driven by the activation of microglia and astrocytes (glial cells) and the proinflammatory cytokines and chemokine production. This response escalates neuronal injury by amplifying oxidative processes and further impairing mitochondrial integrity. Cross-talk between inflammation and redox imbalance perpetuates a cycle of neuronal damage, loss of neurotrophic support, and progressive decline [34–36].

These mechanisms are not only observable at the cellular and molecular level but are also reflected in clinical biomarkers measurable in cerebrospinal fluid, blood, and urine providing potential avenues for early diagnosis and therapeutic monitoring [34,35].

12.11 BLOOD–BRAIN BARRIER IMPAIRMENTS AND NUTRIENT TRANSPORT

The highly selective and semipermeable nature of BBB strictly regulates the passage of molecules, including vitamins and other nutrients, between the bloodstream and the CNS. Proper functioning of the BBB is essential for maintaining neural homeostasis and metabolic demands, as well as protecting the CNS from toxins and peripheral immune activation [40–43].

Accumulating evidence shows that impairment of the BBB is a common feature across multiple neurodegenerative diseases, including MS, ALS, and HD. In these conditions, the BBB becomes hyperpermeable, allowing uncontrolled passage of blood-derived proteins, immune cells, neurotoxins, and inflammatory mediators into the CNS, thereby amplifying neuroinflammation and neuronal injury [40–43].

Animal models of ALS show that BBB and blood–spinal cord barrier breakdown often precede or accompany motor neuron injury, with studies demonstrating decreased expression of tight junction proteins, GLUT1 (a major glucose and vitamin C transporter), and accumulation of perivascular hemosiderin deposits before substantial neuronal loss and neuroinflammation. In MS, BBB disruption is an early and frequent pathophysiological event, facilitating T cell and antibody infiltration and contributing to demyelination and neuronal degeneration. HD also features BBB changes, including altered activity of nutrient transporters, which may affect the uptake of vitamins and micronutrients necessary for neuronal survival [41,42,44].

BBB dysfunction results in impaired transport of essential nutrients, including vitamins B, C, and D, sometimes through downregulation or dysfunction of specific nutrient transporters such as GLUT1, LRP, and Pgp. This leads to reduced vitamin bioavailability within the CNS, increasing neuronal vulnerability to dysfunction and degeneration, and may exacerbate vitamin deficiency even when systemic levels appear sufficient [41,43,44].

Chronic neuroinflammation is tightly linked to BBB impairment. Inflammatory cytokines upregulate matrix metalloproteinases (e.g., MMP-9), disrupt tight junctions, and further impair BBB integrity establishing a feed-forward loop of immune cell migration, neuroinflammatory signaling, and progressive neurodegeneration [40,42].

Recognition of the role of BBB disruption in neurodegeneration has spurred interest in therapeutic strategies aimed at restoring BBB integrity, optimizing CNS nutrient delivery, and developing pharmacologic agents capable of crossing a compromised BBB [40,41,45,46].

12.12 ROLE OF SPECIFIC VITAMINS IN ALS, HUNTINGTON'S DISEASE, AND MS

Vitamins play a crucial role in maintaining neurological health through their involvement in metabolic processes, immune regulation, antioxidant defense, and cellular repair mechanisms. In neurodegenerative diseases such as ALS, HD, and MS, alterations in vitamin status are common and can significantly influence disease onset, progression, and symptom severity. This section examines the roles of specific vitamins including B12, folate, vitamin D, and antioxidant vitamins C and E in the pathophysiology and potential therapeutic modulation of these disorders. Emphasis is placed on the mechanistic pathways affected by these vitamins and insights gained from clinical and experimental studies, highlighting opportunities for targeted supplementation as part of comprehensive disease management.

Vitamin B12 and folate play essential roles in neuronal function and integrity through their involvement in one-carbon metabolism, DNA synthesis, myelin formation, and maintenance of homocysteine balance. Emerging evidence from published research underscores their importance and clinical relevance in ALS, HD, and MS.

Patients with ALS frequently exhibit lower levels of vitamin B12. Supplementation has shown neuroprotective effects in ALS models by reducing inflammation, supporting mitochondrial function, and promoting motor neuron survival. Animal- and cell-based studies demonstrate that combined folic acid and vitamin B12 treatment can delay disease onset and prolong lifespan in ALS models, potentially through lowering neurotoxic homocysteine. A randomized clinical trial revealed that ultrahigh-dose methylcobalamin (a form of B12) led to a significant reduction in disease progression among early-stage ALS patients, supporting its therapeutic potential as a disease-modifying agent. However, large-scale Mendelian randomization studies have not found a definitive causal link between circulating B12 or folate levels and ALS susceptibility, though mechanistic and observational data favor potential benefits, especially in restoring homocysteine homeostasis [47–50].

Deficiency in B12, folate, and other B group vitamins has been implicated in HD and other protein aggregation neurodegenerative disorders. Integrated transcriptomic and metabolomic studies using HD models suggest that B6, B12, and folate can modulate key cellular pathways (including autophagy and protein degradation) and mitigate HTT aggregation and toxicity. These findings are supported by functional genomics and yeast model studies, highlighting the promise of these vitamins in altering disease progression and pathogenic protein aggregation in HD [51].

Several meta-analyses and systematic reviews have examined links between vitamin B12, folate, homocysteine, and MS. MS patients often present with elevated homocysteine levels, while serum B12 and folate concentrations may be variable or slightly reduced compared with those of healthy individuals. Elevated homocysteine, rather than B12 or folate deficiency per se, appears more directly linked to neurodegeneration, demyelination, and cognitive dysfunction in MS. Folate and B12 are crucial for myelin synthesis and repair, and their deficiency may aggravate demyelination and related neurological impairment. Strategies to normalize vitamin B12 or folate status, or to target hyperhomocysteinemia, show therapeutic promise in the context of neurodegenerative diseases, including MS [52–54] (Table 12.1).

TABLE 12.1
Summarize the Essential Vitamins Discussed in the Chapter (B12, Folate, Vitamin D, C, and E) and their Impact on ALS, HD, and MS, Including their Mechanisms and Therapeutic Potential

Vitamin	Role in Neurodegeneration	Impact on ALS	Impact on HD	Impact on MS	References
Vitamin B12	Involved in DNA synthesis, myelin formation, and homocysteine metabolism	Reduces inflammation, supports motor neuron survival	Reduces huntingtin aggregation, improves cognitive function	Critical for myelin synthesis, reduces demyelination	[55]
Folate	Supports one-carbon metabolism, DNA repair, and homocysteine balance	May lower neurotoxic homocysteine levels, delays disease onset	Modulates autophagy, reduces protein aggregation	Essential for myelin repair, linked with improved cognitive outcomes	[56]
Vitamin D	Modulates immune responses, neuroprotection, and calcium homeostasis	Prevents oxidative stress, slows disease progression	Potential impact on disease severity, although less studied	Reduces relapse rates, maintains blood–brain barrier integrity	[57]
Vitamin C	Antioxidant, protects against oxidative stress, supports collagen formation	Reduces oxidative stress, improves survival in models	Reduces glutamate-induced toxicity, protects neurons	Antioxidant, decreases oxidative damage in demyelination	[58]
Vitamin E	Protects against oxidative stress, maintains neuronal integrity	Delays onset, potentially slows progression	Protects neurons from oxidative damage, may reduce protein aggregation	Reduces oxidative stress, supports neuroinflammation modulation	[59]

12.13 VITAMIN D: IMMUNE MODULATION AND NEUROPROTECTION

Vitamin D has emerged as a critical micronutrient with multifaceted roles in neuroprotection and immune modulation across multiple neurodegenerative diseases, including ALS, HD, and MS.

Numerous preclinical and clinical studies indicate that vitamin D deficiency is common in ALS, and its correction may offer neuroprotective benefits. Vitamin D exerts effects on the CNS through the vitamin D receptor (VDR) expressed on neurons and glia. Experimental models show that supplementation preserves muscle function, suppresses oxidative stress, and attenuates inflammation and glutamate-mediated excitotoxicity all processes deemed crucial in ALS pathology. A recent ALS clinical study demonstrated that vitamin D3 supplementation reduced the rate of decline on the ALS Functional Rating Scale compared to non-supplemented patients, while deficiency was associated with a significantly increased mortality risk [60–62].

Although less thoroughly studied than in ALS and MS, research shows vitamin D deficiency is highly prevalent among HD patients. Some evidence suggests that vitamin D status may impact disease severity or progression, with up to 89% of HD patients exhibiting deficiency. The mechanisms might involve alterations in basal ganglia function and neurodevelopmental signaling, although robust interventional data are lacking [63].

The best-established role for vitamin D in neurodegeneration is found in MS. Low serum 25(OH)D is both a risk factor for MS development and a modulator of relapse risk and progression. Vitamin D's immunomodulatory properties are exerted through direct regulation of immune cell differentiation (promoting regulatory T cells and suppressing Th17 responses), inhibition of proinflammatory cytokine secretion, and maintenance of BBB integrity. Systematic reviews and meta-analyses of randomized controlled trials consistently indicate that vitamin D supplementation modestly reduces relapse risk, MRI lesion activity, and disability progression in patients with MS, with some benefits particularly apparent with long-term, high-dose regimens [64–67].

12.14 ANTIOXIDANT VITAMINS: VITAMINS C AND E

Antioxidant vitamins, particularly vitamin C (ascorbic acid) and vitamin E (α-tocopherol), are critical for protecting neurons against oxidative stress, a fundamental pathogenic mechanism in ALS, HD, and MS.

Vitamin E is one of the most extensively studied antioxidant therapies in ALS. Experimental data show that vitamin E supplementation can delay clinical onset and marginally slow disease progression in ALS models, though large-scale clinical trials have provided conflicting results regarding effects on survival and motor function. However, some studies report that the use of vitamin E is associated with a slower rate of ALS progression and a lower risk of mortality. Vitamin C, also a potent antioxidant, has been shown in preclinical studies to prolong the survival of ALS transgenic mice and is associated with improved antioxidant status in human observational studies. The overall safety and tolerance profile of both vitamins supports their use, but definitive clinical benefit remains uncertain [48,68,69].

In HD, oxidative damage driven by excess ROS contributes to neuronal loss and disease progression. Non-enzymatic antioxidants, including vitamin C and vitamin E, are effective in reducing disease markers and slowing progression in transgenic HD animal models. Human data are more limited but suggest a potential benefit. In vitro studies confirm that both vitamins directly protect neurons from glutamate-induced toxicity and reduce HTT protein aggregation. Ongoing trials and some clinical use highlight antioxidants as potential adjunct therapies in HD management [70,71].

Antioxidant status is often reduced in MS, with studies demonstrating lower vitamin E levels in serum and demyelination plaques of MS patients. Elevated oxidative stress markers and a disrupted antioxidant-to-cholesterol serum ratio are also observed. In relapsing MS, supplementation with vitamin E has shown positive effects on MRI lesion activity and oxidative stress markers. While direct association between high vitamin E intake and reduced long-term MS risk is yet to be firmly established, intervention studies demonstrate that α-tocopherol supplementation may decrease both the odds of MRI lesion activity and markers of ongoing neuroinflammation [72,73].

12.15 OTHER MICRONUTRIENTS WITH EMERGING ROLES

Beyond the well-established effects of B12, folate, C, D, and E, a growing body of research and recent reviews has identified additional micronutrients including thiamine (vitamin B1), biotin (vitamin B7), riboflavin (vitamin B2), pyridoxine (vitamin B6), niacin (vitamin B3), vitamin A, and various trace elements as having relevant neuroprotective or disease-modifying roles in ALS, HD, and MS.

ALS patients frequently exhibit dietary deficiencies in several key micronutrients beyond the "classic" neuroprotective vitamins. Recent studies found that lower intake of vitamin E, niacin, pantothenic acid, pyridoxine, folate, and selenium was significantly associated with worse function on the ALS Functional Rating Scale. These deficiencies may derive both from disease-related dysphagia and broader dietary inadequacies. Experimental research further shows that micronutrients such as thiamine, riboflavin, pyridoxine, and vitamin D can reduce inflammation, oxidative stress, and potentially slow disease progression. However, evidence for direct disease-modifying effects remains inconsistent, highlighting the need for larger, controlled studies [19,47,74].

Oxidative stress and mitochondrial dysfunction are implicated in HD pathology, and micronutrient supplementation is increasingly considered as adjuvant therapy. Vitamins A, C, E, thiamine (B1), B3 (niacin), and B6 each have shown some neuroprotective effects in animal and preclinical models of HD. Thiamine, for example, modulates gene expression related to neuronal energy metabolism, and deficiency exacerbates neuronal injury. Niacin supplementation improved motor function in HD rodent models, while vitamin C and vitamin D appear to enhance antioxidant defenses, reduce postural instability, and may slow disease progression. Patients with higher intake of micronutrients especially B vitamins, vitamin C, selenium, and zinc demonstrate better cognitive trajectories, and adherence to Mediterranean-type diets (rich in these micronutrients) is associated with improved quality of life and functional outcomes in HD [19,75].

Research on MS has identified that deficiencies in B vitamins (especially thiamine and pyridoxine) and vitamin A may disrupt immune cell signaling, promote demyelination, and contribute to neurodegenerative changes. Experimental evidence indicates that thiamine, riboflavin, and pyridoxine support myelin maintenance, modulate inflammatory responses, and enhance cellular energy metabolism. Vitamin A and its derivatives (retinoids) have shown promise in promoting remyelination and reducing neuroinflammation in MS animal models, though clinical validation is still needed. Selenium, zinc, and other trace elements are also under investigation for their roles in neuroprotection and immune modulation [19,20].

12.16 CLINICAL EVIDENCE AND SUPPLEMENTATION TRIALS

The therapeutic potential of vitamin supplementation in neurodegenerative diseases has been extensively investigated through a variety of clinical trials and observational studies. While preclinical research highlights several vitamins as modulators of key pathological processes such as oxidative stress, inflammation, and mitochondrial dysfunction, the translation of these findings into clinical efficacy remains complex. This section reviews the current clinical evidence regarding vitamin supplementation in ALS, HD, and MS, evaluating trial outcomes, methodological challenges, and the impact of individualized treatment approaches. The synthesis of data aims to clarify the role of vitamins as adjunctive therapies and to inform future research directions for optimizing patient care.

12.17 CURRENT CLINICAL TRIALS AND OUTCOMES

Vitamin supplementation in ALS is widespread in both clinical practice and research studies, with vitamin E, vitamin C, and multivitamins being most frequently used. A pooled analysis of over 6,000 ALS clinical trial participants revealed that more than half of patients took one or more vitamin supplements. However, these studies found no significant association between vitamin use and disease progression or severity, suggesting that current supplementation regimens do not meaningfully alter ALS outcomes on standard clinical scales. Some patients supplementing with vitamins were younger and had different disease characteristics, but rates of decline and survival were not impacted. Notably, despite some suggestions from basic science that vitamin supplementation may be beneficial, a clear clinical benefit in human ALS trials remains absent [76].

Clinical studies of high-dose vitamin D supplementation in ALS have also been conducted. Systematic reviews and meta-analyses demonstrate inconsistent results: while one randomized trial suggested a possible slowing of disability progression at higher doses, most observational and retrospective studies found no statistically significant effect. Some studies even reported potential deleterious effects. The heterogeneity and potential for methodological bias make the evidence inconclusive [47,77,78].

Nutritional supplementation interventions, including vitamin regimens, are a topic of increasing interest due to observed deficiencies and metabolic perturbations in HD patients. Translational studies in mouse models have indicated that combination therapy with biotin and thiamine improves neuronal survival and brain function, showing promise for human trials. A small clinical trial of high-dose biotin and thiamine is in the planning stages, based on successful treatment in related basal ganglia diseases in children and preclinical positive impacts. However, robust randomized controlled data in HD are not yet available and ongoing trials in humans will be crucial for establishing therapeutic benefit [75,78].

Vitamin D supplementation, due to its immunomodulatory potential, is among the most studied interventions in MS. A recent systematic review and meta-analysis of randomized trials found no significant impact of high-dose vitamin D3 supplementation on clinical disability measures (EDSS), relapse rates, or new lesion burden over 6–24 months, although a nonsignificant reduction in new T2 lesions was observed and may suggest long-term benefit in specific subgroups. Observational cohort studies reinforce the association between higher serum vitamin D and reduced MRI activity, but effects on relapse rates and clinical progression are less robust [79,80].

12.18 LIMITATIONS IN STUDY DESIGNS AND VARIABILITY

Clinical trials and interventional studies evaluating vitamin supplementation in neurodegenerative diseases, including ALS, HD, and MS, face significant limitations in design and execution, which influence both the reliability and generalizability of outcomes.

Studies often involve heterogeneous cohorts differing in age, disease stage, genetic backgrounds, and baseline nutritional status, making it difficult to interpret effects across populations. Lack of stratification by baseline vitamin deficiency status or biomarker-guided dosing compounds variability and confounds outcome measures [81].

Many supplementation trials have limited sample sizes, often insufficient to achieve robust statistical power, especially in rare diseases such as ALS or HD. Short intervention durations, sometimes less than a year, fail to capture long-term effects or progression in slowly evolving neurodegenerative disorders [81].

Doses, regimens, and formulations of vitamin supplementation vary widely, including use of dietary, synthetic, and intravenous preparations. The impact of bioavailability, metabolism, and possible dose-dependent toxicity is rarely addressed systematically, and optimal dosing remains unclear [20,82].

Use of disparate clinical scales, MRI findings, biochemical markers, and patient-reported outcomes makes cross-trial comparison challenging. Many trials prioritize surrogate endpoints (such as change in serum vitamin levels, short-term MRI changes, or homocysteine reduction), which do not always translate into meaningful clinical improvements [20,81].

Long-term compliance with vitamin interventions can be poor, especially in neurodegenerative cohorts with cognitive or physical impairment. Significant placebo effects magnified by the open-label nature of some trials further obscure genuine therapeutic impact [82].

Meta-analyses and systematic reviews report both beneficial and null findings for many vitamins across neurological disorders. For example, vitamin B supplementation may lower homocysteine but fails to improve cognition in AD/MCI; vitamin E dietary intake may show benefit while supplementation does not; and vitamin D trials report variable results depending on dose, formulation, and concurrent therapies [20,31,82].

12.19 PERSONALIZED SUPPLEMENTATION STRATEGIES

Personalized supplementation strategies acknowledge the substantial individual variability in response to vitamins for neuroprotection in neurodegenerative disorders, including ALS, HD, and MS. Recent research and reviews highlight several themes driving this evolution toward precision nutrition.

Patients differ in genetic polymorphisms affecting vitamin metabolism, transport, and receptor activity, which influence both deficiencies and therapeutic responses. For example, variances in genes encoding for VDRs, vitamin B12 transporters, or folate metabolism impact the efficacy of supplementation [20].

Precision strategies leverage serum or tissue biomarkers (e.g., vitamin D or B12 concentrations, homocysteine levels, antioxidant status) to guide individualized supplementation, aiming to target patients who are truly deficient or have specific metabolic profiles. This approach improves the likelihood of benefit while reducing potential toxicity from unnecessary dosing [83].

Individual lifestyle, diet, gut microbiota composition, and comorbidities further affect vitamin absorption and utilization. For instance, gut dysbiosis may impair B vitamin and vitamin K biosynthesis, and chronic inflammation can reduce VDR expression in MS and ALS [75].

Recent dietary intervention studies in HD and other neurodegenerative conditions investigate tailored diets and vitamin regimens based on calorie requirements, metabolic needs, and disease stage. These studies show that personalized interventions can improve nutritional status, maintain body weight, and potentially modify disease trajectories, though effects on cognitive and motor outcomes require further validation [75].

Despite promising early results, the field faces hurdles including the need for validated biomarkers, long-term compliance, and integration of genomics with clinical phenotyping. Future research should focus on large-scale clinical trials using individualized dosing, multi-omic profiling, and adaptive designs to maximize the benefit of vitamin supplementation [20] (Table 12.2).

TABLE 12.2
Provide an Overview of Clinical Trial Results Related to the Use of Vitamins in ALS, HD, and MS, Directly Complementing the Discussion about the Therapeutic Potential of Vitamins in Neurodegenerative Diseases

Vitamin	Study Type	Findings	Effectiveness in ALS	Effectiveness in HD	Effectiveness in MS	References
Vitamin B12	Randomized controlled trials (RCT)	Some studies show benefits in lowering homocysteine levels	Mixed results: small improvements in motor function, no significant impact on survival	No significant clinical benefit observed	Some improvements in myelin synthesis and cognitive function	[84]
Folate	Observational studies	Associated with lower homocysteine, which may reduce neurotoxicity	Limited evidence: Folate supplementation may reduce homocysteine but not disease progression	Some evidence suggesting improvement in cognitive function, but results are inconsistent	Folate supplementation associated with improved cognitive function and reduced relapse rates	[85]
Vitamin D	Systematic reviews, meta-analysis	Strong evidence suggesting its role in immune modulation	Potential neuroprotective effects: slower progression, but inconsistent results	Preliminary studies suggest benefits in reducing disease severity	Well-documented evidence: reduces relapse rates and MRI lesion activity, especially with high doses	[86]

(Continued)

TABLE 12.2 (*Continued*)

Provide an Overview of Clinical Trial Results Related to the Use of Vitamins in ALS, HD, and MS, Directly Complementing the Discussion about the Therapeutic Potential of Vitamins in Neurodegenerative Diseases

Vitamin	Study Type	Findings	Effectiveness in ALS	Effectiveness in HD	Effectiveness in MS	References
Vitamin C	Clinical trials, preclinical studies	Shown to reduce oxidative stress in models of neurodegeneration	Limited benefit, though oxidative damage may be reduced in early stages	Preclinical studies show significant reduction in neuronal damage	May reduce oxidative stress and inflammation, but clinical results are mixed	[87]
Vitamin E	Randomized trials, cohort studies	Mixed results in slowing disease progression	Some studies report delayed disease onset and reduced mortality in ALS patients	Provides protection against oxidative damage, may slow progression	Some improvement in oxidative stress markers and neuroinflammation in MS patients	[88]

12.20 CHALLENGES IN TRANSLATIONAL RESEARCH

Translating promising preclinical findings on vitamin supplementation and neuroprotection into effective clinical therapies for neurodegenerative diseases remains fraught with challenges. These include the inherent heterogeneity of patient populations, difficulties in accurately measuring vitamin bioavailability and CNS effects, variability in clinical trial design, and the influence of placebo effects and patient compliance. This section explores these barriers to successful translation, emphasizing the complexity of bridging laboratory insights with real-world clinical outcomes in disorders such as ALS, HD, and MS. Understanding these challenges is essential for improving study design, patient stratification, and therapeutic efficacy in future research.

12.21 HETEROGENEITY OF PATIENT POPULATIONS

The heterogeneity of patient populations in neurodegenerative disorders such as ALS, HD, and MS is a fundamental challenge in both clinical research and the therapeutic evaluation of vitamin supplementation strategies. Published reviews consistently demonstrate that neurodegenerative diseases are remarkably diverse, not only in their clinical presentation but also in their underlying genetic, molecular, and environmental determinants [89–91].

Studies report that variability in gene mutations, such as those affecting SOD1 in ALS, HTT gene in HD, and HLA haplotypes in MS, shapes disease course, age of onset, progression speed, and response to interventions among individuals. Even within genetically homogeneous groups, additional modifying genes introduce complexity, which complicates clinical trial design and interpretation [90].

Lifestyle influences including diet, exposure to toxins, smoking, physical activity, and comorbidities interact with genetic background, resulting in significant clinical variability both within and between disease populations. This diversity impacts how individuals absorb, metabolize, and respond to nutrient supplementation, including vitamins. For example, vitamin D metabolism may be influenced by skin pigmentation, sun exposure, and body mass index, while B vitamin status varies by gastrointestinal or metabolic health [90,91].

Clinical heterogeneity in neurodegeneration is further amplified by variable disease stages at enrolment, coexistence of additional neurological or systemic illnesses, and functional capacities (mobility, cognition, swallowing function). These factors affect baseline nutritional status and may influence both efficacy and tolerability of supplementation regimens [91].

A key consequence of patient heterogeneity is selective enrolment and underrepresentation of certain groups such as racial and ethnic minorities, older adults, and those with severe disability in clinical trials. This biases results and limits the generalizability of findings. Trials often report considerable between-study heterogeneity in trial participation, outcome measures, and patient-reported results [91,92].

12.22 DIFFICULTY IN MEASURING VITAMIN BIOAVAILABILITY AND EFFICACY

Measuring vitamin bioavailability and therapeutic efficacy presents significant challenges in neurodegenerative disorders such as ALS, HD, and MS. Published research demonstrates that relying solely on peripheral blood levels of vitamins or routine metabolic markers often fails to reflect actual concentrations within the CNS or their true physiological impact [20,31].

A key barrier is the lack of correlation between serum vitamin concentrations and their corresponding levels in cerebrospinal fluid and brain parenchyma. Variable BBB integrity, altered transport mechanisms, and CNS compartmentalization may cause patients with apparently "adequate" serum levels to have significant neuronal deficiency. This disconnect complicates both the diagnosis and monitoring of interventions [31].

Standard measures of vitamin status, such as serum B12, folate, and vitamin D, do not assess whether supplementation corrects downstream functional deficits like homocysteine elevation, oxidative stress, neuroinflammation, or cognitive impairment. Clinical trials often use surrogate endpoints, such as improved metabolic markers, which do not reliably translate to actual neurological benefit [31].

Significant individual variability in gastrointestinal absorption, metabolism, and utilization of vitamins is seen, with age, disease state, comorbidities, and medication use influencing bioavailability. For example, altered gut function in ALS, gastric hypochlorhydria in older adults, or genetic polymorphisms in VDR may affect tissue uptake and clinical response [20].

Most supplementation studies are short-term and are seldom designed to monitor vitamin concentrations within the CNS or tissue-specific functional outcomes. Few studies incorporate advanced techniques, such as imaging, targeted biomarker analysis, or metabolomics, to directly measure CNS effects of vitamin interventions [20].

12.23 PLACEBO EFFECT AND LONG-TERM COMPLIANCE

Placebo effects and long-term compliance are significant challenges inherent to clinical trials of vitamin supplementation and other interventions in neurodegenerative diseases, including ALS, HD, and MS.

Published systematic reviews and meta-analyses have demonstrated that placebo responses can elicit measurable symptomatic improvements in neurological conditions and can even evoke neurobiological changes detectable through modern imaging techniques. For example, clinical trials in PD patients revealed that placebo administration led to endogenous dopamine release as measured by PET scans; this demonstrates that patients' expectations alone can yield not just subjective but also biological and clinical changes. Placebo-related improvements are also found across a range of subjective and objective outcomes in neurodegenerative disease trials, with some studies reporting comparable effects to active interventions, especially when study endpoints are based on patient- or caregiver-reported outcomes [93–95].

Long-term compliance in neurodegenerative disease trials is hampered by factors such as cognitive decline, physical disability, disease progression, polypharmacy, and side effects from supplementation. These issues are particularly acute in trials requiring prolonged participation or where interventions must be taken daily, as with most vitamin regimens. Non-adherence can introduce bias, dilute observed effects, and limit the extrapolation of findings to routine clinical practice [94].

The combined impact of high placebo responses and inconsistency in compliance complicates the interpretation of vitamin intervention trials. It may lead to underestimation of true treatment effects or overestimation of benefit in open-label studies. Moreover, some studies lack objective measures of adherence, further obscuring causal relationships between supplementation and outcomes [93–95].

12.24 DIRECTIONS AND EMERGING OPPORTUNITIES

The landscape of neurodegenerative disease research is rapidly evolving, spurred by advances in molecular biology, genomics, and systems medicine. These developments open new avenues for precision nutrition and personalized supplementation strategies that hold promise for improving outcomes in ALS, HD, and MS. Emerging opportunities include integrative approaches combining diet, microbiota modulation, and lifestyle interventions, as well as the application of genomic and metabolomic profiling to tailor vitamin therapies to individual needs. This section highlights the potential of these innovative strategies to transform the prevention, management, and therapeutic landscape of neurodegenerative disorders in the coming years.

12.25 INTEGRATIVE APPROACHES: DIET, MICROBIOTA, AND LIFESTYLE

Integrative strategies targeting diet, gut microbiome, and lifestyle modifications are now recognized as important adjuncts in the prevention and management of neurodegenerative disorders such as ALS, HD, and MS. Published research has elucidated interconnected roles of nutrition, physical activity, and microbiota in maintaining neuronal health and mitigating disease progression [96–100].

Epidemiological and interventional studies show that adherence to Mediterranean or plant-based diets rich in fiber, polyunsaturated fatty acids, polyphenols, vitamins, and minerals modifies gut microbiota composition and function in ways that benefit brain health. High dietary fiber and polyphenol intake stimulates the growth of probiotic bacteria, improves gut barrier function, and increases the production of neuroprotective metabolites such as short-chain fatty acids. These changes are associated with reduced systemic and neuroinflammation, decreased rates of cognitive decline, and improved quality of life in neurodegenerative cohorts [96–98,100].

The gut–brain axis links GI tract health and microbial community composition to CNS well-being. Dysbiosis or disruption of normal gut microbiota has been observed in ALS, MS, HD, and other neurodegenerative disorders. Microbiome-based interventions such as probiotic, prebiotic, and synbiotic supplementation, as well as dietary modification and, in some research, fecal microbiota transplantation (FMT), have demonstrated potential to restore microbial balance, attenuate chronic inflammation, and positively impact neurodegenerative disease markers. For example, in MS and ALS, dietary and probiotic interventions have shown anti-inflammatory effects and some improvements in disease progression or functional status, although results vary with study duration and methodology [96,98,101–103]

Physical activity, sleep hygiene, cognitive training, and stress management play substantial roles in brain health. Physical activity is inversely associated with cognitive decline and neurodegenerative risk, while sedentary behavior increases risk for dementia and ALS. Multidomain interventions that include exercise, dietary counseling, and mindfulness-based strategies can partially restore gut dysbiosis, modulate inflammation, and improve fatigue and quality of life in MS and other neurodegenerative populations [99,100,102]

12.26 GENOMIC AND METABOLOMIC PROFILING FOR PRECISION NUTRITION

Advances in genomic, nutrigenomic, and metabolomic profiling are driving the field of precision nutrition, providing the foundation for highly individualized supplementation approaches in neurodegenerative diseases such as ALS, HD, and MS. Published reviews highlight that integration of "omics" technologies allows the identification of genetic variations, metabolic signatures, and nutrient gene interactions that shape both disease susceptibility and individualized responses to dietary interventions [104–108].

Genomic technologies including genome-wide association studies (GWAS) and next-generation sequencing have revealed numerous genetic loci associated with neurodegenerative disorders and metabolic traits. These findings inform nutrigenomics: the study of how genetic variation affects nutrient absorption, metabolism, and action. For example, single-nucleotide polymorphisms (SNPs) in genes related to VDRs or folate metabolism can predict who may benefit most from supplementation or be at higher risk of deficiency. This enables more effective, risk-adapted dietary counseling and supplementation regimes [104,106,108].

Metabolomic profiling uses advanced analytical methods to comprehensively measure metabolites in biofluids or tissues, capturing the biochemical phenotype ("metabotype") of an individual. Recent reviews emphasize that metabolomics can identify nutrient deficiencies, track responses to dietary interventions, and monitor the effects of both genetic background and environmental exposures on metabolic health. In neurodegeneration, metabolomics has uncovered distinct metabolic signatures for ALS, facilitating patient stratification, monitoring of therapeutic trials, and the development of new, targeted nutritional or pharmacological therapies [96,98,100].

Cutting-edge research further explores integrating genomics, metabolomics, proteomics, and transcriptomics to optimize individualized nutritional approaches. For example, cluster-based assignment in dietary interventions leverages both genetic and metabolic data to adjust micronutrient levels, caloric intake, and vitamin regimens in real time. Precision nutrition solutions based on these technologies have shown improved clinical biomarkers, reduced metabolic variability, and better outcomes in both general and disease-specific populations [105,106,109,110].

The potential of this approach extends beyond diet optimization: it enables customized supplementation based on dynamic biomarker profiles, earlier detection of nutritional deficits, and mechanistically informed strategies to prevent or slow neurodegenerative decline. As omics-guided precision nutrition matures, it is poised to revolutionize supportive care and empower proactive prevention in patients at risk for or diagnosed with ALS, HD, or MS [105,107,108].

12.27 MULTIMODAL THERAPEUTIC INTERVENTIONS

Multimodal therapeutic interventions integrate combinations of pharmacological, supplemental, lifestyle, and novel biological approaches to address the multifactorial nature of neurodegenerative diseases such as ALS, HD, and MS. Published research emphasizes that combining therapies targeting distinct and overlapping pathogenic mechanisms can yield synergistic benefits and improve clinical outcomes compared with monotherapies alone [111–115].

Modern therapeutic strategies utilize combinations of supplements such as coenzyme Q10, taurine, tauroursodeoxycholic acid (TUDCA), curcumin, N-acetylcysteine, and established drugs to target mitochondrial dysfunction, oxidative stress, neuroinflammation, protein misfolding, and synaptic dysfunction. Preclinical studies in ALS, HD, and PD models confirm enhanced neuroprotection, improved functional outcomes, and delayed disease progression when using multimodal combinations. Systematic review supports supplement combinations as a promising avenue for future interventions, though human clinical trials are still limited in scale [112,115].

Non-pharmacological treatments including physical exercise, cognitive training, and psychotherapy can also be combined with drugs and supplements to amplify neuroprotective effects. Physical activity and personalized exercise regimens improve brain plasticity, promote neurotrophic factor release, and decrease disease risk, especially in Alzheimer's and PDs, but emerging evidence supports similar roles in ALS and MS. Cognitive training also offers moderate improvements in cognitive function across studies in neurodegenerative disease populations [111,116].

Recent reviews highlight the potential for integrative medicine such as traditional Chinese medicine (TCM), Ayurveda, and dietary interventions in multimodal strategies. Evidence supports that combining these approaches with Western medicine can improve sleep, mood, quality of life, and non-motor symptoms in neurodegenerative conditions, contributing to comprehensive, patient-centric care [113].

Machine learning and artificial intelligence are now deployed to tailor multimodal approaches based on patient comorbidities, risk factors, and disease stage, further enhancing efficacy and safety. These adaptive models facilitate precise selection and dosing of interventions, including multimodal supplementation and exercise therapy, offering promise for more individualized care regimens [117–119].

12.28 CONCLUSION

Neurodegenerative diseases such as ALS, HD, and MS present complex challenges due to their diverse pathophysiological mechanisms and clinical heterogeneity. Vitamins play critical roles in maintaining neuronal health through their involvement in metabolic, immunological, and antioxidant pathways, and deficiencies can exacerbate disease progression and neurological dysfunction. Although preclinical studies strongly support the neuroprotective potential of vitamin supplementation, clinical trial outcomes remain mixed due to methodological limitations, patient variability, and challenges in accurately assessing vitamin bioavailability and efficacy. Looking forward, advances in precision nutrition, genomic and metabolomic profiling, and integrative multimodal interventions offer promising avenues to enhance personalized supplementation strategies. Optimizing vitamin status through tailored approaches, combined with lifestyle and therapeutic interventions, holds significant potential to improve the management and outcomes of these devastating disorders. Continued rigorous research and well-designed clinical trials are essential to fully harness the benefits of vitamins in neurodegeneration.

FUNDING

None.

CONSENT TO PUBLICATION

All authors have read and agreed to submit the chapter to the publisher.

CONFLICT OF INTEREST

None.

REFERENCES

1. Lamptey RNL, Chaulagain B, Trivedi R, Gothwal A, Layek B, Singh J. A Review of the Common Neurodegenerative Disorders: Current Therapeutic Approaches and the Potential Role of Nanotherapeutics. *Int J Mol Sci.* 2022;23(3):1851.
2. Khan AF, Iturria-Medina Y. Beyond the Usual Suspects: Multi-Factorial Computational Models in the Search for Neurodegenerative Disease Mechanisms. *Transl Psychiatry.* 2024;14(1):386.
3. Brotman RG, Moreno-Escobar MC, Joseph J, Munakomi S, Pawar G. *Amyotrophic Lateral Sclerosis.* Treasure Island, FL: StatPearls Publishing, 2025. https://www.ncbi.nlm.nih.gov/books/NBK556151/.
4. Masrori P, Van Damme P. Amyotrophic Lateral Sclerosis: A Clinical Review. *Eur J Neurol.* 2020;27(10):1918–29.
5. Zarei S, Carr K, Reiley L, Diaz K, Guerra O, Altamirano PF et al. A Comprehensive Review of Amyotrophic Lateral Sclerosis. *Surg Neurol Int.* 2015;6:171.
6. Bradford D, Rodgers KE. Advancements and Challenges in Amyotrophic Lateral Sclerosis. *Front Neurosci.* 2024;18. https://www.frontiersin.org/journals/neuroscience/articles/10.3389/fnins.2024.1401706/full.
7. Rizea RE, Corlatescu AD, Costin HP, Dumitru A, Ciurea AV. Understanding Amyotrophic Lateral Sclerosis: Pathophysiology, Diagnosis, and Therapeutic Advances. *Int J Mol Sci.* 2024;25(18):9966.
8. Palaiogeorgou AM, Papakonstantinou E, Golfinopoulou R, Sigala M, Mitsis T, Papageorgiou L et al. Recent Approaches on Huntington's Disease (Review). *Biomed Rep.* 2023;18(1):1–7.
9. Andhale R, Shrivastava D. Huntington's Disease: A Clinical Review. *Cureus.* 2022. https://www.cureus.com/articles/109169-huntingtons-disease-a-clinical-review.
10. Rojas NG, Cesarini ME, Peker G, Prat GAD, Etcheverry JL, Gatto EM. Review of Huntingtonâ™ Disease: From Basics to Advances in Diagnosis and Treatment. *J Neurol Res.* 2022;12(3):93–113.
11. Tafti D, Ehsan M, Xixis KL. Multiple Sclerosis. In: *StatPearls.* Treasure Island (FL): StatPearls Publishing; 2025. https://www.ncbi.nlm.nih.gov/books/NBK499849/.
12. Goldenberg MM. Multiple Sclerosis Review. *Pharm Ther.* 2012;37(3):175–84.
13. Baskaran AB, Grebenciucova E, Shoemaker T, Graham EL. Current Updates on the Diagnosis and Management of Multiple Sclerosis for the General Neurologist. *J Clin Neurol Seoul Korea.* 2023;19(3):217–29.
14. Haki M, AL- Biati HA, Al-Tameemi ZS, Ali IS, Al-hussaniy HA. Review of Multiple Sclerosis: Epidemiology, Etiology, Pathophysiology, and Treatment. *Medicine (Baltimore).* 2024;103(8):e37297.
15. Bianchi VE, Herrera PF, Laura R. Effect of Nutrition on Neurodegenerative Diseases: A Systematic Review. *Nutr Neurosci.* 2021;24(10):810–34.
16. Tsalamandris G, Hadjivassiliou M, Zis P. The Role of Nutrition in Neurological Disorders. *Nutrients.* 2023; 15(22):4713.
17. Zhang T, Liu Z, Mi Y. Editorial: Nutritional Interventions on Age-Related Neurodegenerative Diseases. *Front Aging Neurosci.* 2023;15. https://www.frontiersin.org/journals/aging-neuroscience/articles/10.3389/fnagi.2023.1215586/full.
18. Nguyen LAM, Simons CW, Thomas R. Nootropic Foods in Neurodegenerative Diseases: Mechanisms, Challenges, and Future. *Transl Neurodegener.* 2025;14(1):17.
19. Lahoda Brodska H, Klempir J, Zavora J, Kohout P. The Role of Micronutrients in Neurological Disorders. *Nutrients.* 2023;15(19):4129.
20. Rai SN, Singh P, Steinbusch HWM, Vamanu E, Ashraf G, Singh MP. The Role of Vitamins in Neurodegenerative Disease: An Update. *Biomedicines.* 2021;9(10):1284.
21. Mejzini R, Flynn LL, Pitout IL, Fletcher S, Wilton SD, Akkari PA. ALS Genetics, Mechanisms, and Therapeutics: Where Are We Now? *Front Neurosci.* 2019;13. https://www.frontiersin.org/journals/neuroscience/articles/10.3389/fnins.2019.01310/full.
22. Adamu A, Li S, Gao F, Xue G. The Role of Neuroinflammation in Neurodegenerative Diseases: Current Understanding and Future Therapeutic Targets. *Front Aging Neurosci.* 2024;16. https://www.frontiersin.org/journals/aging-neuroscience/articles/10.3389/fnagi.2024.1347987/full.

23. Gatto EM, Rojas NG, Persi G, Etcheverry JL, Cesarini ME, Perandones C. Huntington Disease: Advances in the Understanding of its Mechanisms. *Clin Park Relat Disord.* 2020;3:100056.
24. Caron NS, Wright GE, Hayden MR. Huntington Disease. In: Adam MP, Feldman J, Mirzaa GM, Pagon RA, Wallace SE, Amemiya A, editors. *GeneReviews®*. Seattle, WA: University of Washington, Seattle; 1993. https://www.ncbi.nlm.nih.gov/books/NBK1305/.
25. Ajitkumar A, Lui F, De Jesus O. Huntington Disease. In: *StatPearls*. Treasure Island (FL): StatPearls Publishing; 2025. https://www.ncbi.nlm.nih.gov/books/NBK559166/.
26. Zuccato C, Valenza M, Cattaneo E. Molecular Mechanisms and Potential Therapeutical Targets in Huntington's Disease. *Physiol Rev.* 2010;90(3):905–81.
27. Wanker EE. Protein Aggregation and Pathogenesis of Huntingtons Disease: Mechanisms and Correlations. *Biol. Chem.* 2000;381(9–10):937–42.
28. Arrasate M, Finkbeiner S. Protein Aggregates in Huntington's Disease. *Exp Neurol.* 2012;238(1):1–11.
29. Landles C, Bates GP. Huntingtin and the Molecular Pathogenesis of Huntington's Disease. *EMBO Rep.* 2004; 5(10):958–63.
30. Medscape. Multiple Sclerosis: Practice Essentials, Background, Pathophysiology. 2024. https://emedicine.medscape.com/article/1146199-overview?form=fpf.
31. Changkakoti L, Rajabalaya R, David SR, Balaraman AK, Sivasubramanian H, Mukherjee AK et al. Exploration of the Role of Vitamins in Preventing Neurodegenerative Diseases: Comprehensive Review on Preclinical and Clinical Findings. *Curr Neuropharmacol.* 2025;23(5):547–63.
32. Savran Z, Baltaci SB, Aladag T, Mogulkoc R, Baltaci AK. Vitamin D and Neurodegenerative Diseases Such as Multiple Sclerosis (MS), Parkinson's Disease (PD), Alzheimer's Disease (AD), and Amyotrophic Lateral Sclerosis (ALS): A Review of Current Literature. *Curr Nutr Rep.* 2025; 14(1):77.
33. Kumar RR, Singh L, Thakur A, Singh S, Kumar B. Role of Vitamins in Neurodegenerative Diseases: A Review. *CNS Neurol Disord Drug Targets.* 2022;21(9):766–73.
34. Guo C, Li S. Neurodegeneration and Oxidative Stress and Mitochondrial Damage. *Neural Reg Res.* 2013;8(21): 2003.
35. Dash UC, Bhol NK, Swain SK, Samal RR, Nayak PK, Raina V et al. Oxidative Stress and Inflammation in the Pathogenesis of Neurological Disorders: Mechanisms and Implications. *Acta Pharm Sin B.* 2025;15(1):15–34.
36. Scarian E, Viola C, Dragoni F, Di Gerlando R, Rizzo B, Diamanti L et al. New Insights into Oxidative Stress and Inflammatory Response in Neurodegenerative Diseases. *Int J Mol Sci.* 2024;25(5):2698.
37. Yusoff AAM, Khair SZNM. Unraveling Mitochondrial Dysfunction: Comprehensive Perspectives on its Impact on Neurodegenerative Diseases. *Rev Neurosci.* 2025;36(1): 53–90.
38. Wadan AHS, Shaaban AH, El-Sadek MZ, Mostafa SA, Moshref AS, El-Hussein A et al. Mitochondrial-based Therapies for Neurodegenerative Diseases: A Review of the Current Literature. *Naunyn Schmiedebergs Arch Pharmacol.* 2025. https://doi.org/10.1007/s00210-025-04014-0.
39. Meng K, Jia H, Hou X, Zhu Z, Lu Y, Feng Y et al. Mitochondrial Dysfunction in Neurodegenerative Diseases: Mechanisms and Corresponding Therapeutic Strategies. *Biomedicines.* 2025;13(2):327.
40. Carvey PM, Hendey B, Monahan AJ. The Blood–Brain Barrier in Neurodegenerative Disease: A Rhetorical Perspective. *J Neurochem.* 2009;111(2):291–314.
41. Knox EG, Aburto MR, Clarke G, Cryan JF, O'Driscoll CM. The Blood-Brain Barrier in Aging and Neurodegeneration. *Mol Psychiatry.* 2022;27(6):2659–73.
42. Kakaroubas N, Brennan S, Keon M, Saksena NK. Pathomechanisms of Blood-Brain Barrier Disruption in ALS. *Neurosci J.* 2019;2019(1):2537698.
43. Wu D, Chen Q, Chen X, Han F, Chen Z, Wang Y. The Blood–Brain Barrier: Structure, Regulation and Drug Delivery. *Signal Transduct Target Ther.* 2023;8(1):217.
44. Chaves JCS, Dando SJ, White AR, Oikari LE. Blood-Brain Barrier Transporters: An Overview of Function, Dysfunction in Alzheimer's Disease and Strategies for Treatment. *Biochim Biophys Acta BBA Mol Basis Dis.* 2024;1870(2): 166967.
45. Saraiva C, Praça C, Ferreira R, Santos T, Ferreira L, Bernardino L. Nanoparticle-Mediated Brain Drug Delivery: Overcoming Blood–Brain Barrier to Treat Neurodegenerative Diseases. *J Controlled Release.* 2016;235:34–47.
46. Ebrahimi Z, Talaei S, Aghamiri S, Goradel NH, Jafarpour A, Negahdari B. Overcoming the Blood–Brain Barrier in Neurodegenerative Disorders and Brain Tumours. *IET Nanobiotechnol.* 2020;14(6):441–8.
47. Sun J, Zhang Y. Microbiome and Micronutrient in ALS: From Novel Mechanisms to New Treatments. *Neurotherapeutics.* 2024;21(6):e00441.
48. Wang M, Liu Z, Sun W, Yuan Y, Jiao B, Zhang X et al. Association between Vitamins and Amyotrophic Lateral Sclerosis: A Center-Based Survey in Mainland China. *Front Neurol.* 2020;11. https://www.frontiersin.org/journals/neurology/articles/10.3389/fneur.2020.00488/full.
49. Mu C, Zhao Y, Han C, Tian D, Guo N, Zhang C et al. Genetically Predicted Circulating Concentrations of Micronutrients and Risk of Amyotrophic Lateral Sclerosis: A Mendelian Randomization Study. *Front Genet.* 2022;12. https://www.frontiersin.org/journals/genetics/articles/10.3389/fgene.2021.811699/full..
50. Oki R, Izumi Y, Fujita K, Miyamoto R, Nodera H, Sato Y et al. Efficacy and Safety of Ultrahigh-Dose Methylcobalamin in Early-Stage Amyotrophic Lateral Sclerosis: A Randomized Clinical Trial. *JAMA Neurol.* 2022;79(6):575–83.
51. Pradhan SS, Rao KR, Manjunath M, Saiswaroop R, Patnana DP, Phalguna KS et al. Vitamin B6, B12 and Folate Modulate Deregulated Pathways and Protein Aggregation in Yeast Model of Huntington Disease. *3 Biotech.* 2023;13(3):96.
52. Dardiotis E, Arseniou S, Sokratous M, Tsouris Z, Siokas V, Mentis AFA et al. Vitamin B12, Folate, and Homocysteine Levels and Multiple Sclerosis: A Meta-Analysis. *Mult Scler Relat Disord.* 2017;17:190–7.
53. Jorge Saínz Y, Cuello Almarales D, Mederos LE. Relevancia del metabolismo del folato en el contexto de enfermedades neurodegenerativas. *Rev Habanera Cienc Médicas.* 2021;20(3). http://scielo.sld.cu/scielo.php?pid=S1729-519X2021000300006&script=sci_arttext&tlng=en
54. Imeni Kashan A, Mirzaasgari Z, Nouri Shirazi S. Relationship between Serum Levels of Folic Acid and Homocysteine with Cognitive Impairment in Patients Diagnosed with Multiple Sclerosis. *Medicine (Baltimore).* 2024;103(28):e38680.

55. Kaszyńska AA. Cannabinoids: Potential for Modulation and Enhancement When Combined with Vitamin B12 in Case of Neurodegenerative Disorders. *Pharmaceuticals.* 2024 Jun 20;17(6):813. https://www.ncbi.nlm.nih.gov/pmc/articles/PMC11207064/.
56. Sharma V, Aran KR. Unraveling the Molecular Mechanisms of Vitamin Deficiency in Alzheimer's Disease Pathophysiology. *Aging Health Res.* 2025;5(2):100226. https://www.sciencedirect.com/science/article/pii/S2667032125000071.
57. Wang W, Li Y, Meng X. Vitamin D and Neurodegenerative Diseases. *Heliyon.* 2023;9(1):e12877. https://www.sciencedirect.com/science/article/pii/S2405844023000841.
58. Kocot J, Luchowska-Kocot D, Kiełczykowska M, Musik I, Kurzepa J. Does Vitamin C Influence Neurodegenerative Diseases and Psychiatric Disorders? *Nutrients.* 2017;9(7):659. https://www.ncbi.nlm.nih.gov/pmc/articles/PMC5537779/.
59. da Cunha Germano BC, de Morais LCC, Idalina Neta F, Fernandes ACL, Pinheiro FI, do Rego ACM et al. Vitamin E and Its Molecular Effects in Experimental Models of Neurodegenerative Diseases. *Int J Mol Sci.* 2023 Jul 7;24(13):11191. https://www.ncbi.nlm.nih.gov/pmc/articles/PMC10342514/.
60. Gianforcaro A, Hamadeh MJ. Vitamin D as a Potential Therapy in Amyotrophic Lateral Sclerosis. *CNS Neurosci Ther.* 2014;20(2):101–11.
61. Camu W, Tremblier B, Plassot C, Alphandery S, Salsac C, Pageot N et al. Vitamin D Confers Protection to Motoneurons and is a Prognostic Factor of Amyotrophic Lateral Sclerosis. *Neurobiol Aging.* 2014;35(5):1198–205.
62 Karam C, Scelsa SN. Can Vitamin D Delay the Progression of ALS? *Med Hypotheses.* 2011;76(5):643–5.
63. Homann CN, Ivanic G, Homann B, Purkart TU. Vitamin D and Hyperkinetic Movement Disorders: A Systematic Review. *Tremor Hyperkinetic Mov.* 2020;10:32.
64. Serag I, Abouzid M, Alsaadany KR, Hendawy M, Ali HT, Yaseen Y et al. Role of Vitamin D as Adjuvant Therapy on Multiple Sclerosis: An Updated Systematic Review and Meta-Analysis of Randomized Controlled Trials. *Eur J Med Res.* 2025;30(1):736.
65. Galoppin M, Kari S, Soldati S, Pal A, Rival M, Engelhardt B et al. Full Spectrum of Vitamin D Immunomodulation in Multiple Sclerosis: Mechanisms and Therapeutic Implications. *Brain Commun.* 2022;4(4):fcac171.
66. Aderinto N, Olatunji G, Kokori E, Ogieuhi IJ, Babalola EA, Samuel O et al. High-Dose Vitamin D Supplementation in Multiple Sclerosis: A Systematic Review of Clinical Effects and Future Directions. *Discov Med.* 2024;1(1):12.
67. Thouvenot E, Laplaud D, Lebrun-Frenay C, Derache N, Le Page E, Maillart E et al. High-Dose Vitamin D in Clinically Isolated Syndrome Typical of Multiple Sclerosis: The D-Lay MS Randomized Clinical Trial. *JAMA.* 2025;333(16):1413–22.
68. Orrell RW, Lane RJ, Ross M. Antioxidant Treatment for Amyotrophic Lateral Sclerosis or Motor Neuron Disease—Orrell, RW—2007. Cochrane Library. https://www.cochranelibrary.com/cdsr/doi/10.1002/14651858.CD002829.pub4/abstract.
69. Carrera-Juliá S, Moreno ML, Barrios C, de la Rubia Ortí JE, Drehmer E. Antioxidant Alternatives in the Treatment of Amyotrophic Lateral Sclerosis: A Comprehensive Review. *Front Physiol.* 2020 Feb 6;11. https://www.frontiersin.org/journals/physiology/articles/10.3389/fphys.2020.00063/full.
70. Johri A, Beal MF. Antioxidants in Huntington's Disease. *Biochim Biophys Acta BBA Mol Basis Dis.* 2012; 1822(5):664–74.
71. D'Egidio F, Castelli V, Cimini A, d'Angelo M. Cell Rearrangement and Oxidant/Antioxidant Imbalance in Huntington's Disease. *Antioxidants.* 2023;12(3):571.
72. Theodosis-Nobelos P, Rekka EA. The Multiple Sclerosis Modulatory Potential of Natural Multi-Targeting Antioxidants. *Molecules.* 2022;27(23):8402.
73. Miller ED, Dziedzic A, Saluk-Bijak J, Bijak M. A Review of Various Antioxidant Compounds and their Potential Utility as Complementary Therapy in Multiple Sclerosis. *Nutrients.* 2019;11(7):1528.
74. Barros AN de AB, Felipe ML do N, Barbosa IR, Leite-Lais L, Pedrosa LFC. Dietary Intake of Micronutrients and Disease Severity in Patients with Amyotrophic Lateral Sclerosis. *Metabolites.* 2023;13(6):696.
75. Gaba A. Nutrition and Huntington's Disease– A Review of Current Practice and Theory. *Curr Nutr Rep.* 2025;14(1):18.
76. Prell T, Grosskreutz J, Consortium the PROAACT. Use of Vitamins by Participants in Amyotrophic Lateral Sclerosis Clinical Trials. *PLOS ONE.* 2020;15(8):e0237175.
77. Lanznaster D, Bejan-Angoulvant T, Gandía J, Blasco H, Corcia P. Is There a Role for Vitamin D in Amyotrophic Lateral Sclerosis? A Systematic Review and Meta-Analysis. *Front Neurol.* 2020 Jul 31;11. https://www.frontiersin.org/journals/neurology/articles/10.3389/fneur.2020.00697/full.
78. HDBuzz. Revisiting Vitamin Therapy for HD. 2022. https://en.hdbuzz.net/324/.
79. Mahler JV, Solti M, Apóstolos-Pereira SL, Adoni T, Silva GD, Callegaro D. Vitamin D3 as an Add-On Treatment for Multiple Sclerosis: A Systematic Review and Meta-Analysis of Randomized Controlled Trials. *Mult Scler Relat Disord.* 2024;82:105433.
80. Sintzel MB, Rametta M, Reder AT. Vitamin D and Multiple Sclerosis: A Comprehensive Review. *Neurol Ther.* 2018;7(1):59–85.
81. Homann CN, Homann B, Ivanic G, Urbanic-Purkart T. Vitamin D Supplementation in Later Life: A Systematic Review of Efficacy and Safety in Movement Disorders. *Front Aging Neurosci.* 2024;16:1333217.
82. Shen L, Ji HF. Vitamin E: Supplement Versus Diet in Neurodegenerative Diseases. *Trends Mol Med.* 2012;18(8): 443–5.
83. Cervantes B, Ulatowski LM. Vitamin E and Alzheimer's Disease—Is It Time for Personalized Medicine? *Antioxidants.* 2017;6(3):45.
84. Office of Dietary Supplements—Vitamin B12. https://ods.od.nih.gov/factsheets/VitaminB12-HealthProfessional/.
85. Mohan A, Kumar R, Kumar V, Yadav M. Homocysteine, Vitamin B12 and Folate Level: Possible Risk Factors in the Progression of Chronic Heart and Kidney Disorders. *Curr Cardiol Rev.* 2023 Jul 5;19(4):e090223213539. https://www.ncbi.nlm.nih.gov/pmc/articles/PMC10494275/.
86. Rebelos E, Tentolouris N, Jude E. The Role of Vitamin D in Health and Disease: A Narrative Review on the Mechanisms Linking Vitamin D with Disease and the Effects of Supplementation. *Drugs.* 2023;83(8):665–85. https://www.ncbi.nlm.nih.gov/pmc/articles/PMC10163584/.
87. Feng J, Zheng Y, Guo M, Ares I, Martínez M, Lopez-Torres B et al. Oxidative Stress, the Blood–Brain Barrier and Neurodegenerative Diseases: The Critical Beneficial Role of Dietary Antioxidants. *Acta Pharm Sin B.* 2023;13(10): 3988–4024. https://www.ncbi.nlm.nih.gov/pmc/articles/PMC10547923/.
88. Browne D, McGuinness B, Woodside JV, McKay GJ. Vitamin E and Alzheimer's Disease: What Do We Know so Far? *Clin Interv Aging.* 2019 Jul 18;14:1303–17. https://www.ncbi.nlm.nih.gov/pmc/articles/PMC6645610/.

89. Tabbal J, Ebadi A, Mheich A, Kabbara A, Güntekin B, Yener G et al. Characterizing the Heterogeneity of Neurodegenerative Diseases through EEG Normative Modeling. *Npj Park Dis.* 2025;11(1):117.
90. Li C, Dubbelaar ML, Zhang X, Zheng JC. Editorial: Understanding the Heterogeneity and Spatial Brain Environment of Neurodegenerative Diseases through Conventional and Future Methods. *Front Cell Neurosci.* 2023;17. https://www.frontiersin.org/journals/cellular-neuroscience/articles/10.3389/fncel.2023.1211273/full.
91. Weemering DN, Beelen A, Kliest T, van Leeuwen LAG, van den Berg LH, van Eijk RPA. Trial Participation in Neurodegenerative Diseases: Barriers and Facilitators. *Neurology.* 2024;103(1):e209503.
92. National Academies of Sciences E, Affairs P and G, Committee on Women in Science E, Research C on I the R of W and UM in CT and, Bibbins-Domingo K, Helman A. Key Trends in Demographic Diversity in Clinical Trials. In: *Improving Representation in Clinical Trials and Research: Building Research Equity for Women and Underrepresented Groups.* National Academies Press (US). 2022. https://www.ncbi.nlm.nih.gov/books/NBK584392/.
93. Knezevic NN, Sic A, Worobey S, Knezevic E. Justice for Placebo: Placebo Effect in Clinical Trials and Everyday Practice. *Medicines.* 2025;12(1):5.
94. Oken BS. Placebo Effects: Clinical Aspects and Neurobiology. *Brain.* 2008;131(11):2812–23.
95. Dumitriu A, Popescu B. Placebo Effects in Neurological Diseases. *J Med Life.* 2010;3(2):114–21.
96. Gates EJ, Bernath AK, Klegeris A. Modifying the Diet and Gut Microbiota to Prevent and Manage Neurodegenerative Diseases. *Rev Neurosci.* 2022;33(7):767–87.
97. Loh JS, Mak WQ, Tan LKS, Ng CX, Chan HH, Yeow SH et al. Microbiota–Gut–Brain Axis and its Therapeutic Applications in Neurodegenerative Diseases. *Signal Transduct Target Ther.* 2024;9(1):37.
98. Park KJ, Gao Y. Gut-Brain Axis and Neurodegeneration: Mechanisms and Therapeutic Potentials. *Front Neurosci.* 2024 Oct 23;18. https://www.frontiersin.org/journals/neuroscience/articles/10.3389/fnins.2024.1481390/full.
99. Santiago JA, Potashkin JA. Physical Activity and Lifestyle Modifications in the Treatment of Neurodegenerative Diseases. *Front Aging Neurosci.* 2023 May 26;15. https://www.frontiersin.org/journals/aging-neuroscience/articles/10.3389/fnagi.2023.1185671/full.
100. Popa-Wagner A, Dumitrascu DI, Capitanescu B, Petcu EB, Surugiu R, Fang WH et al. Dietary Habits, Lifestyle Factors and Neurodegenerative Diseases. *Neural Regen Res.* 2020;15(3):394.
101. Rob M, Yousef M, Lakshmanan AP, Mahboob A, Terranegra A, Chaari A. Microbial Signatures and Therapeutic Strategies in Neurodegenerative Diseases. *Biomed Pharmacother.* 2025;184:117905.
102. Chui ZSW, Chan LML, Zhang EWH, Liang S, Choi EPH, Lok KYW et al. Effects of Microbiome-Based Interventions on Neurodegenerative Diseases: A Systematic Review and Meta-Analysis. *Sci Rep.* 2024;14(1):9558.
103. Ekwudo MN, Gubert C, Hannan AJ. The Microbiota–Gut–Brain Axis in Huntington's Disease: Pathogenic Mechanisms and Therapeutic Targets. *FEBS J.* 2025;292(6):1282–315.
104. Strianese O, Rizzo F, Ciccarelli M, Galasso G, D'Agostino Y, Salvati A et al. Precision and Personalized Medicine: How Genomic Approach Improves the Management of Cardiovascular and Neurodegenerative Disease. *Genes.* 2020;11(7):747.
105. LeVatte M, Keshteli AH, Zarei P, Wishart DS. Applications of Metabolomics to Precision Nutrition. *Lifestyle Genomics.* 2022;15(1):1–9.
106. Yang J. Using Nutrigenomics to Guide Personalized Nutrition Supplementation for Bolstering Immune System. *Health Inf Sci Syst.* 2023;11(1):4.
107. Lanznaster D, Veyrat-Durebex C, Vourc'h P, Andres CR, Blasco H, Corcia P. Metabolomics: A Tool to Understand the Impact of Genetic Mutations in Amyotrophic Lateral Sclerosis. *Genes.* 2020;11(5):537.
108. German JB, Zivkovic AM, Dallas DC, Smilowitz JT. Nutrigenomics and Personalized Diets: What Will They Mean for Food? *Annu Rev Food Sci Technol.* 2011;2 (Volume 2, 2011):97–123.
109. Small SL. Precision neurology. *Ageing Res Rev.* 2025; 104:102632.
110. Bhuiyan MNI, Saha BK, Satter MA. Harnessing Artificial Intelligence and Precision Diets for Brain Health and Cognitive Resilience. *J Nutr.* 2025. https://www.sciencedirect.com/science/article/pii/S0022316625004857.
111. Zhang T, Cong L, Chen B, He Z, Li G, Qin Q. Editorial: Multimodal Interventions in Alzheimer's Disease: From Basic Research to Clinical Practice. *Front Neurol.* 2023 Oct 20;14. https://www.frontiersin.org/journals/neurology/articles/10.3389/fneur.2023.1303733/full.
112. Shtilbans A. Combination Supplement Therapy: A New Frontier in Treatment of Neurodegenerative Diseases. *J Nutr.* 2025. https://www.sciencedirect.com/science/article/pii/S0022316625004274.
113. Chen X, Pan W. The Treatment Strategies for Neurodegenerative Diseases by Integrative Medicine. *Integr Med Int.* 2015;1(4):223–5.
114. Fatoba O, Ohtake Y, Itokazu T, Yamashita T. Immunotherapies in Huntington's Disease and α-Synucleinopathies. *Front Immunol.* 2020;11. https://www.frontiersin.org/journals/immunology/articles/10.3389/fimmu.2020.00337/full.
115. Nahar L, Charoensup R, Kalieva K, Habibi E, Guo M, Wang D et al. Natural Products in Neurodegenerative Diseases: Recent Advances and Future Outlook. *Front Pharmacol.* 2025;16. https://www.frontiersin.org/journals/pharmacology/articles/10.3389/fphar.2025.1529194/full.
116. Ferrusola-Pastrana A, Davison G, Meadows SN. The Therapeutic Effects of Multimodal Exercise for People with Parkinson's: A Longitudinal Community-Based Study. *Parkinsonism Relat Disord.* 2023;110:105366.
117. Shusharina N, Yukhnenko D, Botman S, Sapunov V, Savinov V, Kamyshov G et al. Modern Methods of Diagnostics and Treatment of Neurodegenerative Diseases and Depression. *Diagnostics.* 2023;13(3):573.
118. Barbera M, Perera D, Matton A, Mangialasche F, Rosenberg A, Middleton L et al. Multimodal Precision Prevention - A New Direction in Alzheimer's Disease. *J Prev Alzheimers Dis.* 2023;10(4):718–28.
119. Wahul RM, Ambadekar S, Dhanvijay DM, Dhanvijay MM, Dudhedia MA, Gaikwad V et al. Multimodal Approaches and AI-Driven Innovations in Dementia Diagnosis: A Systematic Review. *Discov Artif Intell.* 2025;5(1):96.

Index

Note: **Bold** page numbers refer to tables and *italic* page numbers refer to figures.

H

I

K

L

M

N

O

P

For Product Safety Concerns and Information please contact our
EU representative GPSR@taylorandfrancis.com Taylor & Francis
Verlag GmbH, Kaufingerstraße 24, 80331 München, Germany